Advances in the Use of NOAA AVHRR Data for Land Applications

EURO
_____

COURSES

A series devoted to the publication of courses and educational seminars organized by the Joint Research Centre Ispra, as part of its education and training program.
Published for the Commission of the European Communities, Directorate-General Telecommunications, Information Industries and Innovation, Scientific and Technical Communications Service.

The EUROCOURSES consist of the following subseries:

- Advanced Scientific Techniques
- Chemical and Environmental Science
- Energy Systems and Technology
- Environmental Impact Assessment
- Environmental Management
- Health Physics and Radiation Protection
- Computer and Information Science
- Mechanical and Materials Science
- Nuclear Science and Technology
- Reliability and Risk Analysis
- Remote Sensing
- Technological Innovation

REMOTE SENSING

Volume 5

The publisher will accept continuation orders for this series which may be cancelled at any time and which provide for automatic billing and shipping of each title in the series upon publication. Please write for details.

Advances in the Use of NOAA AVHRR Data for Land Applications

Edited by

Giles D'Souza

Alan S. Belward

and

Jean-Paul Malingreau

Commission of the European Communities,
Joint Research Centre,
Institute for Remote Sensing Applications, Ispra, Italy

KLUWER ACADEMIC PUBLISHERS

DORDRECHT / BOSTON / LONDON

Based on the lectures given during the Eurocourse on
Advances in the Use of NOAA AVHRR Data for Land Applications
held at the Joint Research Centre, Ispra, Italy,
December 6–12, 1993

A C.I.P. Catalogue record for this book is available from the Library of Congress

ISBN-13: 978-94-010-6575-7   e-ISBN-13: 978-94-009-0203-9
DOI: 10.1007/978-94-009-0203-9

Publication arrangements by
Commission of the European Union
Directorate-General Telecommunications, Information Industries and Innovation,
Scientific and Technical Communications Unit, Luxembourg

EUR 16325
© 1996 ECSC, EEC, EAEC, Brussels and Luxembourg
Softcover reprint of the hardcover 1st edition 1996

Published by Kluwer Academic Publishers,
P.O. Box 17, 3300 AA Dordrecht, The Netherlands.

Kluwer Academic Publishers incorporates the publishing programmes of
D. Reidel, Martinus Nijhoff, Dr W. Junk and MTP Press.

Sold and distributed in the U.S.A. and Canada
by Kluwer Academic Publishers,
101 Philip Drive, Norwell, MA 02061, U.S.A.

In all other countries, sold and distributed
by Kluwer Academic Publishers Group,
P.O. Box 322, 3300 AH Dordrecht, The Netherlands.

Table of Contents

PREFACE

Mapping and monitoring of the fragile and increasingly pressured land surface that we live on is becoming more and more important. Regular observation from space has obvious advantages and it is in this perspective that a whole series of specially-designed new Earth-observation sensors are soon to be commissioned (*e.g.* ATSR-2, MODIS, MERIS, VEGETATION, *etc*). However, although the NOAA-AVHRR sensor was initially designed for meteorological purposes, for many years its data have been (and continue to be) used in a whole range of land-based applications, often in an operational manner. Indeed, a recent survey of data requests from the international data providers indicates that as many as 35% of the requests may be for terrestrial applications. Although not ideal, the AVHRR system offers a potentially powerful configuration of large area of coverage with spatial, temporal and spectral resolution characteristics suitable for many land applications. Added to this is the relative cheapness, long-term availability and easy accessibility of the data.

The increased usage of AVHRR data since their potential was first realised (in 1981) has led to several significant developments in their acquisition, pre-treatment and application, but results, algorithms and coefficients for many of the new techniques, have tended to be become available only slowly, and then in a diverse range of special publications, technical reports and papers. For several users of the data, it has often proved difficult to collect and use the latest and most relevant algorithms, methodologies and coefficients.

The Institute for Remote Sensing Applications of the Euopean Union's Joint Research Centre in Ispra, Italy, is one of the largest users of NOAA AVHRR data. Recognising the advances made in the use of these data for land applications, it organised a well-attended training course in November 1993, entitled "Recent advances in the use of AVHRR data for land applications". One main objective of the training course was to assemble, in one set of comprehensive notes, the latest available methods and materials for the pre-processing, utilisation and understanding of NOAA AVHRR imagery. The success of that course, and feedback from it, prompted us to prepare this book, which forms a slightly updated version of what was presented in the course. In compiling this book we hope to provide an up-to-date, detailed set of notes covering all aspects of NOAA AVHRR data collection, pre-processing, analysis and application. Bearing in mind the rapid rate of change of Science and technology in this field, we have endeavoured to include as many FTP sites, email addresses and URL locations wherever possible, to enable readers to access the most up-to-date information electronically wherever possible.

Some chapters of the book address particular aspects of the NOAA AVHRR system (*e.g.* radiometric calibration and geometric correction), but other chapters provide general and important information of interest to any remote sensing studies (*e.g.* the chapters on radiative transfer modelling and atmospheric correction). The book is basically organised in five parts. The first part (chapter 1) relates the history of the use of AVHRR for terrestrial applications. The second part of the book (chapters 2-8) deals with recent advances in pre-processing methods (geometric and radiometric calibration, cloud masking, radiative transfer modelling and atmospheric correction with respect to the channels 1 and 2 of the AVHRR), and the third part (chapters 9-10) covers new methods and models for the retrieval of important surface parameters from the imagery. Chapters 11-16 review examples of land applications for which AVHRR data have proved to be an invaluable source, and illustrate the effect that the new pre-processing methods and algorithms have had in the advancement of the understanding and application of the imagery. The third part of the book also covers other aspects such as the integration of AVHRR imagery with data of higher spatial resolution. The final part of the book (chapters 17-19) deals with aspects of data availability and provision by the international space agencies and others. The final chapter (chapter 19) reviews current large data holdings of global-scale AVHRR data and access to them.

Like many of the other contributions, this last chapter also makes reference to the IGBP-DIS 1km global land cover project- an ambitious international project aimed at collecting and processing daily, full-resolution NOAA-AVHRR data for all the Earth's land surfaces from 1992 onwards. The availability of data from that project will undoubtedly promote even more interest in the use of AVHRR data for land applications, so we feel that the publication of this book, covering all important aspects of treatment and understanding of the data in one volume, is very timely, and we hope it will prove to be a convenient, informative and favourite "recipe book" for many users of NOAA AVHRR data.

The authors of this book are drawn from a range of European and American establishments that are actively engaged in the research, development and use of NOAA AVHRR imagery for land applications. The time and effort put into their contributions is gratefully acknowledged. Others who deserve much gratitude for the successful organisation and execution of the training course include several staff of the EUROCOURSES secretariat at Ispra and of the Institute of Remote Sensing Applications. Vivienne Coleman deserves special thanks for her brilliant drawing of many of the figures, proof-reading and final type-setting of many of the contributions.

List of contributors

F. Achard. Institute for Remote Sensing Applications, Commission of the European Union Joint Research Centre, 21020 Ispra (VA), Italy. Email: frederic.achard@jrc.it

O. Arino. ESA/ESRIN, C.P.64, via Galileo Galilei, 00044 Frascati, Italy. Email: Olivier.Arino@mail.esrin.esa.it

A.S. Belward. Institute for Remote Sensing Applications, Commission of the European Union Joint Research Centre, 21020 Ispra (VA), Italy. Email: alan.belward@jrc.it

G. Dedieu. LERTS, Unité mixte CNES-CNRS, 18 avenue Edouard Belin, 31055 Toulouse Cedex, France.

G. D'Souza. Institute for Remote Sensing Applications, Commission of the European Union Joint Research Centre, 21020 Ispra (VA), Italy. Now at: Geographic Data Support Ltd., 11 Fir Tree Close, Flitwick, Bedfordshire MK45 1NY, UK. Email: 100255.604@compuserve.com

N. El Saleous. NASA/Goddard Space Flight Center, Greenbelt, Maryland 20771, USA. Email: nazmi@kratmos.gsfc.nasa.gov

C. Estreguil. Institute for Remote Sensing Applications, Commission of the European Union Joint Research Centre, 21020 Ispra (VA), Italy. Email: christine.estregiul@jrc.it

S. Flasse. Natural Resources Institute, Central Avenue, Chatham Maritime, Kent, ME4 4TB, UK. Email: stephane.flasse@nri.org

J-M. Grégoire. Institute for Remote Sensing Applications, Commission of the European Union Joint Research Centre, 21020 Ispra (VA), Italy. Email: jean-marie.gregoire@ jrc.it

B.N. Holben. NASA/Goddard Space Flight Center, Greenbelt, Maryland 20771, USA. Email: brent@kratmos.gsfc.nasa.gov

x

K.B. Kidwell.   National Oceanic and Atmospheric Administration, National Environmental Satellite Data and Information Service, National Climatic Data Center, Satellite Data Services Division, Washington, D.C. 20233, USA.
Email: kkidwell@ncdc.noaa.gov

K.T. Kriebel.   Deutsche Forschungsanstakt für Luft- und Raumfahrt, Institut für Physik der Atmosphäre, Postfach 11 16, D- 82230 Wessling, Oberpfaffenhofen, Germany.

R.M. Lucas.   Department of Geography, University College of Swansea, Singleton Park, Swansea, SA2 8PP, UK.
Email: R.Lucas@swansea.ac.uk

J.P. Malingreau.   Institute for Remote Sensing Applications, Commission of the European Union Joint Research Centre, 21020 Ispra (VA), Italy.
Email: jean-paul.malingreau@jrc.it

J.C. Roger.   NASA/Goddard Space Flight Center, Greenbelt, Maryland 20771, USA.

G. Saint.   Centre National d'Etudes Spatiales, 18 avenue Edouard Belin, 31055 Toulouse Cedex, France.

T.D.G. Sandford.   Department of Computing, University of Bradford, Bradford BD7 1DP, United Kingdom.
Email: t.d.g.sandford@computing.bradford.ac.uk

B. Seguin.   INRA Bioclimatologie Avignon, BP 91, F.84 143 Montfavet (Cedex), France.

H.J. Stibig.   Institute for Remote Sensing Applications, Commission of the European Union Joint Research Centre, 21020 Ispra (VA), Italy.

C.J. Tucker.   Laboratory for Terrestrial Physics - Code 923, NASA/Goddard Space Flight Center, Greenbelt, Maryland 20771, USA.
Email: compton@kratmos.gsfc.nasa.gov

E. Vermote.   NASA/Goddard Space Flight Center, Greenbelt, Maryland 20771, USA.
Email: eric@kratmos.gsfc.nasa.gov

M.M. Verstraete.   Institute for Remote Sensing Applications, Commission of the European Union Joint Research Centre, 21020 Ispra (VA), Italy. Email: michel.verstraete@jrc.it

J.V. Vogt.   Institute for Remote Sensing Applications, Commission of the European Union Joint Research Centre, 21020 Ispra (VA), Italy. Email: juergen.vogt@jrc.it

P. Vossen.   Agriculture Information Systems Unit, Institute for Remote Sensing Applications, Commission of the European Union Joint Research Centre, 21020 Ispra (VA), Italy. Email: paul.vossen@jrc.it

HISTORY OF THE USE OF AVHRR DATA FOR LAND APPLICATIONS

C.J. TUCKER
Laboratory for Terrestrial Physics, Code 923,
NASA/Goddard Space Flight Center,
Greenbelt,
Maryland 20771 U.S.A.

ABSTRACT. Changes made to the National Oceanic and Atmospheric Administration's (NOAA) TIROS-N satellite series Advanced Very High Resolution Radiometer (AVHRR) sensor's channel 1, narrowing it from 0.55-0.90 μm to 0.55-0.70 μm on the NOAA-6 satellite in 1979, made possible coarse-resolution time-series monitoring of green vegetation at regional, continental, and global scales for the first time. Since 1980, an ever-increasing use of NOAA AVHRR data for land applications has resulted. The principal use of AVHRR data for terrestrial applications has been for vegetation index studies at local, regional, and global scales; secondary use has made use of the 3.5-3.9 μm channel for monitoring biomass burning and detection of large-scale tropical deforestation and disturbance.

1. NOAA AVHRR Satellite Data

1.1. BACKGROUND

The Advanced Very High Resolution Radiometer (AVHRR) sensor was first flown on the TIROS-N polar-orbiting meteorological satellite in 1978. TIROS-N's AVHRR was configured with 4 channels for meteorological applications: 0.55-0.90 μm; 0.73-1.1 μm; 3.5-3.9 μm; and 10.5-11.5 μm. Before the launch of TIROS-N, scientists (led by Stanley Schneider and Dave McGinnis) in the National Oceanic and Atmospheric Administration's National Earth Satellite Service (NOAA-NESS) succeeded in having future AVHRRs modified. This began with the NOAA-6 satellite, where the first channel was narrowed to 0.55-0.70 μm (Schneider and McGinnis 1981). The principal reason for confining channel 1 to the visible was to increase AVHRR effectiveness for snow mapping (Schneider, personal communication).

The NOAA-series of sun-synchronous meteorological satellites polar-orbit the Earth at an altitude of about 850 km. Most odd-numbered satellites in this series (NOAA-7, -9, -11 and NOAA-14 more recently) have a daytime overpass time of about 14.30 hours local solar time (LST) immediately after launch, moving later and later with time (Price 1991). Most satellites in the NOAA series with even numbers (NOAA-6, -8, -10, and -12) have equatorial overpass times of about 07.30 and 19.30 hours LST, which is maintained more or less throughout the life of the satellite. The AVHRR sensor scans nearly ±55° from nadir and complete coverage of the Earth is potentially available twice daily with two spatial resolutions at the satellite subpoint: 1.1 km and a spatially-degraded resolution of

1

G. D'Souza et al. (eds.), Advances in the Use of NOAA AVHRR Data for Land Applications, 1–19.
© 1996 *ECSC, EEC, EAEC, Brussels and Luxembourg.*

approximately 5.5 x 3.3 km. The approximately 5.5 x 3.3 km data are formed as a partial average of a 5 by 3 element block of 1.1 km pixels. The first four 1.1 km pixels are averaged in the first scan line, the fifth pixel in that scan line is skipped, and the next two 5-pixel blocks are also skipped. Thus, the approximately 5.5 x 3.3 km data form a $^4/_{15}$ sample of the area in question and represent a 15-fold reduction in data volume. These sampled data are referred to as global area coverage (GAC).

Beginning with the launch of NOAA-6 in 1979, the calculation of vegetation indices from AVHRR data became possible, as channel 1 was confined to the upper portion of the visible spectrum, while channel 2 continued to sense at 0.73-1.1 μm. Most of the vegetation indices derived from NOAA AVHRR data use the normalized difference vegetation index (NDVI), which is calculated from channels 1 (0.55-0.70 μm) and 2 (0.73-1.1 μm), often with channel 5 (11.5-12.5 μm) used as a thermal cloud mask. The NDVI is calculated as (channel 2 - channel 1)/(channel 2 + channel 1). Beginning with NOAA-7, and continuing with all the odd-numbered NOAA satellites, the AVHRRs are 5-channel instruments. A fifth channel from 11.5-12.5 μm was added to enable improved sea-surface temperature determinations (Kidwell 1988). NOAA-12's AVHRR is also a 5-channel instrument.

Several people, working independently, but all with an interest in terrestrial or land applications, began to use NOAA-6 AVHRR data in 1980 and early 1981. I first became interested in using NOAA AVHRR data in the course of a study described in Tucker *et al.* (1981). In this study, we commented that the only operational satellite sensor system at that time which might be used to make estimates of primary production using a vegetation index approach from space was the AVHRR on the NOAA series of satellites. Our enthusiasm was somewhat dampened by the AVHRR's wide field of view (±55 °) and the wide bandwidths of channel 1 and channel 2. The day time overpass time of NOAA-6 (07.30 hours LST) was thought to be too early in the morning, while the day time overpass time for NOAA-7 (14.30 hours LST) was thought to be too late in the afternoon because of cloud build-up. At about the same time in 1980, I was approached by Bill Stroud of NASA/Goddard, who was concerned that Landsat-3 would fail. He suggested investigating the possible use of the NOAA AVHRR, should this Landsat multi-spectral scanner failure occur. I was subsequently able to begin work with J. A. Gatlin of NASA/Goddard and S. R. Schneider of NOAA/NESDIS on the processing and interpretation of NOAA-6, and later, NOAA-7, AVHRR data.

One of the first projects we worked on was to look at AVHRR 1-km data from the Nile Delta. The U.S. Air Force was acquiring daily local area coverage (LAC) or tape-recorded 1-km data over this part of the Mediterranean, and we were able to purchase these data from NOAA. The Nile Delta work involved selecting the best cloud-free data from near-daily overpass (Tucker *et al.* 1984). It was apparent to us that NOAA AVHRR data offered an excellent satellite data source to study large areas. Furthermore, the data were inexpensive and were readily available for many areas of interest.

Beginning with NOAA-7's AVHRR, launched in June 1981 with an equatorial overpass time of 02.30 and 14.30 hours LST, other workers also began evaluating these data for land vegetation purposes. Gray and McCrary (1981a 1981b), Schneider *et al.* (1981), Townshend and Tucker (1981), Greegor and Norwine (1981), Schneider and McGinnis

(1982), Ormsby (1982), Cicone and Metzler (1982), Yates and Tarpley (1982), Brown *et al.* (1982), Barnett and Thompson (1982), Duggin *et al.* (1982), and Tucker *et al.* (1983) all reported on early attempts using NOAA-6 and NOAA-7 AVHRR data. Initially, almost all of the emphasis was on using selected cloud-free 1-km data and most of the topics investigated involved agricultural monitoring. Agricultural applications (crop yield forecasts, droughts, *etc.*) were viewed at that time as being the principal rationale for satellite remote sensing (Gray and McCrary 1981a, Schneider and McGinnis 1982, Cicone and Metzler 1982, Yates and Tarpley 1982, Brown *et al.* 1982, Barnett and Thompson 1982, and Duggin *et al.* 1982).

In late summer and early fall of 1981, co-workers in Senegal and I attempted to compare 1-km NDVI data from NOAA-7's AVHRR instrument to ground-collected total above-ground herbaceous dry biomass measurements. A request had been submitted to NOAA to tape record 1-km data for West Africa, centred over Senegal. We were simply lucky that all of our overpasses, scheduled about once every 9-10 days, were largely cloud-free for our area of interest (Tucker *et al.* 1983). However, in 1982 we were very unsuccessful in obtaining cloud-free NOAA-7 AVHRR 1-km data over Senegal as scheduled once every 9-10 days during the growing season. This prompted us to investigate different compositing approaches, atmospheric effects, cloud detection, directional reflectance effects, *etc.* (Kimes 1983, Kimes *et al.* 1984, Gatlin *et al.* 1984, Holben and Fraser 1984, Holben 1986, Holben *et al.* 1986). These investigations continued into 1983 and 1984. By early 1983, it was common operating procedure to form maximum value composite (MVC) images and process daily Level 1b global area coverage (GAC) data for Africa. Our group at NASA/Goddard then provided these to FAO in near-real-time for famine early warning and desert locust control.

The work of Holben and Fraser (1984) and Holben (1986) established the scientific basis for forming maximum value normalized difference composite images; that is, selecting the highest normalized difference vegetation index value for a given location from several successive observations. In 1983, NOAA began producing the global vegetation index, GVI (Tarpley *et al.* 1984, Kidwell 1990) and several uses of the data for large-area study were subsequently reported (Townshend and Tucker 1984, Justice *et al.* 1985, Goward *et al.* 1985, Tucker *et al.* 1985a and 1985b, Tucker *et al.* 1986, Malingreau 1986). In 1986 the Global Inventory Monitoring and Modeling Study (GIMMS) of NASA/Goddard Space Flight Center published a special edition entitled *"Monitoring the grasslands of semi-arid Africa using NOAA-AVHRR data"* (Justice 1986), which summarised several years of AVHRR 1-km and 4-km work from Sahelian Africa, with emphasis on rangeland monitoring. A subsequent special edition entitled *"Coarse Resolution Remote Sensing of the Sahelian Environment"* was published in 1991 (Prince and Justice 1991). These two special editions are excellent references on AVHRR land applications, though the focus is on African rangelands.

Simultaneously to research using AVHRR data for various land applications, other researchers have investigated important AVHRR considerations. These include within- and between-satellite channel 1 and channel 2 calibration (Brest and Rossow 1992, Holben *et al.* 1990, Kaufman and Holben 1993, Los 1993, Smith *et al.* 1988, Staylor *et al.* 1990, and Teillet *et al.* 1990), spatial resolution questions (Belward and Lambin 1990,

Justice *et al.* 1987 and 1988, and Townshend and Justice 1988 and 1990); navigation considerations (Brush 1988, and Emery *et al.* 1989); and cloud screening and atmospheric effects (Eck and Kalb 1991, Holben *et al.* 1991, Justice *et al.* 1991, Soufflet *et al.* 1991, and Tanré *et al.* 1992). Goward *et al.* (1990 and 1993), Gutman (1991), and Singh (1988) have investigated considerations associated with existing NOAA AVHRR vegetation index data sets.

My review of the history of AVHRR data for land applications will be that of an overview, hopefully citing most of the work in several areas. I will first review some of the background about vegetation indices, what they mean biophysically, and how they can be used. I will then review the several areas, many of them unique among satellite applications, where AVHRR data have been used for land applications. These areas are: grassland/steppe biomass estimation; regional, continental, and/or global studies; and large-scale deforestation determination and fire detection. I encourage anyone to bring to my attention references or papers which are not cited, and apologise in advance for my unintended oversights.

Before proceeding further, the unique capabilities of the AVHRR data must be discussed, as these relate directly to many of the various land applications. While AVHRR data are recorded at 1-km and at the so-called "4-km" spatial resolutions at the satellite sub-point, they are available for the entire terrestrial surface at least once a day (more frequent as latitude increases toward the poles). Therefore, AVHRR data provide the potential ability to image a given area every day, thus providing the means to overcome, or at least potentially minimise, cloud cover or other unfavourable atmospheric conditions. Most importantly, this inherent frequency of observation provides the means to study the vegetation index time history of many areas of interest. What the AVHRR loses in spatial detail, it makes up in the possibility of temporal information. Used together with Landsat or SPOT data, which provide spatial detail, AVHRR data are complimentary and unique for temporal considerations. In addition, the AVHRR's channel 3 (3.5-3.9 µm) provides information about temperature differences in the -50 °C to +50 °C range (Kaufman *et al.* 1990).

1.2. USE OF VEGETATION INDICES

Red and near-infrared remote sensing data have been used to study green leaf vegetation density in a wide variety of settings. Such data can be combined into what are called "*vegetation indices*", which a number of researchers have shown to be non-destructive estimates of the intercepted photosynthetically active radiation (0.4-0.7 µm) (Hatfield *et al.* 1984, Asrar *et al.* 1984, 1985, and 1986, Daughtry *et al.* 1983,Gallo *et al.* 1985, Wiegand and Richardson 1984). The ability to non-destructively determine the intercepted photosynthetically active radiation fraction permits remote sensing studies of primary production (Kumar and Monteith 1982, Asrar *et al.* 1985, Sellers 1985 and 1987, Tucker and Sellers 1986). Daughtry *et al.* (1983) and others have used vegetation indices to study biomass production from ground and satellites from a wide range of vegetation types from grasslands to salt marshes to agricultural crops (Hardisky *et al.* 1984, Steinmetz *et al.* 1990, Steven *et al.* 1983, Tucker *et al.* 1981, 1983, and 1985, Prince 1991a).

Ground-based vegetation index measurements have been made using spectrometers or simple multi-channel radiometers coupled with detailed biological sampling of the exact areas spectrally measured (Pearson and Miller 1972, Tucker 1979, Asrar *et al.* 1985, Tucker *et al.* 1981, Holben *et al.* 1980, Kimes *et al.* 1981). Similar studies have been made using aircraft, Landsat, and NOAA AVHRR data where the same relationships found on the ground were demonstrated to hold using aircraft and satellite data. These relationships have been extensively tested with ground data from a wide variety of settings (summarised in Perry and Lautenschlager 1984, Tucker 1980, Curran 1983, Asrar *et al.* 1985).

2.   Grassland/Steppe Herbaceous Biomass Estimation

One of the areas where AVHRR data have proven to be very useful is in the area of herbaceous total biomass accumulation estimation in grasslands and steppes. This work follows from the basic relationship of NDVI to intercepted PAR, and this relationship to photosynthesis. Summed over the growing season, PAR and NDVI are highly related to total photosynthesis, and hence to total dry biomass production.

A body of work has accumulated, using first ground-based red and near-infrared spectral measurements, and later satellite data, to estimate total biomass production from time-series vegetation indices. The satellite application of this technique was first tested in northern Senegal with C. Vanpraet and co-workers in 1981-1983 (Tucker *et al.* 1983 and 1985). Subsequently, many other investigators expanded this work throughout the Sahelian zone and elsewhere, and reported similar results (Diallo *et al.* 1991, Hiernaux and Justice 1986, Justice and Hiernaux 1986, Prince 1991a, 1991b, Prince and Astel 1986, Prince and Tucker 1986, and Wylie *et al.* 1991) (see Figure 1).

Figure 1 is rather remarkable. It represents data collected at 363 ground locations, by large numbers of different people over eight years, from Senegal, Mali, and Niger. The ground data are end-of-season total above-ground dry biomass. A substantial fraction of the total biomass production is below ground, but this is not sampled because of logistic difficulties. Furthermore, there is a mismatch between the destructive sampling, where 10-40 1-m^2 plots are clipped at the end of the growing season, and the satellite data, in terms of actual area measured/sampled by both: 10 to 40 m^2 versus 1,000,000 m^2. This is compounded by navigation errors of ±1 pixel (±1 km at best), atmospheric effects not completely removed by compositing, calibration differences, and a host of other problems. In spite of qualifications associated with the ground data used in Figure 1, every point in this figure represents hard-won ground data, collected under Sahelian conditions.

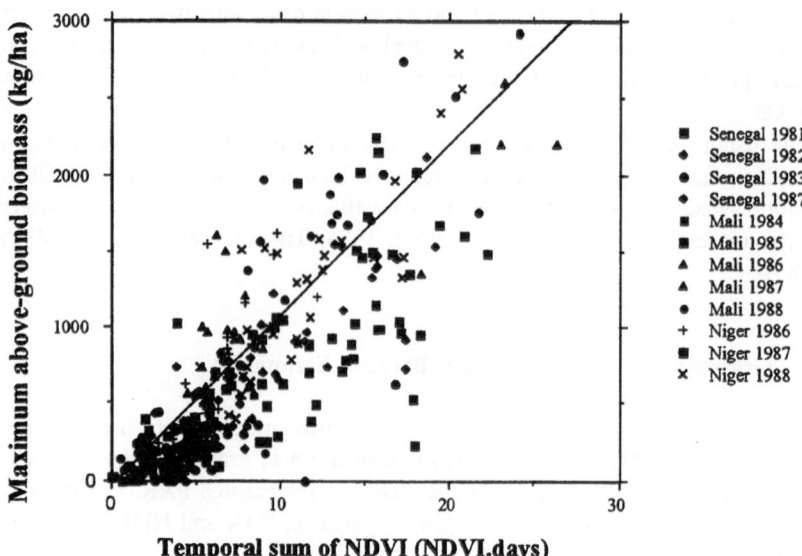

Figure 1. Summary figure for NOAA AVHRR 1-km NDVI-Sahelian biomass relationship from 1981-1988. This figure represents the specific comparisons between ground-sampled above-ground total dry herbaceous biomass sampled at the end of the growing season and the integrated or summed NDVI data from the same growing season for these specific locations. From Prince (1991) [biomass (kg/ha) = -86+114*ndvi-days; Confidence Intervals: @3 ndvi-days, ±61 kg/ha; @10 ndvi-days, ±51 kg/ha].

3. Regional, Continental, and Global Studies

3.1. INTRODUCTION

AVHRR data have been used in a variety of studies at regional, continental, and global scales. All of these types of studies have used time-series of AVHRR data, usually derived from daily GAC data. Examples of these include land cover studies, large-scale primary production studies, and studies of linkages between vegetation phenology and animals, such as the desert locust, the mosquito which carries Rift Valley fever, and the African weaver bird.

3.2. CONTINENTAL LAND COVER CLASSIFICATION AND MONITORING

AVHRR data have been formed into time-series of the NDVI and used to investigate the possibility of continental land cover determination (Townshend *et al.* 1985 and 1987, Tucker *et al.* 1985, Tucker *et al.* 1991, Tateishi and Kajiwara 1992, Eastman and Fulk 1993). The assumption in all of the AVHRR continental land cover studies is that the various large aggregations of land cover have different NDVI responses through time. The different NDVI magnitudes and time variations are the means to discriminate or classify one cover type, say desert, semi-arid steppe, savanna, forest, *etc.* from another.

Problems arise when more and more land cover aggregations are desired, or when larger and larger areas are studied. Problems also occur when multiple continents are studied together. These problems stem from an uncertainty of clear boundaries between different vegetation aggregations. This difficulty makes exact boundary delineation impossible in many situations.

An example of where this has been overcome has been reported by Tucker *et al.* (1991) and involves determination of the area of the Sahara Desert from 1980 to 1993. In this study, precipitation data were correlated with coincident NDVI data for 42 Sahelian station locations (Figure 2). This provided the means to use a specific NDVI value as the threshold corresponding to a specific precipitation amount. Tucker *et al.* (1991) used the 200 mm/yr precipitation isoline as their boundary between desert and non-desert, on the south side of the Sahara. This approach was then used from 1980 through to the present

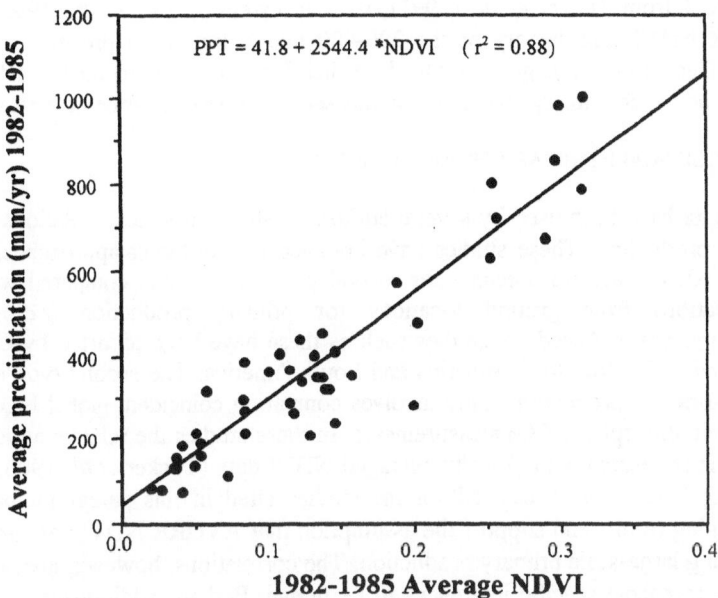

Figure 2. Comparison between the average precipitation from 1982-1985 for 24 stations from West Africa and the coincident normalized difference vegetation index averaged from 1982-1985. A similar relationship exists for the 0-600 mm/yr zone between precipitation and the normalized difference vegetation index. From Tucker *et al.* (1991).

to document the expansion and contraction of the Sahara in the south. Other land cover determinations must use specified thresholds for assigning areas studied to specific land cover aggregations. Otherwise, it is difficult to assign specific boundaries between the various cover types which have meaning.Figures 1 and 2 represent two methods for determining the accuracy of NOAA AVHRR satellite NDVI data in terms of coincident ground observations from specified areas. Other relationships from different locales have been reported by Nicholson and her students (Davenport and Nicholson 1993, Malo and Nicholson 1990, and Nicholson *et al.* 1990).

Associated with land cover classification is land cover monitoring, where time-series AVHRR NDVI data are used to document conditions in arid and semi-arid areas and elsewhere. With the AVHRR NDVI time history, going back to 1981, comparisons can be made to historical conditions, and inferences drawn about the present situation. This is the rationale for the Famine Early Warning project sponsored by the U. S. Agency for International Development (Hutchinson 1991). The monitoring potential of AVHRR data is substantial, as many workers have reported (Goward 1989, Henrickson and Durkin 1986, Hielkema *et al.* 1986, Justice *et al.* 1991, Justice and Hiernaux 1986, Justice *et al.* 1986, Justice *et al.* 1985, Malingreau 1986, Malingreau and Tucker 1987, and Townshend and Justice 1990).

Figure 3 from Tucker *et al.* (1991), with an extension through to 1992, presents an AVHRR NDVI time history of the 200-400 mm/yr long-term precipitation zone from Africa. This corresponds generally to the Sahel Zone, and documents the ebb and flow of rainfall and hence primary production in this semi-arid pastoral zone of Africa.

3.3. LARGE-SCALE PRIMARY PRODUCTION STUDIES

NDVI data have been used by several authors to study large-scale relations to terrestrial primary production. These studies have involved two different approaches. In the first, satellite NDVI data are obtained for several yearly cycles and compared with published extrapolations from ground locations for primary production (*i.e.* accumulated carbon/area/year). Anecdotal studies such as these have been reported by Goward *et al.* (1985 and 1987) for North America and South America. The second type of large-scale satellite-primary production study involves comparing coincident global NDVI data with coincident atmospheric CO_2 measurements. In these studies the relative atmospheric CO_2 content is correlated with globally-averaged NDVI data (Tucker *et al.* 1986, and Fung *et al.* 1986, Box *et al.* 1989). All of the studies cited in this paragraph have provided encouraging results and support the assumption that AVHRR NDVI data are very useful in studying large-scale primary production. The correlations, however, are not sufficiently rigorous to enable specific inferences to be drawn. Perhaps additional work in this area will enable more quantitative conclusions to be drawn from these types of studies.

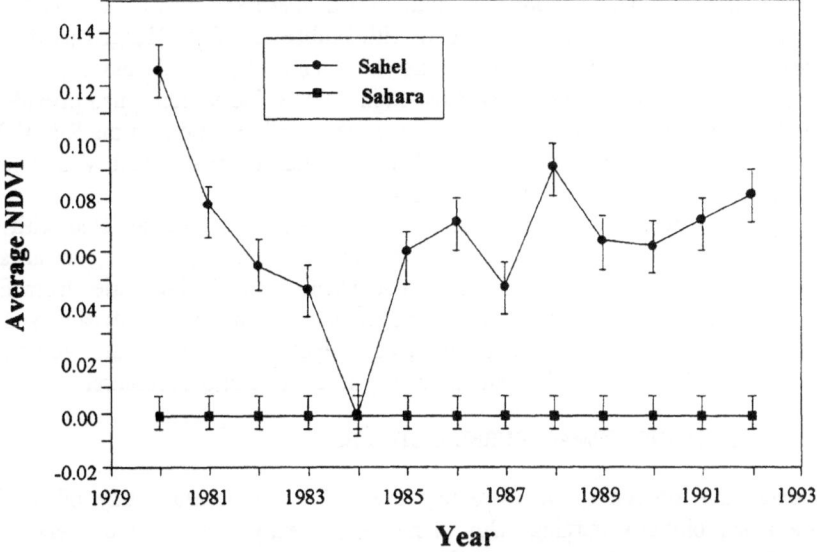

Figure 3. Average vegetation index values for the approximate 200-400 mm/yr mean precipitation zone and a 2,500,000 km^2 area of the central hyper-arid Sahara Desert, both plotted against time, from 1980-1992. There is a direct correlation between primary production and the vegetation index (Figure 1), precipitation and primary production, and hence precipitation and the vegetation index (Figure 2). From Tucker *et al.* (1991) with extension to 1991 and 1992.

4. Linking Africa NDVI Dynamics with Ecological Phenomena

4.1. GENERAL

Interesting relationships have been reported between NDVI time-series data and the life cycles of insects and birds in Africa. This is not surprising, as green vegetation dynamics are a major driving ecological variable in many arid and semi-arid systems. Tucker *et al.* (1985) and Hielkema *et al.* (1986) have reported on the close association between green vegetation development in the 17,000,000 km^2 desert locust recession area and potential upsurges of these pests. AVHRR NDVI data are now routinely used to identify areas for closer inspection, very close to real-time. This minimises desert locust plagues, while also minimising the use of dangerous insecticides when desert locust plagues happen. NASA/Goddard Space Flight Center, FAO, and the U.S. Agency for International Development are presently co-operating in continuing the desert locust monitoring program.

NOAA AVHRR NDVI data have also been shown to be highly correlated to outbreaks of the mosquito which is responsible for the transmission of the virus known to cause Rift

Valley fever, which can be fatal in humans. High NDVIs in areas of East Africa where Rift Valley fever is endemic have been correlated with outbreaks of this dangerous disease which threatens man and domesticated animals (Linthicum *et al.* 1987 and 1990). This results from heavy rains which indirectly cause high NDVIs. Heavy rains also provide an environment for the mosquitoes to prosper. Hence, very wet conditions in the Rift Valley fever area, as manifested in very high NDVI values, indicate that a Rift Valley fever outbreak is likely.

Green vegetation dynamics in East Africa are also known to be associated with the movement of the *Quela quela* weaver bird into agricultural areas, resulting in sizeable grain losses. Wallin *et al.* (1992) have shown that AVHRR NDVI time-series from East Africa are highly correlated to *Quela quela* movements and population dynamics. It is assumed that other couplings exist between green vegetation dynamics and ecological and/or biomedical phenomena. This is an interesting area for additional research.

4.2. DEFORESTATION AND BIOMASS BURNING STUDIES

AVHRR data have proven to be extremely useful for the general study of tropical deforestation and biomass burning. Our planet is presently facing the conversion of thousands of square kilometres each year from tropical forest to other types of land use. Not only is there concern over the possible greenhouse gas contributions from tropical forest conversion, but there is urgency over tremendous losses of plant and animal species. Associated with tropical deforestation are two adverse effects on biological diversity: (1) direct habitat destruction and (2) indirect adverse effects on remaining habitat, through the isolation of forest fragments, and from "*edge effects*" where deforested areas are next to remaining forest. The combination of these effects leads to species loss - directly in the former case, indirectly in the latter case. AVHRR data are useful for an early warning system, to detect where tropical deforestation is occurring, so to focus more detailed studies with Landsat or SPOT data in those locales.

Research by Malingreau *et al.* (1989) has demonstrated the unique ability of NOAA AVHRR data to provide global evaluation of tropical deforestation. Two elements related to vegetation dynamics are readily obtained from AVHRR data: undisturbed primary forest can be discriminated from secondary forest or non-forest; and disturbance, drought and fire can also be readily detected. Use of the combination of visible, near-infrared, and thermal AVHRR channels makes this possible. The highly repetitive coverage (*i.e.* every day) usually results in the acquisition of adequate data sets over extensive areas of tropical forest. Time-series analyses can also be a useful tool.

A global program of AVHRR tropical forest monitoring is affordable. It also provides important information on extensive portions of the tropics where land transformations are under way. When combined with high-resolution satellite data and field observations, AVHRR data can form the basis for a global early warning system for the conservation of tropical forests and their associated biological diversity (Malingreau *et al.* 1989). Much of this is described in the chapter by Malingreau *et al.* later in this book.

Deforestation and biomass burning studies make use of the 3.5-3.9 µm channel of the AVHRR. This results from a high sensitivity in the 3.5-3.9 µm channel to temperatures in the -50 °C to +50 °C range (Schutt and Kerber 1986, Kaufman *et al.* 1990). The 3.5- 3.9 µm channel saturates at around 320 °K (Matson *et al.* 1987). This means that a fire with a temperature of greater than 250 °C will saturate the channel 3 response if it occupies only 0.1% of the pixel, while having a small effect in channel 4 or channel 5 responses. Furthermore, flaming fires (temperatures greater than 500 °C) which occupy only 0.01% of the pixel can also be observed. As a result, it is possible to count fires but it is not possible to determine the size of individual fires (Kaufman *et al.* 1990). Fire counts have been reported by Setzer *et al.* (1988) and Setzer and Pereira (1991). Much of this work with fire detection and monitoring is covered in a later chapter by Grégoire *et al.*

The high sensitivity of the 3.5-3.9 µm AVHRR channel to slight temperature variations also makes it rather suitable to distinguish areas of deforestation and areas of disturbed forest (areas which have been thinned, *etc.*) from areas of intact forest (Malingreau *et al.* 1989). Areas where roads are being opened into the forest, together with activities such as mining, logging, burning, clearing, *etc.* are all readily detected by the 3.5-3.9µm AVHRR channel because of their resulting higher brightness temperatures. Thus the use of data from this channel is useful in identifying areas where deforestation is proceeding.

AVHRR 1-km data are capable of providing accurate estimates of deforestation extent in areas where the deforested areas are large. In places where the deforestation is more mixed, or where the areas of deforestation are much smaller, AVHRR deforestation accuracies fall off (Nelson and Holben 1986, Malingreau and Tucker 1988). However, this should not detract from the unique ability of AVHRR data to supply a global evaluation of tropical deforestation.

5.   Conclusions

Since 1981 NOAA AVHRR data have found a tremendous use for the study of terrestrial vegetation. This was unexpected as the AVHRR was designed as a meteorological instrument. Had Schneider and McGinnis of NOAA/NESS not prevailed in the mid-1970s in restricting channel 1 to the upper portion of the visible spectrum, we would not have the opportunity to use this important source of terrestrial imagery.

References and Bibliography

Asrar, G., Fuchs, M., Kanemasu, E.T., and Hatfield, J.L., 1984. Estimating absorbed photosynthetically active radiation and leaf area index from spectral reflectance in wheat. Agronomy Journal, 76:300-306.

Asrar, G., Kanemasu, E.T., Jackson, R.D., and Pinter, P.J., 1985. Estimation of total above-ground phytomass production using remotely sensed data. *Remote Sensing of Environment*, 17:211-220.

Asrar, G., Kanemasu, E.T., Miller, G.P., and Weiser, R.L., 1986. Light interception and leaf area estimates from measurements of grass canopy reflectance. *IEEE Transactions Geoscience Remote Sens.*, **GE24**:76-82.

Barnett, T.L. and Thompson, D.R., 1982. Large area relation of satellite spectral data to wheat yields. In, *Proc. 8th Int. Symp. Machine Processing Remotely Sensed Data*, Purdue University, Indiana, USA, pp. 213-219.

Belward, A.S. and Lambin, E., 1990. Limitations to the identification of spatial structures from AVHRR data. *International Journal of Remote Sensing*, **11**:921-927.

Box, E.O., Holben, B.N., and Kalb, V., 1989. Accuracy of the AVHRR vegetation index as a predictor of biomass, primary productivity and net CO_2 flux. *Vegatatio*, **80**:71-89.

Brest, C.L., and Rossow, W.B., 1992. Radiometric calibration and monitoring of NOAA AVHRR data for ISCCP. *International Journal of Remote Sensing*, **13**:235-273.

Brown, R.J., Bernier, M., and Fedosejevs, G., 1982. Geometric and radiometric considerations of NOAA AVHRR imagery. In, *Proc. 8th Int. Symp. Machine Processing Remotely Sensed Data*, Purdue University, Indiana, USA, pp. 374-381.

Brush, R.J.H., 1988. The navigation of AVHRR imagery. *International Journal of Remote Sensing*, **9**:1491-1502.

Cicone, R.C., and Metzler, S.D., 1982. Comparisons of Landsat MSS, Nimbus-7 CZCS, and NOAA-6/7 AVHRR sensors for land use analysis. In, *Proc. 8th Int. Symp. Machine Processing Remotely Sensed Data*, Purdue University, Indiana, USA, pp. 291-297.

Curran, P.J., 1983. Multi-spectral remote sensing for the estimation of green leaf area index. *Phil. Trans. Royal Soc.*, **A309**:257-270.

Daughtry, C.S.T.,Gallo, K.P., and Bauer, M.E., 1983. Spectral estimates of solar radiation intercepted by corn canopies. *Agronomy Journal*, **75**:527-531.

Davenport, M.L., and Nicholson, S.E., 1993. On the relationship between rainfall and the normalized difference vegetation index for diverse vegetation types in East Africa. *International Journal of Remote Sensing (in press)*.

Diallo, O., Diouf, A., Hanan, N.P., Ndiaye, A., and Prevost, Y., 1991. AVHRR monitoring of savanna primary production in Senegal, West Africa: 1987-1988. *International Journal of Remote Sensing*, **12**:1259-1280.

Duggin, M.L., Piwinski, D., Whitehead, V., and Tyland, G., 1982. Evaluation of NOAA AVHRR data for crop assessment. *Applied Optics*, **21**:1873-1875.

Eastman, R.R., and Fulk, M., 1993. Long sequence time-series evaluation using standard principal components. *Photogrammetric Engineering and Remote Sensing*, **59**:991-996.

Eck, T.F., and Kalb, V.L., 1991. Cloud-screening for Africa using a geographically and seasonally variable infrared threshold. *International Journal of Remote Sensing*, **12**:1205-1222.

Emery, W.J., Brown, J., and Nowak, Z.P., 1989. AVHRR image navigation: summary and review. *Photogrammetric Engineering and Remote Sensing*, **55**:1175-1183.

Fung, I.Y., Tucker, C.J., and Prentice, K.C., 1986. On the applicability of the AVHRR vegetation index to study the atmosphere-biosphere exchange of CO_2. *Journal of Geophysical Research*, **92**:2999-3015.

Gallo, K.P., Daughtry, C.S.T., and Bauer, M.E., 1985. Spectral estimation of absorbed photosynthetically active radiation in corn canopies. *Agronomy Journal*, **78**:752-756.

Gatlin, J.A., Sullivan, R.J., and Tucker, C.J., 1984. Considerations of and improvements to large-scale vegetation monitoring. *IEEE Trans. Geosci. Remote Sens.*, **GE22**:496-502.

Goward, S.A., 1989. Satellite bioclimatology. *Journal of Climate*, **7**:710-720.

Goward, S.A., Tucker, C.J., and Dye, D., 1985. North American vegetation patterns observed with the NOAA-7 advanced very high resolution radiometer. *Vegetatio*, **64**:3-14.

Goward, S.A., Kerber, A., Dye, D., and Kalb, V., 1987. Comparison of North and South American biomes from AVHRR observations. *Geocarto*, **2**:27-40.

Goward, S.N., Markham, B., Dye, D.G., Dulaney, W., and Yang, J., 1990. Normalized difference vegetation index measurements from the advanced very high resolution radiometer. *Remote Sensing of Environment*, **35**:257-277.

Goward, S.N., Dye, D.G., Turner, S., and Yang, J., 1993. Objective assessment of NOAA global vegetation index data product. *International Journal of Remote Sensing* (in press).

Gray, T.I., and McCrary, D.G., 1981a. Meteorological Satellite Data - A Tool to Describe the Health of the World's Agriculture. AgRISTARS Report EW-NI-04042, Johnson Space Center, Houston, Texas, 7 p.

Gray, T.I., and McCrary, D.G., 1981b. The environmental vegetation index, a tool potentially useful for arid land management. In, *Proc. 15th Conf. on Agriculture and Forest Meteorology and 5th Conf. on Biometeorology*, American Meteorological Soc., pp. 205-207.

Greegor, D.H., and Norwine, J., 1981. A Gradient Model of Vegetation and Climate Utilizing NOAA Satellite Imagery - Phase I: Texas Transect. AgRISTARS Report JSC-17435, FC-J1-04176, Johnson Space Center, Houston, Texas, 58 p.

Gutman, G.G., 1991. Vegetation indices from AVHRR: An update and future prospects. *Remote Sensing of Environment*, **35**:121-136.

Hanan, N.P., Prince, S.D., and Hiernaux, P.H.Y., 1991. Spectral modelling of multi-component landscapes in the Sahel. *International Journal of Remote Sensing*, **12**:1243-1259..

Hardisky, M.A., Daiber, F.C., Roman, C.T., and Klemas, V., 1984. Remote sensing of biomass productivity of a salt marsh. *Remote Sensing of Environment*, **16**:91-106.

Hatfield, J.L., Asrar, G., and Kanemasu, E.T., 1984. Intercepted photosynthetically active radiation in wheat canopies estimated by spectral reflectance. *Remote Sensing of Environment*, **14**:65-76.

Henrickson, B.L., and Durkin, J.W., 1986. Growing period and drought early warning in Africa using satellite data. *International Journal of Remote Sensing*, **7**:1515-1532.

Hielkema, J.U., Prince, S.D., and Astle, W.L., 1986. Rainfall and vegetation monitoring in the Savanna Zone of the Democratic Republic of Sudan using NOAA advanced very high resolution radiometer. *International Journal of Remote Sensing*, **7**:1499-1514.

Hielkema, J.U., Roffey, J., and Tucker, C.J., 1986. Assessment of ecological conditions associated with the 1980/1981 desert locust plague upsurge in West Africa using environmental satellite data. *International Journal of Remote Sensing*, 7:1609-1622.

Hiernaux, P.H.Y., and Justice, C.O., 1986. Suivi du developpement vegetal au cours de l'ete 1984 dans le Sahel Malien. *International Journal of Remote Sensing*, 7:1515-1532.

Holben, B.N., 1986. Characteristics of maximum-value composite images from temporal AVHRR data. *International Journal of Remote Sensing*, 7:1417-1434.

Holben, B.N., Eck, T.F., and Fraser, R.S., 1991. Temporal and spatial variability of aerosol optical depth in the Sahel region in relation to vegetation remote sensing. *International Journal of Remote Sensing*, 12:1147-1164.

Holben, B.N., and Fraser, R.S., 1984. Red and near-infrared sensor response to off-nadir viewing. *International Journal of Remote Sensing*, 5:145-160.

Holben, B.N., Kaufman, Y.J., and Kendall, J., 1990. NOAA-11 AVHRR visible and near-infrared inflight calibration. *International Journal of Remote Sensing*, 11:1511-1519.

Holben, B.N., Kimes, D., and Fraser, R.S., 1986. Directional reflectance response in AVHRR red and near-ir bands for three cover types and varying atmospheric conditions. *Remote Sensing of Environment*, 19:213-236.

Huete, A.R., Post, D.F., and Jackson, R.D., 1985. Soil spectral effects on 4-space vegetation discrimination. *Remote Sensing of Environment*, 15:155-165.

Huete, A.R., and Jackson, R.D., 1988. Soil and atmosphere influences on the spectra of partial canopies. *Remote Sensing of Environment*, 25:89-105

Huete, A.R., and Tucker, C.J., 1991. Investigation of soil influences in AVHRR red and near-infrared vegetation index imagery. *International Journal of Remote Sensing*, 12:1223-1242.

Hutchinson, C.F., 1991. Use of satellite data for famine early warning in sub-Saharan Africa. *International Journal of Remote Sensing*, 12:1405-1421.

Justice, C.O., 1986. Special edition entitled *Monitoring the Grasslands of Semi-Arid Africa Using NOAA AVHRR Data*. *International Journal of Remote Sensing*, 7:1383-1622.

Justice, C.O., Dugdale, G., Townshend, J.R.G., Narracott, A.S., and Kumar, M., 1991. Synergism between NOAA AVHRR and Meteosat data for studying vegetation development in semi-arid West Africa. *International Journal of Remote Sensing*, 12:1349-1368.

Justice, C.O., Eck, T.F., Tanré, D., and Holben, B.N., 1991. The effect of water vapor on the normalized difference vegetation index derived for the Sahelian regions from NOAA AVHRR data. *International Journal of Remote Sensing*, 12:1165-1188.

Justice, C.O., and Hiernaux, P., 1986. Monitoring the grasslands of the Sahel using NOAA AVHRR data: Niger 1983. *International Journal of Remote Sensing*, 7:1475-1498.

Justice, C.O., Holben, B.N., and Gwyne, M.D., 1986. Monitoring East African vegetation using AVHRR data. *International Journal of Remote Sensing*, 7:1453-1474.

Justice, C.O., Markham, B.L., Townshend, J.R.G., and Kennard, R., 1988. Spatial degradation of satellite data. *International Journal of Remote Sensing*, 10:1539-1561.

Justice, C.O., Townshend, J.R.G., Holben, B.N., and Tucker, C.J., 1985. Analysis of the phenology of global vegetation using meteorological satellite data. *International Journal of Remote Sensing*, **6**:1271-1318.

Justice, C.O., Townshend, J.R.G., and Markham, B.L., 1987. MODIS spatial resolution study, *International Journal of Remote Sensing*, **8**:1271-1318.

Kaufman, Y.J., and Holben, B.N., 1993. Calibration of the AVHRR visible and near-IR bands by atmospheric scattering, ocean glint, and desert reflection. *International Journal of Remote Sensing*, **12**:21-52.

Kaufman, Y.J., Tucker, C.J., and Fung, I.Y., 1990. Remote sensing of biomass burning in the tropics. *Journal of Geophysical Research*, **95**:9927-9939.

Kidwell, K.B., 1988. *NOAA Polar Orbital Data Users Guide* (NOAA National Climate Data Center, Washington, D.C).

Kidwell, K.B., 1990. *Global Vegetation Index Users Guide* (NOAA National Climate Data Center, Washington, D.C).

Kimes, D.S., 1983. Dynamics of directional reflectance factor distributions for vegetation canopies. *Applied Optics*, **22**:1364-1372.

Kimes, D.S., Holben, B.N., Tucker, C.J., and Newcomb, W.W., 1984. Optimal directional view angles for remote sensing missions. *International Journal of Remote Sensing*, **5**:877-891.

Kimes, D.S., Markham, B.L., Tucker, C.J., and McMurtrey, J.E., 1981. Temporal relationships between spectral response and agronomic variables of a corn canopy. *Remote Sensing of Environment*, **11**:401-411.

Kumar, M., and Monteith, J.L., 1982. Remote sensing of plant growth, In, *Plants and the Daylight Spectrum* (H. Smith, ed.), Pitman, London, pp. 133-144.

Linthicum, K.J., Bailey, C.L., Davies, F.G., and Tucker, C.J., 1987. Detection of Rift Valley Fever viral activity in Kenya by satellite remote sensing imagery. *Science*, **235**:1656-1659.

Linthicum, K.J., Bailey, Tucker, C.J., Mitchell, K.D., Logan, T.M., Davies, F.G., Kamu, C.W., Thande, P.C., and Wagateh, J.N., 1990. Application of polar-orbiting, meteorological satellite data to detect flooding of Rift Valley Fever virus vector mosquito habitats in Kenya. *Med. Vet. Entom,*. **4**:433-438.

Los, S.O., 1993. Calibration adjustment of the NOAA AVHRR normalized difference vegetation index without recourse to component channel 1 and 2 data. *International Journal of Remote Sensing*, **14**:1907-1917.

Malingreau, J.P., 1986. Global vegetation dynamics: satellite observations over Asia. *International Journal of Remote Sensing*, **7**:1121-1146.

Malingreau, J.P., Stephens, G., and Fellows, L., 1985. Remote sensing of forest fires: Kalimantan and North Borneo in 1982-1983. *Ambio*, **14**:314-321.

Malingreau, J.P., and Tucker, C.J., 1987. Measuring tropical forest disturbances using NOAA satellite data. The case of the Amazon Basin. *Proc. IGARSS '87 Symp.*, Ann Arbor, Michigan, 18-21 May 1987, (Piscataway, New Jersey: Inst. Electrical and Electronic Engineers), pp. 443-448.

Malingreau, J.P., and Tucker, C.J., 1987. La vegetation vue de l'espace. *La Recherche*, **18**:180-189.

Malingreau, J.P., and Tucker, C.J., 1988. Large-scale deforestation in the Southeastern Amazon Basin of Brazil. *Ambio*, **17**:49-55.

Malingreau, J.P., Tucker, C.J., and Laporte, N,. 1989. AVHRR for monitoring global tropical deforestation. *International Journal of Remote Sensing*, **10**:855-867.

Malo, A.R., and Nicholson, S.N., 1990. A study of rainfall and vegetation dynamics in the African Sahel using normalized difference vegetation index. *Journal of Arid Environments*, **19**:1-24..

Matson, M., and Dozier, J., 1981. Identification of sub-resolution high temperature sources using thermal IR sensor. *Photogrammetric Engineering and Remote Sensing*, **47**:131-138.

Matson, M., Stephens, G., and Robinson, J., 1987. Fire detection using data from the NOAA-n satellites. *International Journal of Remote Sensing*, **8**:961-970.

Nelson, R., and Holben, B.N., 1986. Identifying deforestation in Brazil using multi-resolution satellite data. *International Journal of Remote Sensing*, **7**:429-448.

Nicholson, S.E., Davenport, M.L., and Malo, A.R., 1990. A comparison of the vegetation response to rainfall in the Sahel and East Africa, using normalized difference vegetation index from NOAA AVHRR. *Climate Change*, **17**:209-242.

Nicholson, S.E., 1989. African drought: characteristics, causal theories, and global teleconnections. In, *Understanding Climate Change* (Berger, A., Dickinson, R.E., and Kidson, J.W., editors), American Geophysical Union, Washington, D.C., pp. 79-100.

Ormsby, J.P., 1982. Classification of simulated and actual NOAA-6 AVHRR data for hydrological land-surface feature definition. *IEEE Trans. Geosci. Remote Sens.*, **GE-20**:262-268.

Otterman, J., and Tucker, C.J., 1984. Satellite measurements of surface albedo and temperature in semi-desert. *Journal of Climate and Applied Meteorology*, **24**:228-235.

Pearson, R.L., and Miller, L.D., 1972. Remote mapping of standing crop biomass for estimation of the productivity of the shortgrass prairie. In, *Proc. 8th Intl. Symp. Remote Sens. Environ.*, Univ. Michigan, Ann Arbor, pp. 1357-1381.

Perry, C.R., and Lautenschlager, L.F., 1984. Functional equivalence of spectral vegetation indices. *Remote Sensing of Environment*, **14**:169-182

Price, J., 1987. Calibration of satellite radiometers and the comparison of vegetation indices. *Remote Sensing of Environment*, **21**:15-27. .

Price, J., 1991, Timing of the NOAA afternoon passes. *International Journal of Remote Sensing*, **12**:193-198.

Prince, S.D., 1991a., Satellite remote sensing of primary production: comparison of results for Sahelian grasslands 1981-1988. *International Journal of Remote Sensing*, **12**:1301-1312.

Prince, S.D., 1991b A model of regional primary production for use with coarse-resolution satellite data. *International Journal of Remote Sensing*, **12**:1313-1330.

Prince, S D., and Astel, W.L., 1986. Satellite remote sensing of rangelands in Botswana. I. Landsat MSS and herbaceous vegetation. *International Journal of Remote Sensing*, **7**:1533-1554.

Prince, S.D. and Justice, C.O., 1991. Special edition entitled *"Coarse Resolution Remote Sensing of the Sahelian Environment"*. *International Journal of Remote Sensing*, **12**:1133-1421.

Prince, S.D., and Tucker, C.J., 1986. Satellite remote sensing of rangelands in Botswana. II. NOAA AVHRR and herbaceous vegetation. *International Journal of Remote Sensing*, **7**: 1555-1570.

Schneider, S.R., McGinnis, D.F., and Gatlin, J.A., 1981. *Use of NOAA AVHRR Visible and Near-Infrared Data for Land Remote Sensing*. NOAA Technical Report NESS 84, NOAA National Earth Satellite Service, Washington, D.C., 48 p.

Schneider, S.R., and McGinnis, D.F., 1982. The NOAA AVHRR: A new satellite sensor for monitoring crop growth. In, *Proc. 8th Int. Symp. Machine Processing Remotely Sensed Data*, Purdue University, Indiana, USA, pp. 281-290.

Schutt, J., and Kerber, A., 1986. Utility of AVHRR channels 3 and 4 in land cover mapping. *Photogrammetric Engineering and Remote Sensing*, **52**:1877-1883.

Sellers, P.J., 1985. Canopy reflectance, photosynthesis, and transpiration. *International Journal of Remote Sensing*, **6**: 1335-1372.

Sellers, P.J., 1987. Canopy reflectance, photosynthesis, and transpiration. II. The role of biophysics in the linearity of their interdependence. *Remote Sensing of Environment*, **21**:143-183.

Setzer, A.W., and Pereira, M.C., 1991. Amazon biomass burning in 1987 and their tropospheric emissions. *Ambio*, **20**:19-27.

Setzer, A.W. *et al.*, 1988. Relatorio de atividades do projeto IBDF-INPE "SEQE"-ANO 1987, Publ. INPE-4534-RPE/565, Inst. de Pesqui. Espacias, Sao Jose dos Campos, Sao Paulo, Brazil.

Singh, S.M., 1988. Simulation of solar zenith angle effect on vegetation index from AVHRR. *International Journal of Remote Sensing*, **9**:237-248.

Smith, G.R., Levin, R.H., Abel, P., and Jacobowitz, H., 1988. Calibration of the solar bands of the NOAA-9 AVHRR using high-altitude aircraft measurements. *J. Atmos. and Oceanogr. Tech.*, **5**:631-639.

Soufflet, V., Tanré D., Begue, A., Podaire, A., and Deschamps, P.Y., 1991. Atmospheric effects on NOAA AVHRR data over Sahelian regions. *International Journal of Remote Sensing*, **12**:1189-1203.

Steven, M.D., Biscoe, P.V., and Jaggard, K.W., 1983. Estimation of sugar beet productivity from reflection in the red and near-infrared spectral bands. *International Journal of Remote Sensing*, **4**:325-334.

Staylor, W.F., 1990. Degradation rates of the AVHRR visible channel from the NOAA-6, NOAA-7, and NOAA-9 spacecraft. *Journal of Atmosphere and Oceanogr. Tech.*, **7**:411-423.

Steinmetz, S., Gucrif, M., Delccollc, R., and Baret, F., 1990. Spectral estimates of the absorbed photosynthetically active radiation and light use efficiency of a winter wheat canopy subjected to nitrogen and water deficiencies. *International Journal of Remote Sensing*, **11**: 1797-1808.

Tanré, D., Holben, B.N., and Kaufman, Y.J., 1992. Atmospheric correction algorithm for NOAA-AVHRR products: theory and application. *IEEE Trans. Geosci. Remote Sens.*, **30**:231-248.

Tarpley, J.P., Schneider, S.R., and Money, R.L., 1984. Global vegetation indices from NOAA-7 meteorological satellite. *Journal of Climate and applied Meteorology*, **23**:491-494.

Tateishi, R., and Kajiwara, K., 1992. Global land cover monitoring by AVHRR NDVI data. *Earth Environment*, **7**:4-14.

Teillet, P.M., Slater, P.N., Ding, Y., Santer, R.P., Jackson, R.D., and Moran, M.S., 1990. Three methods for the absolute calibration of the NOAA AVHRR sensors, in-flight. *Remote Sensing of Environment*, **31**:105-120.

Townshend, J.R.G. and Tucker, C.J., 1981. Utility of AVHRR NOAA-6 and -7 for vegetation mapping. In, *Matching Remote Sensing Technologies and Their Applications Proceedings*, Remote Sens. Soc., London, pp 97-109.

Townshend, J.R.G., Justice, C.O., and Kalb, V., 1987. Characterization and classification of South American land cover types using satellite data. *International Journal of Remote Sensing*, **8**:1189-1207.

Townshend, J.R.G., and Justice, C.O., 1986. Analysis of the dynamics of African vegetation using the normalized difference vegetation index. *International Journal of Remote Sensing*, **7**:1435-1446.

Townshend, J.R.G., and Justice, C.O., 1988. Selecting the spatial resolution of satellite sensors required for global monitoring of land transformations. *International Journal of Remote Sensing*, **9**:187-236.

Townshend, J.R.G., and Justice, C.O., 1990. The spatial variation of vegetation changes at very coarse scales. *International Journal of Remote Sensing*, **11**:149-157.

Townshend, J.R.G., Goff, T.E., and Tucker, C.J., 1985. Multitemporal dimensionality of images of normalized difference vegetation index at continental scales. *IEEE Trans. Geosci. Remote Sens.*, **GE-23**:888-895.

Townshend, J.R.G., and Tucker, C.J., 1984. Objective assessment of AVHRR for land cover mapping. *International Journal of Remote Sensing*, **5**:497-504.

Tucker, C.J., 1979. Red and photographic infrared linear combinations for monitoring vegetation. *Remote Sensing of Environment*, **8**:127-150.

Tucker, C.J., 1980. A critical review of remote sensing and other methods for non-destructive estimation of standing crop biomass. *Grass and Forage Science*, **35**:177-182.

Tucker, C.J., 1989. Comparing SMMR and AVHRR data for drought monitoring. *International Journal of Remote Sensing*, **10**:1663-1672.

Tucker, C.J., Dregne, H.E., and Newcomb, W.W., 1991. Expansion and contraction of the Sahara Desert from 1980 to 1990. *Science*, **253**:299-301.

Tucker, C. J., Fung, I. Y., Keeling, C. D., and Gammon, R. H., 1986. Relationship between atmospheric CO_2 variations and a satellite-derived vegetation index. *Nature*, **319**:195-199.

Tucker, C. J., Gatlin, J. A., and Schneider, S. R., 1984. Monitoring vegetation in the Nile Delta with NOAA-6 and NOAA-7 AVHRR imagery. *Photogrammetric Engineering and Remote Sensing,* **50**:53-61.

Tucker, C.J., Hielkema, J.U., and Roffey, J., 1985. The potential of satellite remote sensing of ecological conditions for survey and forecasting desert-locust activity. *International Journal of Remote Sensing,* **6**:127-138.

Tucker, C.J., Holben, B.N., Elgin, J.H., and McMurtrey, J.E., 1981. Remote sensing of total dry matter accumulation in winter wheat. *Remote Sensing of Environment,* **11**:171-189.

Tucker, C.J., Holben, B.N., and Goff, T.E., 1984. Intensive forest clearing in Rondonia, Brazil, as detected by satellite remote sensing. *Remote Sensing of Environment,* **15**:255-262.

Tucker, C.J., and Matson, M., 1985. Determination of volcanic dust deposition of El Chichon from ground and satellite data. *International Journal of Remote Sensing,* **6**:619-628.

Tucker, C.J., and Sellers, P.J., 1986. Satellite remote sensing of primary production. *International Journal of Remote Sensing,* **7**:1395-1416.

Tucker, C.J., Vanpraet, C.L., Boerwinkel, E., and Gaston, A., 1983. Satellite remote sensing of total dry matter production in the Senegalese Sahel: 1980-1984. *Remote Sensing of Environment,* **13**:461-474.

Tucker, C.J., Vanpraet, C. L., Sharman, M. J., and Van Ittersum, G., 1985. Satellite remote sensing of total herbaceous biomass production in the Senegalese Sahel: 1980-1984. *Remote Sensing of Environment,* **17**:233-249.

Wallin, D.O., Elliott, C.C.H., Shugart, H.H., Tucker C.J., and Wilhelmi, F., 1992. Satellite remote sensing of breeding habitat for an African weaver-bird. *Landscape Ecology,* **7**:87-99.

Wiegand, C.L., and Richardson, A.J., 1984. Leaf area, light interception, and yield estimates from spectral components analysis. *Agronomy Journal,* **76**:543-548.

Wylie, B.K., Harrington, J.A., Prince, S.D., and Denda, I., 1991. Satellite and ground-based pasture production assessment in Niger: 1986-1988. *International Journal of Remote Sensing,* **12**:1281-1300.

Yates, H.W,. and Tarpley, J.D., 1982. The role of meteorological satellites in agricultural remote sensing. In, *Proc. 8th Int. Symp. Machine Processing Remotely Sensed Data,* Purdue University, Indiana, USA, pp. 23-32.

BASIC RADIOMETRIC CALIBRATION OF NOAA-AVHRR DATA

G. D'SOUZA
TREES Project
Institute of Remote Sensing Applications
CEU Joint Research Centre
I-21020 Ispra (VA)
Italy

ABSTRACT. All quantitative analysis of NOAA-AVHRR data should be based on real physical parameters derived from the imagery. In this chapter we cover the basic steps involved in the conversion of the raw digital numbers from the NOAA-AVHRR data stream into real physical quantities such as radiance, top-of-atmosphere reflectance and brightness temperature. Both preflight and inflight calibration of all NOAA AVHRR channels are covered in detail.

1. Introduction

As seen in the previous chapter by Tucker, the redesignation of the visible channel on the NOAA-AVHRR sensors led to the rapid development of a number of land applications, especially the monitoring of vegetation dynamics using vegetation indices. In general, vegetation indices are based on a ratio of *reflectance*, as derived from channels 1 and 2 of the AVHRR sensor. Cloudy pixels are often excluded by examining *brightness temperatures* calculated from data from the thermal channels 4 and 5. In this chapter we review the basic methodology for converting the raw AVHRR 10-bit digital numbers (DN) into more meaningful physical quantities such as reflectance, brightness temperature and radiance. Much of the information presented in this chapter is extracted from various technical manuals and reports including Planet (1988), Rao (1987) and Kidwell (1991).

AVHRR data themselves are transmitted in two different formats: Analogue Picture Transmission (APT) and High-Resolution Picture Transmission (HRPT). APT form a simple analogue representation, which is useful for qualitative examination of imagery, but for any detailed quantitative work the digital HRPT data should be used. These consist of 10-bit data from all the AVHRR channels for 2048 pixels per line of data, and are compressed into 16-bit words, and appended with other ancillary information. Since the sensor was initially designed to view a wide variety of environmental conditions, the range of sensitivity for each channel is very large so in practice only a small part of the available 1024 level range is occupied in any one image.

Although the raw digital counts do provide some information about the relative behaviour of various surfaces (relative brightness and temperatures), for any detailed application, the 10-bit digital numbers must be extracted and converted to real physical parameters. This is the process of data calibration, which is particularly important in

21

G. D'Souza et al. (eds.), Advances in the Use of NOAA AVHRR Data for Land Applications, 21-48.
© 1996 ECSC, EEC, EAEC, Brussels and Luxembourg.

applications where data from different sensors, or data from the same sensor at different times, are used.

The AVHRR sensors basically function in a similar way to any other spectroradiometers on Earth-observing satellites. Radiant energy is guided to the sensor through the optical system where some selective filtering is made on the basis of wavelength. The Instantaneous Field Of View (IFOV) for the different channels varies from about 1.39 to 1.51 milliradians, and the chosen IFOV leads to a resolution at the satellite subpoint of 1.1 km for a nominal altitude of 833 km. Photons incident on the detector from the column being viewed at any one instant, induce minute electrical currents through the photoelectric effect. The resulting analogue current signal is then quantised on-board the satellite at a rate of 39,936 samples per second per channel. Each sample step corresponds to an angle of scanner rotation of 0.95 milliradians so, at this sampling rate, there are 1,362 samples per IFOV. The resulting analogue electrical signal is then amplified and transformed into a digital number (DN). This DN therefore corresponds to the raw measurement recorded by the sensor, *i.e.* its quantised value, which is proportional to the number of photons and thus to the amount of energy that entered the sensor aperture and reached the detector. The overall correspondence between DNs and the amount of incoming radiation is complex and may be altered by a number of factors including signal to noise effects in the induction of electrical currents in the sensor; the loss of information due to the discretisation into a DN; degradation of the optical components; changes in the gain characteristics of the electrical amplification components, and even noise arising from various other electronic systems on the platform.

It is the radiometric calibration process that establishes a relationship between incoming radiant energy at the satellite and the recorded values. In general, radiometric calibration involves exposing a radiometer to known or well-documented sources of radiation (perhaps those that have been calibrated against primary or secondary standards) and then determining a direct relationship between the output of the radiometer and the intensity of the incident radiation.

2.   Pre-flight Calibration

2.1.   MIDDLE AND THERMAL-INFRARED CHANNELS

2.1.1   Normalised sensor response function. In order to characterise their spectral and sensitivity performance, all the radiometers flown on the NOAA series of satellites are extensively tested and radiometrically calibrated before launch. One of the most fundamental parameters derived during the pre-launch calibration is the spectral response function of all the various channels. This is normally determined for fine samples of wavelength (about 0.03 μm), and is related to the relative performance at peak response. The results are therefore normally referred to as normalised spectral response data and may be presented in graphs such as those shown in Figure 1.

The middle and thermal-infrared channels of the AVHRR (channels 3, 4 and 5) are calibrated against precision blackbody sources whose calibration is traceable to the

National Institute of Standards Technology (NIST) of the USA (which was previously known as the National Bureau of Standards (NBS)). The sensitivity of each of these channels is determined for 60 steps of about 0.03-0.06 μm each. By NOAA's conventions, the sensitivity is expressed as a function of wave number (which is the inverse of wavelength). For example, the spectral sensitivity of channel 4 of NOAA-11 is expressed in terms of the normalised spectral response functions provided for 60 steps of 2.505 cm^{-1}, starting at a wave number of 854.7 cm^{-1}. It is important to note that the values used differ slightly for starting wave number, wave number per step and normalised sensor response for each channel of each AVHRR. The values for NOAA's 6-9 are provided in Planet (1979), and the values for subsequent satellites are provided in a number of appendices, published shortly after launch of each new satellite. An example of the normalised spectral responses for channels 3, 4 and 5 for NOAA-11 is provided in Figure 1. Note that NOAA only supply these spectral response function values in terms of wavenumber, so the initial calculations must be made in these units.

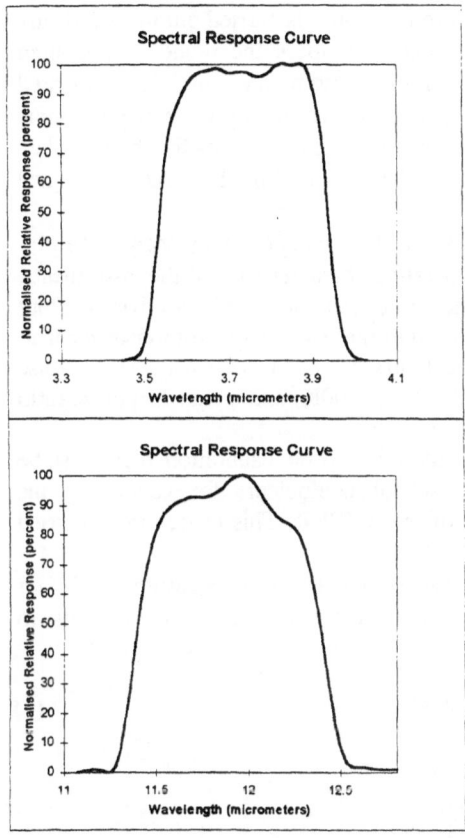
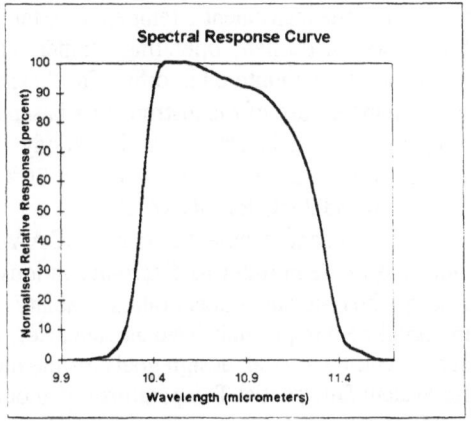

Figure 1.

Normalised sensor response functions for channels 3, 4 and 5 for NOAA 11.

Once the normalised sensor response functions across the wavelength range of sensitivity have been determined, the responsivity of the sensor to different target temperatures has to be established. In a thermally-controlled vacuum chamber where actual space conditions are simulated and the absorption of radiation in the path between the source and radiometer is minimised, the radiometer is made to view in turn a warm calibrated laboratory blackbody (representing the Earth "target"), then a blackbody cooled to approximately 77K (representing the cold space view) and finally its own internal blackbodies. In laboratory testing and inflight, temperatures of all the internal blackbodies are sensed with thermistors (for TOVS-HIRS channels) and platinum resistance thermometers (PRTs) for TOVS-MSU and AVHRR channels.

During the laboratory calibration phase, the warm calibrated laboratory blackbody is cycled through a sequence of temperature plateaux approximately 10K apart between 175 and 320K, which spans the entire range of possible Earth target temperatures (including clouds and oceans). From the laboratory data, for each of the target temperature plateaux, radiances for each channel can then be computed using the inverse Planck law and related to the 0-1023 levels yielded by the sensor. However, as the responsivity is partly a function of the instrument's temperature, the entire procedure is carried out independently for several instrument operating temperatures that encompass the range of operating temperatures encountered in orbit. The "instrument operating temperature" is represented by the temperature of the instrument's baseplate, which is kept at approximately the same temperature as its internal warm blackbody. The operating temperatures for the baseplate used in calibration tests are 10, 15 and 20°C for AVHRRs on-board NOAA's 6-11 and 10, 15, 20 and 25°C for NOAA 12.

The instrument manufacturers and NOAA-NESDIS then independently analyse the pre-launch test data in order to determine various operating characteristics of the instruments, such as their signal-to-noise ratios, stability, linearity of response, and sensitivity (output in digital counts per unit incident radiance). Although these will vary from instrument to instrument, the overall design goals for the thermal infrared channels were a NEDT (Noise Equivalent Differential Temperature) of about 0.12K (at 300K) and a signal to noise ratio (S/N) of 3:1 at 0.5 percent albedo.

Another set of important parameters necessary for in-orbit calibration that must be derived from the pre-launch analysis is the set of coefficients for calibrating the temperature sensors in the internal blackbodies of the AVHRR. This procedure is carried out as follows (Kidwell 1991):

The AVHRR has a single internal blackbody, whose temperature is measured with four PRTs, for each of which there is a set of polynomials to convert its outputs to temperature, $i.e.$

$$T = \sum_{j=0}^{4} a_j x^j \qquad (1)$$

where   X is the PRT output in digital counts,

              T is the temperature in degrees Kelvin and

              a_0 - a_4 are the coefficients that are specific to each PRT.

A set of coefficients for each PRT, determined by the calibration against a thermometric standard, is normally provided by the instrument manufacturers, so there are four sets of five coefficients for each AVHRR sensor. The set of coefficients for the four PRTs from the AVHRR sensor on-board NOAA-14 are provided in Table 1 below for illustration. The average temperature of the internal target is then calculated (normally as the average of the temperatures of the four PRTs).

Because the in-orbit calculation is to be traceable first to the laboratory blackbody and then to the NIST standard, the above coefficients really only relate brightness temperature (as opposed to kinetic temperature) to counts. For the NOAA instruments, the conversion from one to the other is achieved by using the instrument itself to transfer the calibration of the laboratory blackbody to the PRTs. This process utilises data from the pre-launch tests as follows (Kidwell 1991):

- The radiometer calibration, *i.e.* the relationship between target radiance and DN output of the AVHRR in digital counts, is derived from data collected when the radiometer viewed the calibrated blackbody.
- The radiometer's outputs on viewing its internal blackbody are then converted from counts to radiances and then to equivalent brightness temperatures. In other words, the radiometer itself is used to measure the brightness temperature of its internal blackbody. A data set of internal target brightness temperatures versus the DN outputs, in digital counts, of the internal blackbody's PRTs is thereby assembled. Then the number of samples is determined by the number of laboratory blackbody temperature plateaux and the number of instrument operating temperature plateaux.
- Finally, a fifth-order polynomial relating brightness temperature of the internal blackbody to PRT, in counts, is fitted to the data by least-squares regression, yielding the five a_i coefficients for equation (1).

In summary then, the pre-launch calibration relates the AVHRR channel 3, 4 and 5 outputs in digital counts to the radiances of their target. In pre-launch tests, the target used is the laboratory blackbody, and the calibration relationship is calculated as a function of the channel and baseplate temperature. However, one cannot expect the sensitivity to temperature relationships to stay the same in orbit as they were before

Table 1. Coefficients for conversion of PRT counts to temperature for AVHRR on-board NOAA-14.

PRT	a_0	a_1	a_2	a_3	a_4
1	276.597	0.051275	1.363E-06	0.0	0.0
2	276.597	0.051275	1.363E-06	0.0	0.0
3	276.597	0.051275	1.363E-06	0.0	0.0
4	276.597	0.051275	1.363E-06	0.0	0.0

launch. Among the reasons that these may change is the shock and possible outgassing of the filters during and after launch. Another reason is that the thermal environment varies with position in the orbit, resulting in orbitally-varying sensitivity. Furthermore, instrument components age and degrade in the several years that usually elapse between the time of the pre-launch tests and the actual launch itself, and the aging process continues during the two or more years that the instrument typically operates in orbit.

For these reasons, on-board calibration is also provided for all the NOAA-AVHRR radiometers (at least those for channels 3, 4 and 5). In consistency with their performance during the laboratory tests, the thermal and middle-infrared radiometers have been designed to view, in turn, cold space and one or more internal warm blackbodies as part of each of their normal scan sequences in orbit. The temperatures of these blackbodies are regularly sensed by PRTs, so this provides data in the relevant channels for determining signal-to-noise ratios and radiometric slopes and intercepts, as will be described in the section 3.1. Unfortunately, there are no on-board calibration sources for the visible and near-infrared channels. Also, calibration by observation of only two calibration targets (space and internal blackbody/baseplate) provides only two points for calibration and therefore allows only a linear relationship to be derived. This has been found to be sufficiently precise for channel 3 detectors, but not for channels 4 and 5.

2.1.2 Non-linearity corrections. As described above, all the inflight calibration is based on observation of two extrema, so this provides only for a linear calibration correction, which is an inadequate representation of the non-linear relationship of channels 4 and 5 (Rao *et al* 1993b, Weinreb *et al* 1990). Thus, in order to account for the non-linearity in the AVHRR's response in these channels, it must first be determined as follows:

- using the calibration test data for targets at different temperatures, a quadratic equation is fitted by the least squares method to the scene radiance versus AVHRR output count data.
- This quadratic equation is applied to the AVHRR response, in counts, when it viewed its internal blackbody. This determines the radiance of the internal blackbody. In effect, the AVHRR itself is used to transfer the calibration of the laboratory blackbody to the internal blackbody. By carrying out the procedure in this way, no assumptions need be made about the actual emissivity of the internal blackbody.
- Then, using data from the space-view cold target (whose radiance is assumed to be zero) and the internal target, the linear calibration relationship is obtained
- This linear calibration is then applied to the AVHRR output, in counts, obtained when the AVHRR viewed the laboratory blackbody. This produces radiances, one for each of the temperature plateaux of the laboratory blackbody, which are then converted to brightness temperatures by the method outlined in section 3.1.
- Finally, the brightness temperatures thus calculated are subtracted from the actual temperatures of the laboratory blackbody, determined from its PRTs, and the differences so determined are the correction terms.

Note that in this procedure the calibration of the laboratory blackbody is transferred directly to the internal blackbody, and the radiance of the internal blackbody is computed without recourse to the measurements by its four PRTs. However, for in-orbit calibrations, the radiances of the internal blackbody must be based on the measurements by the PRTs. This is why it is important that the calibration of the laboratory blackbody must be transferred to the PRTs of the internal blackbody, as described above.

To account for non-linearities NESDIS conveniently provides a set of corrections for both channel 4 and 5 of each AVHRR sensor (where appropriate). These corrections are tabulated against scene temperature, and there is a separate table for each channel and each baseplate temperature. Users are recommended to interpolate values for actual baseplate and target temperatures. For illustration of the magnitudes of the amounts involved, the non-linear corrections as published for the sensor on-board NOAA-11 are provided in Table 2, and the non-linear corrections are shown plotted against temperature in Figure 2. As can be seen, corrections of up to two degrees are possible.

However, NOAA have recently adjusted their approach to the non-linear correction process, and are now implementing a new procedure which both corrects the non-linearity of channels 4 and 5, and incorporates an adjustment in order to account for the NOAA-14 AVHRR channel 3 space view response (which apparently does not lie on the measured calibration curve derived from measurements of the laboratory target). This new approach results in a linear correction for channel 3 and quadratic approach for channels 4 and 5, and means that non-linear correction tables will not be required for the AVHRRs from NOAA-14 onwards. More details about the new procedure are provided in section 3.1.

Table 2a. Non-linearity correction tables for channel 4.

TARGET TEMP	INTERNAL	BLACKBODY	TEMPERATURE	(deg C)
(K)	10	15	20	25
205	-1.540000e+00	-1.760000e+00	-1.980000e+00	-1.980000e+00
215	-1.820000e+00	-2.020000e+00	-2.200000e+00	-2.200000e+00
225	-1.900000e+00	-2.140000e+00	-2.360000e+00	-2.360000e+00
235	-1.870000e+00	-2.100000e+00	-2.280000e+00	-2.280000e+00
245	-1.700000e+00	-1.960000e+00	-2.220000e+00	-2.220000e+00
255	-1.410000e+00	-1.720000e+00	-2.030000e+00	-2.030000e+00
265	-1.060000e+00	-1.370000e+00	-1.660000e+00	-1.660000e+00
275	-4.500000e-01	-7.900000e-01	-1.150000e+00	-1.150000e+00
285	2.400000e-01	-2.100000e-01	-6.700000e-01	-6.700000e-01
295	1.050000e+00	6.800000e-01	2.200000e-01	2.200000e-01
305	2.230000e+00	1.730000e+00	1.320000e+00	1.320000e+00
310	2.850000e+00	2.330000e+00	1.910000e+00	1.910000e+00
315	3.500000e+00	2.980000e+00	2.550000e+00	2.550000e+00
320	4.290000e+00	3.710000e+00	3.250000e+00	3.250000e+00

Table 2b. Non-linearity correction tables for channel 5.

TARGET TEMP	INTERNAL	BLACKBODY	TEMPERATURE	(deg C)
(K)	10	15	20	25
205	-1.150000e+00	-1.270000e+00	-1.230000e+00	-1.230000e+00
215	-1.120000e+00	-1.240000e+00	-1.160000e+00	-1.160000e+00
225	-9.400000e-01	-1.060000e+00	-1.160000e+00	-1.160000e+00
235	-8.400000e-01	-1.020000e+00	-1.000000e+00	-1.000000e+00
245	-7.200000e-01	-9.000000e-01	-9.200000e-01	-9.200000e-01
255	-6.000000e-01	-7.700000e-01	-7.800000e-01	-7.800000e-01
265	-3.700000e-01	-5.100000e-01	-6.000000e-01	-6.000000e-01
275	-1.900000e-01	-3.400000e-01	-4.700000e-01	-4.700000e-01
285	7.000000e-02	-7.000000e-02	-2.300000e-01	-2.300000e-01
295	4.300000e-01	2.800000e-01	9.000000e-02	9.000000e-02
305	8.500000e-01	6.400000e-01	4.700000e-01	4.700000e-01
310	1.050000e+00	8.400000e-01	7.000000e-01	7.000000e-01
315	1.230000e+00	1.030000e+00	8.900000e-01	8.900000e-01
320	1.430000e+00	1.260000e+00	1.120000e+00	1.120000e+00

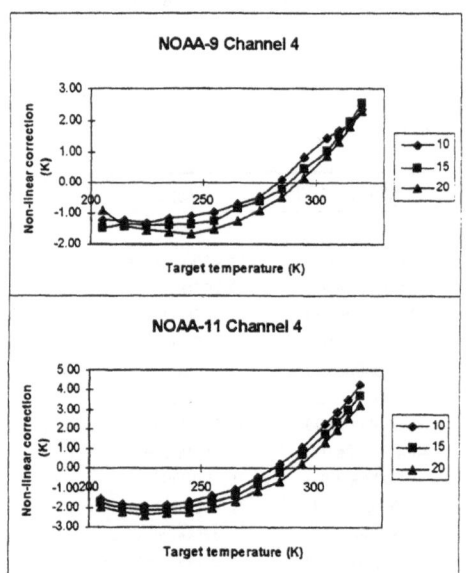
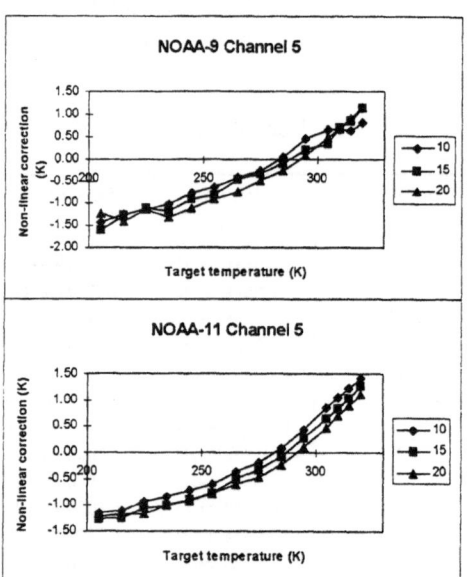

Figure 2. Non-linear corrections for Channels 4 and 5 for NOAA's 9, and 11, at internal target temperatures of 10, 15 and 20 degrees Centigrade.

2.2.   VISIBLE AND NEAR-INFRARED CHANNELS

Similarly to the radiometric calibration for the thermal and mid-infrared channels, the first thing to be determined for the visible and near-infrared channels is their spectral sensitivity. This is provided diagramatically in Kidwell (1991), and an example of the spectral sensitivity curves for channel 1 and 2 of the NOAA-11 AVHRR is provided in Figure 3 below.

Next, the actual pre-launch procedure for visible and near-infrared channel calibration is carried out. This is done at ambient temperature and involves the observation of an aperture cut into an internally illuminated sphere with optically diffusing walls. The value of the spectral radiance emerging through the aperture shows strong uniformity across the aperture, and is traceable to the NIST standard of spectral radiance in the visible and near-infrared region of the spectrum (Rao 1987). The calibration source used in this case is a large aperture integrating sphere equipped with 12 calibrated quartz-halogen lamps. These lamps are carefully selected to match each other as closely as possible in spectral output and operating current. The sphere is calibrated with all 12 lamps on, against a NIST secondary standard of spectral irradiance. Then the ratio of the output of n lamps to that of the 12 lamps is also determined. This yields the spectral output of the sphere when any number of lamps, n, is on. By varying the number of bulbs which are turned on, a calibration curve from the dark level to a maximum of 12 lamp outputs can be obtained. As defined in Kidwell (1991) the following calculations must be made to present the calibration in terms of percent albedo against radiometer output. First, the spectral output of the sphere is integrated with the spectral response function of the AVHRR channel to yield an effective radiance for the spectral band for 12 lamps operating. This is then multiplied by the appropriate K_n factor to convert n lamps, as described by the following equation:

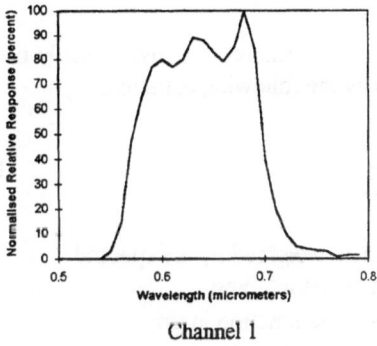
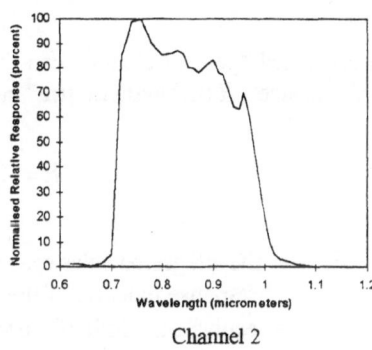

Channel 1   Channel 2

Figure 3. Spectral Sensitivity Curves for Channels 1 and 2 of NOAA-11

$$N_L = K_n \int_{\lambda_1}^{\lambda_2} C(\lambda)\phi(\lambda)d\lambda \tag{2}$$

where N_L = effective radiance as seen by the channel in the appropriate spectral band
K_n = the factor to convert to radiance for n lamps
$C(\lambda)$ = calibrated spectral radiance of the sphere with 12 lamps on
λ = wavelength, in the spectral region λ_1 to λ_2
$\phi(\lambda)$ = the measured spectral response of the channel being calibrated

Similarly, if one takes the solar irradiance at the top of the atmosphere and performs a similar calculation, the results are:

$$N_s = \frac{1}{\Pi} \int_{\lambda_1}^{\lambda_2} S(\lambda)\phi(\lambda)d\lambda \tag{3}$$

where N_S = effective radiance of the radiometer viewing reflected sunlight
$S(\lambda)$ = spectral irradiance viewed at the top of the atmosphere
$\phi(\lambda)$ = spectral response function of the channel

Thus the resultant N_S represents what would be "seen" from space with a 100% reflecting, diffuse surface when the solar zenith angle is zero, *i.e.* normalised. The percent "albedo", A, then is calculated using the following equation:

$$A = \frac{N_L}{N_S} x100 \tag{4}$$

This calculated "percent albedo" can then be related to the more conventionally used "spectral radiance", R (in Watts/m^2 μm^{-1} steradian^{-1}) by the following equation:

$$R = A\left[\frac{F}{100.W.\Pi}\right] \tag{5}$$

where F = the integrated solar spectral irradiance, weighted by the spectral response function of the channel in Watts/m^2, and
W = equivalent width of the spectral response function in μm.

The values for F and W for all the relevant NOAA AVHRR instruments are provided in Table 3.

Table 3. Values for W and F for AVHRR Channels 1 and 2.

Satellite	W_1	F_1	W_2	F_2
TIROS-N	0.325	443.3	0.303	313.5
NOAA-6	0.109	179.0	0.223	233.7
NOAA-7	0.108	177.5	0.249	261.9
NOAA-8	0.113	183.4	0.230	242.8
NOAA-9	0.117	191.3	0.239	251.8
NOAA-10	0.108	178.8	0.222	231.5
NOAA-11	0.113	184.1	0.229	241.1
NOAA-12	0.124	200.1	0.219	229.9
NOAA-14	0.136	221.4	0.245	252.3

As the visible and near-infrared sensors are also assumed to behave in a linear manner, so the final pre-launch procedure aims to determine *slope* and *intercept* calibration coefficients such that:

$$A = M\,(DN) + I, \tag{6}$$

where A = "albedo" (as referred to by NOAA).

M = the channel slope

and I = the channel intercept.

The slope and intercept values for all channels have been published by NOAA as pre-launch calibration coefficients and these are summarised in Table 4. To convert DNs to radiances, the user must multiply the DN by the appropriate slope value provided in Table 4, add the intercept value, then apply equation 5 using appropriate values for F and W.

However, sensor performance degradation during storage, during launch and post-launch also occurs in visible and near-infrared sensors, but unfortunately no on-board

Table 4. Pre-launch slopes and intercepts for NOAA AVHRR channels 1 and 2.

Satellite	M_1	I_1	M_2	I_2
TIROS-N	0.1071	-3.9	0.1051	-3.5
NOAA-6	0.1071	-4.1	0.1058	-3.5
NOAA-7	0.1068	-3.4	0.1069	-3.5
NOAA-8	0.1060	-4.2	0.1060	-4.2
NOAA-9	0.1063	-3.8	0.1075	-3.9
NOAA-10	0.1059	-3.7	0.1058	-3.6
NOAA-11	0.0950	-3.8	0.1061	-3.6
NOAA-12	0.1042	-4.4	0.1014	-4.0
NOAA-14	0.1081	-3.9	0.1090	-3.7

calibration is provided for these channels. Hence, much research has since been carried out on the derivation of post-launch calibration coefficients using Earth targets. This research has indicated that the amount of degradation is considerable, rendering the use of pre-launch calibration coefficients inappropriate for a large number of applications. The definition and derivation of post-launch calibration information, and more appropriate visible and near-infrared calibration coefficients are provided in the section 3.2 below.

3.   In-flight calibration: practical considerations

3.1.   THERMAL CHANNELS

3.1.1. NOAAs 6-12.   As already mentioned, one cannot expect radiometric characteristics of radiometers to be the same in orbit as they were before launch. One reason is that the instrument components age, and their performance deteriorates in the several years that usually elapse between the time of the pre-launch tests and launch itself, and the aging process continues during the two or more years that the instrument typically operates in orbit. Also, the thermal environment varies with position in the orbit, causing sensitivities to vary orbitally. Therefore, the TIROS/NOAA radiometers have been designed to view cold space and one or more internal warm blackbodies as part of their normal scan sequences in orbit. The temperatures of these blackbodies are sensed by thermistors or PRTs, which provide data in the middle and thermal infrared channels for determining signal-to-noise and radiometric slopes and intercepts, as will be described in the following sections.

However, as only two points on the calibration curve are derived, these are sufficient to determine only a straight-line approximation for the calibration. The linear assumption is appropriate for channel 3 which has an indium antimonide (InSb) detector, but is not adequate for channels 4 and 5 which are based on mercury cadmium telluride (HgCdTe) detectors. Thus, for accurate brightness temperature derivations from channels 4 and 5, non-linear corrections have to be incorporated. In practice, the calculations are made as if all sensors behaved linearly and non-linear corrections are added subsequently.

The information required for producing AVHRR middle and thermal infrared channel calibration coefficients is located in the 103-word HRPT header. Header words 18, 19 and 20 contain a five-point subcommutation of the outputs of the four PRTs that monitor the temperature of the internal blackbody. Any of these words, when extracted from five consecutive HRPT minor frames, produces a reference value (REF) and one sample of each of the four PRTs. The pattern is as follows:

n	REF
n+1	PRT1
n+2	PRT2
n+3	PRT3
n+4	PRT4
n+5	REF

The reference value is easily identified as it is the only output having a count value of less than 10. NESDIS recommends the averaging of 10 samples from each PRT to produce a mean PRT count value for conversion to temperature units. The 30 words of internal target data (header words 23-52) provide 10 samples each for channels 3, 4 and 5. The 50 words of space view data (header words 53-102) provide 10 samples each for all five AVHRR channels. NESDIS itself averages 50 samples of space and internal target radiometric data per channel to produce mean count values. To calculate the internal blackbody radiance and corresponding temperature, it is necessary to compute the target temperature, which can be done using the appropriate PRT conversion coefficients as described in section 2.1.1 and equation 1.

Next, the slope and intercept values are derived such the digital counts (DN) of each channel is a linear function of the observed radiance according to:

$$R = M(DN) + I \tag{7}$$

where M = the channel slope
and I = the channel intercept.

The quantity M (in units of radiance/count) is calculated for each channel from the equation:

$$M = (R_t-R_{sp})/(DN_t-DN_{sp}) \tag{8}$$

where R_t = is the average radiance (of 50 consecutive samples) when the instrument views its internal radiance calibration target
 R_{sp} = is the average radiance of deep space (of 50 consecutive samples)
 DN_t = is the average digital count of the internal target (of 50 consecutive samples).
and DN_{sp} = is the average digital count of deep space (of 50 consecutive samples).

The intercept can then be calculated for each channel by:

$$I = R_{sp} - M(DN_{sp}) \tag{9}$$

Typical values for the slopes for channels 3, 4 and 5 are of the order of -0.0015, -0.16 and -0.15 respectively, and typical values for the intercepts are of the order of 1.5, 160.0 and 150.0 for channels 3, 4 and 5 respectively. Note that the slope values are all negative, so that in uncalibrated imagery low DN values represent pixels of high radiance and pixels yielding high digital numbers represent areas of low radiance. The slope and intercept values are unlikely to vary a great deal over a few minutes of AVHRR HRPT data, so most users find adequate the use a single value per image for slope and intercept (perhaps

most users find adequate the use a single value per image for slope and intercept (perhaps an average over the whole image). For those users who obtain SHARP-format data from ESA or NOAA-Level 1b format data from NOAA, derived slope and intercept values for all three channels are already calculated and included in the data stream (see chapters by Arino and Kidwell).

3.1.2. Digital Count to Radiance to Temperature Conversions:

NOAA method 1 (Central Wavenumber approach): One recommended way of converting derived radiances from channels 3, 4 and 5 to brightness temperature estimates is provided in Kidwell (1991). The scaled thermal channel slope values derived by the methods shown above are in units of radiance per count, and the intercept values are in radiance units. Thus, the amount of radiant energy measured by the sensor in any one channel (say channel i) can then be computed as a linear function of the input digital numbers as follows:

$$R_i = M_i(DN) + I_i \tag{10}$$

where R_i is the radiant energy value in $mWm^{-2}sr^{-1}\mu m^{-1}$, and DN is the input digital count (ranging from 0-1023). Then, to convert the value of radiance into a brightness temperature estimate, Kidwell (1991) recommends the use of the non-standard formulation of the Planck radiation equation:

$$T(R) = \frac{c_2 v}{ln(1 + \frac{c_1 v^3}{R})} \tag{11}$$

where T is the temperature in degrees Kelvin for the radiance value R,

         v is the central wave number of the channel filter (in cm^{-1}),

and c_1 and c_2 are the Planck constants ($c1 = 1.1910659 \times 10^{-5}\ mWm^{-2}sr^{-1}cm^4$

and $c2$ $= 1.438833\ cmK$).

Note that the brightness temperatures obtained by this procedure are not corrected for atmospheric attenuation, *etc.* but represent "Top Of Atmosphere" brightness temperatures. Also, non-linear corrections for channels 4 and 5 are still required. Furthermore, the use of this simplified Planck formulation is valid only over limited temperature ranges, and requires the calculation of a number of "central wave numbers" which minimise the error introduced over the temperature range considered. NOAA have calculated and published the central wave numbers (in cm^{-1}) for channels 3, 4 and 5 as a function of temperature range for each satellite (see Kidwell 1991). As an example those for NOAA-11 are included in Table 5. Although simple to apply, this method of conversion from radiance to brightness temperature does require that the user has a specific temperature range of interest for his own particular application.

Table 5. Central wavenumber values (v_i) for the channels 3, 4 and 5 of the NOAA-11 AVHRR sensor (from Wooster *et al* 1995).

Valid temperature range (K)	Central Wave Numbers		
	Channel 3 (cm^{-1})	Channel 4 (cm^{-1})	Channel 5 (cm^{-1})
180-225	2663.50	926.81	841.40
225-275	2668.15	927.36	841.81
275-320	2671.40	927.83	842.20
270-310	2670.96	927.75	842.14

ESA and NOAA method 2 (Sensor spectral sensitivity weighted method) : The method of converting radiance to brightness temperature estimates described above assumes a perfect spectral response over a limited range of wavelength centered around the central wave number. Obviously this is a simplification which is only valid over a limited temperature range. Planet (1979, 1988) and Buongiorno (1992) describe a more precise method of converting radiance to brightness temperature values. This method accounts for each channel's normalised spectral responses over the whole wavelength range, and yields a result valid for any temperature range.

In the first step of this method, the radiance is calculated from the DN as before (*i.e.* by application of equation (10)). Then, under the assumption that the Earth emits as a blackbody in the spectral wavelength of interest, the radiance R_i derived above can be considered as the radiance, sensed in the channel i, from a blackbody at temperature T_i. Thus R_i corresponds to the weighted mean of the Planck function over the spectral response function of the channel i. In terms of discrete steps for the spectral response function:

$$L_i(T_i) = \Sigma_n B(v_{i,n}, T_i) * \Phi(v_{i,n}) / \Sigma_n \Phi(v_{i,n}), \quad \text{for i=3,5 and n=0,59} \qquad (12)$$

where $B(v_{i,n}, T_i)$ is the Planck's function for a blackbody, given in terms of wavenumber as:

$$B(v_{i,n}, T_i) = c_1 v_{i,n}^3 / \left[exp(c_2 v_{i,n} / T_i) - 1 \right] \text{ in mWm}^{-2}\text{sr}^{-1}\text{cm}$$

and

$v_{i,n}$	$= vI_i + n* \Delta v_i,$	(cm^{-1})
	the wave number in the spectral bandwidth of channel i	
vI_i	= starting wave number of channel i	(cm^{-1})
Δv_i	= wave number increment for channel i	(cm^{-1})
$\Phi(v_{i,n})$	= band i spectral response function at the wave number $v_{i,n}$	
T_i	= band i brightness temperature	(K)

The values vl_i, Δv_i and the band spectral responses $\Phi(v_{i,n})$ for 60 values of $v_{i,n}$ are available in Planet (1979 and updates) and reproduced in Buongiorno (1992). Because these values are only supplied in terms of wavenumber and wavenumber increments, only the Planck formulation provided above can be used conveniently. As this equation is not easily reversible, in practice, users construct a look-up table of radiances for each channel for the range and precision of brightness temperatures they require for their application. This look-up table need be constructed only once, but it is important to note that the non-linear corrections have not yet been included.

Figure 4 shows the range of temperature differences that may arise from the use of the two different methods for radiance to brightness temperature conversion. As can be seen, the amount of temperature difference error is unlikely to be more than 0.1 degrees, unless inappropriate central wavenumbers are used. The likely error is larger for channel 3 than for the thermal channels. Errors resulting from the different methods of radiance to brightness temperature conversion are very small in comparison with errors of non-linear correction for channels 4 and 5.

3.1.3. Non-linearity corrections. Once the radiance values have been converted to brightness temperatures, the non-linear corrections should be applied (for pre-NOAA 14 sensors). This requires the use of the non-linear correction tables, interpolating values according to the calculated brightness temperature, and the derived temperature of the internal calibration target.

3.1.4. NOAA-14 changes. Tests on the AVHRR sensor flown on-board NOAA-14 have shown that the channel 3 space view response (which should be zero) does not lie on the measured (linear) calibration curve derived from measurements of the laboratory target. Thus, from this sensor onwards, NESDIS will implement a new radiance correction procedure which corrects both the non-linearity of channels 4 and 5 and the observed effect of the space point from the calibration curve for channel 3. As was the case with previous instruments, gain and slope calibration coefficients are derived which allow the user to perform a linear calibration. However, these coefficients are derived using a non-zero radiance for space, R_{SP}, which is a constant for each channel and which makes the radiance corrections *independent of the internal calibration target temperature*. The gain and intercept coefficients are derived with the following equations:

$$M = (R_T - R_{SP})/(DN_T - DN_{SP})$$
$$I = R_{SP} - M \times DN_{SP} \tag{13}$$

where
DN_{SP}	=	counts for space
M	=	gain
I	=	intercept
R_T	=	radiance for target
R_{SP}	=	radiance for space (as given in Table below)
DN_T	=	counts for target

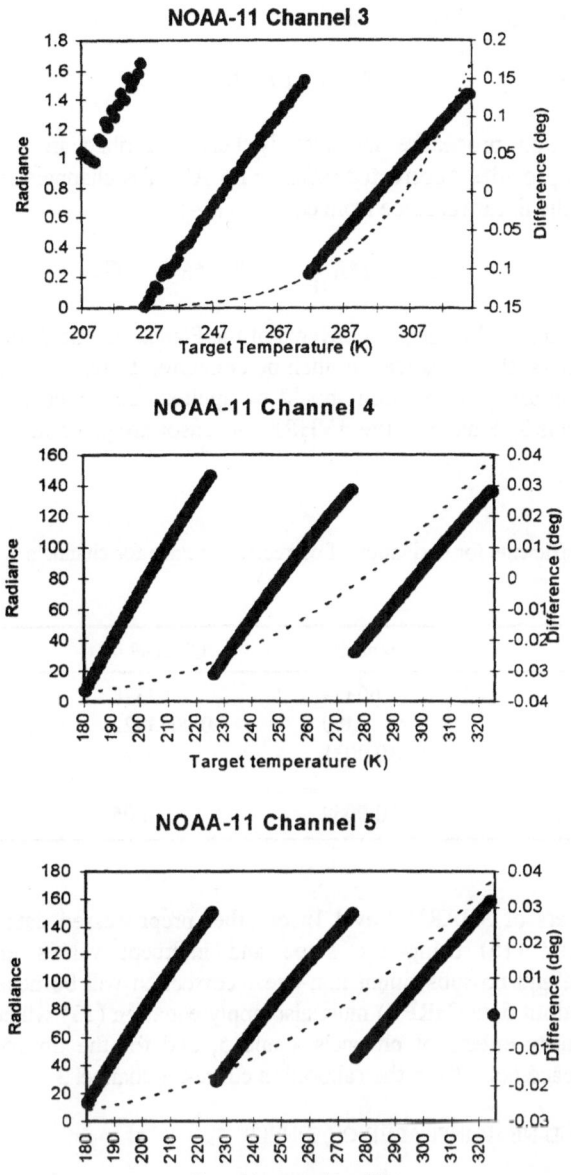

Figure 4. Radiance values for ranges of brightness temperature targets for channels 3, 4 and 5 of NOAA-11 (dashed lines). Also shown are the temperature differences that could arise using the two different radiance to temperature conversion methods (black circles). The temperatures have been calculated using the central wave numbers for the 180-225, 225-275 and 275-320K ranges.

These coefficients provide a linear radiance:

$$R_{LIN} = M \times DN + I \tag{14}$$

This linear calibration procedure has already been described in section 3.1.1. The following equation provides a corrected radiance for AVHRR channels 3, 4 and 5 which is a function only of the linear radiance from equation (14):

$$RAD = A*R_{LIN} + B*R_{LIN}*R_{LIN} + C \tag{15}$$

With the energy tables, or by using the inverse of the Planck function with the appropriate central wave numbers, this radiance can then be converted to temperature. No additional temperature (non-linearity) correction should be applied. The coefficients to apply this equation for channels 3, 4 and 5 of the AVHRR-14 sensor are given in Table 6.

Table 6. Coefficients for derivation of corrected radiance for channels 3, 4 and 5 for NOAA-14.

	Channel 3	Channel 4	Channel 5
Coefficient A	1.00359	0.92378	0.96194
Coefficient B	0.0	0.0003822	0.0001742
Coefficient C	-0.0031	3.72	2.00
Space Radiance R_{SP}	0.0069	-4.05	-2.29

In summary, users of SHARP, Level 1b or other preprocessed data need only apply equations (14) and (15) using the slope and intercept values embedded in the preprocessed data, and no subsequent non-linear correction will be necessary. However, users of direct readout data (HRPT) must also apply equation (13) which will correct for both the non-linearity effects of channels 4 and 5, and for the observed effect of the departure of the space point from the calibration curve for channel 3.

3.2   VISIBLE AND NEAR-INFRARED CHANNELS

The fact that the visible and near-infrared channels became so widely used for land applications makes the lack of on-board calibration for these channels particularly unfortunate. Additionally, the performance of the visible and near-infrared radiometers has been seen to deteriorate considerably during and post launch. This makes the use of pre-flight calibration coefficients particularly unsuitable, especially when long time-series of data are being processed. In order to overcome the limitations, a number of investigations have been carried out to derive more suitable inflight calibration coefficients from observations of light and dark targets on the Earth's surface and atmosphere. These investigations have shown that pre-flight calibrations can be very inappropriate for post-launch data acquired by nearly all AVHRR instruments, which have demonstrated substantial decreases in sensitivity over time after launch. Most calibration approaches have been developed and applied to the afternoon-pass NOAA satellites, as the light levels and thermal contrasts at that time of day are more suitable for land-based applications.

Several visible and near-infrared inflight calibration attempts have been made, some using observations of standard targets (*e.g.* Manore and Brown, 1986, Frouin and Gautier 1987); cloud radiance (Justus 1989); simultaneous airborne and satellite-based measurements (Abel *et al* 1988, Smith *et al* 1988); and desert, cloud and ocean observations (Vermote *et al* 1992, Kaufman and Holben 1992, Teillet *et al* 1990). The desert approach (in which deserts are observed regularly and assumed to be time invariant targets) has proved to be one of the most satisfactory approaches, since a large number of observations can be made under the right viewing conditions. Frouin and Gautier (1987) compared NOAA-derived reflectances with those derived from laboratory measurements of the White Sands reflectance. Teillet *et al* (1990) used ground measurements of the reflectance of the alkali flats area at the White Sands missile range and the nearly simultaneous image data from high-resolution Landsat Thematic Mapper imagery to derive the actual surface reflectance during the AVHRR observation.

In a simpler approach, Holben *et al* (1990) and Kaufman and Holben (1992) used a series of NOAA image observations of the Libyan Desert, restricting illumination and viewing geometries over carefully selected, very bright sites. Very bright sites were chosen because the variable atmospheric contribution to the upwelling radiance at the top of the atmosphere is small compared with the contribution of the radiation from the bright desert surface (Brest and Rossow 1991, Kaufman and Holben 1992, Rao and Chen 1994). Furthermore, only near-nadir viewing geometry (about every nine days) was employed to remove sun-scanner azimuthal dependencies, and image data were screened carefully to minimise cloud and aerosol contamination.

Most of the studies cited above all indicated that the in-orbit performance of the AVHRRs visible and near-infrared channels degraded in time, initially during storage, then because of outgassing (*e.g.* water vapour loss from filter interstices) and other launch-associated contamination (*e.g.* rocket exhaust contamination), and subsequently because of the continued exposure to the harsh space environment. However, several authors reported different degradation rates for the two sensors (see Che and Price 1992). It was thus recognised as one priority of the NOAA/NASA Pathfinder program that the in-orbit degradation of the AVHRR visible and near-infrared channels be evaluated and appropriate correction algorithms developed and disseminated. Furthermore, the interrelationships among the radiance measurements made by the three AVHRRs should also be established so that continuity of record of the environmental products for the Pathfinder period , 1981- present, would be ensured.

In 1993, the most active researchers on the subject working together in the AVHRR Pathfinder Calibration Working Group carried out a statistical analysis of a long time-series of data from NOAAs -7, -9 and -11 over the southeastern Libyan desert. The choice of the southeastern part of the Libyan desert as a time-invariant target was governed by the following considerations: long-term stability, high reflectance combined with low solar zenith angles, surface uniformity, low cloudiness and low precipitation and sufficiently large spatial extent (of the order of a few hundred square kilometres) (Rao *et al* 1993a).

On the basis of previous work, it was assumed that the radiometer response degraded in time exponentially. Then, only the upwelling radiances confined to small satellite zenith angles were considered in the analysis, thereby effectively eliminating the azimuth angle dependency of the upwelling radiance. Under these conditions, the empirical bidirectional reflection model used, led to the equivalent assumption that the reflectance factor has a power law dependence on a simple function of the cosines of the solar and satellite zenith angles which exhibits reciprocity in the two angular parameters. An iterative procedure, consisting of regressions in succession of (a) the reflectance factor on the simple function of cosines of the satellite and solar zenith angles, and (b) the natural logarithm of the ratio of the measured to the estimated reflectance factor on days from launch, yielded the apparent degradation per day of the reflectance factor of the invariant desert target. This was interpreted as the rate of decrease of the gain in units counts/radiance of the instrument. Using this procedure, the degradation rates of channels 1 and 2 of the AVHRRs on the NOAA-7, -9 and -11 spacecraft were determined, covering the Pathfinder period July 1981 to December 1991 (Rao and Chen 1994).

Next, the availability of the results of several aircraft/satellite congruent path measurements over White Sands, New Mexico employing a well-calibrated radiometer on-board a U-2 aircraft (Smith *et al* 1988), made it possible to translate the *relative* degradation rates into variations in time of the slope parameter of the calibration. Thus, the relative degradations of the two channels of the NOAA-9 AVHRR were anchored to the absolute calibrations based on aircraft/satellite measurements made during October/November 1986, and a couple of time-dependent expressions relating the channel 1 and channel 2 slopes to days after launch were derived. These were then recommended to the user community by the NOAA/NASA AVHRR Pathfinder Calibration working group (Rao *et al* 1993a).

Finally, using regression analyses for 11 sets of matched data for NOAA-7/NOAA-9 and 10 sets of matched data for the NOAA-11/NOAA9 combinations, the AVHRR on NOAA-9 was used as the normalisation standard to establish strong linear interrelationships between the visible and near-infrared radiances measured by the three AVHRRs (see Rao and Chen 1994 for more details on the selection criteria used and regression analysis performed). Rao and Chen (1994) claim that the attainable accuracies in the radiances calculated using the formulae they provide (Table 7) are at best comparable with the estimated accuracies of the order of a few percent of the Fall 1986 aircraft-based absolute calibrations of the AVHRR on NOAA-9, which were used as the normalisation standard in their work. The relative degradation rates and the time-dependent slopes for the two channels of all three AVHRRs are shown in Figure 5 and Table 7.

Table 7. Formulae for the calculation of calibrated radiances (from Rao and Chen 1994)

Spacecraft and channel	Radiance $(W/(m^{-2} sr^{-1} \mu m^{-1}))$
NOAA-7	
Channel 1	$0.5753 \times \exp(1.01 \times 10^{-4} \times d) \times (DN\ -36)$
Channel 2	$0.3914 \times \exp(1.20 \times 10^{-4} \times d) \times (DN\text{-}37)$
NOAA-9	
Channel 1	$0.5406 \times \exp(1.66 \times 10^{-4} \times d) \times (DN\text{-}37)$
Channel 2	$0.3808 \times \exp(0.98 \times 10^{-4} \times d) \times (DN\text{-}39.6)$
NOAA-11	
Channel 1	$0.5496 \times \exp(0.33 \times 10^{-4} \times d) \times (DN\text{-}40)$
Channel 2	$0.3680 \times \exp(0.55 \times 10^{-4} \times d) \times (DN\text{-}40)$

In this table, DN is the digital number in the respective channel, and d is the number of days from launch (zero on the day of launch). Although the digital number for the space view is extractable from the HRPT data stream, users are recommended to use the value provided in the equations above as the effective counts have been normalised to mean Earth-Sun distance in the supplied formulae.

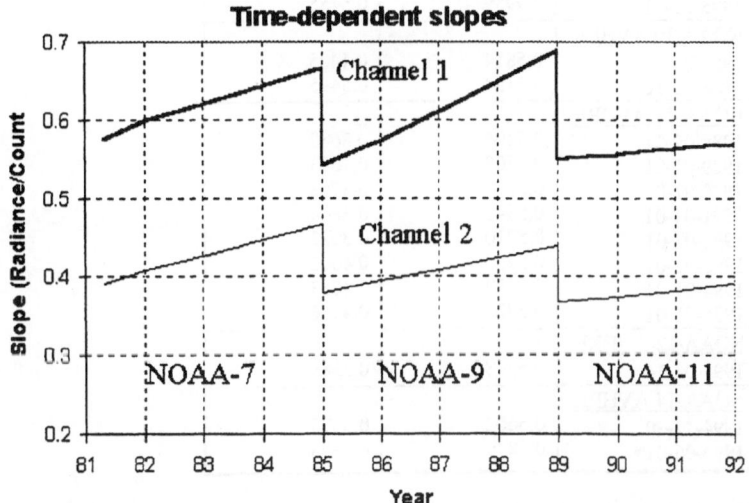

Figure 5. Time-dependent slopes of channels 1 and 2 of the AVHRR's on NOAA-7, -9 and -11 (From Rao and Chen 1994).

Table 8. Slope values for channels 1 and 2 of the NOAA-AVHRR series of sensors. The even-numbered satellite data are taken from Teillet and Holben (1994), and those for NOAA's 7, 9 and 11 are recalculated from the equations of Rao and Chen (1994). NOAA-14 prelaunch and preliminary post-launch calibration coefficients are also included.

Satellite and Date	Channel 1 Slope	Channel 2 Slope
NOAA-6 AVHRR		
1979-06-27	0.5587	0.3534
NOAA-7 AVHRR		
1981-06-23	0.5753	0.3914
1982-01-01	0.5865	0.4004
1983-01-01	0.6085	0.4184
1984-01-01	0.6313	0.4371
1985-01-01	0.6550	0.4567
NOAA-8 AVHRR		
1983-03-28	0.5465	0.3690
NOAA-9 AVHRR		
1984-12-12	0.5406	0.3808
1985-01-01	0.5423	0.3815
1986-01-01	0.5761	0.3954
1987-01-01	0.6121	0.4098
1988-01-01	0.6504	0.4247
1989-01-01	0.6910	0.4401
1990-01-01	0.7342	0.4562
1991-01-01	0.7800	0.4728
1992-01-01[+]	0.8287	0.4900
1993-01-01[+]	0.8805	0.5079
1994-01-01[+]	0.9355	0.5264
1995-01-01[+]	0.9939	0.5455
NOAA-10 AVHRR		
1986-09-17	0.5814	0.4405
1988-02-10	0.7194	0.5495
NOAA-11 AVHRR		
1988-09-24	0.5496	0.3680
1989-01-01	0.5508	0.3693
1990-01-01	0.5574	0.3768
1991-01-01	0.5642	0.3844
1992-01-01	0.5710	0.3922
1993-01-01	0.5780	0.4002
1994-01-01	0.5850	0.4083
1995-01-01	0.5920	0.4166
NOAA-12 AVHRR		
1991-05-14	0.5348	0.3389
NOAA-14 AVHRR		
1994-12-30	0.5602	0.3573
1995-06-01*	0.5849	0.4695

[+]   although NOAA-9 slopes have been interpolated using the supplied equations, the local crossing time had by these dates drifted so much that the values derived are likely to be less reliable and the data less useful for land applications.

*   NOAA-14 post launch calibration coefficients proposed by Vermote (*pers comm*) based on observation of oceans, clouds and sunphotometers.

It is important to note that the radiance values calculated by use of the formulae presented above represent the "Top Of Atmosphere" radiance only. No account has been taken for the effects of the atmosphere (see chapter by Vermote and El Saleous for further discussion of this topic). It is also worthwhile noting that in most of the radiometric calibration work, the authors have assumed the perfect linearity of response of the visible and near-infrared sensors, and have also assumed that the spectral response itself has remained consistent to that measured in the laboratory throughout the working life of the satellite sensor. This may not always be the case (as discussed by Vermote and Holben in a following chapter) but would be very difficult to detect or quantify. In any case, the relative error caused by possible spectral shifts in the sensors is likely to be small in comparison with degradation effects over time.

Figure 6 shows the magnitude of error in calculation of the normalised vegetation index (NDVI) that would occur if prelaunch, instead of inflight calibration coefficients, were used for NOAA-7, -9 and -11. The equation used for the calculation of NDVI difference is the approximation suggested by Che and Price (1992). The figure illustrates the importance of using inflight calibration coefficients wherever possible, and especially for multi-temporal studies and studies where data from more than one instrument are used.

4.   Discussion

Proper radiometric calibration is extremely important in any application which utilises quantitative analysis of NOAA-AVHRR data. Calibration is particularly important when time-series of data, or data from several sensors, are to be used together. The provision of on-board calibration for channels 3, 4 and 5 allows good inflight calibration to be performed, but users must take care to use the recommended procedures for conversion from radiance to brightness temperature. The correct application of the non-linear corrections for channels 4 and 5 is also important.

The visible and near-infrared channels are very important in all land applications of the NOAA-AVHRR, so it is particularly unfortunate that there is no on-board calibration source provided for these channels. However, a significant amount of work has been carried out on inflight calibration using Earth-based targets, and the scientific and operational community now appear to produce converging methodologies and results in reasonable time frames. Use of the prelaunch calibration coefficients can lead to major discrepancies, so post-launch calibration coefficients should always be used wherever practicable. The latest of these have been listed in this chapter. It is hoped that as operational results of inflight calibration for existing and future satellites becomes available, these will be more quickly available to the user community over the INTERNET (see chapter by Kidwell, and URL: http://psbsgi1.fb4.noaa.gov:8080/EBB/pubs/CAL).

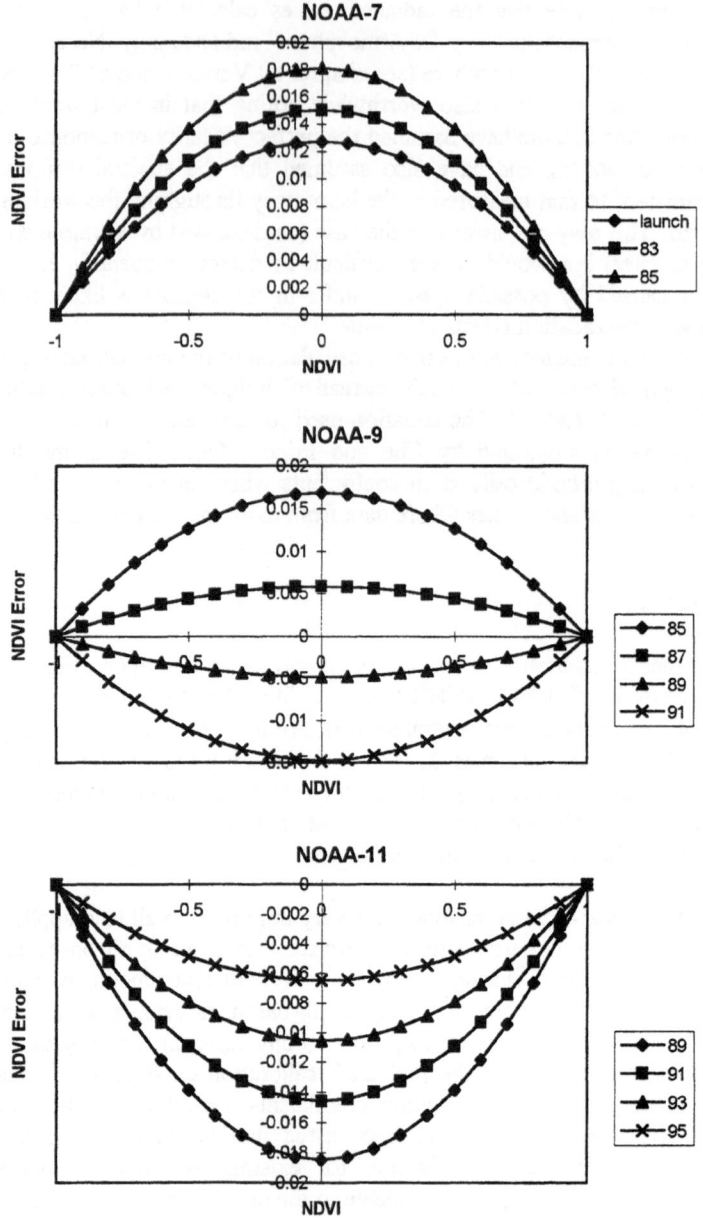

Figure 6. NDVI error as a function of NDVI for NOAA's 7, 9 and 11, using prelaunch instead of recalculated inflight calibration coefficients.

5. Acknowledgements

The kind assistance of Drs. Vermote and Teillet for provision of calibration coefficients is gratefully acknowledged.

6. References and Bibliography

Abel, P., 1990. Prelaunch calibration of the NOAA-11 AVHRR visible and near IR channels. *Remote Sensing of Environment*, **31**: 227-229.

Abel, P., 1991. Clouds as calibration targets for AVHRR reflected-solar channels: results from a two year study at NOAA/NESDIS, *Proceedings of the Society of Photo-optical Instrument Engineers*, **1493**.

Abel, P., Smith, G.R., Levin, R.H. and Jacobowitz, H., 1988. Results from aircraft measurements over White Sands, New Mexico, to calibrate the visible channels of spacecraft instruments. In *Proceedings of the Society of Photo-optical Instrument Engineers*, **924**.

Brest, C.L. and Rossow, W.B., 1990. Radiometric calibration and monitoring of NOAA AVHRR visible data. *7th Conference of Atmospheric Radiation* (Boston, MA: AMS). pp 68-74.

Brest, C.L. and Rossow, W.B., 1991. Radiometric calibration and monitoring of NOAA AVHRR data for ISCCP. *International Journal of Remote Sensing*, **13**: 235-273.

Buongiorno,A., 1992. SHARP Level 2 Development Procedures and Format Specifications, ESA Earthnet Programme Office, Frascati, Italy.

Che,N., and Price,J.C., 1992. Survey of radiometric calibration results and methods for visible and near-infrared channels of NOAA-7, -9 and -11 AVHRRs. *Remote Sensing of Environment*, **41**, 19-27.

Che, N., Grant, B.G., Flitter, D.E. *et al*, 1991. Results of calibrations of the NOAA-11 AVHRR made by reference to calibrated SPOT imagery at White Sands, New Mexico. *Proceedings of the Society of Photo-optical Instrument Engineers*, **1493**.

Fraser, R.S. and Kaufman, Y.J., 1985. The relative importance of scattering and absorption in remote sensing. *IEEE Transactions on Geoscience and Remote Sensing*, **23**: 623-633.

Fraser, R.S. and Kaufman, Y.J., 1986. Calibration of satellite sensors after launch. *Applied Optics*, **25**: 1177-1185.

Frouin, R. and Gautier, C., 1987. Calibration of NOAA-7 AVHRR, GOES-5 and GOES-6 VISSR/VAS solar bands. *Remote Sensing of Environment*, **22**: 73-101.

Holben, B.N., Kaufman, Y.J. and Kendall, J.D., 1990. NOAA-11 AVHRR visible and near IR inflight calibration. *International Journal of Remote Sensing*, **11**, 1511-1519.

Holben. B.N., and Fraser, R.S., 1984. Red and near-IR sensor response to off-nadir viewing. *International Journal of Remote Sensing*, **5**: 145-160.

ITT Aerospace, 1982. AVHRR/2 Advanced Very High Resolution Radiometer Description, ITT Aerospace/Optical Division, NASA Contract No. NAS5-26771, Fort Wayne, Indiana 46801, USA.

Justus, C.G., 1989. An operational procedure for calibrating and assessing the stability and accuracy of shortwave satellite sensors, Final Report to NOAA/NESDIS and USDA/ARC, 99 pp.

Kaufman, Y.J. and Holben, B.N., 1992. Calibration of the AVHRR visible and near-IR bands by atmospheric scattering, ocean glint and desert reflection. *International Journal of Remote Sensing*, **14**: 41-52.

Kaufman, Y.J., Holben, B.N., Kendall, J.D. and Mekler, Y., 1990. Inflight calibration of the NOAA-AVHRR visible and near-IR channels. *IGARSS Symposium Washington D.C.*, (New Jersey: IEEE Inc), pp 511-514.

Kidwell,K., 1991. NOAA Polar Orbiter Data Users Guide. NOAA NESDIS Technical Report, USDoC, Washington, DC.

Lauritson,L, Nelson,G.J. and Porto,F.W., 1988. Data extraction and calibration of TIROS-N/NOAA radiometers, NOAA Technical Memorandum NESS 107, USDoC, Washington, DC.

Manore, M. and Brown, R.J., 1986. Secondary targets for the radiometric correction of AVHRR imagery for crop monitoring, *Proceedings 10th Canadian Symposium on Remote Sensing*, Edmonton, Alberta, Canada, 875-889.

NOAA, 1990. Report of the Workshop on Radiometric Calibration of Satellite Sensors of Reflected Solar Radiation, NOAA Technical Report NESDIS 55, Washington, DC, 33 pp.

Planet, W.G., (Ed) 1979 + several updates. Data extraction and calibration of TIROS-N/NOAA radiometers, NOAA Technical Memorandum NESS 107-Rev. 1, Appendix B.

Price, J.C., 1987a. Radiometric calibration of satellite sensors in the visible and near-IR: history and outlook. *Remote Sensing of Environment*, **23**: 3-9.

Price, J.C., 1987b. Calibration of satellite radiometers and the comparison of vegetation indexes. *Remote Sensing of Environment*, **21**: 15-27.

Price, J.C., 1988. An update on visible and near infrared calibration of satellite instruments. *Remote Sensing of Environment*, **24**: 419-422.

Rao,C.R.N., 1987. Prelaunch calibration of channels 1 and 2 of the Advanced Very High Resolution Radiometer, NOAA Technical Report NESDIS 26, USDoC, Washington, DC.

Rao,C.R.N., and Chen,J., 1994. Post launch calibration of the visible and near-infrared channels of the Advanced Very High Resolution Radiometer on NOAA-7, -9 and -11 spacecraft, NOAA Technical Report NESDIS 78, USDoC, Washington, DC.

Rao. C.R.N., Stowe, L.L. and McClain, E.P., 1989. Remote Sensing of aerosol over ocean using AVHRR data, theory, practice and application. *International Journal of Remote Sensing*, **10**: 743-749.

Rao. C.R.N., Stowe, L.L., McClain, E.P. and Sapper, J., 1988. Development and application of aerosol remote sensing with AVHRR data from the NOAA satellite. In *Aerosol and Climate*, edited by P.V. Hobbs, M.P. McCormick (Hampton, Virginia: Deepak), pp 69-79.

Rao,C.R.N., Chen,J., Staylor,F.W., Abel,P., Kaufman,Y.J., Vermote,E., Rossow,W.R., and Brest,C., 1993a. Degradation of the visible and near-infrared channels of the Advanced Very High Resolution Radiometer on the NOAA-9 Spacecraft: Assessment and recommendations for corrections. NOAA Technical Report NESDIS 70, USDoC, Washington, DC.

Rao,C.R.N., Sullivan,J.T., Walton,C.C., Brown,J.W. and Evans,R.H., 1993b. Nonlinearity corrections for the thermal infrared channels of the AVHRR: Assessment and recommendations, NOAA Technical Report NESDID 69, USDoC, NOAA, Washington,DC.

Slater, P.N., Biggar, S.F., Holm, R.G. *et al.* 1987. Reflectance based and radiance based methods for the inflight absolute calibration of multispectral sensors. *Remote Sensing of Environment*, **22**: 11-37.

Smith, G.R., Levin, R.H., Abel, P. and Jacobowitz, H., 1987. Calibration of the solar bands of the NOAA-9 AVHRR and GOES-6 VISSR using high altitude aircraft measurements. *Presented in the International Union of Geodesy and Geophysics, Vancouver, Canada.*

Smith, G.R., Levin, R.H., Abel, P. and Jacobowitz, H., 1988. Calibration of the solar bands of the NOAA-9 AVHRR using high altitude aircraft measurements. *Journal of Atmosphere and Ocean*, **5**: 631-639.

Staylor, W.F., 1986. Site selection and directional models of desert used for ERBE validation targets. NASA TP-2540.

Staylor, W.F., 1990. Degradation rates of the AVHRR visible channel for the NOAA 6, 7 and 9 spacecraft. *Journal of Atmospheric and Ocean Technology*, 7, 411-423.

Tarpley, J.D., Schneider, S.R. and Money, R.L., 1984. Global vegetation indices from the NOAA-7 meteorological satellite. *Journal of Climate and Applied Meteorology*, **23**: 491-494.

Teillet, P.M. and Holben,B.N., 1994. Towards operational radiometric calibration of NOAA AVHRR imagery in the visible and near-infrared channels, *Canadian Journal of Remote Sensing*, **20**, 1-11.

Teillet, P.M., Slater, P.N., Mao, Y., Ding, Y., Bartell, R.J., Biggar, S.F., Santer, R.P., Jackson, R.D. and Moran, M.S., 1988. Absolute radiometric calibration of the NOAA AVHRR sensors. *Society of Photo-optical Instrument Engineers Conference, Orlando, Florida, April 1988. Society of Photo-optical Instrument Engineers*, **924**: 196-207.

Teillet, P.M., Slater, P.N., Mao, Y., Ding, Y., Bartell, R.J., Biggar, S.F., Santer, R.P., Jackson, R.D. and Moran, M.S., 1990. Three methods for he absolute calibration of the NOAA AVHRR sensors in flight. *Remote Sensing of Environment*, **31**: 105-120.

Vermote,E., Santer,R., Deschamps,P.Y., and Herman,M., 1992. In-flight calibration of large field of view sensors at short wavelengths using Rayleigh scattering, *International Journal of Remote Sensing*, **13**, 3409-3429.

Weinreb,M.P., Hamilton,G., Brown,S. and Koczor,R.J., 1990. Nonlinearity corrections in calibration of AVHRR infrared channels. *Journal of Geophysical Research*, **95**, 7381-7388.

Whitlock, C.H., Staylor, W.F., Smith, G., *et al.* 1990. AVHRR and VISSR satellite instrument calibration results for both cirrus and marine stratocumulus IFO periods. NASA CP 3083.

Whitlock, C.H., Staylor, W.F., Smith, G., Levin, R., Frouin, R., Gautier, C., Teillet, P.M., Slater, P.N., Kaufman, Y.J., Holben, B.N., Rossow, W.B., Brest, C. and LeCroy, S.R., 1988. AVHRR and VISSR satellite instrument calibration results for both cirrus and marine stratocumulus IFO periods. FIRE Science Team Meeting, Vail, Colorado.

Wooster, M.J., Richards, T.S. and Kidwell, K., 1995. NOAA-11 AVHRR/2 thermal channel calibration update, *International Journal of Remote Sensing*, **16**, 359-363.

RADIATIVE TRANSFER MODELLING FOR CALIBRATION AND ATMOSPHERIC CORRECTION

E. VERMOTE AND J.C. ROGER
NASA / Goddard Space Flight Center,
Greenbelt, MD, 20771,
U.S.A.

ABSTRACT. The remotely-sensed signals in the visible and near infrared channels at satellite or airborne platforms are combinations of surface and atmospheric contributions, with relative amounts varying across the two wavelength regions, depending on the condition of the atmosphere. In order to derive accurate sensor calibration and atmospheric correction, the contribution of the atmospheric constituents to the total retrieved signal must be understood and modelled. This chapter reviews the different atmospheric contributors to the signal, the formulation of their effect and their relative effects on the measured signal. In particular, the functionality, precision and accuracy of a widely-used radiation transfer code, 5S (Simulation of Satellite Signal in the Solar Spectrum), and its recent successor, 6S (Second Simulation of Satellite Signal in the Solar Spectrum), which enables accurate simulation and correction for atmospheric effects, are examined.

1. Introduction

As shown by Tucker in an earlier chapter, the visible and near-infrared channels of the Advanced Very High Resolution Radiometers (AVHRR) onboard the NOAA satellite platforms have proven to be valuable tools for terrestrial applications. However, the signal measured in each of these channels represents a combination of surface and atmospheric effects, usually in different proportions depending on the condition of the atmosphere. Therefore, inherent in any study of the Earth's surface or vegetation from space, is the need to extract the surface contribution from the combined surface/atmosphere reflectance received at the sensor (Deschamps *et al.* 1983, Gordon *et al.* 1988, Justice *et al.* 1991).

This so-called *"decoupling"* of the atmosphere and the surface effect, is a challenging problem, and in the past the research community has attempted to avoid the need for precise atmospheric correction by developing vegetation indices such as the Normalised Difference Vegetation Index (NDVI) (Tucker 1979) which significantly reduces the atmospheric effect due to the normalisation involved in its calculation (Kaufman and Tanré 1992). Further reduction of atmospheric effects, such as those caused by dense haze and sub-pixel-sized clouds, is achieved by the adoption of compositing techniques in which several consecutive images are examined and the value corresponding to the highest value of vegetation index for each pixel is chosen to represent the *"correct"* value for the time period considered (Holben 1986, Kaufman 1987, Tanré *et al.* 1992).

G. D'Souza et al. (eds.), Advances in the Use of NOAA AVHRR Data for Land Applications, 49-72.

As well as these pragmatic methods for the removal of unwanted atmospheric effects, there have also been attempts to perform more explicit atmospheric correction by using radiative transfer codes (Moran *et al.* 1990). When such codes are used in conjunction with field measurements of atmospheric optical depth made on the day of satellite overpass, quite accurate atmospheric corrections are possible (Moran *et al.* 1992). However, the acquisition of regular sun-photometer data everywhere is clearly an impossibility. Therefore, simplified methods rely on assumptions, or simulations, of atmospheric conditions, with varying degrees of accuracy (Dozier 1981, Otterman and Fraser 1976, Singh 1988). A major difficulty with these methods is that the highly spatially and temporally variable distribution of the major interfering atmospheric constituents, aerosols and water vapour, cannot be adequately dealt with. Alternatively, and optimally, information about the atmospheric optical properties should be acquired from the satellite scene itself, and combined with suitable radiative transfer models to perform accurate surface/atmospheric decoupling and atmospheric correction. Some methods for the determination of aerosol optical depth directly from the satellite imagery have been developed (Kaufman and Sendra 1988, Holben *et al.* 1992), though these are not yet validated over all terrestrial surfaces.

The need to understand and model the various elements of radiative transfer through the atmosphere as accurately as possible, is obvious. In this chapter we describe various elements of radiative transfer in the wavelength regions of the first two channels of the AVHRR, with reference to one of the most widely-used models for this purpose, 5S (Simulation of Satellite Signal in the Solar Spectrum) (Tanré *et al.* 1983, 1990, 1992), and especially in comparison with the improvements included in its successor, 6S (Second Simulation of Satellite Signal in the Solar Spectrum).

2.   Radiative Transfer Modelling: The 5S code

In most cases, only a fraction of the visible and near-infrared light reflected from a target reaches the sensor. The two most responsible atmospheric processes for this are: absorption by gases (when observation bands overlap with gaseous absorption bands) and scattering by aerosols or molecules in the atmosphere. In the simple case of a lambertian, homogeneous target at sea level viewed by a satellite sensor (under zenith angle of view θ_v, azimuth angle of view ϕ_v) and illuminated by sun (θ_s, ϕ_s), the reflectance received at the sensor may be written as:

$$\rho^*_{Target}(\theta_s, \theta_v, \phi_s-\phi_v) = Tg(\theta_s, \theta_v)\left[\left(\rho^*_{Rayleigh} + \rho^*_{aerosols}\right) + T^{\Downarrow}(\theta_s)T^{\Uparrow}(\theta_v)\frac{\rho_{Target}}{1-S\rho_{Target}}\right] \quad (1)$$

Where Tg is the gaseous transmission (in the visible and infrared atmospheric window) and H_2O, CO_2, O_2 and O_3 are the principal absorbing gases. Over a simple black target, the intrinsic atmospheric reflectance observed, ($\rho^*_{Rayleigh} + \rho^*_{aerosols}$) is written here as the simple sum of reflectance of aerosols and Rayleigh contributions. This simplification, however, is not valid at short wavelengths (less than 0.45 μm) or large sun and view zenith angles (Deschamps *et al.* 1983). The transmission, Tg, is a non-linear function of the effective amount of absorbers in the atmosphere, dependent on the pressure and temperature profiles. In the 5S code this term is computed by a two-band absorption model.

As well as consideration of absorption by atmospheric gases, the 5S code includes routines for a detailed treatment of the scattering process (considering all possible scattering processes along the sun-target-sensor path) and an approximation for the interaction between the absorption and scattering processes (Tanré *et al.* 1992). Since its publication and widespread distribution, the 5S code has been quite widely used in decoupling surface/atmosphere reflectance and atmospheric correction routines. Its success however, has been limited, largely because of a number of shortcomings: the imprecision with which it deals with Rayleigh and aerosol scattering; the limited nature of its inherent spectroscopic data; and it makes no correction for variations in altitude for both sensor and target. Moreover, the assumption of a target with fully lambertian behaviour is a major limitation to its successful application.

In 6S the code accuracy has been improved by addressing all of these specific issues separately. First, the atmospheric reflectance transmission function and spherical albedo aspects of Rayleigh scattering were considered in more detail along with the effects of aerosols, and improvements were also made to the 5S spectroscopic data. Second, the problem of variable altitude for both sensor and target was addressed. Third, bidirectional effects were incorporated with large improvements in the resulting accuracy of the model over the one using the lambertian assumption. In the next sections we provide details concerning the modifications made to 5S that result in the new, improved 6S code, and provide measures of the likely improvements in accuracy and precision.

3.   The 6S Modifications: Code improvements

3.1.   RAYLEIGH

3.1.1. *Atmospheric Reflectance.* One of the first improvements to consider in the 6S code (over the 5S version), is concerned with the treatment of atmospheric reflectance. For isotropic scattering Chandrasekhar (1960) showed how solutions derived for small optical thicknesses may be extended to larger values of optical thickness, τ. He expressed the atmospheric reflectance

$$\rho^a(\mu_s, \mu_v, \phi_v - \phi_s)$$

as:

$$\rho^a_l (\mu_s, \mu_v, \phi_v - \phi_s) = \rho^a_l (\mu_s, \mu_v, \phi_v - \phi_s) + (1 - e^{-\tau/\mu_s})(1 - e^{-\tau/\mu_s}) \Delta(\tau) \qquad (2)$$

where:

$$\rho^a_l (\mu_s, \mu_v, \phi_v - \phi_s)$$

52

is the single-scattering contribution and the second term accounts roughly for higher orders of scattering.

In 6S, we used this approach to compute the molecular scattering reflectance (Vermote and Tanré 1992). This molecular reflectance is shown plotted versus the reflectance computed from the successive order of scattering method (see section 3.2) for $\tau = 0.35$ in Figure 1. Four values of the solar zenith angle (0°, 53°, 66° and 70°), 17 values of the viewing zenith angle (from 0° to 60° with a step of 3.3°) and 19 values of the difference of the azimuth angles (from 0° to 180° with a step of 10°), covering a large range of possible geometrical conditions, have been selected. As can be seen, most points fall on the 45-degree line. The right-hand scale, which gives absolute differences between the two results, shows clearly that the precision of 0.001 is achieved for the full range of geometric conditions.

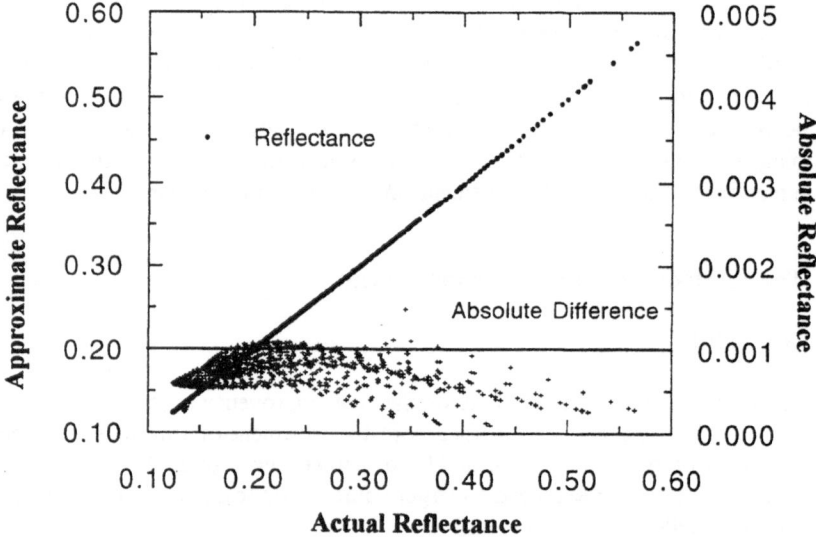

Figure 1. Molecular reflectance versus successive-order-of-scattering reflectance and absolute differences between the two methods for several geometrical conditions with $\tau = 0.35$.

3.1.2. *Transmission Function.* The transmission function refers to the normalised flux measured at the surface. There are several approximate expressions (Joseph *et al.* 1976, Lenoble 1977, Zdunkowski *et al.* 1980) based on the two-stream methods for computing the transmitted flux. The accuracy of these expressions depends on the scattering properties of the atmospheric layer (thick or thin clouds or aerosols) and on the geometrical conditions. The delta-Eddington method (Joseph *et al.* 1976) proved to be well suited for our purposes, and was therefore selected for inclusion in the code. Since molecular scattering is conservative ($\omega_0 = 1$) and the anisotropy factor g is equal to zero, we may write:

$$T(\mu) = \frac{[(2/3) + \mu] + [(2/3) - \mu]e^{-\tau/\mu}}{(4/3) + \tau} , \tag{3}$$

where μ is the cosine of the solar and/or observational zenith angle and τ is the optical thickness.

Figure 2 shows a similar comparison as that in Figure 1, this time for the transmission function Equation (3) for four optical thicknesses (0.05, 0.10, 0.25 and 0.35). For the largest optical thickness, the accuracy remains adequate even for very low observation and/or sun angles of 70° for example. The maximum difference is around 0.005, which means a relative accuracy of better than 0.1%.

3.1.3. *Spherical Albedo.* In conservative cases such as molecular scattering, the spherical albedo s is given by:

$$s = 1 - \int_0^1 \mu T(\mu) d\mu \tag{4}$$

where $T(\mu)$ has been given in Equation (3) above. Using Equations (3) and (4), the spherical albedo can be written as:

$$s = \frac{1}{4 + 3\tau} [3\tau - 4E_3(\tau) + 6E_4(\tau)], \tag{5}$$

where $E_3(\tau)$ and $E_4(\tau)$ are exponential integrals for the argument τ. These functions are easily computable from expressions given in the "Handbook of Mathematical Functions" (Abramowitz and Stegun 1970).

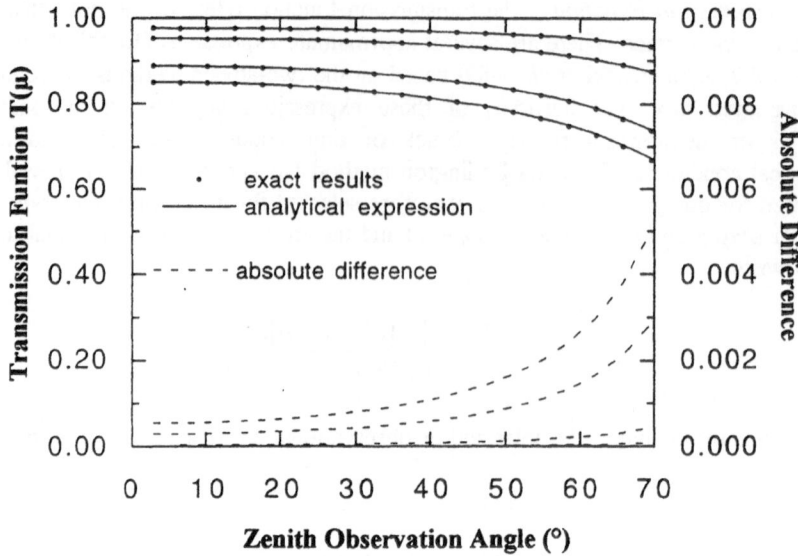

Figure 2. Comparison between actual transmission function and the delta-Eddington approximation used in 6S.

Figure 3 shows results of the expression for the spherical albedo, s. The differences between the exact results and Equation (5) are around only 0.003 for $\tau = 0.35$ which results in an error of only 0.0003 for a surface albedo of 0.10. In the red part of the solar spectrum for which the surface albedo may be larger, the error is still below 0.002.

3.2. AEROSOL

Sobolev's (1975) approximation for the reflectance, Zdunkowsky's method (1980) for the transmittance, and a semi-empirical formula for the spherical albedo. The advantage of this was that users with limited computing resources could still obtain approximations quickly. The drawback was that the accuracy of the computations could be inaccurate by a few percent in reflectance units, especially at large view and sun angles and high optical thickness. In addition, these approximations could be completely inadequate for handling the integration of the downward radiance field with non-lambertian ground conditions, typically the problem of bidirectional reflectance distribution function (BRDF) simulation. The new scheme of computation where aerosol plus Rayleigh contributions are treated as a coupled system relies on the "Successive Order of Scattering" method used by several authors (e.g. Ahmad and Fraser 1982, Deuzé et al. 1989). The accuracy of such a scheme is better than 2×10^{-4} reflectance units.

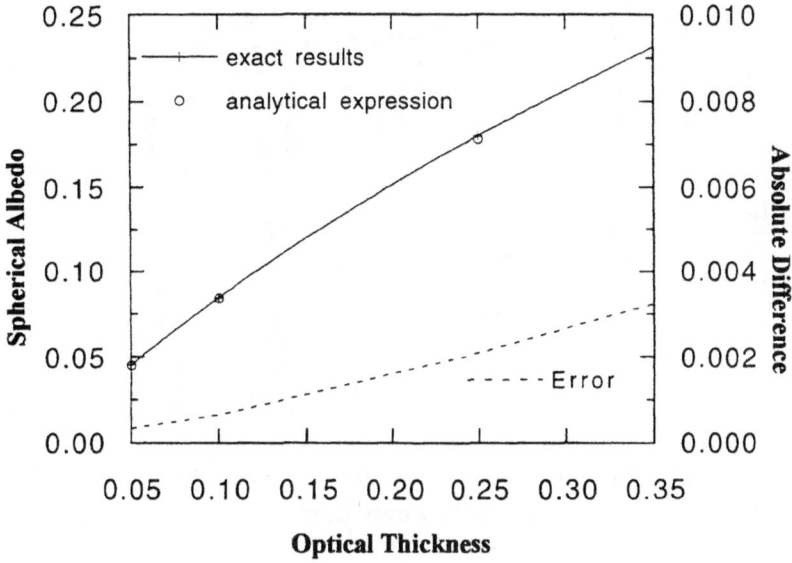

Figure 3. Comparison between actual spherical albedo and the approximation used in 6S.

In our application, the atmosphere is divided into 13 layers, which enables exact simulations of airborne observations. The downward radiation field is also computed for a quadrature of 13 gauss emerging angles which provides the necessary inputs for BRDF simulations (see section 5). The computing time remains reasonable and is in the order of only a few seconds on an HP735 Workstation (approximately 124 MIPS).

3.3.   SPECTROSCOPIC DATA

In relation to spectroscopy, the computation scheme used in 5S has not been changed, but the accuracy and resolution of the spectroscopic data have been improved. The spectral resolution of 6S has been improved to 10 cm⁻¹ with respect to band absorption models - the data having been generated using the HITRAN database. Also, CH_4, CO and NO_2 are now taken into account in the computation of the gaseous transmission. The computation of the water vapour absorption has also been improved according to recent findings, and differences between 5S and 6S values can reach a few percent for this term. Figures 4a-c show that the 6S computations match very well those obtained by MODTRAN (5 cm⁻¹ resolution) in the example of the typical US62 atmosphere.

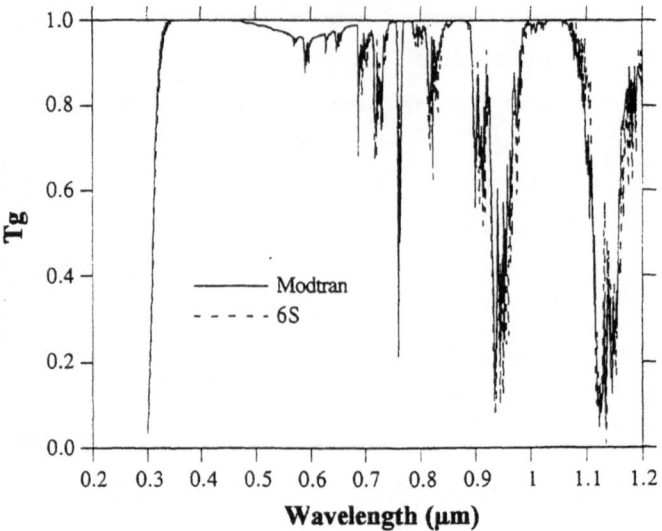

Figure 4a. Comparison of gaseous transmission as computed by MODTRAN and 6S for a typical US62 atmosphere between 0.20 and 1.20 μm.

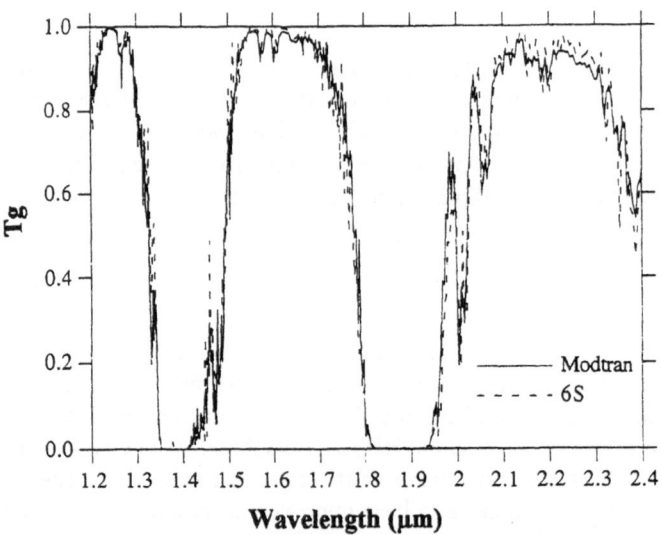

Figure 4b. Comparison of gaseous transmission as computed by MODTRAN and 6S for a typical US62 atmosphere between 1.20 and 2.40 μm.

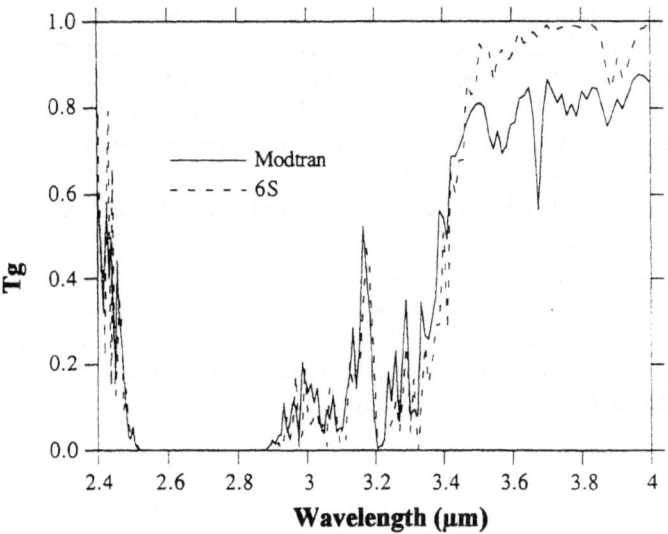

Figure 4c. Comparison of gaseous transmission as computed by MODTRAN and 6S for a typical US62 atmosphere between 2.40 and 4.0 μm.

4.   The 6S Modifications: target and sensor altitude

4.1.   ELEVATED TARGET SIMULATION

Where targets are above sea level, Equation 1 may be modified as follows:

$$\rho^*_{T\arg et}\ (\theta_s,\theta_v,\phi_s\text{-}\phi_v,z_t) =$$

$$Tg(\theta_s,\theta_v,z_t)\ [\rho^*_{Rayleigh}\ (z_t) + \rho^*_{aerosols} + T^{\Uparrow}(\theta_v,z_t)T^{\Downarrow}(\theta_s,z_t)\frac{\rho_{Target}}{1\text{-}S(z_t)\rho_{Target}}], \qquad (6)$$

where z_t is the target altitude.

The target altitude and pressure indicates the amount of scatterers above the target (molecules and aerosols) and the amount of gaseous absorbents. In the 5S code, the amount and type of aerosol is entered as a fixed parameter, thus the aerosol characteristics implicitly depend on target altitude because these are measured at target location. For 6S the target altitude is handled in the following manner: first the atmospheric profile and the target altitude or pressure is selected, then a new atmospheric profile is computed by stripping out the atmospheric level above target altitude and interpolating if necessary. In this way, a more precise computation of the atmospheric parameters is made, without any kind of approximation and taking into account the coupled pressure-temperature effect on absorption. In most cases, only the integrated content may be modified as already pointed out by Teillet and Santer (1991). The user still has the option to enter the total amount of H_2O and O_3, but in this case the quantity entered must be representative of the level measured or estimated at the target location.

The influence of target altitude on Tg for an observation made with solar zenith angle, $\theta_s = 30°$ and view zenith angle $\theta_v = 60°$, is shown in Figure 5. As can be seen, the absorption effect of O_3 is not sensitive to altitude target, because the ozone layer is located in the upper levels of the atmosphere. However, target altitude does have an important effect on absorption by H_2O as the models show. This is because most of the water vapour is located in the lower atmosphere, although it should be noted that the water vapour profile in the atmosphere is highly variable.

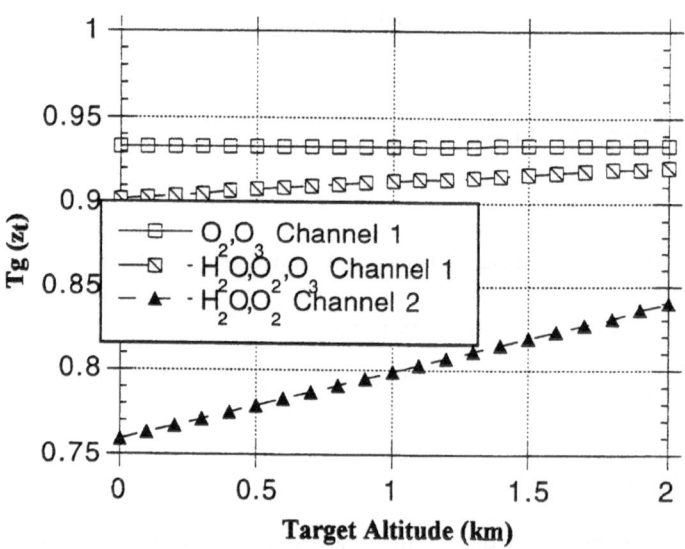

Figure 5. The influence of target altitude on the transmission function, Tg.

The effect of target altitude on molecular optical thickness is also accounted for precisely in 6S. A good approximation, however, is to consider that τ_R is proportional to the pressure at target level. Figure 6, compares, for the case of observation in AVHRR channel 1, the exact computation (derived from modified Mid-latitude Summer profile) to the approximation τ_R proportional to pressure (z_{target}).

Figure 7 illustrates the influence of altitude target, in terms of absolute variation of the Rayleigh reflectance for AVHRR channel 1 for the whole globe, using the 1/3 of degree resolution elevation map. For each 1/3 x 1/3 degree cell, considering a constant view angle of 30°, a map has been computed for solar zenith angle of 30° (backscattering). The Digital Elevation Model used in this simulation is ETOPO5. The error made by neglecting the target altitude in Rayleigh correction can reach 0.016 reflectance units but is usually between 0.001 and 0.01 for most cases.

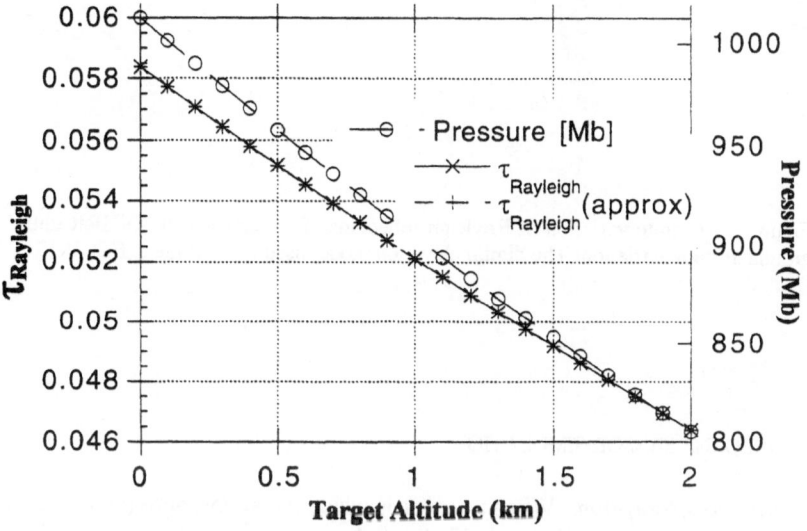

Figure 6. The effect of target altitude on molecular optical thickness as approximated in 6S.

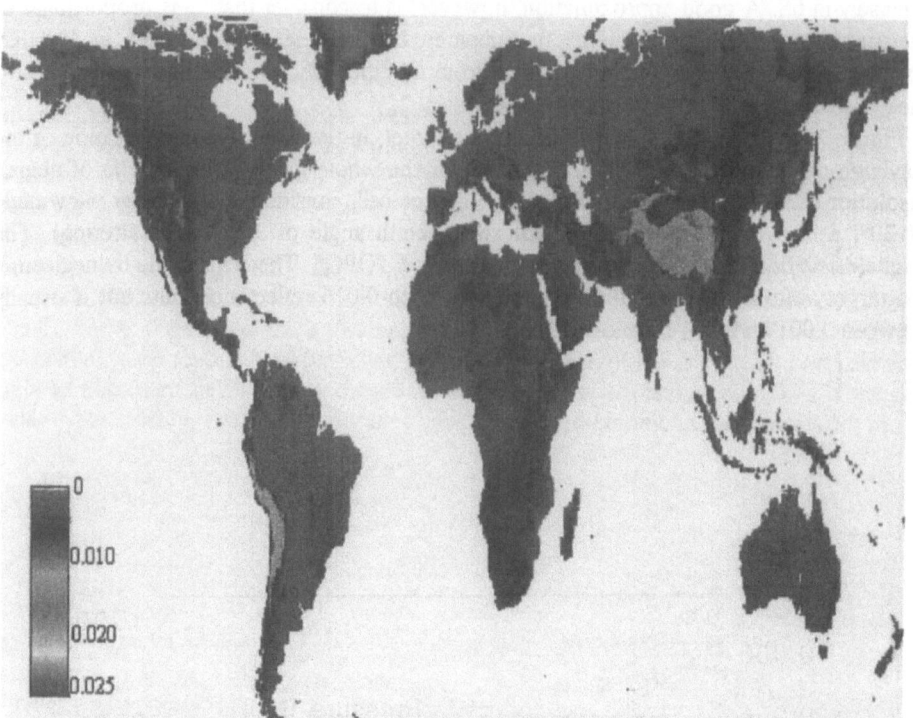

Figure 7. Absolute variation of Rayleigh reflectance for channel 1 of AVHRR due to ground altitude variation. The digital elevation model used for altitude is ETOPO5.

4.2.   AIRBORNE SENSOR SIMULATION

4.2.1. *Gaseous Absorption.* When a sensor is inside the atmosphere (as is the case with airborne sensors), Equation (1) is modified as follows:

$$\rho^*_{Target}\ (\theta_s, \theta_v, \phi_s-\phi_v, z) =$$

$$\mathrm{Tg}\,(\theta_s, \theta_v, z)\ [\rho^*_{Rayleigh}\,(z) + \rho^*_{aerosols}(z) + \mathrm{T}^{\Uparrow}(\theta_v, z)\mathrm{T}^{\Downarrow}(\theta_s, z)\frac{\rho_{Target}}{1 - S(z_t)\rho_{Target}}], \qquad (7)$$

In this case gaseous absorption is computed using a technique similar to that used for targets above sea level. The upward path only is modified: the atmosphere level above the sensor altitude is stripped, so computation is carried out only to the altitude of the sensor (with interpolation of the atmospheric profile as necessary). Figure 8 illustrates the effect of altitude on gaseous transmission computation, for $\theta_s = 30°$, $\theta_v = 60°$. In the case of the spectral wavelength of AVHRR's visible channel, O_3 absorption along the target-sensor path is no longer taken into account because these molecules are only located above the airborne sensor. For H_2O, absorption is highly dependent on altitudes up to 4 km. Thus, if the observed channel is sensitive to water vapour absorption (as is the case with AVHRR channel 2) we recommend that additional measurements of water vapour should be taken from the aircraft. An additional option has been set up in 6S for this purpose and enables

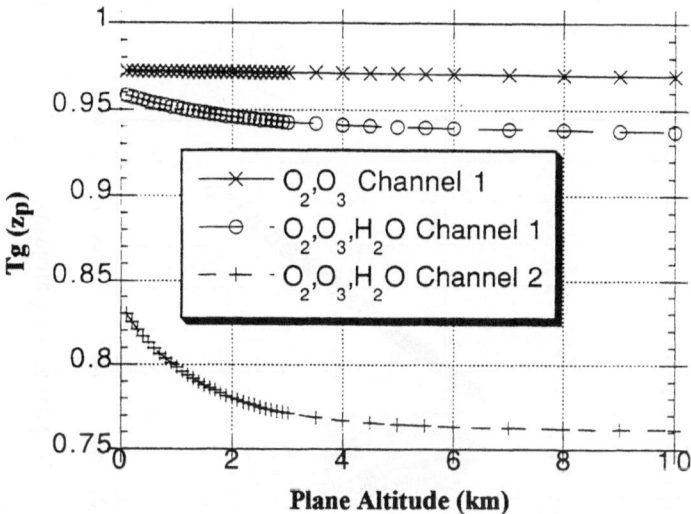

Figure 8. The effect of altitude on gaseous transmission computation for $\theta_s = 30°$ and $\theta_v = 60°$.

the user to enter aerosol, ozone and water vapour content for the portion of the atmosphere located under the aircraft if these are known or can be calculated.

4.2.2. *Atmospheric Reflectance and Transmittance.* In most cases the simple approximation "equivalent atmosphere", for atmospheric reflectance and transmittance is sufficiently accurate, *i.e*:

$$\rho^*_{Rayleigh}(z) \cong \rho^*_{Rayleigh}(z = \infty, \tau_R(0 \rightarrow z)) \tag{8}$$

$$T^{\uparrow}(\tau_{R,}, \theta_v, z) \cong T^{\uparrow}(\tau_R(0 \rightarrow z), \theta_v, z = \infty) \tag{9}$$

In Figures 9 and 10 we present the comparison between approximations using Equation (8) and (9) and exact computation, provided by a radiative code, again based on the Successive Order of Scattering method. As can be seen, the approximations work very well with only small absolute differences.

However, at wavelengths less than 550 nm, as can be seen in Figure 11, the accuracy of Equation 8 is not as good in the case of a mixed Rayleigh-aerosol atmosphere. Therefore, in 6S, the computation is performed exactly by defining one of the multiple layers used in the Successive Order of Scattering at the altitude of the sensor. This enables exact computation of both reflectance and transmission term for a realistic mixing between aerosol and Rayleigh components.

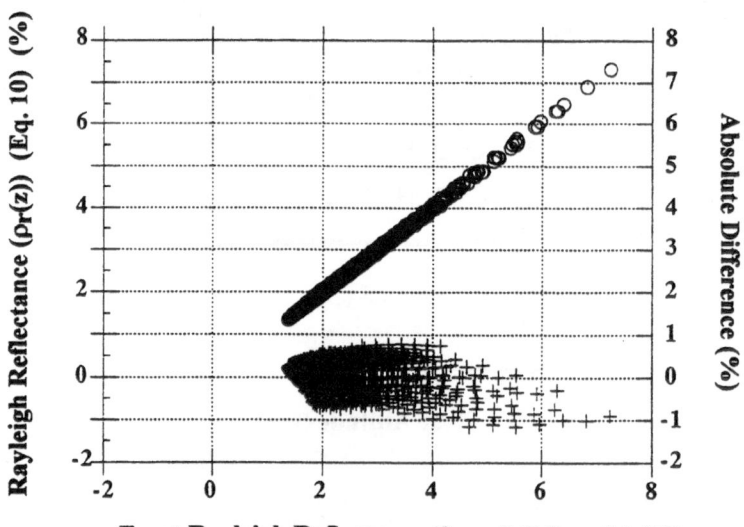

Figure 9. Atmospheric Rayleigh reflectance approximation (Equation 8) as used in 6S versus exact Rayleigh reflectance calculated using the Successive Order of Scattering (S.O.S.) code.

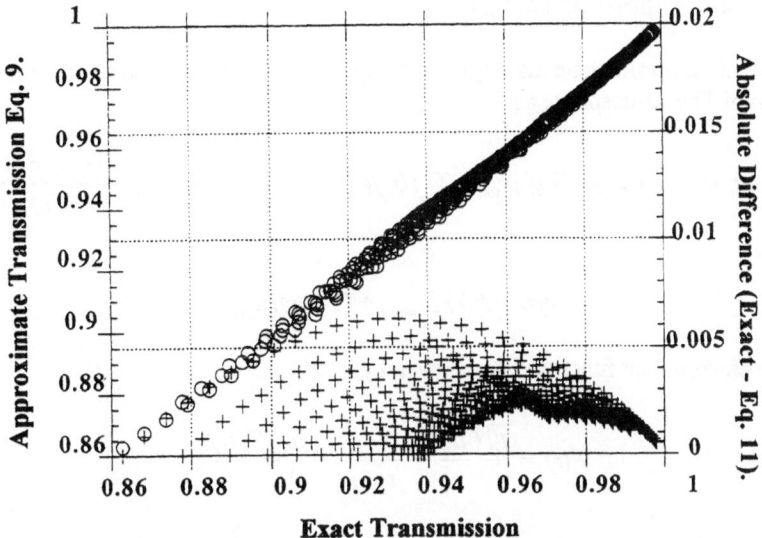

Figure 10. Atmospheric transmission approximation (Equation 9) as used in 6S versus exact transmission calculated using the Successive Order of Scattering (S.O.S.) code

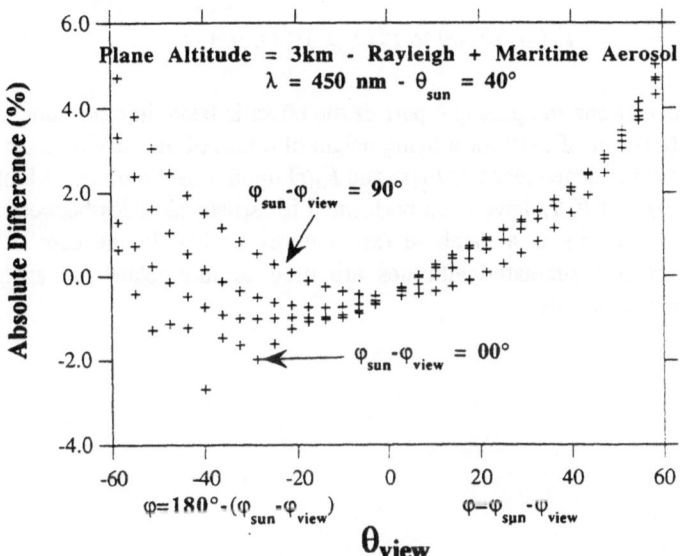

Figure 11. Comparison of Rayleigh reflectances between approximation (Equation 8) and exact computation at 450 nm in case of a moderate maritime aerosol background (Visibility of 23 km).

4.3. NON-HOMOGENEOUS TARGET

In the case of non-homogeneous targets the approach adopted in 5S is to write the signal at the Top of The Atmosphere as:

$$\rho^*_{Target}(\theta_s,\theta_v,\phi_s-\phi_v) = \text{Tg}\left(\rho^*_{atm} + \text{T}^{\parallel}(\theta_s)(\rho_{Target}\,e^{-\tau/\mu_v} + t_d(\vartheta_v))\frac{<\rho>}{1-S(z)<\rho>}\right) \tag{10}$$

where:

$$<\rho> = F(r)\,\rho_{Target} + (1-F(r))\rho_{env} \tag{11}$$

and the environment function $F(r)$ is given by:

$$F(r) = \frac{t_d^R(\mu_v)F_R(r) + t_d^A(\mu_v)F_A(r)}{t_d(\mu_v)} \tag{12}$$

If we consider target and environment to be at the same altitude, the problem of a target above sea level can be solved just by modifying the Rayleigh optical thickness.

In the case of aircraft observations, however, we have to first take into account reduction of the amount of scatterers under the aircraft. This can be done just by adjusting the term,

$$t_d^R(\mu_v)F_R(r) \text{ to } t_d^R(z,\mu_v)t_d^A(z,\mu_v)$$

Once this has been done the principal part of the effect is taken into account, that is, a global reduction (a factor of 5-10 for a flying height of 6 km) of the "environment effect". The second effect is the dependence of $F_R(r)$ and $F_A(r)$ upon sensor altitude. Monte Carlo simulations of $F_R(r)$ and $F_A(r)$ have been performed for sensor altitudes between 0.5 and 12 km and included in 6S as a database (see Figures 12-13). In the case of aircraft observation, the closest simulated altitudes are used to interpolate the environment function at the aircraft altitude.

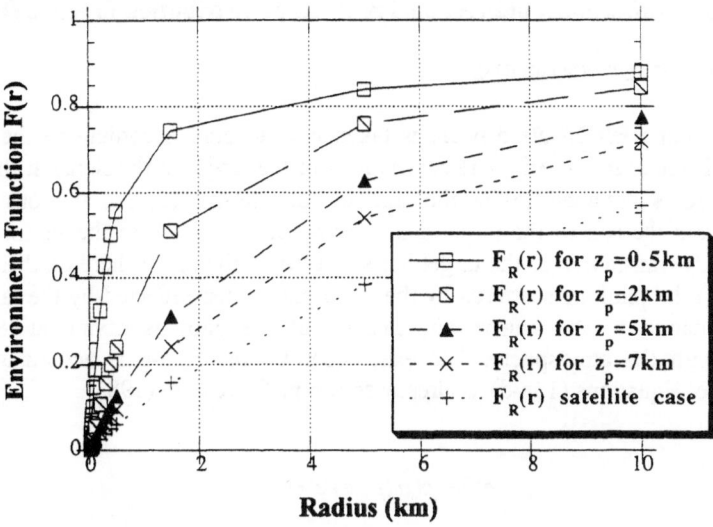

Figure 12. Variation of the environment functions, Rayleigh (F_R) for different altitudes of the sensor.

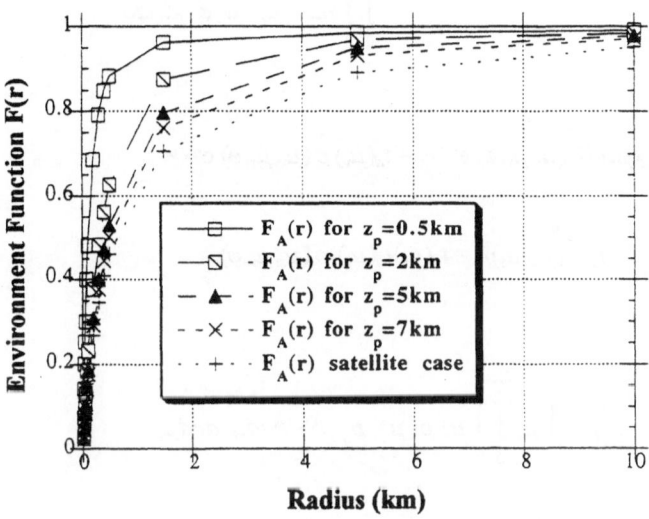

Figure 13. Variation of the environment functions for aerosol (F_A) for different altitudes of the sensor.

5. The 6S modifications: bidirectional reflectance distribution function (BRDF)

5.1. THEORETICAL BACKGROUND

In 6S, the coupling BRDF atmosphere is taken into account according to the scheme presented in Tanré *et al.* (1986). The contribution of the target to the signal at the top of the atmosphere is assumed to be the sum of four terms: (a) the photons directly transmitted from the sun to the target and directly reflected back to the sensor, (b) the photons directly transmitted to the target but scattered by the atmosphere on their way to the sensor, (c) the photons scattered by the atmosphere then reflected by the target and directly transmitted to the sensor, and finally, (d) the photons having at least two interactions with the atmosphere. The exact contribution of a-d is according to the following set of Equations (13a-d) as already shown in Tanré *et al* (1983).

$$e^{-\tau/\mu_s} \, \rho(\mu_s, \mu_v, \phi) \, e^{-\tau/\mu_v} \tag{13a}$$

$$t_d(\mu_s) \, \overline{\rho}(\mu_s, \mu_v, \phi) \, e^{-\tau/\mu_v} = t_d(\mu_s) \frac{\displaystyle\int_0^{2\pi}\int_0^1 \mu L^{\Downarrow}(\mu_s, \mu_v, \phi') \rho(\mu, \mu_v, \phi' - \phi) d\mu d\phi}{\displaystyle\int_0^{2\pi}\int_0^1 \mu L^{\Downarrow}(\mu_s, \mu, \phi') d\mu d\phi} e^{-\tau/\mu_v} \tag{13b}$$

$$t_d(\mu_v) \, \overline{\rho}'(\mu_s, \mu_v, \phi) \, e^{-\tau/\mu_s} = t_d(\mu_v) \, \overline{\rho}(\mu_v, \mu_s, \phi) \, e^{-\tau/\mu_s} \tag{13c}$$

$$t_d(\mu_v) t_d(\mu_s) \overline{\overline{\rho}} = t_d(\mu_v) t_d(\mu_s) \, \rho'(\mu_s, \mu_v, \phi), \tag{13d}$$

where

$$\overline{\overline{\rho}} \cong \int_0^1 \mu_s \int_0^{2\pi}\int_0^1 \mu_v \rho(\mu_s, \mu_v, \phi' - \phi) d\mu_v \, d\phi \, d\mu_s \tag{13e}$$

In 6S, the first three contributions are computed exactly using a downward radiation field as obtained by the successive order of scattering method. The fourth contribution which involves at least two interactions between the atmosphere and the BRDF

(Equations 13d and e) is approximated by taking $\bar{\bar{\rho}}$ equal to the hemispherical albedo of the target. This approximation is necessary because the exact computation would require a double integration, and it is justified by the limited impact on the total signal of this last contribution relative to $t_d(\mu_v)t_d(\mu_s)$, and also because multiple scattering tends to be isotropic.

5.2. VALIDATION

Thus, the only approximation in the computing scheme of 6S for BRDF effects is in the estimation of multiple interaction between target and atmosphere. Figure 14 shows that the effect of this approximation is only small for a typical BRDF signature, the clover patch measured by Woessner and Hapke (1987), and Figure 15 shows that the approximation works quite well in a range of condition as compared with the results obtained from the Successive Order of Scattering method. The ground BRDF in the example is from Kime's measurements over a ploughed field fitted with the Hapke BRDF model.

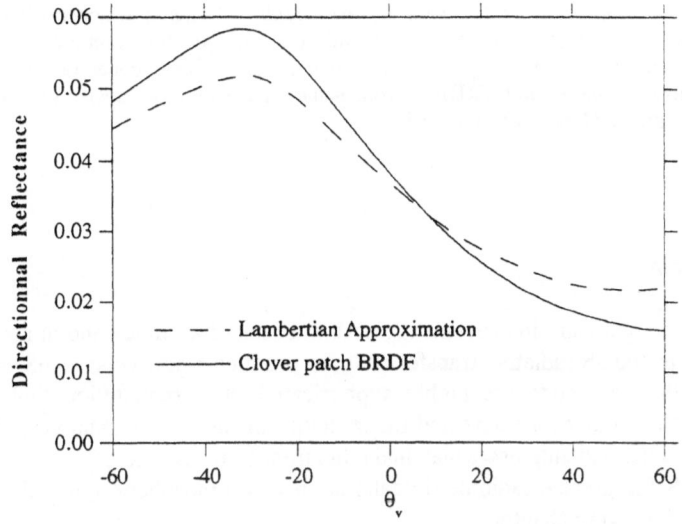

Figure 14. Comparison of the exact signal with the approximation given by Equation 13d. The target is a clover patch at 450 nm, the atmosphere is clear (visibility of 100 km) and the solar zenith angle is 30 degrees.

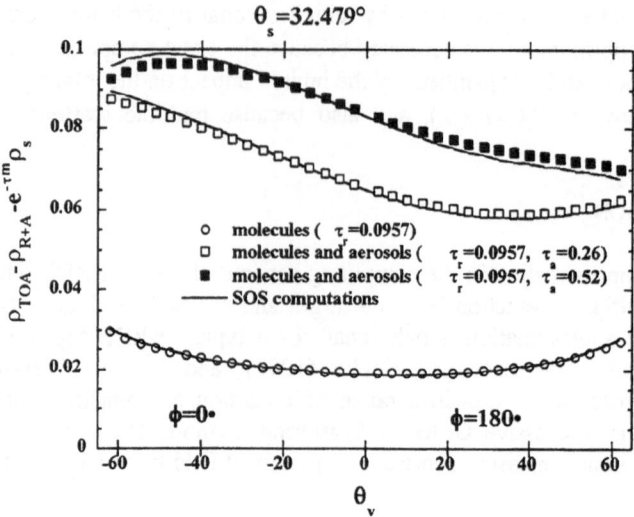

Figure 15 Comparison of the sum of the coupling terms atmosphere-BRDF: $\rho_{TOA} - (\rho_{R+A} - e^{-\tau m}\rho s)$; computed by 6S with the same quantity computed by the Successive Order of Scattering code for different atmospheric conditions (clear, average, turbid). The ground BRDF is from Kime's measurements over a ploughed field fitted with the Hapke BRDF model

6. Conclusions

Modifications and significant improvements, both in terms of accuracy and in application, have been made to the 5S radiative transfer code, to produce a new version, 6S. Although some parts of the new code are highly sophisticated, the computation time remains reasonable, and the input parameters and the structure of the code remain very similar to the 5S version, so that existing users may make the transition easily to 6S.

The use of 6S for precise radiometric calibration and atmospheric correction will be illustrated in the following chapters.

Glossary

θ_S	Solar zenith angle
μ_S	Cosine of solar zenith angle
θ_V	View zenith angle
μ_S	Cosine of view zenith angle
ϕ_S	Solar azimuth angle
ϕ_V	View azimuth angle
ρ	reflectance (unitless)
Tg	gaseous transmission
T	total scattering transmission (diffuse+direct)
t_d	diffuse transmittance factor
τ	optical thickness (unitless)
s	spherical albedo
z	altitude of the target
μm	micrometer
F(r)	environment function

References and Bibliography

Abel, P., Guenther, B., Galimore, R. N. and Cooper, J. W., 1993, Calibration Results for NOAA-11 AVHRR Channels 1 and 2 from Congruent Path Aircraft observations. *Journal of Atmospheric and Oceanic Technology*, **10**, 4, 493-508.

Abramowitz, M. and Stegun, I., A, 1970, *Handbook of Mathematical Functions* (New-York: Dover Publications,Inc).

Ahmad, Z. and Fraser, R. S., 1982, An iterative radiative transfer code for ocean-atmosphere system. *Journal of Atmospheric Science*, **39**, 656-665.

Brest, C. L. and Rossow, W. L., 1991, Radiometric calibration and monitoring of NOAA AVHRR data for ISCCP. *International Journal of Remote Sensing*, **13**, 235-273.

Chandrasekhar, S., 1960, *Radiative Transfer* (Dover: New York).

Che, N. and Price, J. C., 1992, Survey of Radiometric Calibration Results and Methods for Visible and Near Infrared Channels of NOAA-7,-9, and -11 AVHRR's. *Remote Sensing of Environment*, **41**, 19-27.

Cox, C. and Munk, W., 1965, Slopes of the sea surface deduced from photographs of sun glitter. *Bulletin of Scripps Institute of Oceanography of University of California*, **6**, 401-488.

Deschamps, P. Y., Herman, M. and Tanré, D., 1983, Modeling of the atmospheric effects and its application to the remote sensing of ocean color. *Applied Optics*, **22**, 23, 3751-3758.

Deuzé, J. L., Herman, M. and Santer, R., 1989, Fourier series expansion of the transfer equation in the atmosphere-ocean system. *Journal of Quantitative Spectroscopy and Radiative Transfer*, **41,** 6, 483-494.

Dozier, J., 1981, A method for satellite identification of surface temperature fields of subpixel resolution. *Remote Sensing of Environment*, **11,** 221-229.

Fraser, R. S. and Kaufman, Y. J., 1986, Calibration of satellite sensors after launch. *Applied Optics*, **25,** 1177-1185.

Frouin, R. and Gautier, C., 1987, Calibration of NOAA-7 AVHRR, GOES-5 and GOES-6 VISSR/VAS solar channels. *Remote Sensing Environment*, **22,** 73-101.

Gordon, H. R., Brown, J. W. and Evans, R. H., 1988, Exact Rayleigh scattering calculations for use with the Nimbus-7 Coastal Zone Color Scanner. *Applied Optics*, **27,** 5, 862-871.

Holben, B. N., 1986, Characteristics of maximum-value composite images for temporal AVHRR data. *International Journal of Remote Sensing*, **7,** 11, 1435-1445.

Holben, B. N., Kaufman, Y. J. and Kendall, J. D., 1990, NOAA-11 AVHRR visible and near-IR inflight calibration. *International Journal of Remote Sensing*, **11,** 8, 1511-1519.

Holben, B. N., Vermote, E., Kaufman, Y. J., Tanré, D. and Kalb, V., 1992, Aerosol retrieval over Land from AVHRR data-application for atmospheric correction. *IEEE Transaction on Geoscience and Remote Sensing*, **30,** 2, 212-222.

Joseph, J. H., Wiscombe, W. J. and Weinman, J. A., 1976, Solar Flux Transfer Through Turbid Atmospheres Evaluated by the Delta-Eddington Approximation. *Journal of Atmospheric Science*, **33,** 2452-2459.

Justice, C. O., Eck, T. F., Tanré, D. and Holben, B. N., 1991, The effect of water vapour on the normalized difference vegetation index derived for the Sahelian region from NOAA AVHRR data. *International Journal of Remote Sensing*, **12,** 6, 1165-1187.

Kaufman, Y., J, 1987, The effect of subpixel cloud on remote sensing. *International Journal of Remote Sensing*, **8,** 839-856.

Kaufman, Y. J. and Holben, B. N., 1993, Calibration of the AVHRR visible and near-IR bands by atmospheric scattering, ocean glint and desert reflection. *International Journal of Remote Sensing*, **14,** 21-52.

Kaufman, Y. J. and Sendra, C., 1988, Algorithm for automatic atmospheric corrections to visible and near-IR satellite imagery. *International Journal of Remote Sensing*, **9,** 8, 1357-1381.

Kaufman, Y. J. and Tanré, D., 1992, Atmospherically resistant vegetation index (ARVI) for EOS-MODIS. *IEEE Transactions on Geoscience and Remote Sensing*, **30,** 2, 261-270.

King, M., Harshvardhan, D. and Arking, A., 1984, A model of the radiative properties of the El Chichon Stratospheric Aerosol layer. *Journal of Climate and Applied Meteorology*, **23,** 7, 1121-1137.

Koepke, P., 1982, Vicarious satellite calibration in the solar spectral range by means of calculated radiances and its application to Meteosat. *Applied Optics*, **21**, 15, 2845-2854.

Koepke, P., 1984, Effective reflectance of oceanic white caps. *Applied Optics*, **20**, 24,

Lenoble, J., 1977, Standard procedures to compute Atmospheric Radiative Transfer in a Scattering Atmosphere Radiation Commission, IAMAP, National Center for Atmospheric Research, Boulder,CO,

London, J., Bojkov, D. R., Oltmans, S. and Kelly, J. L., 1976, Atlas of the Global Distribution of the Total Ozone July 1957-June 1967 NCAR/TN/113+STR, NCAR, Boulder, Colorado, USA.

McCormick, M. P. and Veiga, R. E., 1992, SAGE II Measurements of early Pinatubo aerosols. *Geophysical Research letters*, **19**, 2, 155-158.

Mitchell, R. M., O'Brien, D. M. and Forgan, B. W., 1992, Calibration of the NOAA AVHRR Short-wave Channels Using Split Pass Imagery: I. Pilot Study. *Remote Sensing of Environment*, **40**, 57-65.

Moran, M. S., Jackson, R. D., Hart, G. F., Slater, P. N., Bartell, R. J., Biggar, S. F., Gellman, D., I. and Santer, R. P., 1990, Obtaining surface factors from atmospheric and view angle corrected SPOT-1 HRV data. *Remote Sensing of Environment*, **32**, 203-214.

Moran, M. S., Jackson, R. D., Slater, P. N. and Teillet, P. M., 1992, Evaluation of simplified procedures for retrieval of land surface reflectance factors from satellite sensor output. *Remote Sensing of Environment*, **41**, 169-184.

Otterman, J. and Fraser, R., 1976, Earth-atmosphere system and surface reflectivities in arid regions from Landsat MSS data. *Remote Sensing of Environment*, **5**, 247-266.

Price, J. C., 1987, Calibration of satellite radiometers and the comparison of vegetation indexes. *Remote Sensing Environment*, **21**, 15-27.

Price, J. C., 1988, An update on visible and near IR calibration of satellite instruments. *Remote Sensing Environment*, **24**, 419-422.

Singh, S. M., 1988, Simulation of solar zenith angle effect on global vegetation index (GVI) data. *International Journal of Remote Sensing*, **9**, 2, 237-248.

Smith, G. R., Levin, R. H., Abel, P. and Jacobowitz, H., 1988, Calibration of the solar channels of the NOAA-9 AVHRR using high altitude aircraft measurements. *Journal of Atmospheric and Oceanic Technology*, **5**, 631-639.

Sobolev, V. V., 1975, *Light scattering in Planetry Atmospheres*, Pergamon Press, New York.

Staylor, W. F., 1990, Degradation rates of the AVHRR visible channel for the NOAA 6,7 and 9 Spacecraft. *Journal of Atmospheric and Oceanic Technology*, 7, 411-423.

Stowe, L. L., Carey, R. M. and Pellegrino, P. P., 1992, Monitoring the Mt. Pinatubo aerosol layer with NOAA/11 AVHRR data. *Geophysical Research Letter*, **19**, 2, 159-162.

Stowe, L. L., McClain, E. P., Carey, R., Pellegrino, P., Gutman, G. G., Davis, P., Long, C. and Hart, S., 1991, Global Distribution of Cloud Cover derived from

NOAA/AVHRR operational Satellite Data. *Advances in Space Research / COSPAR,* 11 COSPAR), pp. 51-54.

Tanré, D., Herman, M. and Deschamps, P. Y., 1983, Influence of the atmosphere on space measurements of directional properties. *Applied Optics,* **21,** 733-741.

Tanré, D., Deroo, C., Duhaut, P., Herman, M., Morcette, J. J., Perbos, J. and Deschamps, P. Y., 1990, Description of a computer code to simulate the satellite signal in the solar spectrum: 5S code. *International Journal of Remote Sensing,* **11,** 659-668.

Tanré, D., Holben, B. N. and Kaufman, Y. J., 1992, Atmospheric Correction algorithm for NOAA-AVHRR Products: Theory and Application. *IEEE Transaction on Geoscience and Remote Sensing,* **30,** 2, 231-248.

Teillet, P. M. and Santer, R. P., 1991, Altitude dependence in a semi-analytical atmospheric code. *Physical Measurements and Signatures in Remote Sensing, France,* ESA, pp. 36-44.

Teillet, P. M., Slater, P. N., Ding, Y., Santer, R. P., Jackson, R. D. and Moran, M. S., 1990, Three Methods for the Absolute Calibration of the NOAA AVHRR Sensors In-Flight. *Remote Sensing of Environment,* **31,** 105-120.

Tucker, C. J., 1979, Red and photographic infrared linear combinations monitoring vegetation. *Remote Sensing Environment,* **8,** 127-150.

Tucker, C. J., 1986, Maximum normalized difference vegetation index images for sub-Saharan Africa for 1983-1985. *International Journal of Remote Sensing,* **7,** 1383-1384.

Vermote, E., El Saleous, N. and Holben, B. N., 1993, Atmospheric Correction of AVHRR visible and Near Infrared Data. *Workshop on atmospheric correction of Landsat Data.*

Vermote, E., Santer, R., Deschamps, P. Y. and Herman, M., 1992, In-flight calibration of large field of view sensors at short wavelengths using Rayleigh scattering. *International Journal of Remote Sensing,* **13,** 18, 3409-3429.

Vermote, E. F. and Tanré, D., 1992, Analytical Expressions for Radiative Properties of Planar Rayleigh Scattering Media Including Polarization Contribution. *Preprint for Journal Of Quantitative Spectroscopy and Radiative Transfer,* **47,** 4, 305-314.

Whitlock, C. H., Staylor, W. F., Smith, G., Levin, R., Frouin, R., Gautier, C., Teillet, P. M., Slater, P. N., Kaufman, Y. J., Holben, B. N., Rossow, W. B., Brest, C. and LeCroy, S. R., 1988, AVHRR and VISSR satellite instrument calibration results for both cirrus and marine stratocumulus IFO periods. *FIRE Science Team Meeting, Vail, Colorado*

Woessner, P. and Hapke, B., 1987, Polarization of light scattered by clover. *Remote Sensing of Environment,* **21,** 243-261.

Zdunkowski, W. G., Welch, R. M. and Korb, G., 1980, An investigation of the Structure of Typical Two-stream-methods for the Calculation of Solar Fluxes and Heating Rates in Clouds. *Contributions to Atmospheric Physics,* **53,** 2, 147-165.

ABSOLUTE CALIBRATION OF AVHRR CHANNELS 1 AND 2

E. VERMOTE AND NAZMI EL SALEOUS
NASA / Goddard Space Flight Center,
Greenbelt, MD, 20771,
U.S.A.

Abstract. This chapter considers the calibration of channels 1 and 2 of the Advanced Very High Resolution Radiometers (AVHRR), a crucial process for the quantitative interpretation of remotely sensed data acquired through time. In this chapter, a new method of absolute calibration of the first two AVHRR channels based on observation of oceans and deep thick clouds is presented, along with sensitivity tests on the possible error sources. Calibration results obtained by this method for NOAA- 7, -9 and -11 from 1981 to the present are also included, as is a comparison of the derived NOAA-9 values with other recently published ones.

1. Introduction

As was shown in the previous chapter, separating the atmospheric and ground signals in Advanced Very High Resolution Radiometer (AVHRR) data is an important step towards the use of these data in quantitative studies. This chapter deals with a related issue, radiometric calibration. The AVHRR data, in common with those from other Earth observation satellite sensors, are originally recorded as digital counts or numbers (DNs). These may be converted to radiance values through the use of appropriate calibration coefficients. These coefficients (gain and offset) are usually supplied from pre-flight calibration of the sensor. However, large changes can, and do, occur to the optical system, both during and post-launch. Onboard calibration devices are designed to take these into account, but unfortunately no such devices are available for the AVHRR visible and near-infrared sensors. Thus, comparison of data from different AVHRR sensors, or time-series from a single sensor, should ideally incorporate the use of absolute calibration coefficients which take into account the post-launch changes in sensor performance.

Several methods, both relative and absolute, have been developed for this purpose over both land and ocean surfaces (Frouin and Gautier 1987, Smith *et al.* 1988, Whitlock *et al.* 1988, Holben *et al.* 1990, Staylor 1990, Teillet *et al.* 1990, Brest and Rossow 1991, Abel *et al.* 1993, Kaufman and Holben 1993). Relative calibration methods do show relative sensor degradation but they also need to be related to some absolute value. Aircraft calibration (*e.g.* as carried out by Abel *et al.* 1993) is expensive and cannot be applied to historical datasets. Other absolute calibration methods combining surface and atmospheric ground-based measurements (Teillet *et al.* 1990) suffer from problems in sampling of the surface and uncertainty in accounting for bidirectional relectance distribution function (BRDF) effects.

This chapter describes a recent and robust method for absolute sensor calibration based on the approach of Kaufman and Holben (1993). This method, first introduced by

73

G. D'Souza et al. (eds.), Advances in the Use of NOAA AVHRR Data for Land Applications, 73–92.
© 1996 ECSC, EEC, EAEC, Brussels and Luxembourg.

Vermote *et al.* (1992) for the calibration of the SPOT/HRV radiometer, unlike previous methods, takes into account the actual precipitable water vapour and aerosol optical thickness qualities of the atmosphere at the time of imaging.

2.   Sensor calibration

Ocean areas have often been used to calibrate satellite sensors. For example, Koepke (1982) used ground and atmospheric measurements to calibrate the METEOSAT sensor and Fraser and Kaufman (1986) developed a method for the calibration of the Visible Infrared Spin Scan Radiometer (VISSR with effective wavelength of 0.61 μm), carried onboard GOES-5 and 6. The latter method concerned the relating of the observed digital count to the actual reflectance by assuming mean conditions both for the atmosphere and the ocean - the dominant signal being modelled for areas outside the ocean glint pattern. This method was then adapted also to calibrate AVHRR channel 1 (Kaufman and Holben 1993).

In this chapter we report on developments of the Kaufman and Holben (1993) approach. Correction of channel 2 is introduced, and channel 1 and 2 intercalibration used to reduce the uncertainty due to variability in surface and atmospheric conditions, which relate principally to wind speed and aerosol concentration respectively. The theoretical background and accuracy of this method is described and examined, and the method is then applied to the AVHRR sensors on NOAA-7, -9 and -11, from 1981 to the present. Split-window water vapour estimates form part of the output of this method, and these are then validated by ancillary data from the SSMI sensor. Finally, the results obtained for NOAA-9 are validated by comparison with other recently published values.

3.   Intercalibration between channel 1 and 2

3.1.   METHOD

The method of calibration described in this chapter is based on AVHRR observations of high clouds. These may be considered lambertian reflectors and, if they are sufficiently thick, the effects of the atmosphere and surface below them can be ignored. Therefore, the signal at the top of the atmosphere can be written as a function of the cloud top altitude z:

$$\rho^* (z) = T_g [\rho_a(z) + T_a \frac{\rho_{cloud}}{1 - S(z)\rho_{cloud}}] \qquad (1)$$

where T_g represents the gaseous transmission, $\rho_a(z)$ is the intrinsic reflectance of the atmosphere above the cloud, $T_a(z)$ is the transmission of the atmosphere above the cloud, and $S(z)$ is the albedo of the atmosphere above the cloud.

If the cloud is high enough (above 4 km), then the aerosol and water vapour influence on the signal can also be ignored because these are normally present only in the lower layers of the atmosphere. Therefore, the signal could be written for both channels 1 and 2 of the AVHRR as:

$$\rho_1 = T_{goz1}\, T_{gox1}\, [\rho_{r1}(z) + T_{r1}(z)\ \frac{\rho_{cloud}}{1 - S_{r1}(z)\rho_{cloud}}\ ] \tag{2a}$$

$$\rho_2 = T_{goz2}\, T_{gox2}\, [\rho_{r2}(z) + T_{r2}(z)\ \frac{\rho_{cloud}}{1 - S_{r2}(z)\rho_{cloud}}\ ] \tag{2b}$$

In practice we have:

$$\rho_{m1} = \rho_1\, r_1 = T_{goz1}\, T_{gox1}\, [\rho_{r1}(z) + T_{r1}(z)\ \frac{\rho_{cloud}}{1 - S_{r1}(z)\rho_{cloud}}\ ]\, r_1 \tag{3a}$$

$$\rho_{m2} = \rho_2\, r_2 = T_{goz2}\, T_{gox2}\, [\rho_{r2}(z) + T_{r2}(z)\ \frac{\rho_{cloud}}{1 - S_{r2}(z)\rho_{cloud}}\ ]\, r_2 \tag{3b}$$

where ρ_{m1}, ρ_{m2} are the measured AVHRR signals with pre-flight calibration for channels 1 and 2 (Price 1987, Price 1988) and r_1 and r_2 are the respective degradation coefficients (Kaufman and Holben 1993).

Therefore, assuming $\rho_r(z)\, r \cong \rho_r(z)$, it can be shown that:

$$\rho'_1 = \frac{\dfrac{\rho_{m1}}{T_{goz1}\, T_{gox1}} - \rho_{r1}(z)}{T_{r1}(z)} \tag{4a}$$

$$\rho'_2 = \frac{\dfrac{\rho_{m2}}{T_{goz2}\, T_{gox2}} - \rho_{r2}(z)}{T_{r2}(z)} \tag{4b}$$

where ρ'_1 and ρ'_2 are the corrected signals for molecular scattering and gaseous absorption. Then, correcting for the atmospheric albedo, $S_r(z)$, it can be shown that:

$$\rho''_1 = \frac{\rho'_1}{1 + S_{r1}(z)\rho'_1} \cong r_1\, \rho_{clouds} \tag{5a}$$

$$\rho''_2 = \frac{\rho'_2}{1 + S_{r2}(z)\rho'_2} \cong r_2\, \rho_{clouds} \tag{5b}$$

So, $\rho''_1/ \rho''_2 = r_1/r_2 = r_{12}$ assuming that the cloud reflectance is constant for all wavelengths in the region of 0.60 to 1.0 μm.

3.2.   ERROR BUDGET

3.2.1 *Lambertian Clouds.* In order to minimise the gaseous absorption and scattering effects, the clouds chosen for use in this method are restricted to those within angles of view of between 0 and 10 degrees. Furthermore, because it is the ratio of cloud reflectances that is utilised, any remaining errors resulting from the assumption of lambertian behaviour are expected to cancel out.

3.2.2. *Aerosol and Water Vapour.*   Clouds are chosen relative to their apparent brightness temperature in AVHRR Channel 4 in the range from 220 K to 225 K at tropical latitudes. This corresponds to an altitude range of 8 km to 13 km, which is well above the layers at which water vapour or aerosols are present. Figure 1 shows the variation of the signal observed over clouds for both channels versus the temperature in channel 4 (10.8 μm). Both channel 1 and 2 signals increase when the temperature decreases due to the decreasing amount of water vapour and other molecules above the cloud. Figure 2 shows the calibration ratio, r_1/r_2, versus temperature. The clouds selected to determine the ratio are located at the lower end of the temperature range where the water vapour effect is minimal and the ratio therefore reaches a plateau. By running a gaseous aborption model, it can be shown that the residual absorption by water vapour at this range of altitude is only of the order of 2-3% (in channel 2) depending on the atmospheric profile. Since our clouds are also situated in the moist tropical convective zone it was consequently felt appropriate to adopt a general correction factor of 0.97.

3.2.3. *Rayleigh Correction.*   For clouds of apparent reflectance greater than 0.5 and higher than 8 km, for view zenith angle lower than 30°, the terms ρ_r, T_r and S_r only influence about 1% of the total signal. Therefore, using pre-flight calibration coefficients to correct for Rayleigh scatter, this will lead to a subsequent error of only 0.3%.

3.2.4. *Ozone and Oxygen Correction.*   The overall amount of oxygen absorption is weak in both channels and depends mainly on altitude. As the uncertainty regarding cloud altitude should be less than 1 km, so the uncertainty concerning total oxygen amount should be less than 5%. Thus, the variation of the transmission in channels 1 and 2 due to the variation of oxygen amount is negligible.

The corrections used for ozone absorption are based on the ozone amounts taken from the latitudinal monthly climatological data published by London *et al.* (1976). The relative accuracy of these included ozone amounts should be in the order of 10%, which translates, for an airmass of 2, to an uncertainty of only 1.0% in the radiance for channel 1, and therefore to a similarly low uncertainty value for the ratio r_{12}.

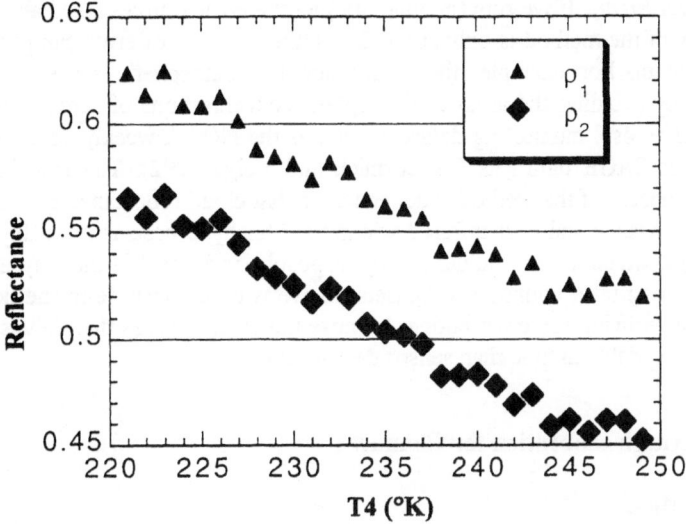

Figure 1. Cloud reflectances observed in channel 1 and 2 of the AVHRR as a function of T4.

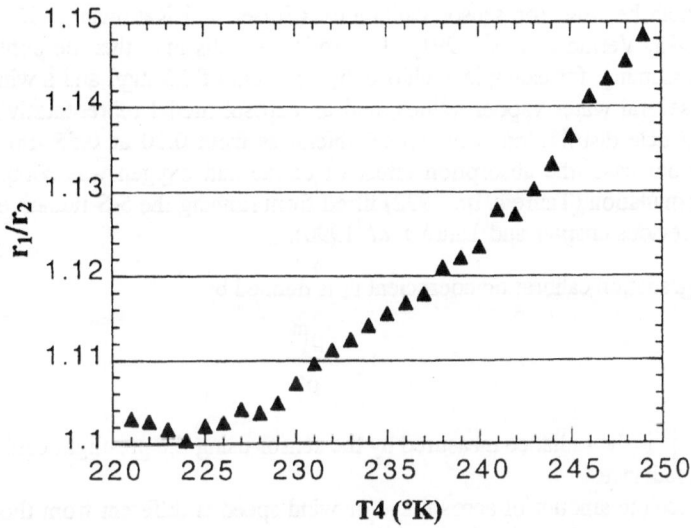

Figure 2. The ratio r_1/r_2 as a function of the temperature in channel 4.

3.2.5. *Total Error.* If we sum (in quadratic) all the error sources listed above, the overall uncertainty of the method is only about 2%. Other sources of error can probably be kept to a minimum. For example, the errors due to stratospheric aerosol effects can be minimised by avoiding the areas of the highest concentrations of aerosols as depicted by operational aerosol monitoring datasets, such as the NOAA weekly composite (Stowe *et al.* 1992) or SAGE data sets (McCormick and Veiga 1992). The main assumption on which the success of the method is dependent is that cloud reflectance does not vary much between the visible and near-infrared windows. This hypothesis is problably true for high thick clouds, which are not influenced by tropospheric aerosol in the way they form. The analysis of long-term continuous datasets, as we will demonstrate in the next section, is one way of verifying the error budget because the error sources themselves will probably be far more variable in time than sensor deterioration.

4. Rayleigh calibration for Channel 1

4.1. METHOD

For a cloudless air mass over the ocean with only a small amount of haze and a reasonable distance from regions of sun glint, the major contribution to the upward radiance in the visible part of the spectrum is from molecular scattering (in the order of 70-80 %). This amount can be accurately computed from a radiative transfer model (Deuzé *et al.* 1989) and can then be used for sensor calibration (Fraser and Kaufman 1986, Kaufman and Holben 1993, Vermote *et al.* 1991). Theoretical signals may then be computed in both channels assuming, for example, a chlorophyll content of 0.3 mg/l and a wind speed of 10 ms^{-1}, for several water vapour values, and an aerosol model representative of maritime-aerosol particle distribution with optical thickness from 0.10 at 0.55 μm (Deuzé *et al.* 1989). In our case, the absorption effect of ozone and oxygen was computed using an explicit formulation (Tanré *et al.* 1992) fitted from running the 5-S radiative transfer code (see the previous chapter and Tanré *et al.* 1990).

The degradation calibration coefficient r_1 is defined by:

$$r_1 = \frac{\rho_1^m}{\rho_1^r} \tag{6}$$

where ρ_1^m is the radiance measured by the sensor using the pre-flight calibration and ρ_1^r is the real radiance.

Since the amount of aerosol and/or wind speed is different from those used in the theoretical computations, the theoretical radiance, ρ^t has to be corrected by an offset $\delta\rho_1$

$$\rho_1^r = \rho_1^t + \delta\rho_1 \tag{7}$$

At satellite level this equation then becomes:

$$\rho_1^m = r_1 \rho_1^t + r_1 \delta\rho_1 \tag{8}$$

A discrepancy will also be observed in channel 2, that is:

$$\rho_2^m \ r_2 \rho_2^t + r_2 \delta\rho_2 \tag{9}$$

where $\delta\rho_1$ is related to $\delta\rho_2$ through the spectral dependence of this perturbation I_{12}:

$$\delta\rho_1 = I_{12}\delta\rho_2 \tag{10}$$

I_{12} is derived from simulations of the signal for both channels 1 and 2 with an aerosol optical thickness of 0.15 and 0.05.

Therefore, ρ_1^m may be written as :

$$\rho_1^m = r_1 \rho_1^t + r_1 I_{12} (\rho_2^m - r_2 \rho_2^t) / r_2 \tag{11}$$

and, solving for r_1 it was found that:

$$r_1 = \frac{\rho_1^m - r_{12}I_{12}\rho_2^m}{\rho_1^t - I_{12}\rho_2^t} \tag{12}$$

where r_{12} is the intercalibration coefficient computed with the method previously outlined and is equal to r_1/r_2.

In practice, the calculation of r_1 is based on more than one measurement. The data are extracted from the Pacific Ocean data set referred to above. For a period of nine days, rigourous cloud screening is performed using the methods of Stowe et al. (1991). Then, a composite of the non-cloudy pixels and non-cloud-shadowed pixels is produced using the minimum value in channel 1. The result of this "geometrical composite" is then screened manually to select a zone of 25 scans where the cloud amount is low and the Rayleigh contribution significant (in practice as close to the principal plane as possible). An average value for non-cloudy pixels is then calculated for each view direction for both channels. The values, after subtracting the deep-space count, are then converted to reflectance units using the pre-flight calibration coefficient (Figure 3). For each pixel of the anti-specular portion of the scan (angle of view between 40-70 degrees) the numerator of Equation (12) is shown plotted against the denominator of the same Equation 12 (Figure 4) (with the water vapour amount as given by the split window technique) (Vermote et al. 1993). Then the slope of the linear regression of the numerator (measured reflectance) versus the denominator (predicted reflectance) is the degradation coefficient r_1. Both the intercept (which should be small) and the correlation coefficient (which should be high) provide good indicators of the accuracy of the calibration.

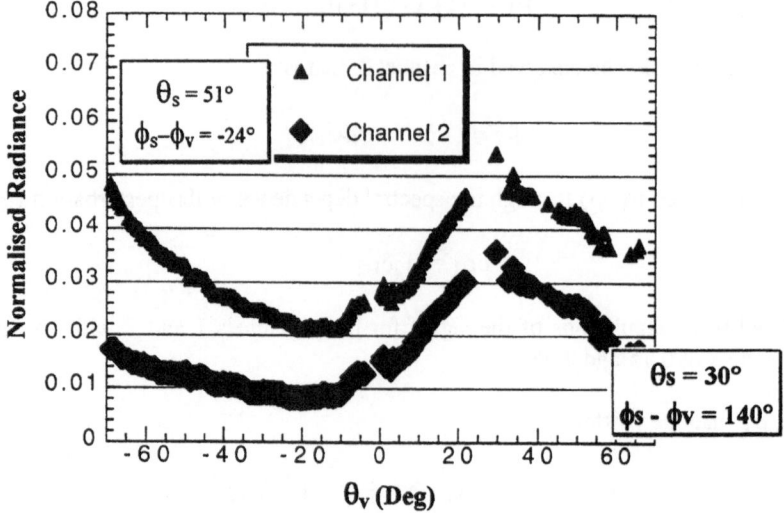

Figure 3. Radiances observed in channel 1 and 2 of the AVHRR over ocean.

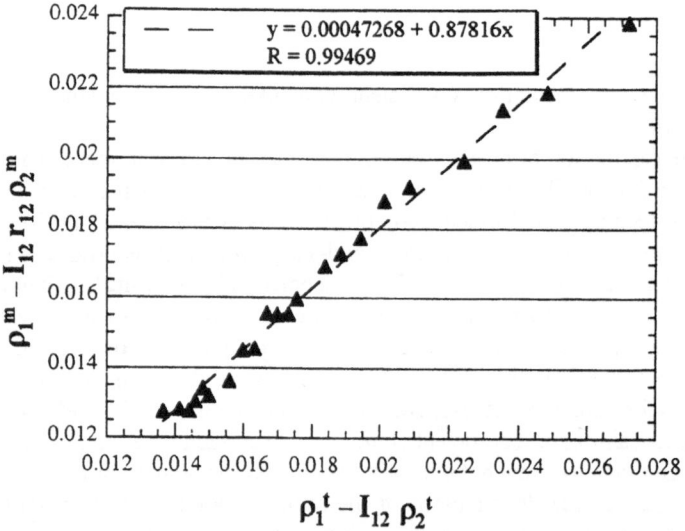

Figure 4. Linear regression between the numerator of Equation 12 *(y)* and the demoninator of Equation 12 *(x)*, that leads to an estimate of degradation (r_1), the slope of the linear regression.

4.2.   ERROR BUDGET

Even after we assume perfect linearity for the instrument, and stability in the spectral response, several other sources of uncertainty remain. Basically, these are uncertainties in the radiative computation of the model, and errors in: input parameters for ozone and water vapour amount, ocean colour, pressure and, finally, errors in the estimation of I_{12} because of an assumed aerosol type and constant wind speed. The relative magnitude of these error sources is further discussed below.

4.2.1. *Intercalibration Coefficient (r_{12})*.   The error on r_{12} estimated at 2% will be directly related to r_1 (Equation 12) as described above (section 3.2.4).

4.2.2. *Radiative Transfer Computation*.   If the sun is never lower than 75° over the horizon, the accuracy of the radiative transfer calculation based on plane parallel approximation is better than 1×10^{-3} in reflectance units (Vermote and Tanré 1992). Thus for the given level of radiance (2×10^{-2}) and sun angle (75°), of the pixels we are using for calibration, there will be a relative uncertainty due to the radiative transfer computation in the order of only 1%.

4.2.3. *Gaseous Absorbers*.   The ozone amount used in the computation, which primarily affects the signal in channel 1, is the latitudinal monthly climatological data (London *et al.* 1976). The assumed relative accuracy of the ozone amount should be of the order of 10%. That translates, for an airmass of 4, to an uncertainty of only 1.5% in the radiance for channel 1. This uncertainty is translated to an uncertainty of 2% for the calibration coefficient (see Equation 12), as a typical ratio from channel 1 to channel 2 over clear water is 2.
    The expected accuracy in the determination of the total column water vapour amount is 0.5 cm. Water vapour is mainly distributed in the lower layer of the atmosphere (0-2 km) and most of the water vapour absorption occurs in the longer spectral wavelength region of channel 2. Therefore, uncertainty in the determination of the total column water vapour amount will translate directly to an uncertainty in the aerosol correction process. The radiation transfer code 5S was again used to determine the amount of possible uncertainty on transmission; this was found to be in the order of 2%. For a typically clear day with an aerosol amount of 0.1 m this translates to a residual of 2×10^{-4} in reflectance units (assuming that 0.1 m gives 0.01 reflectance units). That leads to an uncertainty of only 0.1% in r_1 for a typical value of the numerator of Equation 12.

4.2.4. *Ocean Colour*.   The effect of ocean colour was determined using data from an area over the Pacific Ocean well away from the coast to avoid turbidity effects. According to data from the Coastal Zone Color Scanner (CZCS), if the chlorophyll content is lower than 0.3 mg/l and known with an accuracy better than 0.1 mg/l, this would lead to an uncertainty of only 1×10^{-4} in reflectance units for channel 1, which translates to an uncertainty of only 0.1% for the calibration coefficient.

4.2.5. *Pressure.* The data set used in this study was completely free of clouds, typical of a high-pressure weather system, and a pressure of 1010 millibars (Mb) was consequently assumed. If the pressure is known to within an accuracy of 10 millibars (Mb), then the Rayleigh optical thickness can be determined to within 1% in both channels. This would lead to an uncertainty of only 1% for the calibration coefficient.

4.2.6. *Wind Speed.* A simulation was performed in order to assess the uncertainty induced by wind speed. The wind speed was taken as 10 m/s and a simulation of the measured radiance was made at 5 m/s and 15 m/s with all the other parameters remaining fixed (Pressure = 1010 Mb, maritime aerosol model, aerosol optical thickness = 0.1). The wind speed affects the signal at the top of the atmosphere in two ways: on the sun glint by changing the distribution of wave slopes (Cox and Munk 1965) and by changing the area covered by foam (Koepke 1984). The latter is the most important effect because the calibration is done in the backscattering direction where the direct sunglint influence is low.

Figure 5 gives the values observed in channel 1 and 2 for the nomimal wind speed as well as at 5 m/s and 15 m/s. It also gives the values of the calibration coefficient (equal to 1 for the nominal wind speed). It can be deduced that the effect of wind speed introduces a relative uncertainty of about 2%.

4.2.7. *Aerosol Type.* A simulation was performed for two aerosol types in addition to the maritime model: for continental aerosol at an optical thickness of 0.1 and 0.25 and for stratospheric aerosol (King *et al.* 1984) at optical thicknesses of 0.1, 0.2 and 0.3. In each case, the error is significant, ranging from 5% to 15%. We do not expect continental aerosol to be present all the time over the ocean, especially after the compositing process used. In cases where stratospheric aerosol concentration is significant, for example following the Pinatubo or El Chichon volcanic eruptions, the method presented is not expected to work, at least in its current state. In the total error budget, any uncertainty due to aerosol type is not reported. This assumes that there is no stratospheric aerosol contamination, that the composite eliminates continental aerosol outbreaks, and also that the haze is perfectly representative of the mean aerosol background over the Pacific Ocean.

4.2.8. *Total Error.* Adding all the effects listed above, the theoretical error budget shows that, under reasonable aerosol loading, the intercalibration coefficient can be determined to within a precision of 4%. This figure relies largely on the fact that strong aerosol contamination is avoided. Any dubious results could be filtered out by inspecting the correlation and intercept of the regression (see Figure 4). The consistency of the result over a month (four weekly results) should also confirm the estimate and the size of its likely error.

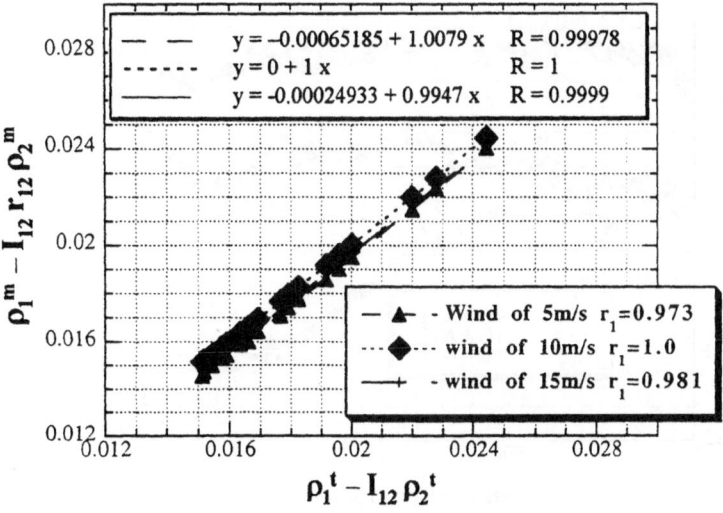

Figure 5. Effect of wind speed on the determination of r_1.

5. Calibration results

Figure 6 shows intercalibration coefficient estimates for NOAA-7, -9 and -11 obtained using the cloud technique. The spread of values is very low (less than 2%).

Figure 7 shows the estimate of the degradation in channel 1, using the ocean technique. Points where the intercept was found to be larger than 2×10^{-3} were eliminated. The results were also averaged over a period of 36 days (four composites) to minimise noise. Once again, the results are very encouraging; the stability of the derived coefficient, 4-5%, is inside our error budget estimate. There is a seasonal variation of about ±2% which can be attributed to the ozone variability. The data are actually reprocessed to address that point.

Figure 8a shows the comparison between U2 aircraft calibration (Abel *et al.* 1993) and our result from the cloud calibration method for NOAA-9. Figure 8b shows the same comparison for NOAA-11. The temporal trends from both methods, and even the absolute values, agree reasonably well. The figures also show values from Che and Price (1992) obtained from a compilation of various previously published methods.

For the Ocean method, Figures 9a-b show comparisons for both NOAA-9 and NOAA-11 respectively between U2 estimates and our results. In both cases, the value for deterioration observed over the ocean is well below the value determined by the U2 or others. However, this degradation has also been reported for NOAA-11 by Mitchell *et al.* (1992). The values from their paper are also plotted on the figure for comparison.

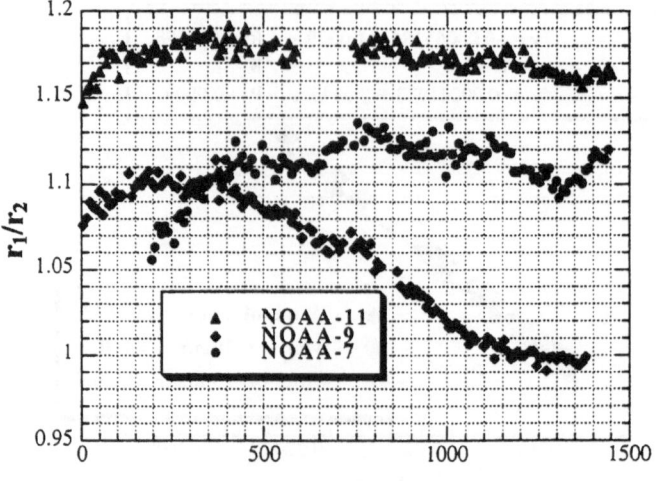

Figure 6. Ratio between the deterioration of channel 1 and 2, as observed over high reflective clouds for NOAA-7-9 and -11, as a function of the number of days since the beginning of launch year.

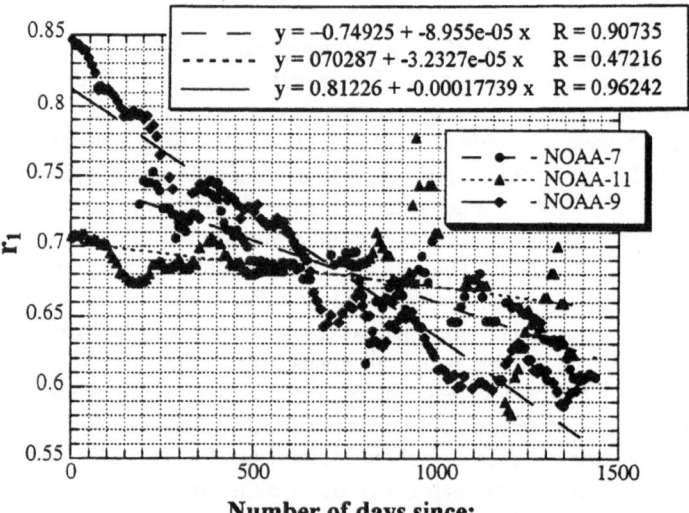

Figure 7. Deterioration of channel 1 of the AVHRR as observed using an ocean target and the method described in 3.2 for NOAA-7-9 and -11.

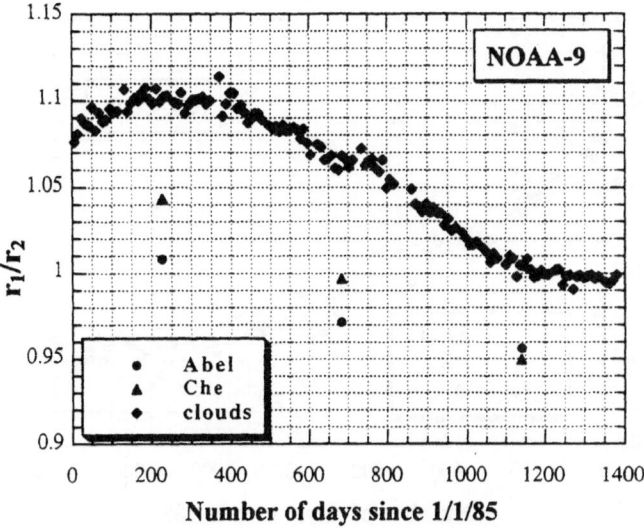

Figure 8a. Comparison of the r_1/r_2 derived using clouds and other methods for
NOAA-9.

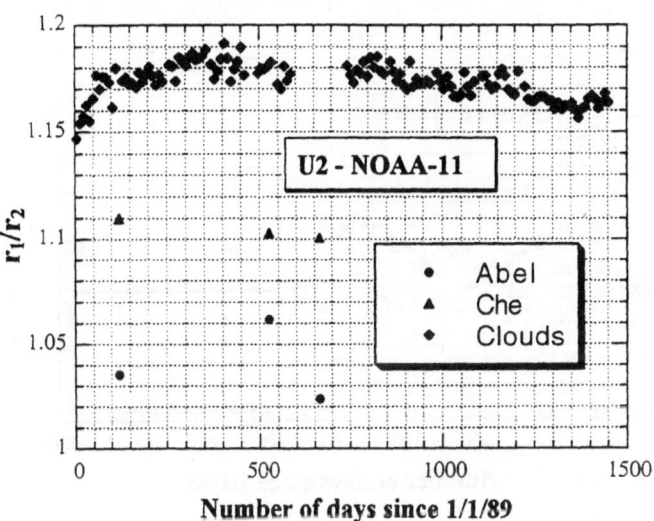

Figure 8b. Comparison of the r_1/r_2 derived using clouds and other methods for
NOAA-11.

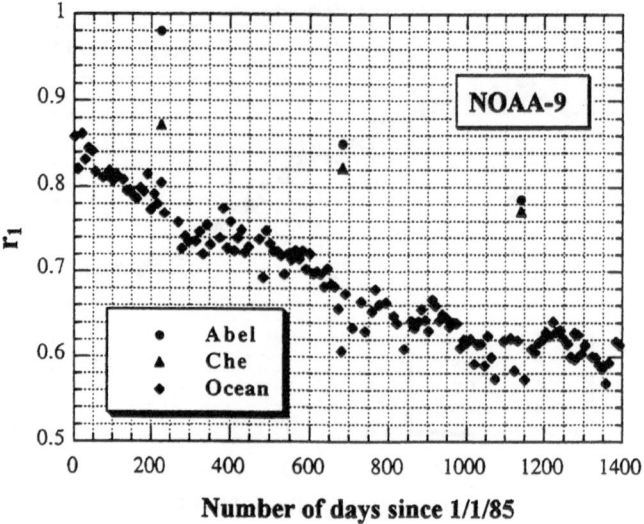

Figure 9a. Comparison of the r_1 derived using ocean and other methods for NOAA-9.

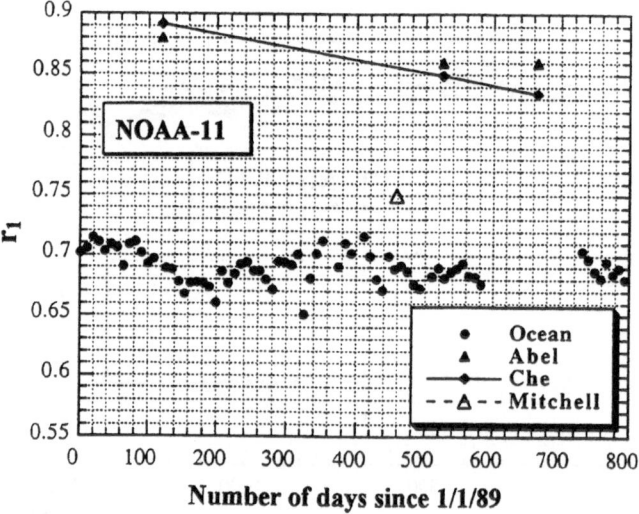

Figure 9b. Comparison of the r_1 derived using ocean and other methods for NOAA-11.

To try and explain this apparent degradation, we analysed data recorded by NOAA-9 over the coast of Tasmania where we had measurements of the optical thickness. Using the U2 absolute calibration, we derived optical thickness over the ocean in both AVHRR channels, and compared these with the actual measurements. Table 1 shows the derived values. The agreement is reasonable in channel 2 but underestimated in channel 1. Because we are in relatively good agreement with U2 concerning the ratio between the two channels, we put the result down to a problem in channel 1. Our hypothesis is that there has been a spectral shift in the response of Channel 1 to a slightly longer wavelength. Our results suggest a shift in the order of 17 nm. Such a shift due to the outgassing of the filter is reasonable as shown by Dinguirard (*personal communication*) for the SPOT-HRV instrument.

Table 1. Analysis of the optical depth values recorded over the coast of Tasmania during 1988 with those derived from NOAA-9 using calibration hypothesis.
(1) U2 calibration.
(2) U2 Calibration with a shift of 17 nm toward the red of channel 1 central wavelength.

Julian Day	τ 0.5 μm	τ 0.86 μm	τ ch1 (1) $r_1 = 0.78$	τ ch2 (1) $r_1 = 0.82$	τ ch1 (2) $r_1 = 0.78$
19	0.04	0.04	0.00	0.05	0.05
28	0.02	0.03	-0.01	0.03	0.04
31	0.10	0.09	0.06	0.12	0.10
40	0.07	0.07	-0.02	0.03	0.02
46	0.055	0.055	0.02	0.06	0.06
47	0.02	0.02	-0.03	0.02	0.02
48	0.03	0.03	-0.05	0.03	0.01
55	0.03	0.03	-0.03	0.02	0.02
57	0.06	0.06	0.05	0.11	0.11
58	0.04	0.04	0.03	0.11	0.08
65	0.03	0.03	0.01	0.06	0.06
66	0.035	0.04	-0.03	0.04	0.02
76	0.05	0.05	0.02	0.08	0.08
77	0.05	0.05	0.03	0.06	0.07
Mean	0.042	0.045	0.005	0.059	0.053

Figures 10a-b, show the results using the hypothesis of a 17 nm spectral shift in channel 1, for both NOAA-9 and 11. According to these results, this hypothesis seems reasonable for both AVHRRs. This is far from unexpected as both have similar filters and both are submitted to the same conditions during post-launch.

6.   Conclusions

A new method for absolute calibration of channel 1 and 2 of AVHRR has been presented here. It consists of observations in these channels of thick high clouds and clear ocean areas. Because the clouds observed are thick and high, water vapour and gaseous absorption can be ignored, and lambertian reflectance can also be reliably assumed. Error sources are further minimised by restricting cloud observations to angles of view of between 0 and 10 degrees, and by utilising an intercalibration coefficient between channels 1 and 2, which reduces the uncertainty due to variability in surface and atmospheric conditions.

The results of the method compare well with previously published methods such as those of the Holben *et al.* (1990) desert calibration method. The ocean-cloud method we propose has the advantage that no additional measurements are necessary (although some ancillary data may be necessary to avoid areas of high and variable stratospheric aerosol content). The results overall are stable and within an acceptable error budget, yet the new approach can be used to determine sensor deterioration on a more frequent basis than other methods.

However, there are still some questions that need to be answered before the absolute calibration of AVHRR channel 1 and 2 data is completely satisfactory. Most pressing perhaps is the question concerning the possible spectral shift in channel 1. We are currently working on a combined approach based on sensor intercomparison with Peter Abel (*personal communication*) to address this point. On the other hand, the sensitivity of the Ocean method to apparent spectral shift could be seen as an advantage for characterising the calibration of ocean colour sensors such as the soon to be launched SeaWIFS.

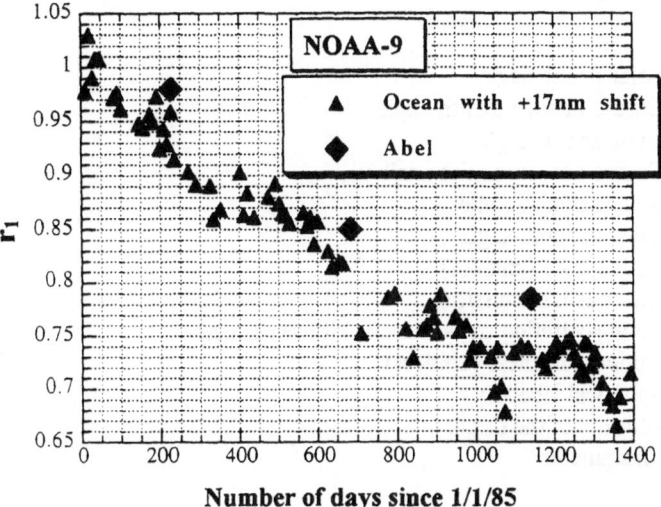

Figure 10a Comparison of the r_1 derived using ocean and other methods for NOAA-9, assuming a shift of 17 nm of the central wavelength of channel 1 toward the red.

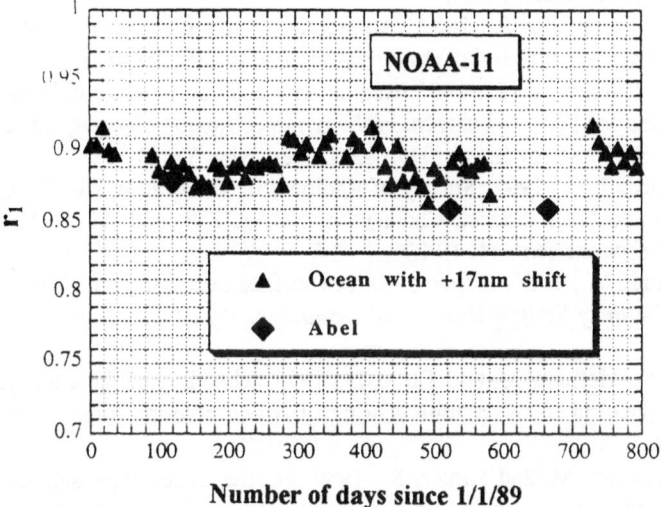

Figure 10b. Comparison of the r_1 derived using ocean and other methods for NOAA-11, assuming a shift of 17 nm of the central wavelength of channel 1 toward the red.

Glossary

θ_S	Solar zenith angle
μ_S	Cosine of solar zenith angle
θ_V	View zenith angle
μ_S	Cosine of view zenith angle
ϕ_S	Solar azimuth angle
ϕ_V	View azimuth angle
ρ	Reflectance (unitless)
Tg	Gaseous transmission
T	Total scattering transmission (diffuse+direct)
τ	Optical thickness (unitless)
s	Spherical albedo
z_t	Altitude of the target
μm	Micrometer

References and Bibliography

Abel, P., Guenther, B., Galimore, R. N. and Cooper, J. W., 1993, Calibration Results for NOAA-11 AVHRR Channels 1 and 2 from Congruent Path Aircraft observations. *Journal of Atmospheric and Oceanic Technology*, **10**, 4, 493-508.

Abramowitz, M. and Stegun, I, A., 1970, *Handbook of Mathematical Functions* (New-York: Dover Publications,Inc).

Ahmad, Z. and Fraser, R. S., 1982, An iterative radiative transfer code for ocean-atmosphere system, *Journal of Atmospheric Science*, **39**, 656-665.

Brest, C. L. and Rossow, W. L., 1991, Radiometric calibration and monitoring of NOAA AVHRR data for ISCCP. *International Journal of Remote Sensing*, **13**, 235-273.

Chandrasekhar, S., 1960, *Radiative Transfer* (Dover: New York).

Che, N. and Price, J. C., 1992, Survey of Radiometric Calibration Results and Methods for Visible and Near Infrared Channels of NOAA-7,-9, and -11 AVHRRs. *Remote Sensing of Environment*, **41**, 19-27.

Cox, C. and Munk, W., 1965, Slopes of the sea surface deduced from photographs of sun glitter. *Bulletin of Scripps Institute of Oceanography of University of California*, **6**, 401-488.

Deschamps, P. Y., Herman, M. and Tanre, D., 1983, Modeling of the atmospheric effects and its application to the remote sensing of ocean colour. *Applied Optics*, **22**, 23, 3751-3758.

Deuzé, J. L., Herman, M. and Santer, R., 1989, Fourier series expansion of the transfer equation in the atmosphere-ocean system. *Journal of Quantitative Spectroscopy and Radiative Transfer*, **41**, 6, 483-494.

Fraser, R. S. and Kaufman, Y. J., 1986, Calibration of satellite sensors after launch. *Applied Optics*, **25**, 1177-1185.

Frouin, R. and Gautier, C., 1987, Calibration of NOAA-7 AVHRR, GOES-5 and GOES-6 VISSR/VAS solar channels. *Remote Sensing Environment,* 22, 73-101.

Holben, B. N., Kaufman, Y. J. and Kendall, J. D., 1990, NOAA-11 AVHRR visible and near-IR inflight calibration. *International Journal of Remote Sensing,* 11, 8, 1511-1519.

Joseph, J, H., Wiscombe, W, J. and Weinman, J, A., 1976, Solar Flux Transfer Through Turbid Atmospheres EValuated by the Delta-Eddington Approximation. *Journal of Atmospheric Science,* 33, 2452-2459.

Kaufman, Y. J. and Holben, B. N., 1993, Calibration of the AVHRR visible and near-IR bands by atmospheric scattering, ocean glint and desert reflection. *International Journal of Remote Sensing,* 14, 21-52.

King, M., Harshvardhan D. and Arking, A., 1984, A model of the radiative properties of the El Chichon Stratospheric Aerosol layer. *Journal of Climate and Applied Meteorology,* 23, 7, 1121-1137.

Koepke, P., 1982, Vicarious satellite calibration in the solar spectral range by means of calculated radiances and its application to Meteosat. *Applied Optics,* 21, 15, 2845-2854.

Koepke, P., 1984, Effective reflectance of oceanic white caps. *Applied Optics,* 20, 24,

Lenoble, J., 1977, Standard procedures to compoute Atmospheric Radiative Transfer in a Scattering Atmosphere Radiation Commission, IAMAP, National Center for Atmospheric Research, Boulder, Colorado, USA.

London, J., Bojkov, D. R., Oltmans, S. and Kelly, J. L., 1976, Atlas of the Global Distribution of the Total Ozone July 1957-June 1967 NCAR/TN/113+STR, NCAR, Boulder, Colorado, USA.

McCormick, M, P. and Veiga, R, E., 1992, SAGE II Measurements of early Pinatubo aerosols. *Geophysical Research Letters,* 19, 2, 155-158.

Mitchell, R. M., O'Brien, D. M. and Forgan, B. W., 1992, Calibration of the NOAA AVHRR Shortwave Channels Using Split Pass Imagery: I. Pilot Study. *Remote Sensing of Environment,* 40, 57-65.

Price, J. C., 1987, Calibration of satellite radiometers and the comparison of vegetation indexes. *Remote Sensing of Environment,* 21, 15-27.

Price, J. C., 1988, An update on visible and near IR calibration of satellite instruments. *Remote Sensing of Environment,* 24, 419-422.

Smith, G. R., Levin, R. H., Abel, P. and Jacobowitz, H., 1988, Calibration of the solar channels of the NOAA-9 AVHRR using high altitude aircraft measurements. *Journal of Atmospheric and Oceanic Technology,* 5, 631-639.

Sobolev, V. V., 1975, *Light scattering in Planetry Atmospheres ,* Pergamon Press, New York.

Staylor, W. F., 1990, Degradation rates of the AVHRR visible channel for the NOAA 6,7 and 9 Spacecraft. *Journal of Atmospheric and Oceanic Technology,* 7, 411-423.

Stowe, L. L., Carey, R. M. and Pellegrino, P. P., 1992, Monitoring the Mt. Pinatubo aerosol layer with NOAA/11 AVHRR data. *Geophysical Research Letter,* 19, 2, 159-162.

Stowe, L. L., McClain, E. P., Carey, R., Pellegrino, P., Gutman, G. G., Davis, P., Long, C. and Hart, S., 1991, Global Distribution of Cloud Cover derived from NOAA/AVHRR operational Satellite Data. *Advances in Space Research / COSPAR*, 11 COSPAR), pp. 51-54.

Tanré, D., Deroo, C., Duhaut, P., Herman, M., Morcette, J. J., Perbos, J. and Deschamps, P. Y., 1990, Description of a computer code to simulate the satellite signal in the solar spectrum: 5S code". *International Journal of Remote Sensing*, **11**, 659-668.

Tanré, D., Herman, M. and Deschamps, P. Y., 1983, Influence of the atmosphere on space measurements of directional properties. *Applied Optics*, **21**, 733-741.

Tanré, D., Holben, B. N. and Kaufman, Y. J., 1992, Atmospheric Correction algorithm for NOAA-AVHRR Products: Theory and Application. *IEEE Transaction on Geoscience and Remote Sensing*, **30**, 2, 231-248.

Teillet, P. M. and Santer, R. P., 1991, Altitude dependence in a semi-analytical atmospheric code. *Physical Measurements and Signatures in Remote Sensing, France*, ESA), pp. 36-44.

Teillet, P. M., Slater, P. N., Ding, Y., Santer, R. P., Jackson, R. D. and Moran, M. S., 1990, Three Methods for the Absolute Calibration of the NOAA AVHRR Sensors In-Flight. *Remote Sensing of Environment*, **31**, 105-120.

Vermote, E., El Saleous, N. and Holben, B, N., 1993, Atmospheric Correction of AVHRR visible and Near Infrared Data. *Workshop on Atmospheric Correction of Landsat Data*. pp. 21-25, Torrance, CA, June 29-July 1, 1993.

Vermote, E., Santer, R., Deschamps, P, Y. and Herman, M., 1991, In-flight calibration of large field of view sensors at short wavelengths using Rayleigh scattering. *International Journal of Remote Sensing*, **13**, 18, 3409-3429.

Vermote, E. F. and Tanré, D., 1992, Analytical Expressions for Radiative Properties of Planar Rayleigh Scattering Media Including Polarisation Contribution. *Preprint for Journal Of Quantitative Spectroscopy and Radiative Transfer*, **47**, 4, 305-314.

Whitlock, C. H., Staylor, W. F., Smith, G., Levin, R., Frouin, R., Gautier, C., Teillet, P. M., Slater, P. N., Kaufman, Y. J., Holben, B. N., Rossow, W. B., Brest, C. and LeCroy, S. R., 1988, AVHRR and VISSR satellite instrument calibration results for both cirrus and marine stratocumulus IFO periods. *FIRE Science Team Meeting, Vail, Colorado, USA*.

Woessner, P. and Hapke, B., 1987, Polarisation of light scattered by clover. *Remote Sensing of Environment*, **21**, 243-261.

Zdunkowski, W. G., Welch, R. M. and Korb, G., 1980, An investigation of the Structure of Typical Two-stream-methods for the Calculation of Solar Fluxes and Heating Rates in Clouds. *Contributions to Atmospheric Physics*, **53**, 2, 147-165.

AEROSOL RETRIEVAL AND ATMOSPHERIC CORRECTION

E. VERMOTE, NAZMI EL SALEOUS AND B. N. HOLBEN
NASA / Goddard Space Flight Center,
Greenbelt, MD, 20771,
U.S.A.

ABSTRACT.   Remote monitoring of global-scale parameters and their change requires the correction of atmospheric effects. Apart from clouds, water vapour and aerosols constitute the primary limitations for the remote sensing of surface features in the AVHRR visible and near-infrared channels. The following chapter presents methods for correcting data in these channels for water vapour and aerosol interference using AVHRR data alone. Included are two methods for aerosol retrieval which are applicable for AVHRR visible and near-infrared radiance data. The first method is the *"dark target"* approach that currently has applicability over oceans and dense dark vegetation. The method has been shown to have an accuracy of approximately 0.1 for optical thickness estimations. The second method, the *"contrast reduction"* method, has a similar accuracy and is complimentary to the first method in that it can be applied in regions of invariant surface cover, such as those where little or no green vegetation grows.

The dark target retrieval is demonstrated on a global-type data set over oceans and land targets. Included is a discussion of calibration, water vapour correction and cloud-screening methods. A method of atmospheric correction for the Pinatubo stratospheric aerosol is then given and applied to part of the same global data set. As validation of the method, data from years after the Pinatubo eruption are seen to compare well with data from years before the eruption, once the Pinatubo stratospheric aerosol correction has been applied.

1.   Aerosol Retrieval: A review of two different approaches

Aerosols have two opposite effects on visible and near-infrared satellite signals. First, they increase the signal measured at the sensor by adding a contribution called the "intrinsic atmospheric reflectance" (caused by photons that are reflected to the atmosphere before interacting with the target). Second, they decrease the amount of light reaching the target, the so-called "transmission effect" (caused by absorption and scattering of photons en route to the target or sensor).

There are basically two methods of determining and isolating the effects of aerosols from the visible and near-infrared satellite signals. In the first method, retrieval of aerosol amount may be achieved from visible (red wavelength) observations of dark targets such as still water bodies or dark dense vegetation, hence its name *"dark target approach"*. In the second method, the variation in the amount of transmission may be estimated by considering an invariant target which will allow a relative correction to be applied. This latter approach is called *"contrast reduction"*.

Both approaches have been successfully applied to data from the Landsat TM sensor and have shown good potential for determining the optical thickness with an accuracy of

93

G. D'Souza et al. (eds.), Advances in the Use of NOAA AVHRR Data for Land Applications, 93–124.
© *1996 ECSC, EEC, EAEC, Brussels and Luxembourg.*

up to 0.1 in the red portion of the electromagnetic spectrum (specifically at 0.55μm). The application of these methods to global data sets implies the establishment of: (1) an operational criterium to identify dark targets, and (2) operational criteria to identify and check the spatial homogeneity and invariance of reference targets.

In this chapter we present the first application of both atmospheric correction methods to global-scale full-resolution and GAC AVHRR data, and further define the criteria required for their global operational application (as well as an estimate of their accuracy).

2.   Conceptualization of aerosol retrievals

The methods for atmospheric retrievals are based on the determination of the concentration of atmospheric aerosol (or the aerosol optical thickness, τ_a) from radiances taken from the image itself. The atmospheric effect on the image may be described by:

$$L(\tau_a) = L_o(\tau_a) + F_d(\tau_a)T(\tau_a)\rho/(1-s(\tau_a)\rho), \tag{1}$$

and it is composed of two parts: the atmospheric path radiance L_o due to photons scattered by the atmosphere to the sensor without being reflected by the surface, and the atmospheric effect on the transmission of the downward flux, F_d, and the upward transmission, T. $s(\tau_a)$ describes the multiple interaction between the ground and atmosphere. Both atmospheric effects have been used in the past to determine the aerosol optical thickness, and their derivation from AVHRR data is described in the following sections.

2.1.   DETERMINATION OF AEROSOL OPTICAL THICKNESS USING PATH RADIANCE

2.1.1. *General.*   In order to determine the aerosol optical thickness from the path radiance, the second term in Equation 1 must be small, so that the uncertainty in the surface reflectance, ρ, will have a minimal effect on the determination of τ_a. A variety of forests, forming dense, dark green canopies have low reflectance, only 1-2% in the red channel, *e.g.* deciduous forest (Kaufman and Sendra 1988), coniferous forest (Deering and Eck 1991, Kriebel 1977), hardwood and pine forests (Kimes *et al.* 1986, Kleman 1987) and tropical forests (Kaufman *et al.* 1992). Therefore, large patches of forests in generally urban or rural areas can be used to derive the aerosol optical thickness and subsequently to apply atmospheric correction. This method we term DDV (dense dark vegetation) and it appears appropriate for the coarse resolution AVHRR data in areas where there are large expanses of dense dark targets such as vegetation (*i.e.* forest) or water.

The accuracy of the determination of the aerosol optical thickness using this path radiance method depends on the accuracy of the assumed reflectance of the dark objects (*e.g.* vegetation or water) and on the ability of estimating the aerosol scattering phase function and single scattering albedo. The total error inherent in this method (including the effect of uncertainty in the surface reflectance and assuming a good calibration) was estimated as $\Delta\tau_a = \pm0.10$ (Kaufman and Sendra 1988) for Landsat MSS data. In the process of correcting the image for the atmospheric effects, the errors due to uncertainty in the aerosol phase function partially cancel out (Kaufman and Sendra 1988) since the same phase function that is used to derive the aerosol optical thickness is also used in the correction process. As a result, the error in the derived surface reflectance is mainly affected by the accuracy of the assumed reflectance of the dark surface ($\Delta\rho = \pm0.01$). Extrapolation of the optical thickness to other channels (from the red to the near-infrared) can be done, but with an uncertainty of $\Delta\tau_a = \pm0.05$, and for the path radiance an uncertainty of $\Delta L = \pm0.002 \cdot F_0/\pi$ (Kaufman 1991).

An aerosol retrieval scheme is possible with one of two methods provided that surfaces with known characteristics can be identified. The required characteristics are: very low surface reflectance for the path radiance method, and a number of invariant contrasting targets for the contrast reduction method. The following sections review ways of delineating such suitable areas.

2.1.2. *Dense dark vegetation.* This method can be applied to images for which there is *a priori* knowledge that dense vegetation is present, taking into account the geographic location and season in which the image was taken (for geographic distribution see Figure 12 of Kaufman and Sendra (1988)). Some minimal fraction of pixels covered by the dense vegetation has to be assumed (the actual fraction may be larger). It is further assumed that the spatial distribution of these pixels is dense enough to cover any spatial variability in the aerosol concentration. Two approaches are possible to determine, on a statistical basis, which pixels are covered by dense dark vegetation.

Determination based on vegetation index examination. This method is described in detail by Kaufman and Sendra (1988) in which they used Landsat MSS data. The following is a brief summary of the method modified for the AVHRR Channel 1. The vegetation index, NDVI, can be used to determine pixels that have the densest, greenest and darkest vegetation in the image. The absolute value of the NDVI depends, amongst other things, on the atmospheric effect (Holben 1986, Tanré *et al.* 1992), viewing and illumination geometry (Holben and Fraser 1984) and surface bidirectional reflectance (Deering and Eck 1991, Kimes *et al.* 1986, Lee and Kaufman 1986, Tanré *et al.* 1992). As long as the image to be analysed is small (*e.g.* less than 200 km by 200 km) aerosol concentration, and the resulting atmospheric effect is more likely to be uniform, geometrical effects will be minimised and pixels that correspond to the densest vegetation should have the highest NDVI values. A small size image is also more likely to yield uniformity in water vapour concentrations. Pixels that correspond to dense, dark vegetation (DDV) are then separated from other vegetated pixels, *i.e.* those having the lowest reflectance in the near-infrared channel. These pixels are also expected to have low, well-defined reflectance values in the visible channel that can be used to determine the aerosol optical thickness. However, large variations in aerosol concentrations from one part of the image to another may also cause large variations in NDVI values across the image, and may therefore lead to the erroneous identification of dense dark vegetation areas. Consequently, this method should not be used in regions with highly variable aerosol concentrations (Kaufman and Sendra 1988).

Determination based on radiance at 3.75 μm. Radiance at 3.75 μm, the wavelength at which the AVHRR channel 3 is sensitive, is affected both by thermal emission and by reflection of the weak solar irradiance at this wavelength. Depending on the value of the surface reflectance and temperature, either of these two effects can dominate. The complex characteristics of this channel make it hard to use, but some widely different applications do emerge. For example, this channel on the AVHRR has been used to determine the size/phase of cloud drops (Arking and Childs 1985), has been shown to be very sensitive to variation in the surface temperature (Kerber and Schutt 1986), and is the most sensitive channel to detect the difference between mature forest and vegetated but deforested areas (Malingreau and Tucker 1988, Tucker *et al.* 1984, and Malingreau *et al.* later in this book).

This latter quality suggests application to aerosol retrieval using the DDV. In an image of mixed surface cover, forest is expected to have the lowest radiance in this channel (except for water, clouds and snow). Due to evaporation, forest is usually cooler than regions with lower density vegetation or bare soil. Also, vegetation has low reflection in this channel due to the absorption of liquid water in live vegetation, and forest decreases this reflection even further by trapping incoming sunlight due to the multilayer canopy structure. These characteristics, as depicted in the radiance of the 3.75 μm channel, in contrast to the visible channel and NDVI depictions, are illustrated in a transect taken from a remarkably cloud-free AVHRR scene over the Amazon forest (Figure 1). It can be seen that variations in the reflectivity of the red channel, caused by variation in the surface cover, are highly correlated with the vegetation index, and are even more apparent in the value of the 3.75 μm channel which has a smaller scan angle effect.

Since most aerosol types have a small particle size, and high transmissivity index in the 3.75 μm wavelength range (Holben *et al.* 1990, Kaufman *et al.* 1992, Shettle and Fenn 1979, Whitby 1978) the aerosol layer is mostly transparent in this channel. Therefore, the selection of dense vegetation pixels is expected to be nearly independent of the aerosol loading and the 3.75 μm channel should be better than the NDVI in selecting the dense dark vegetation pixels. However, it is important to bear in mind that desert dust may behave differently in this respect due to its larger particle size.

2.2. DETERMINATION BASED ON ATMOSPHERIC TRANSMISSION

Determination of the aerosol optical thickness from the atmospheric transmission (second term in Equation 1) is based on the ratio of the transmission among several images and is termed contrast reduction. Tanré *et al.* (1988) suggested and applied this method to several Landsat TM images taken over an arid region. The variation in the transmission is determined from the variation of the difference between the radiance from pixels located a specified distance apart. From Equation 1, the difference of apparent radiance ΔL^*_{ij} between two adjacent pixels (i,j) and $(i,j+1)$, where i and j are the geographical coordinates expressed in line and column numbers, may be related to the actual ground reflectance difference $\Delta\rho_{ij}$ by

$$\Delta L_{ij}^{*}(\tau) \cong \Delta\rho_{ij} \frac{T(\tau_a,\mu_v)F_d(\tau_a,\mu_s)}{1-<\rho>s\cdot2} \tag{2}$$

where $<\rho>$ is the mean reflectance of the two pixels.

If the method is applied to a group of images which includes a relatively clear image (for which the optical thickness is available or estimable), the actual $\Delta\rho_{ij}$ can be derived and the optical thickness for each of the images can then be estimated from Equation 2 (Tanré *et al.* 1988). The derived optical thickness is independent of the aerosol scattering phase function, but depends on the single scattering albedo, ω_0, and the asymmetry parameter of the aerosol.

BRAZIL, JULY 23, 1990.

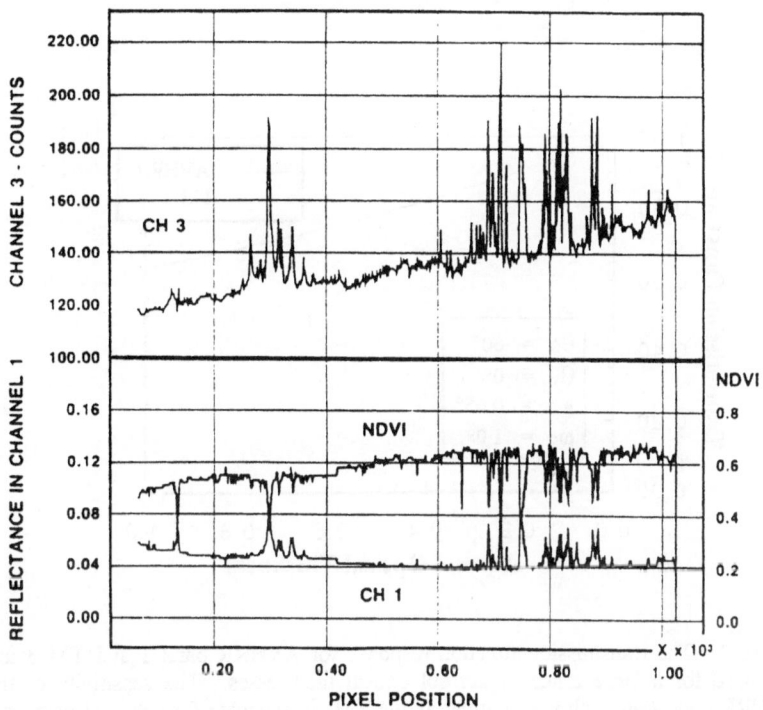

Figure 1. AVHRR bands 1 and 3 are plotted versus scan angle for a transect across the southern Amazon forest dominated by Dense Dark Vegetation. Note the increased sensitivity in band 3 relative to the NDVI for surface variations.

The reduction in contrast for Landsat TM and AVHRR resolution as a function of the aerosol optical thickness, for $\theta_S = 60°$, a nadir observation and a continental aerosol model was simulated (and is shown in Figure 2). The contrast reduction is 40% lower for the AVHRR but still large enough for hazy or dusty conditions in order to be applied in arid or semi-arid regions where large and variable aerosol concentrations prevail (Holben et al. 1991).

To express the contrasted character of the target, $(\Delta\rho_{ij})$, we use the structure function concept, noted Fs(d), and defined by:

$$F*s(d)^2 = \frac{1}{n*(m-d)} \sum_{i=1}^{n} \sum_{j=1}^{m-d} (\rho_{i,j} - \rho_{i,j} + d)^2 \qquad (3)$$

where d is the distance between the pixels and n*(m-d) is the total number of pixels within the target for the structure function Fs(d). From Equations (2) and (3), the structure function observed from the satellite F*s(d) and the actual structure of the surface Fs(d) are related by :

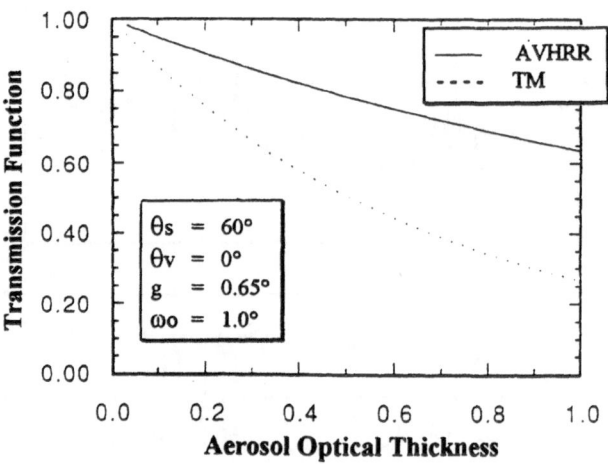

Figure 2. The transmission function response for AVHRR band 1 and TM 3 are simulated for a large range of aerosol optical thicknesses. The sensitivity of the AVHRR band (although not as great as the TM) is adequate for retrieval using the transmission function method.

$$F*s(d) = Fs(d) \frac{T(\tau, \mu_v)Fd(\tau, \mu_s)}{1 - A * s} \qquad (4)$$

where A is the mean albedo of the target .

Provided that Fs(d) is known and invariant, the satellite measurements allow us to estimate the aerosol optical thicknesses by means of the transmissions functions of Equation 4.

3.   Demonstration of aerosol optical thickness retrieval methods

The two methods described above (DDV and Contrast Reduction) generally apply to different cover types. Obviously, the DDV approach is most appropriate in areas where dense dark vegetation is evenly distributed across a scene, in patches of at least more than one pixel size. The contrast reduction method applies more easily to areas devoid of any vegetation as long as there are large contrasts in the surface reflectance in single scenes, which remain invariant between scenes. Therefore, we have chosen two sites to illustrate the two techniques: the Amazon basin for the DDV approach and the Sahel in Mali for the contrast reduction approach. Some ground information was available at both sites.

3.1.   THE DENSE DARK VEGETATION METHOD

This method was applied to four 1989 AVHRR 2 x 2 degree scenes with a pixel resolution of 1 km over the Amazon basin in Brazil. The aerosol optical thickness was measured by the authors on days 247 and 249. Two Amazon Basin sites which both contain large tracts of dense forest canopies were chosen, Alta Floresta and Santarem. In the area of our ground observations, significant deforestation by burning had resulted in large areas of non-forest canopies and bare ground. A third site of intensive forest burning was chosen over the state of Rondonia to illustrate differences in the methods under conditions of spatially variable concentrations of aerosols. Both DDV methods previously described (determination by VI examination, and by examination of radiance at 3.75 μm) were applied to delineate the most suitable areas of dense dark vegetation. Ground observations were made from a five-band sun photometer taken at a single location within the scenes analysed, and were used to verify the AVHRR retrieved aerosol optical thicknesses.

The retrieved aerosol optical thickness in Channel 1 was computed for the NDVI and 3.75 μm techniques according to the method described in the previous section. Cloud screening was applied to images from all dates using a simple interactive thermal 11.0 μm threshold and a primitive cloud shadow approximation based on an assumed cloud height and illumination direction. This eliminated about 20 to 80% of all pixels from the various scenes due to the afternoon broken cumulus cloud layer.

The optical thicknesses measured for each site and the mean values retrieved by each method (days 247 and 249) generally agree to within 0.2. For the uniformly distributed optical thickness (standard deviation in the 3.75 μm method is < 0.05), the NDVI and 3.75 μm methods agree within $\Delta\tau_a = 0.03$. For a day when the optical thickness was very non-uniform and smoke plumes were also observed (252, 1987- σ3.75 = 0.27), τ_a3.75 is larger than τ_aNDVI by 0.24 (Table 1). For day 247 the mean value of the NDVI method is within 0.01 of the ground observation and the 3.75 μm method exceeds the ground observation by 0.11. The standard deviation for the 3.75 μm method, however, is large (Table 1). This is illustrated by plotting the difference of the mean retrieved aerosol optical thickness by the two methods against the standard deviation from the 3.75 μm method (Figure 3). A clear trend of increasing inhomogeneity (larger σ3.75) results in a larger disparity between the methods. A frequency histogram of day 249 at Alta Floresta indicates that the 3.75 μm method retrieves a much broader range of aerosol optical thicknesses than does the NDVI on this date, suggesting a lack of uniformity in the aerosol optical thickness across the scene, and a violation of the basic assumption of the NDVI method (Figure 4a). In contrast, the frequency histograms for day 259 are nearly identical and the range in retrieved aerosol is small (Figure 4b). In this case, the techniques appear equivalent.

The sensitivity of the two methods to aerosol non-uniformity is further illustrated by plotting the brightness temperature at 3.75 μm and the NDVI for the selected DDV pixels in each method against the retrieved aerosol optical thickness for day 249 (Figures 5 and 6). By relaxing the thresholds for the NDVI technique, the range of the retrieved optical thicknesses was matched to the range in retrieved values using the 3.75 μm technique. While the NDVI decreases as a function of τ_a, (Figure 6) the T3.75 is independent of τ_a (Figure 5) which shows the advantage of the 3.75 μm technique for non-uniform aerosol layers. The results clearly demonstrate the importance of the assumption of a uniformly distributed aerosol optical thickness over the scene in question, in order to successfully retrieve the aerosol optical thickness by the NDVI method.

Table 1. Comparison between aerosol optical thickness derived from each of the methods and ground measurements.

Day/Year	Geometry (°) $\theta s/\theta v/\phi v$	Aerosol Optical Thicknesses (τ_a)				Location/ comments
		NDVI Method	3.75 μm method	Ground obs.	Cloud fraction	
247/89	40.9/-43.4/139	0.39±0.02	0.51±0.08	0.40	0.70	Alta Floresta
249/89	35.9/-17.1/137	0.40±0.02	0.58±0.13	0.52	0.26	Alta Floresta
249/89	35.0/-34.2/148	0.36±0.04	0.33±0.02	0.16	0.67	Santarem
250/89	33.6/-1.0/111	0.51±0.02	0.58±0.11	ND	0.68	Alta Floresta
259/89	34.2/-9.4/143	0.33±0.02	0.36±0.05	ND	0.79	Alta Floresta

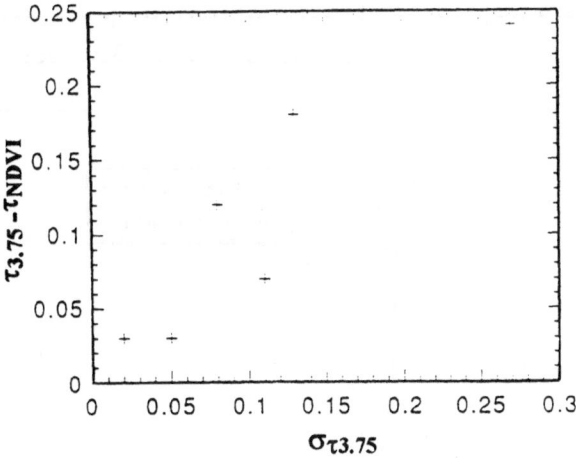

Figure 3. The τ_a difference between the 3.75 μm and NDVI methods is plotted against the standard deviation ($\sigma_{\tau 3.75}$) of the aerosol optical thickness for scenes analysed. As the aerosol inhomogeneity increases (high ($\sigma_{\tau 3.75}$) the difference between the two methods increases illustrating the utility of the 3.75 μm method under these conditions.

The last site considered is in Santarem, Brazil. Ground observations made at the airport approximately 1.5 hours after the AVHRR overpass showed an aerosol optical thickness of 0.16 for AVHRR Channel 1 and the visibility was observed to be greater than 15 km for the region, which had numerous small cumulus clouds present at the time. The retrieved values for both methods are similar, near 0.3, and nearly double the measured value (Table 1). Several factors may be responsible for the disparity between the retrieved and measured values: (i) the measured value at the airport is simply one observation that may not be representative of the retrieval area; (ii) the AVHRR viewed in the hot spot direction for the Santarem scene and this would cause an overestimation of the retrieved aerosol amount (although the *magnitude* of the effect is not known); (iii) because the area surrounding Santarem is populated, we may well expect cultural activities to increase the surface reflectance by slightly more than our assumed 0.02, which would also increase our value of the retrieved aerosol optical thickness; and (iv) the Santarem image (as with many AVHRR afternoon images in this area) is contaminated with cumulus cloud formations which increases the likelihood of incorporating subpixel clouds in the assumed "cloud free" radiances used for the retrievals, and this could also lead to overestimation of the retrieved aerosol optical thickness. With regard to the final factor, it is nonetheless encouraging to retrieve reasonable (though probably high) τ_a values even between the clouds. Recognizing the effect of sub-pixel clouds on the retrieved values, we are investigating further methods of minimising it.

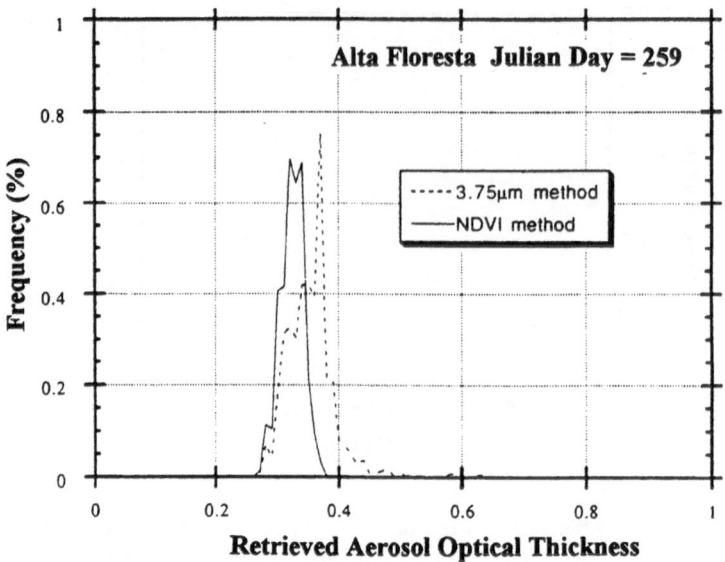

(a)

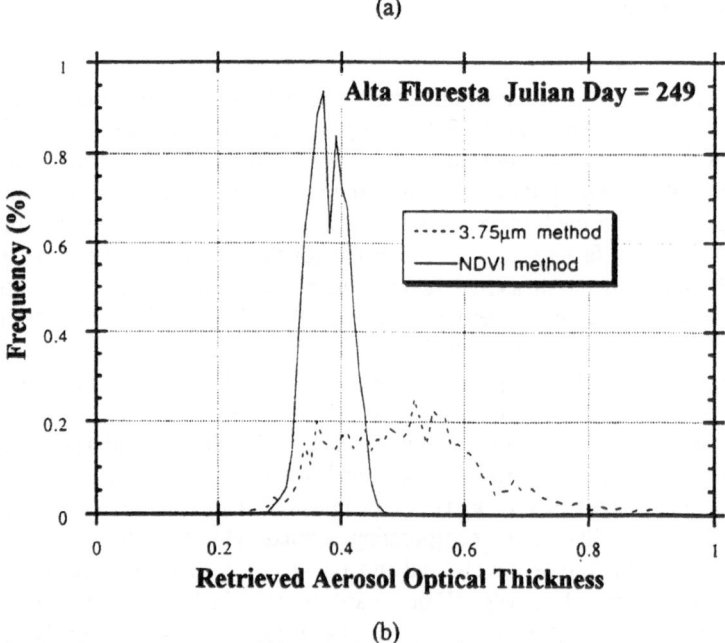

(b)

Figure 4. Frequency histograms of NDVI and channel 3 methods for aerosol retrieval on a hazy day (a) and clear day (b) in Alta Floresta, Brazil.

3.75 µm Method

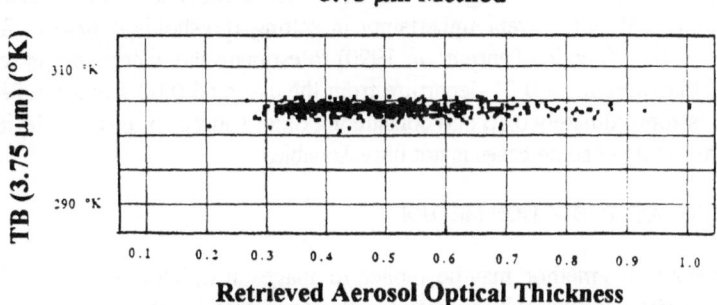

Figure 5. The sensitivity of the 3.75 µm methods to aerosol heterogeneous distributions. For optical thicknesses from 0.25 to 1.0 the 3.75 µm method varies by only 4 degrees K.

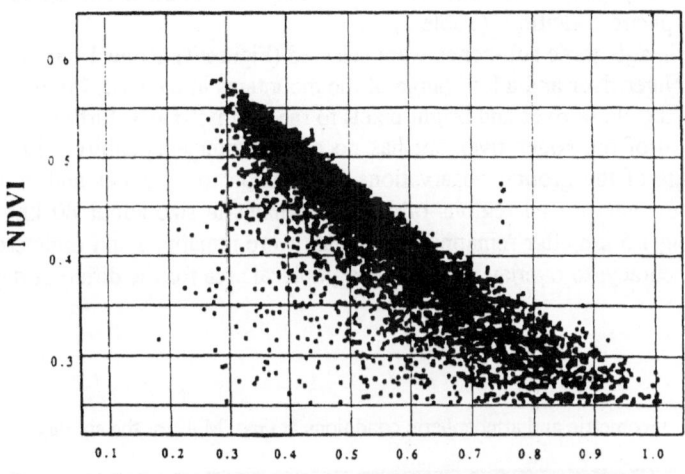

Retrieved Aerosol Optical Thickness

Figure 6. The sensitivity of the NDVI methods to aerosol heterogeneous distributions is illustrated. For optical thicknesses from 0.25 to 1.0 the NDVI varies by a factor of two.

Considering our lack of knowledge of the true surface reflectance in Channel 1 for various conditions and our current uncertainty in setting thresholds *a priori* for both methods, we used the 5S code (Tanré *et al.* 1990) to estimate the uncertainty in retrieved aerosol optical thickness for a 0.01 departure from the assumed 0.02 surface reflectance. The difference is approximately 0.15 and considering other sources of error, a deviation of 0.2 from the true value in some cases is not unreasonable.

3.2.   THE CONTRAST REDUCTION METHOD

As described before, this method may be applied to images in which it is possible to find a number of contrasting targets which have a reflectance that is invariant in time. Such conditions occur typically in arid and semi-arid regions where the DDV method cannot be applied, for example in the Sahel of Mali, from where our case study example is drawn. We selected six AVHRR images for which contemporaneous ground measurements were available, one at the end of the growing season (86/10/3, day 276) and 5 within the dry season, 09/12/86 (day 343), 10/12/86 (day 344), 19/12/86 (day 353), 05/01/87 (day 005) and 15/01/87 (day 015). The selections have similar geometrical conditions but widely varying atmospheric conditions (Table 2).

Within the image, three sub-zones were selected (Figure 7). Zone 1 has two prominent features, the Niger river and a low range of the mountains in the east; Zone 2, west of the first, includes the Niger river and bright pixels to the north; Zone 3, further south, includes a small portion of the Niger river but has no other prominent features (Figure 7). Gao airport, the site of the ground observations, is common to all zones and is indicated by (+). The three zones are all regions of 50 by 50 pixels in size about 80 km square, the pixel size being 1.5 km after remapping. The data were remapped and registered with one pixel r.m.s. accuracy, to overlay the same geographical area for the different days.

Table 2. Geometric and atmospheric conditions at Gao, Mali for the six days.

Day of Year	θ_s	θ_v	ϕ_v	τ_a	U_{H_2O}
276	45.6	54.4	50	0.40	4.20
343	14.0	60.0	34	0.11	1.53
344	7.2	58.0	210	0.08	1.44
353	1.6	58.4	210	0.63	0.66
005	30.4	60.0	33	0.40	1.18
015	17.2	56.8	32	0.35	1.27

Figure 7. The NDVI computed from the reference image shows the location of the three zones each 40 LAC pixels square used for the contrast reduction method. The ground observation site at Gao, Mali, is common to all zones.

Results of the structure function (Equation 3) are plotted against distance from any given pixel within the three zones. This illustrates how the structure function changes due to the influence of variable aerosol loading on each day and the importance of selecting an area of sufficient contrast to observe the atmospheric effect. For zone 1 (Figure 8a), the magnitude of the structure functions is in good agreement with ground optical thickness measurements (Table 2). The largest structure is observed for the clearest day (344) with no difference for day 343 which has similar atmospheric conditions. Days 276, 005, and 015 all have lower magnitudes consistent with the measured intermediate aerosol optical thicknesses. Finally, the most turbid day (353) has the lowest structure function. The same conclusions apply to zone 2 except that day 276 appears as the most turbid day (Figure 8b). The magnitude of the structure is slightly smaller (10%) for all days than for zone 1 because the contrasting mountain feature is not present. For zone 3 (Figure 8c), the magnitude is only half that of zone 1, and the different days do not appear to be ordered in relation to the ground aerosol optical thickness.

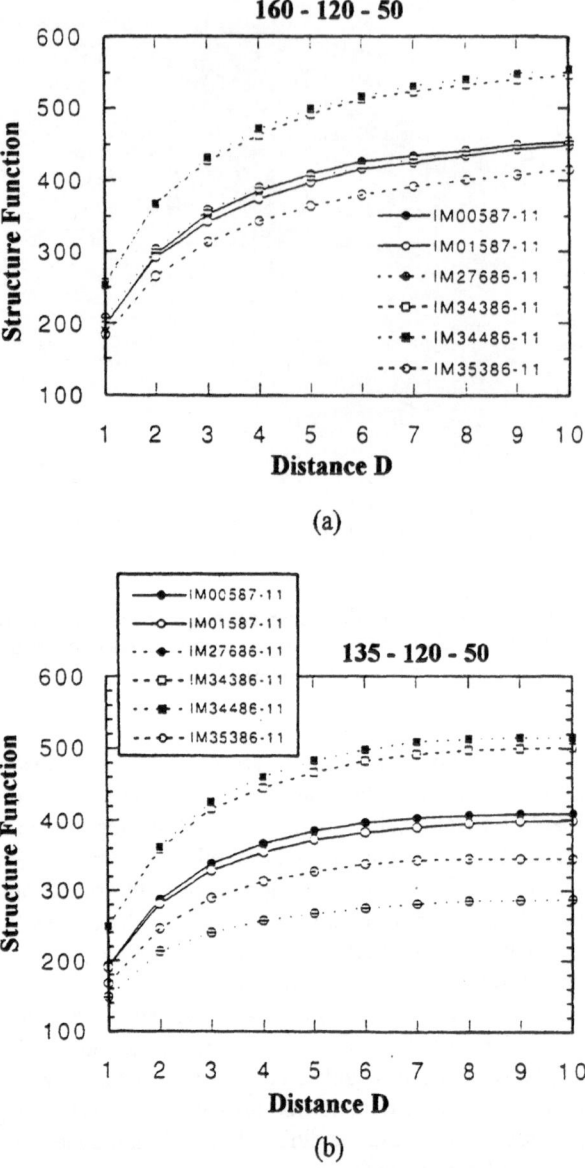

Figure 8. The structure function is plotted against the distance D for zones 1 to 3 (a, b and c respectively). Note the decrease in structure function for zone 3 compared with zones 1 and 2.

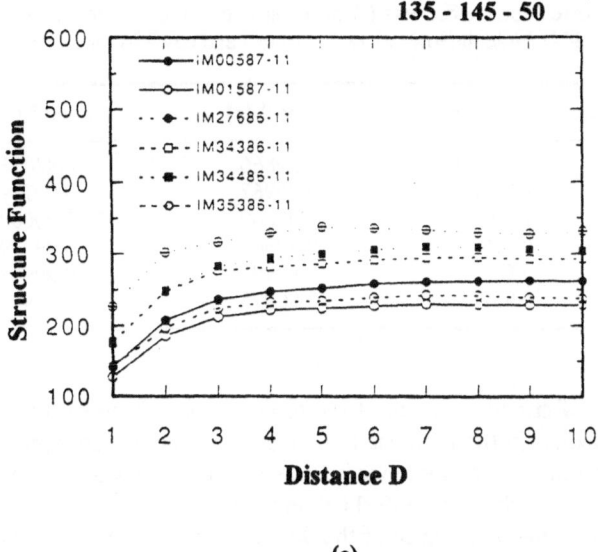

135 - 145 - 50

(c)

Figure 8 continued.

Using the clearest day (344) as the reference day (Table 2), and assuming the surface as invariant, from Equation (4), plots of $F^*s(d)$ as a function of $F^*s(d)$ for day 344 should be straight lines and slopes should be directly related to the function $T(\tau,\mu_V)$ $Fd(\tau,\mu_S)$. Zone 1 and 2 (Figures 9a and 9b) show excellent linearity as indicated by theory, except for day 276. Linear correlation coefficients for each date are greater than 0.9992 for all days (except for 276) for zones 1 and 2 but are generally lower for zone 3 (Table 3). For day 276, starting at $d = 5$, the points deviate from a straight line - a finding more obvious for zone 2 than for zone 1. This probably results from actual surface changes since day 276 corresponds to the end of the rainy season and is about 3 months earlier than the other days. For zone 3, the linearity for any day is lower than for zones 1 and 2 possibly due to the lack of structure, a non-uniform atmosphere and/or to changes in the surface reflectance (Figure 9c) as shown by the lower correlation coefficients. At this point, we do not know exactly the relative contributions of the uncertainties.

The aerosol characteristics are taken from Shettle's (1984) background desert aerosol model. Since the geometrical conditions are known, the only remaining unknown is the aerosol optical thickness, but this can be interpreted from the slopes of the $F^*s(d)$ plots (Figure 9). Comparison of the three zones and the ground measurements show variable results for zone 3 but good overall agreement for zones 1 and 2 (Figure 10). Based on these results we estimate the accuracy to be around 0.1 for the derivation of the aerosol optical thickness.

Table 3. Linear correlation coefficients (r) of the structure function for each date *vs* the reference date, 344. Note the low values for zone 3 and day 276 for all zones.

Day of Year	Zone 1 (r)	Zone 2 (r)	Zone 3 (r)
276	0.99910	0.99866	0.98011
343	0.99983	0.99967	0.99749
344	1.00000	1.00000	1.00000
353	0.99922	0.99968	0.99939
005	0.99974	0.99991	0.99947
015	0.99962	0.99975	0.99916

As well as the lack of structure, some of the discrepancies observed for zone 3 may result from inhomogeneities in the horizontal distribution of the aerosol optical thickness specifically within that zone. Furthermore we have assumed the same aerosol model for all turbidity conditions. D'Almeida *et al.* (1991) found that the radiative properties of the aerosol change relative to the importance of the dust events. They suggest a higher value of the asymmetry parameter resulting from a larger particle size. Considering this, the extreme point of Figure 10 would be closer to the 1:1 line but without more information on the atmosphere and ground conditions it is difficult to draw any further conclusions.

4. Large Area Application

In the following section, a pilot study example of aerosol retrieval and atmospheric correction using AVHRR data for a much larger part of the Earth is provided. It covers the application of an operational atmospheric correction scheme actually tested on the zone covering the area of 60°S to 60°N and 135° West to 30° East. In contrast to the previous section, the analysis is based on GAC data.

4.1. PREPROCESSING

4.1.1 *Data set.* The Global Archive Collection (GAC) is a level 1B product from NOAA consisting of raw data for all AVHRR channels with a nominal spatial resolution 4.4 km x 5 km at nadir, degrading to 20 km x 5 km at the extreme off-nadir view (±55.0°) (see Chapters by Belward and Kidwell for more details about GAC data). This archive represents a large data volume of approximately 200 Megabytes a day for all of the Earth, but one of considerable information content. In this case study, the visible and near-infrared channels are used to derive aerosol optical thickness by correcting for broad-band, water vapour absorption for channel 2 and sensor sensitivity degradation. The preprocessing of the data is described briefly below, and the reader is referred to following chapters for more detailed discussion of preprocessing methodologies and details.

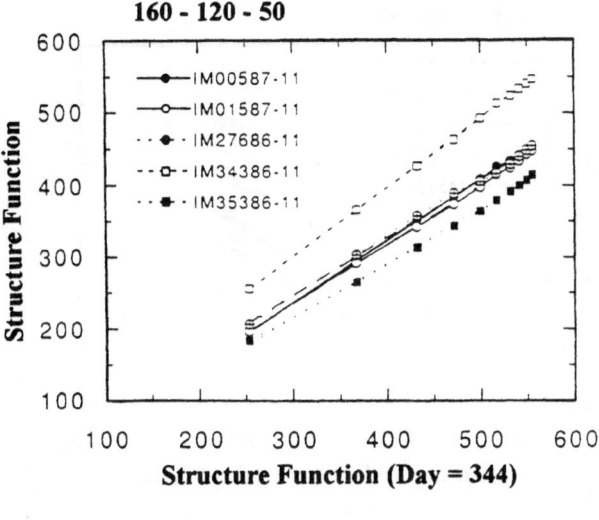

(a)

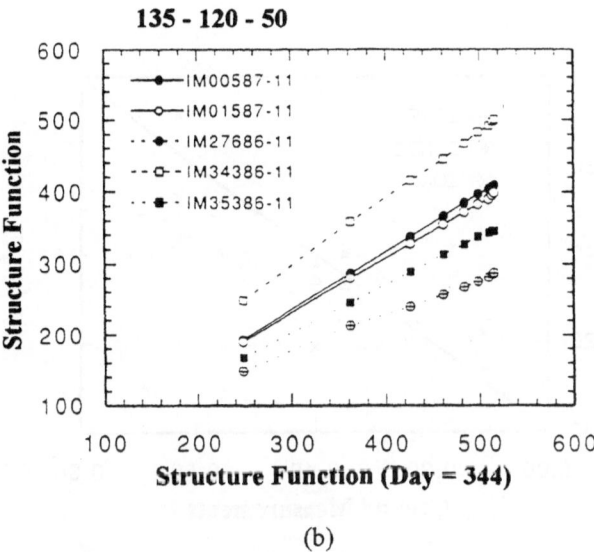

(b)

Figure 9. The structure function for all days is plotted against the reference day, 344, for zones 1, 2 and 3 (a, b and c respectively). The slopes of the lines are a function of the aerosol optical thickness.

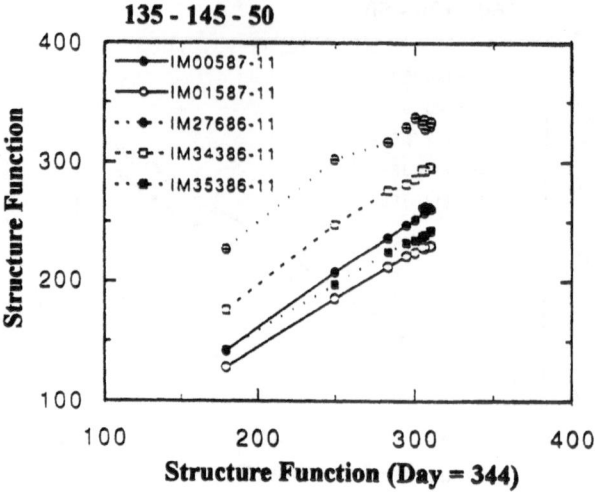

(c)

Figure 9 continued.

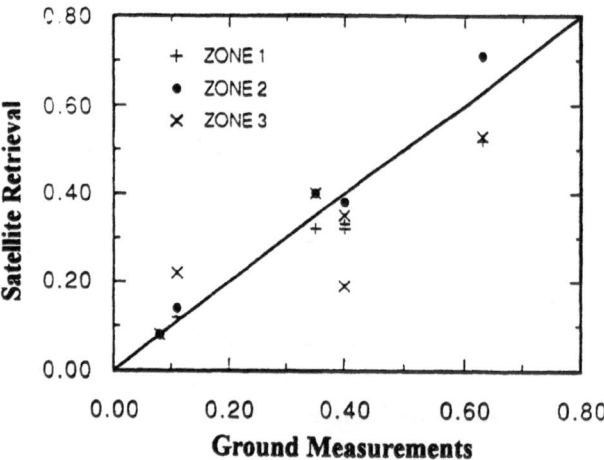

Figure 10. The satellite retrieval from the contrast reduction method is plotted against ground observations for the three zones. Zones 1 and 2 show good agreement with ground truth.

4.1.2. *Calibration.* The GAC data used for this case study were preprocessed in the following way: the reflective channels were unpacked, corrected for dark current responses and scaled to 16-bit digital numbers using pre-flight published calibration coefficients, and the thermal channels were unpacked and converted to temperature values in 16-bits (using on-board information and a non-linearity correction algorithm (Weinreb *et al.* 1990)).

4.1.3. *Cloud Screening.* The CLAVR algorithm developed by Stowe *et al.* (1991) was applied immediately after the unpacking and preflight calibration. The CLAVR algorithm used produces a mask of Ocean/Land/Cloud/DDV/elevation which is used in the next processing step. The DDV is a marker for dense dark vegetation deduced from the channel 3 reflectance which in turn is derived from channel 3, 4 and 5 brightness temperatures assuming an emissivity of 1.0 in each channel. A water vapour correction was applied by using the split-window technique (see Chapter by Vogt). The elevation of the target is used to adjust the Rayleigh correction to the effective amount of scattering by air molecules (see *Rayleigh and Ozone correction* below).

4.1.4. *Geometric Correction.* The output of the preprocessing stage is a linear lat/long grid covering the study area. An inverse navigation scheme was then used to fill this grid with the appropriate pixels extracted from the GAC Level 1B data set. This scheme uses an orbital model based on a Brouwer-Lydanne model to propagate orbital elements supplied by the US Navy ephemeris to the scene time. The propagated orbital elements are then used to resample the satellite data to fit the predefined geographic grid.

4.1.5. *Rayleigh and Ozone correction for Channel 1 and 2.* The consensus of several international working groups on AVHRR products (Faizoun 1992, Townshend 1992) was to make a first-level correction to the visible and near-infrared channel using the reference degradation correction calibration coefficient (Holben *et al.* 1990). The correction terms are the Rayleigh scattering and ozone ($T_{g_{ozone}}$) and oxygen ($T_{g_{oxygen}}$) gaseous transmission (Equation 5). The Rayleigh correction takes into account the elevation of the target and is very accurate (Vermote and Tanré, 1992). The gaseous transmission formulae are simple expressions derived from radiative transfer code runs (Tanré *et al.* 1990, Tanré *et al.* 1992).

$$\rho_{cor} = \frac{Albedo}{Xa} - \rho_{Rayleigh} \frac{P(z)}{P_0} \tag{5}$$

where :
Xa	is the product: d $T_{g_{ozone}} T_{g_{oxygen}}$ cos(Solar Zenith angle) r,
r	is the degradation coefficient (Holben *et al.*, 1991),
d	is the correction term for the variation of the Sun-Earth distance,
$\rho_{Rayleigh}$	is the reflectance due to molecular scattering,
Albedo	is the digital count corrected from dark current and calibrated with pre-flight calibration,
P(z)	is the pressure at target altitude (z) and P_0 is pressure at sea level.

4.2. AEROSOL RETRIEVAL METHODS

4.2.1. *Over water.* Once the ocean signal is corrected for Rayleigh scattering and the gaseous transmission effect, the radiance observed in channel 1 is due mainly to aerosol scattering. Using a radiative model simulating the reflectance above a rough ocean (Deuzé *et al.* 1989), we derived, for several optical thickness values (from 0.0 to 1.0 at 0.55 μm), the reflectance that would be observed by the sensor in case of a maritime model for a wind speed of 10 ms^{-1}. In the glint pattern, simulations are highly dependent on the wind speed so we only inverted data outside a cone of 30 degrees around the specular reflection. We also limited the retrieval to cloud-free pixels only.

In channel 2, the effect of the water vapour should also be corrected for. As its effect varies strongly with time and space, it is important to obtain the water vapour content at the same time and in the same conditions as the radiance. Several published studies have used T_4-T_5 for the retrieval of the amount of water vapour in the atmosphere (*e.g.* Dalu 1986).

Using more than one year of SSMI and AVHRR data over the Pacific Ocean, a relation between the water vapour content and T_4-T_5 has been established. We assume that water vapour content is linearly proportional with T_4-T_5, *i.e.*: $U_{H_2O} = A(\theta)*(T_4$-$T_5)$. A plot of SSMI water vapour content versus nadir measurements of T_4-T_5 is given in Figure 11. The coefficient A can then be deduced $A(0°) = 1.98 \pm 0.5$ g/cm^2/°K. For a viewing angle $\theta = 60°$, this coefficient becomes (see Figure 12) $A(60°) = 1.53 \pm 0.5$ g/cm^2/°K.

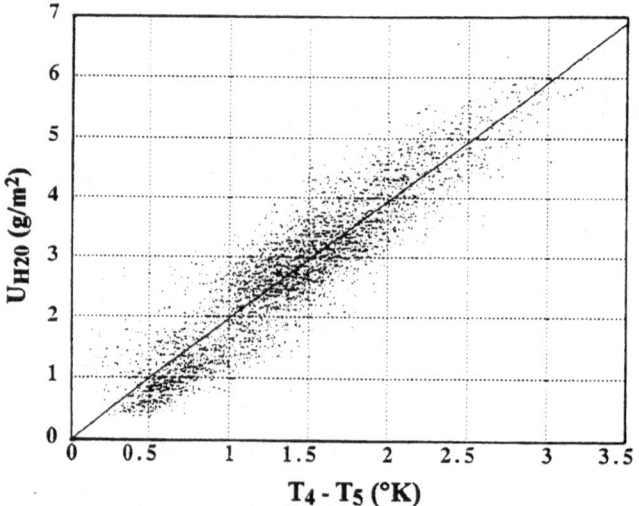

Figure 11. Plot of the water vapour amount (SSMI) versus T_4-T_5 (AVHRR) measured over the Pacific Ocean for a viewing angle of 0°.

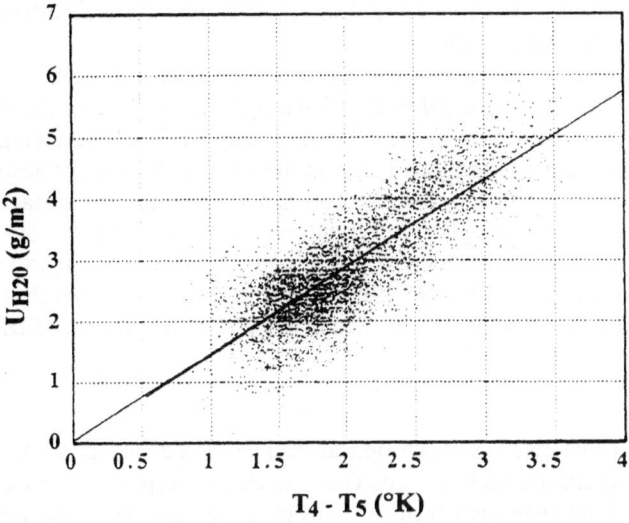

Figure 12. Plot of the water vapour amount (SSMI) versus T_4-T_5 (AVHRR) measured over the Pacific Ocean for a viewing angle of 60°.

The angular dependency for some values of the $\cos(\theta)$ is shown in Figure 13. The retrieved results are then compared with theoretical computations. Defining 36 different atmospheres with a water vapour content ranging from 0 to 6.5 g/cm^2, we computed (with Lowtran 7) the coefficient A for different angles. These computations confirm the angular dependency of the coefficient A (Figure 13) which does not seem to follow a $\cos(\theta)$ law. Once the water vapour content is estimated, the absorption due to it in channel 2 is computed from the radiative transfer model.

4.2.2. *Over Land, the Dark Target concept.* As mentioned in section 2, Dense Dark Vegetation (DDV) targets exhibit very low reflectance in channel 1 of the AVHRR. At this wavelength, 0.65 μm, DDV can be used as a reference dark target (0.01-0.02 reflectance units) to derive the path radiance, and studies of the accuracy of the method shows that the path radiance can be used for the purpose of atmospheric correction with an accuracy of ±0.01. If the aerosol model is known the optical depth can be retrieved with an error of 0.10.

Detection of DDV using the channel 3 brightness temperature as an alternative to channel 1 reflectance has been assessed by Holben *et al.* (1992) where a new method using Channel 3 temperature has been introduced. In this "thermal" channel the DDV pixels are selected because they are cooler, less reflective and more absorbent of water than bare soil or less vegetated pixels.

In this case study we tried to use the *albedo* of channel 3 instead of brightness temperature, as the latter is not suitable for global application. One expression of this albedo can be found in Stowe *et al.* (1991). It is derived from the assumption that emissivity in Channel 4 and 5 is equal to 1, and in this case, water vapour absorption can be corrected by the split-window technique. This is applicable for sea surfaces, but our goal was to give a threshold for DDV detection rather than to try to use the Channel 3 albedo quantitatively. This threshold has been arbitrarily set at 3%. Figure 14 gives a global overview of the aerosol retrieval both on sea and land. The retrieval is limited by the presence of clouds over land and ocean, and over land by the presence of DDV.

4.3. RESULTS

Figure 15 shows a comparison between optical thicknesses derived from AVHRR with sunphotometer measurements taken at Cape Grin (Australia). Agreement between the two is generally good in both channels which validates the retrieval scheme and water vapour correction. However, some discrepancy can occur due to the temporal or spatial variability of the aerosol layer.

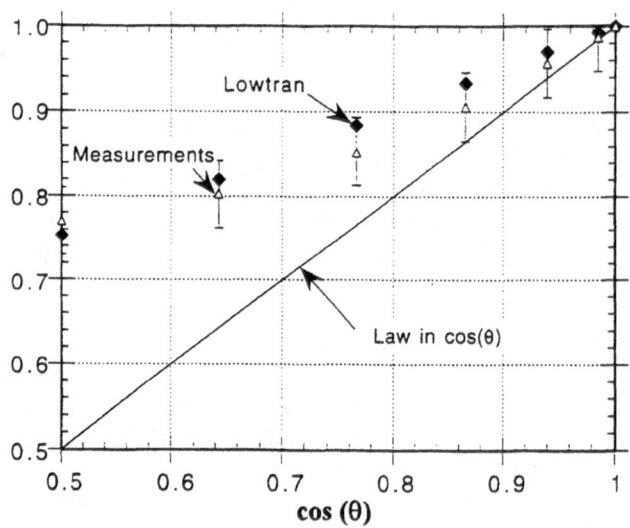

Figure 13. Angular dependency of the coefficient A (normalised to 1) determined both by the measurements and by the computations. Also reported, the law in cos(θ).

Figure 14. 550 nm aerosol depth maximum values retrieved during the 09/03/93 - 09/09/93 period using AVHRR-NOAA-11 data. Values range from 0.1 to 2.0 (over Brazil) due to biomass burning.

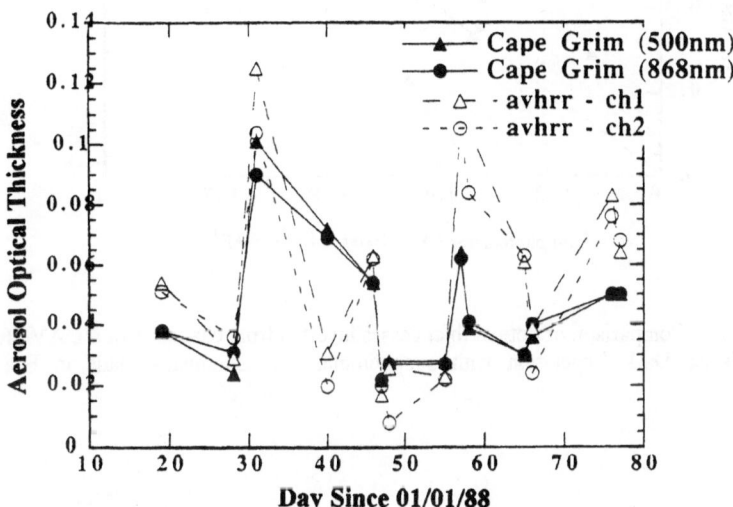

Figure 15. Comparison of optical thicknesses inverted from AVHRR channel 1 and 2 with sunphotometer measurements made at Cape Grim (Australia).

Figure 16 shows a comparison between optical thickness derived from DDV observation in channel 1 with sunphotometer measurements taken in Brazil (1992). The agreement between the two sets is very good considering the uncertainty on the reflectance of DDV in channel 1 (±0.01) which translates directly to an uncertainty of 0.1 in the derived aerosol optical thickness.

4.4   CONCLUSIONS

We have presented a method for atmospheric correction and aerosol retrieval using the visible and near-infrared channels of AVHRR. The method has been applied over a very large area and a first comparison with sunphotometer measurements demonstrates the validity of the approach. Similar algorithms which will be used to correct MODIS data will be tuned using sunphotometer network results, ground measurement campaigns and validated with the current efforts to understand the dynamics of aerosol transport using historical AVHRR data sets.

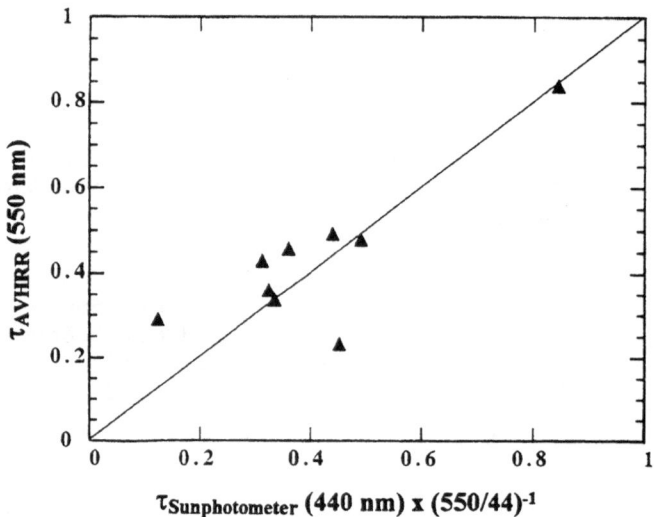

Figure 16. Comparison of optical thicknesses inverted from channel 1 of the AVHRR over Dense Dark Vegetation with sunphotometer measurements made in Brazil (1992).

5.   Stratospheric aerosol perturbations (the Mount Pinatubo eruption)

The second example of the application of the aerosol retrieval and atmospheric correction methods described in this chapter, concerns the operational correction of NDVI for stratospheric aerosol effect, necessitated by the eruption of Mount Pinatubo in the Philippines on 6 June, 1991. Stratospheric aerosols produced by the eruption had a noticeable effect on NOAA-AVHRR data values. Following the eruption, a longitudinally homogeneous dust layer was observed between 20°N and 20°S. The largest optical thickness observed for the dust layer was 0.4-0.6 at 0.5 μm. Due to this layer, the monthly composite Normalized Difference Vegetation Index (NDVI) (generally bounded between -0.1 and 0.6) has shown a systematic decrease of up to 0.15 units, two months after the eruption. Such an atmospheric effect has never previously been observed, and its persistence and spatial extent has seriously compromised the validity of the compositing technique. It stresses that long term monitoring of vegetation using the NDVI necessitates correction of the effect of stratospheric aerosols.

The correction scheme we describe here consists of the examination of a latitudinal profile of optical thickness in channels 1 and 2 over the Pacific Ocean, and use of a radiative transfer code assuming lambertian boundary conditions to correct each pixel in the two channels prior to computing the NDVI.

5.1.   DETERMINATION OF THE LATITUDINAL PROFILE OF STRATOSPHERIC OPTICAL DEPTH.

A few months after Mount Pinatubo injected aerosols into the upper atmosphere, stratospheric winds produced a longitudinally homogeneous aerosol layer (Stowe et al. 1992). This allowed us to derive a stratospheric aerosol optical thickness latitudinal profile by using data from the Pacific Ocean only. This area is known to have a low loading in tropospheric aerosols.

The part of the signal corresponding to stratospheric aerosols can be extracted by subtracting from the total signal, that part produced by the ocean and other atmospheric effects. The latter is determined theoretically by considering a clear ocean with a nominal tropospheric aerosol load. A stratospheric aerosol model (King et al. 1984) is then used to invert the measured signal and deduce the optical depth for different solar and viewing geometry. Figures 17 and 18 give the measured normalised radiance contributed by stratospheric aerosols for channels 1 and 2 (solid line) and theoretical contributions of stratospheric aerosols in these channels computed by King's model for an optical depth of 0.4 (dotted lines). The theoretical plots intercept the measured ones in the same area (latitudinal range) for both channels and for the two different viewing angles considered in these figures. These important features show the spectral and geometric consistency of the model.

With this technique applied to cloud-free weekly composited images of channels 1 and 2 (Figure 19), the values obtained for different zenith angles are averaged to produce the latitudinal profile (Figure 20).

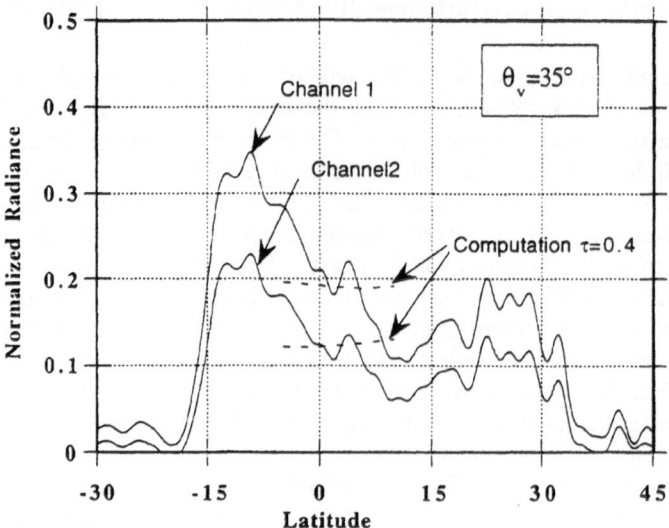

Figure 17. Comparison of the Pinatubo radiances observed at 35° view angle with radiative transfer computation.

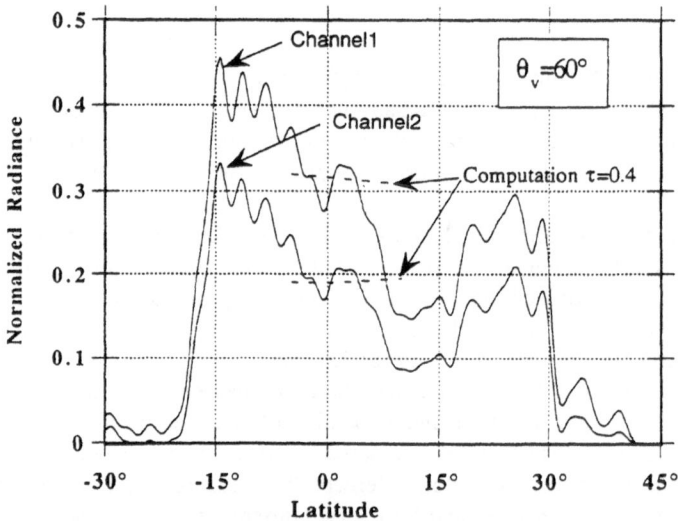

Figure 18. Comparison of the Pinatubo radiances observed at 60° view angle with radiative transfer computation.

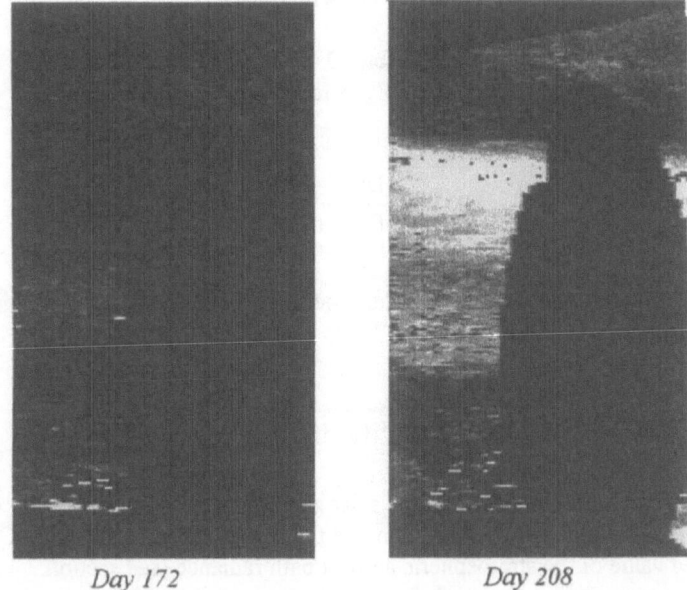

Day 172   Day 208

OPTICAL THICKNESS

Figure 19. "Geometric" composite of the optical thickness observed in channel 1 of the AVHRR before (a) and after the stratospheric aerosol layer formation (b). The top of the image corresponds to a latitude of -60°S, the bottom to +60°N.

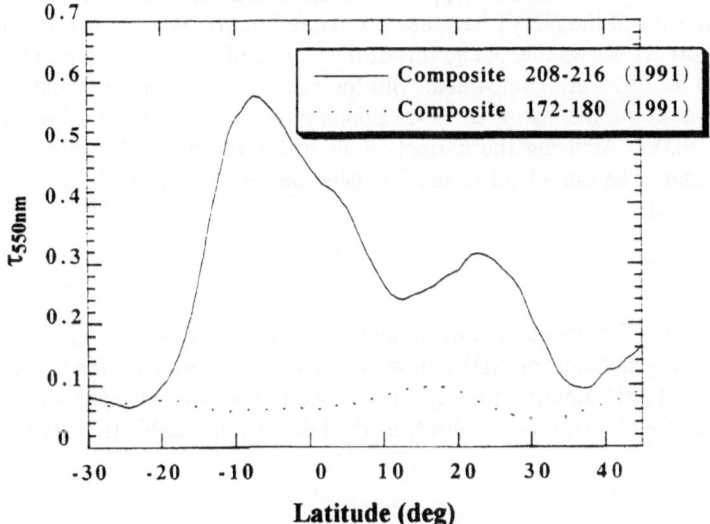

Figure 20. Stratospheric Profiles deduced from Figure 3.

5.2. CORRECTION EQUATION FOR NDVI 30-DAY COMPOSITE

The NDVI is computed from AVHRR visible (0.63 μm) and near infrared (0.87 μm) pre-flight calibrated digital counts, DC_1 and DC_2 as follows:

$$NDVI = \frac{DC_2 - DC_1}{DC_2 + DC_1} \qquad (6)$$

Considering only the effect of stratospheric perturbations, the measured digital counts for channels 1 and 2 can be written as

$$DC^{perturbed} = DC^{stratosphere}(\phi) + T(\mu_s)T(\mu_v)DC^{corrected} \qquad (7)$$

Where μ_s, μ_v and ϕ are the cosine of the solar zenith angle, the cosine of the view angle and the azimuth difference between the Sun and the sensor respectively and $DC^{stratosphere}$ is the measured value of the stratospheric aerosol path radiance (see section 3.1).

$T(\mu)$ is the transmission (upward for μ_v, downward for μ_s) of the aerosol layer computed using King's model (King *et al.* 1984) and optical thickness deduced in section 3.1. The corrected values of channels 1 and 2 are deduced from Equation 7.

5.3. RESULTS AND DISCUSSION

The method described above has been applied over an area in East Africa (-10°S, 10°N). All 10-day composites of the NDVI have been computed for 1989/90/91 before and after the eruption. Figure 21 shows the histograms of the results obtained before eruption. The values computed for 1991 are comparable with the historical data of 1989 and 1990. In Figure 22 (after eruption) the histogram corresponding to the uncorrected values shows a decrease in the NDVI. Applying the correction algorithm reinstates the values back to their expected range. The same kind of test has been performed for 1992, and its results are given in Figure 23.

5.4. CONCLUSIONS

This method has been developed as a quick and effective procedure that can be applied operationally to the generation of NDVI from AVHRR data, corrected for stratospheric aerosol effects. The first results are very encouraging and an extended validation is ongoing as well as the preparation of a direct method directly applicable to NDVI values.

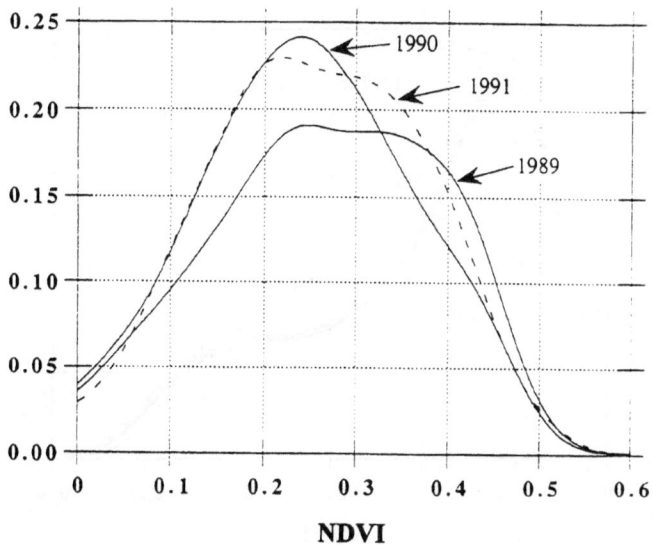

Figure 21. Histogram of NDVI over Africa (-10°S, +10°N) for three years prior to the Pinatubo eruption.

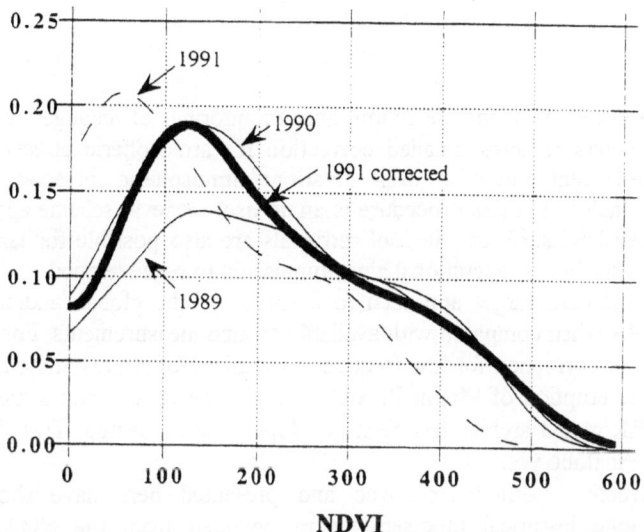

Figure 22. Histogram of NDVI over Africa (-10°.S, +10°N) for three years prior to the Pinatubo eruption. Also included is the post-eruption period (August), and corrected values of NDVI for 1991.

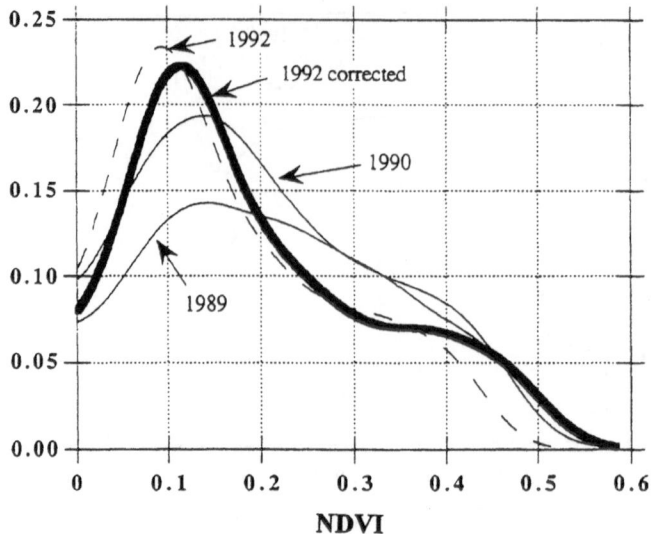

Figure 23. Histogram of NDVI over Africa (-10°.S, +10°N) for two years prior to the Pinatubo eruption. Also included is the post-eruption period (August), and corrected values of NDVI for 1992.

6.   Summary

Analysis of time-series data for detection and monitoring of changes in earth or atmospheric parameters requires detailed correction for atmospheric effects. We have presented here the current state of a quasi-operational atmospheric correction procedure for AVHRR data. Included in this procedure is an aerosol retrieval scheme applicable for oceans and forested land surfaces. Aerosol retrievals are also possible for land surfaces not covered by forests, but an operational algorithm is still to be developed.

Application of the dark target approach to a subset of the global landmass archive showed good results when compared with available ground measurements. For vegetation monitoring, a further correction for low frequency variations of stratospheric aerosol (for example due to the eruption of Mount Pinatubo) is also necessary, and a method using part of the AVHRR global archive has been developed and presented. This shows good agreement to pre-Pinatubo years.

Thus, the corrective methods reviewed and presented here have shown global applicability for long historical time-series data obtained from the NOAA-AVHRR archive, and the application of these techniques are anticipated to provide more realistic and more reliable measurements of earth and atmospheric parameters from long time-series of data. The demonstration of their usefulness to global-scale data sets also encourages us to promote their development and application to forthcoming EOS global-scale data sets.

References

Arking, A. and Childs, J., 1985, Retrieval of cloud cover parameters from multispectral satellite images. *Journal of Climate and Applied Meteorology,* **24,** 322-333.

D'Almeida, G.A., Koepke, P. and Shettle, E.P., 1991. Atmospheric Aerosols Global Climatology and Radiative Characteristics (Hampton: A Deepak Publishers).

Dalu, G., 1986, Satellite remote sensing of atmospheric water vapor. *International Journal of remote Sensing,* **7,** 9, 1089-1097.

Deering, D. W. and Eck, T. F., 1991, Directional Radiance Distributions above and within a forest canopy Ch285-8/90/0000-0879/$01.00, NASA/GSFC,

Deuzé, J. L., Herman, M. and Santer, R., 1989, Fourier series expansion of the transfer equation in the atmosphere-ocean system. *Journal of Quantitative Spectroscopy and Radiative Transfer,* **41,** 6, 483-494.

Faizoun, A. C., 1992, Final Report of the OSS Working Group on NOAA-AVHRR Standard Preprocessing Definition 1, LERTS,

Holben, B., N, Eck, T. and Fraser, R., S, 1991, Temporal and spatial variability of aerosol optical depth in the Sahel region in relation to vegetation remote sensing. *International Journal of Remote Sensing,* **12,** 6, 1147-1163.

Holben, B, N., Kaufman, Y, J., Setzer, A., Tanré, D. and Ward, D. E., 1990, Optical properties of aerosols from biomass burning in the tropics, BASE-A. *Chapman conference on Biomass Burning,*

Holben, B. N., 1986, Characteristics of maximum-value composite images for temporal AVHRR data. *International Journal of Remote Sensing,* **7,** 11, 1435-1445.

Holben, B. N. and Fraser, R. S., 1984, Red and near infrared response to off-nadir viewing. *International Journal of Remote Sensing,* **5,** 145-160.

Holben, B. N., Vermote, E., Kaufman, Y. J., Tanré, D. and Kalb, V., 1992, Aerosol retrieval over land from AVHRR data-application for atmospheric correction. *IEEE Transaction on Geoscience and Remote Sensing,* **30,** 2, 212-222.

Kaufman, Y. J., 1991, Measurements of the Aerosol Optical Thickness and Path Radiance. *Journal of Geophysical Research,* **98,** D2, 2677-2692.

Kaufman, Y. J. and Sendra, C., 1988, Algorithm for automatic atmospheric corrections to visible and near-IR satellite imagery. *International Journal of Remote Sensing,* **9,** 8, 1357-1381.

Kaufman, Y. J., Setzer, A., Ward, D., Tanré, D., Holben, B. N., Menzel, P., Pereira, M. C. and Rasmussen, R, 1992, Biomass Burning Airborne and Space Borne Experiment in the Amazonas (BASE-A). *Journal of Geophysical Research,* **97,** D13, 14581-14599.

Kerber, G. B., and Schutt, J. B., 1986, Utility of AVHRR Channels 3 and 4 in land cover mapping. *Photogrammetry Engineering and Remote Sensing,* **52,** 1877-1883.

Kimes, D. S., Newcomb, W. W., Nelson, R. F. and Schutt, J B., 1986, Directional Reflectance Distributions of a Hardwood and Pine Forest Canopy. *IEEE Transactions on Geoscience and Remote Sensing,* **GE-24,** 2, 281-293.

King, M., Harshvardhan D. and Arking, A., 1984, A model of the radiative properties of the El Chichon Stratospheric Aerosol layer. *Journal of Climate and Applied*

Meteorology, **23,** 7, 1121-1137.

Kleman, J., 1987, Directional Reflectance Factor Distributions for Two Forest Canopies. *Remote Sensing of Environment*, **23,** 83-96.

Kriebel, K., T, 1977, Reflection properties of vegetated surfaces: Tables of measured spectral biconical reflectance factors. DFVLR, Internal Document.

Lee, T. Y. and Kaufman, Y. J., 1986, Non-Lambertian effects on remote sensing of surface reflectance and vegetation index. *IEEE Transactions on Geoscience and Remote Sensing*, **GE-24,** 699-708.

Malingreau, J. P. and Tucker, C. J., 1988, Large-scale deforestation in the southern Amazon Basin of Brazil. *Ambio,* **17,** 49-55.

Shettle, E., P, 1984, Optical and radiative properties of a desert aerosol model. *Symposium of Radiation in the atmosphere,* 1 Deepak Publishing, pp. 74-77.

Shettle, E. P. and Fenn, R. W., 1979, Models for the aerosol of the lower atmosphere and the effect of humidity variations on their optical properties Opt. Phys. Div., Air Force Geoph. Lab., U.S.A.

Stowe, L. L., Carey, R. M. and Pellegrino, P. P., 1992, Monitoring the Mt. Pinatubo aerosol layer with NOAA/11 AVHRR data. *Geophysical Research Letter*, **19,** 2, 159-162.

Stowe, L. L., McClain, E. P., Carey, R., Pellegrino, P., Gutman, G. G., Davis, P., Long, C. and Hart, S., 1991, Global Distribution of Cloud Cover derived from NOAA/AVHRR operational Satellite Data. *Advances in Space Research / COSPAR*, 11 COSPAR), pp. 51-54.

Tanré, D., Deroo, C., Duhaut, P., Herman, M., Morcette, J. J., Perbos, J. and Deschamps, P. Y., 1990, Description of a computer code to simulate the satellite signal in the solar spectrum. 5S code". *International Journal of Remote Sensing*, **11,** 659-668.

Tanré, D., Deschamps, P. Y., Devaux, C. and Herman, M., 1988, Estimation of Saharan aerosol optical thickness from blurring effects in Thematic Mapper data. *Journal of Geophysical Research*, **92,** 15955-15964.

Tanré, D., Holben, B. N. and Kaufman, Y. J., 1992, Atmospheric Correction algorithm for NOAA-AVHRR Products: Theory and Application. *IEEE Transaction on Geoscience and Remote Sensing*, **30,** 2, 231-248.

Townshend, J. R. G., 1992, Improved Global Data for Land Applications: A proposal for a New High Resolution Data Set 20, IGBP,

Tucker, C., J, Holben, B., N and Goff, T., E, 1984, Intensive forest clearing in Rondonia, Brazil, as detected by satellite remote sensing. *Remote Sensing of Environment*, **15,** 255.

Vermote, E. F. and Tanré, D., 1992, Analytical Expressions for Radiative Properties of Planar Rayleigh Scattering Media Including Polarisation Contribution. *Preprint for Journal of Quantitative Spectroscopy and Radiative Transfer*, **47,** 4, 305-314.

Weinreb, M. P., Hamilton, G., Brown, S. and Koczor, R. J., 1990, Nonlinearity Corrections in Calibration of Advanced Very High Resolution Radiometer Infrared Channels. *Journal of Geophysical Research*, **95,** C5, 7381-7388.

Whitby, K., T, 1978, The physical characteristics of sulfur aerosols. *Atmospheric Environment*, **12,** 135-159.

LAND SURFACE TEMPERATURE RETRIEVAL FROM NOAA AVHRR DATA

JÜRGEN V. VOGT
Institute for Remote Sensing Applications,
CEC Joint Research Centre,
I - 21020 Ispra (Varese),
ITALY.

1.   Introduction

Land Surface Temperature (LST) is one of the key parameters for the energy balance at the ground and is also a primary climatological variable. It controls the longwave energy flux towards the atmosphere and depends on the state of other surface parameters, such as albedo, surface moisture and vegetation cover and condition. Knowledge of the spatial distribution of LST and its temporal evolution, therefore, is important for the accurate modelling of energy fluxes between the surface and the atmosphere. Despite the interest in this parameter, however, measurements of LST are not readily available from standard data sources.

In addition, land surfaces are characterised by a strong heterogeneity over short distances. Due to the rapid changes in biophysical surface characteristics, LST tends to change rapidly in space as well as in time. An adequate characterisation of LST distribution and its temporal evolution, therefore, requires measurements with detailed spatial and temporal frequencies. Satellite remote sensing represents a particularly interesting tool in this context, since it offers the only possibility of measuring LST over extended regions with a sufficiently high temporal resolution and - provided no clouds obscure the surface - with a complete spatial sampling.

As a consequence, remote sensing in the thermal infrared has received considerable attention since the beginning of the 1970s. From then on, many attempts have been made to derive surface temperatures from satellite measurements, especially from the Advanced Very High Resolution Radiometer (AVHRR) onboard the NOAA satellites. Currently, this sensor is the only one capable of providing daily measurements of the emitted radiant flux covering the whole globe in at least two wavebands.

Section 2 summarises the most important physical principles underlying remote sensing in the thermal infrared. In section 3 we review the split-window approach to derive surface temperatures from radiances measured at the top of the atmosphere and in section 4 we discuss some of the problems related to the interpretation of satellite-derived LSTs. In order to guide the reader in the practical use of AVHRR data for LST retrieval, the more theoretical development of the subject matter is followed by a discussion of the practical aspects of the actual implementation in section 5. Finally, section 6 summarises our conclusions on the use of AVHRR-derived LSTs.

G. D'Souza et al. (eds.), Advances in the Use of NOAA AVHRR Data for Land Applications. 125–151.
© 1996 *ECSC, EEC, EAEC, Brussels and Luxembourg.*

2.   Remote Sensing in the Thermal Infrared

A sensor onboard a satellite measures the directional radiance L [Wm^{-2}sr^{-1}] of electromagnetic energy reflected or emitted by the target (and the atmosphere) in different wavebands. The basic problem, therefore, is the conversion of this radiance into the desired information, which, in the case of the thermal infrared, is a temperature.

In the thermal infrared part of the spectrum we are concerned with radiation emitted by the Earth's surface and the atmosphere. In order to describe the radiative characteristics of the Earth-atmosphere system, it is convenient to introduce the concept of a blackbody. A blackbody is defined as a perfectly absorbing and emitting object with a constant spectral emissivity (ε_λ) of unity over the full spectrum. This implies that it will absorb all energy radiating onto it and that the radiation it emits in any particular direction will depend entirely on its temperature and can be described according to Planck's radiation law:

$$B_\lambda(T) = \frac{2\,h\,c^2}{\lambda^5\left(e^{hc/\lambda kT} - 1\right)} \tag{1}$$

where:

$B_\lambda(T)$	=	spectral exitance of a blackbody	[Wm^{-2} m^{-1}]
c	=	speed of light in vacuum (2.997 924 580 x 10^8)	[ms^{-1}]
h	=	Planck constant (6.626 076 x 10^{-34})	[Ws^2 = Js]
k	=	Boltzmann constant (1.380 658 x 10^{-23})	[JK^{-1}]
T	=	absolute temperature	[K]
λ	=	wavelength	[m]

Figure 1 shows the spectral distribution of the energy radiated by a blackbody at various temperatures as well as the transmission characteristics of the atmosphere in the thermal infrared part of the spectrum.

From this figure it is evident that in the thermal infrared region only certain spectral intervals are of interest for remote sensing of the Earth's surface. The atmosphere is highly transparent only within these intervals, the so-called *windows*, whereas for other wavelengths it is fairly opaque. The most important windows range from 3.4 to 4.2 μm, from 4.5 to 5 μm and from 8 to 14 μm. Even in these windows, however, the remaining atmospheric influence has a significant influence on the measured signal and requires correction. The AVHRR sensor measures emitted radiation in two of these windows. While channel 3 is centered around 3.7 μm, channels 4 and 5 are centered around 10.8 μm and 11.9 μm, respectively.

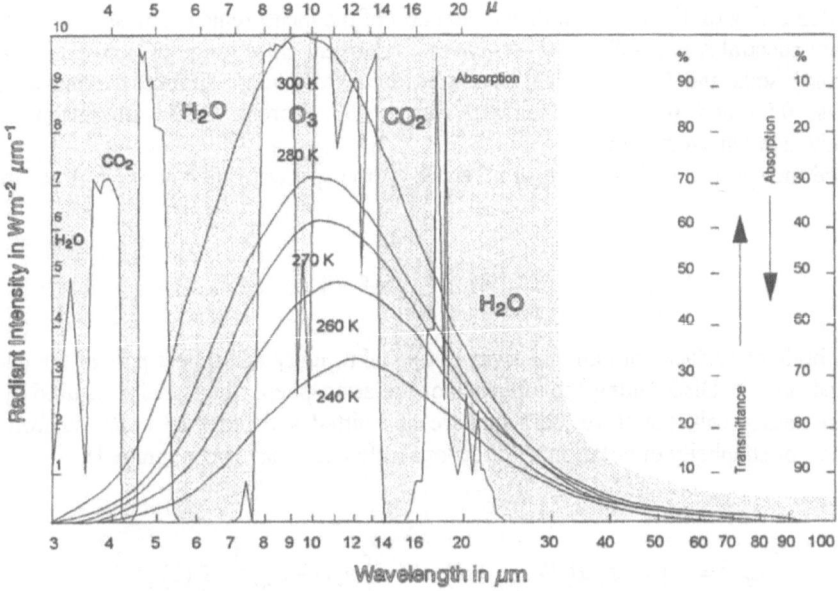

Figure 1: Spectral distribution of the energy radiated by a blackbody at different temperatures, and atmospheric transmittance in the thermal infrared (after Blüthgen and Weischet 1980).

The blackbody, of course, is a theoretical concept and natural surfaces do not behave as blackbodies. The ratio between the radiation emitted by a surface and the radiation emitted by a blackbody of the same temperature is termed the emissivity. It may vary with wavelength (λ) and direction (θ) of the measurement and characterises the efficiency with which a surface or body is able to radiate:

$$\varepsilon_{\lambda,\theta} = \frac{M_{\lambda,\theta}(T)}{B_{\lambda,\theta}(T)} \qquad 0 < \varepsilon_{\lambda,\theta} < 1 \qquad (2)$$

where:

$\varepsilon_{\lambda,\theta}$ = spectral emissivity [dimensionless]

$M_{\lambda,\theta}(T)$ = spectral exitance [$Wm^{-2} \mu m^{-1} sr^{-1}$]

$B_{\lambda,\theta}(T)$ = spectral exitance of a blackbody [$Wm^{-2} \mu m^{-1} sr^{-1}$]

All natural surfaces are characterised by emissivities less than unity, with the absolute value depending on the wavelength and direction of the measurement. Emissivities in the 8 - 14 μm spectral range will, however, cover a relatively close range of between 0.91 for dry sandy soils and 0.98 for fully vegetated areas with most surfaces having a value between 0.94 and 0.97 (Buettner and Kern 1965, Lorenz 1973, Gossmann 1984, Monteith and Unsworth 1990).

Since the spectral emissivity ($\varepsilon_{\lambda,\theta}$) is related to the spectral reflectivity ($\rho_{\lambda,\theta}$) according to:

$$\varepsilon_{\lambda,\theta} = 1 - \rho_{\lambda,\theta} \tag{3}$$

any body or surface with an emissivity not equal to unity will reflect part of the energy radiated onto it. Discounting the influence of the atmosphere, the signal measured by the satellite sensor will, therefore, be a mixture of emitted and reflected radiation from the surface (for simplicity of notation all directional references have been omitted):

$$L_S = \int_{\lambda_1}^{\lambda_2} \phi_\lambda\, \varepsilon_\lambda\, B_\lambda(T_{surface})\, d\lambda \;+\; \int_{\lambda_1}^{\lambda_2} \phi_\lambda (1-\varepsilon_\lambda) \frac{1}{\pi} B_\lambda(T_{sun})\, d\lambda \tag{4}$$

where:
- L_S = radiance received by the sensor   [Wm^{-2} sr^{-1}]
- ϕ_λ = normalised sensor response function   [dimensionless]
- ε_λ = spectral emissivity   [dimensionless]
- $B_\lambda(T)$ = spectral exitance of a blackbody at temperature T   [Wm^{-2} sr^{-1}]

While in the 8 - 13 μm window the reflected part is negligible, up to 50 % of the measured signal in the 3.7 μm window stems from radiation reflected by the surface (Lorenz 1973).

So far we have neglected any perturbations of the radiation due to the atmosphere. While travelling from the surface to the sensor, electromagnetic energy may be absorbed and scattered by the various atmospheric gases and aerosols. Although ozone, carbon dioxide and aerosols play a part, atmospheric water vapour is the main agent of these effects in the thermal infrared. At the same time, the atmosphere emits radiation as a function of its own temperature profile. This energy may reach the sensor directly (upwelling component) or through reflection by the surface (downwelling component). In both cases it will contribute to the signal measured by the sensor. Equation 4, consequently, should be rewritten as follows:

$$L_s = \int_{\lambda_1}^{\lambda_2} \phi_\lambda \, \varepsilon_\lambda \, B_\lambda(T_{surface}) \, \tau_\lambda(z_0) \, d\lambda$$

$$+ \int_{\lambda_1}^{\lambda_2} \phi_\lambda (1 - \varepsilon_\lambda) \frac{1}{\pi} (A_\lambda + B_\lambda(T_{sun})) \, \tau_\lambda(z_0) \, d\lambda \qquad (5)$$

$$+ \int_{\lambda_1}^{\lambda_2} \int_{z_0}^{z_{sat}} \phi_\lambda \, \varepsilon_\lambda \, B_\lambda(T_{air}[z]) \frac{\partial \tau_\lambda}{\partial z} \, dz \, d\lambda$$

with z being the altitude (z_0 = surface, z_{sat} = satellite altitude), τ_λ the total spectral transmission of the atmosphere and A_λ the downwelling spectral radiation from the atmosphere. The first term on the right hand side gives the radiation emitted by the surface, the second term the solar and atmospheric radiation reflected by the surface and the third term characterises the atmospheric contribution to the signal through upwelling radiance. One should note that the bulk of the atmospheric contribution originates from the lower layers of the atmosphere.

Since, in the 8 - 14 μm window, the reflected part is negligible, Equation 5 may be simplified by omitting the second term on the right hand side when analysing data from this part of the spectrum. For the case of the AVHRR this applies to temperature retrievals from channels 4 and 5. In the case of channel 3, located in the 3.4 to 4.2 μm window, this approximation is acceptable only for night-time retrievals.

Equation 5 highlights the influence of the surface emissivity as well as the influence of the atmosphere on the measured signal. The latter through its absorptive characteristics as well as through its own emission.

By inversion of Planck's radiation law (Equation 1) we can retrieve the temperature equivalent to the temperature of a blackbody radiating the measured radiance. This temperature is generally termed *brightness temperature* (also sometimes known as *apparent temperature*). It is important to note that a brightness temperature is neither corrected for atmospheric effects nor for emissivity effects. The influence of these two factors may be of the order of several degrees K, depending on atmospheric and surface conditions. Correction of these effects may be achieved through *a priori* knowledge of the atmospheric state and/or the surface emissivity or through contemporary measurements in several wavebands or from several view angles. Measurements in several wavebands or from several view angles, for example, allow the estimation of the atmospheric contribution to the signal provided that the surface emissivity in the relevant wavebands is known with high precision.

3.   Methods for Surface Temperature Retrieval

3.1.   PRINCIPLES

The theoretical basis for the retrieval of surface temperatures from space has developed since the early 1970s. Since then, various methods have been proposed to correct for the atmospheric contamination of the signal in the thermal infrared. They may be grouped into (1) single channel methods, (2) multi-angle methods, and (3) multi-channel (split-window) methods (Becker and Li 1990a).

Single channel methods take the radiance as measured by a satellite sensor in only one band and correct for residual atmospheric absorption and emission by a radiative transfer code which requires input data on the atmospheric profiles of water vapour and temperature. Clearly, the drawback of these methods is the requirement of additional information which usually is not available with sufficient spatial density or at the time of the passage of the satellite.

Multi-angle methods are based on the differential absorption due to the difference in atmospheric path length when an object is seen from two different view angles. The difference in the strength of the absorption allows for the numerical elimination of the atmospheric effect. However, multi-angle measurements of the same object have not been available in an operational mode until recently. It is only since the launch of ERS-1 in 1991 that such measurements became available through the Along Track Scanning Radiometer (ATSR) onboard this satellite. The main problem for the technique is the availability of measurements from sufficiently different view angles in order to yield a significant difference in the atmospheric absorption even for relatively clear atmospheres. For a more in-depth discussion of the methodology and a comparison with the split-window technique, the reader is referred to Prata (1993).

The *split-window methods* are by far the most widely applied methods, in particular, to analyse AVHRR data. Since the launch of NOAA-7 in June 1981, regular measurements of the upwelling radiation in two narrow spectral bands within the infrared window from 8 to 14 μm have been possible. The concept of the technique goes back to Anding and Kauth (1970) and is based on the different atmospheric transmittance characteristics in two adjacent spectral ranges in the infrared window around 11-12 μm. Hence the name "split-window". This method can take advantage of the two measurements provided by the AVHRR in channels 4 and 5, centred at 10.8 μm and 11.9 μm. Since two satellites are normally in orbit at the same time, data are available at least four times a day for the whole globe.

The two measurements yield a set of two equations in the form of Equation 5. Since the spectral surface emissivities and the atmospheric effect (absorption and emission) are both unknown, however, there is no unique solution to the split-window equations. In fact, any number of wavebands will yield an underdetermined set of equations, since the number of unknowns (the spectral emissivities in each waveband and the atmospheric effect) will always be larger than the number of equations. In order to constrain the problem further, either additional information or certain assumptions will therefore have to be incorporated. These assumptions may include:

(a) The linearisation of the Planck function. This assumption requires that the measurements be made in two spectral ranges close to each other, such that one of the measured radiances may be expressed as a linear function of the other.

(b) The equal magnitude of all temperatures involved (brightness temperatures, air temperature and surface temperature). Under most atmospheric conditions this is true and allows for the elimination of the atmospheric contribution through upwelling radiance.

(c) The spectral invariance of the surface emissivity.

(d) The weak absorption approximation for water vapour absorption. This assumption requires a relatively small total water vapour content of the atmosphere. A linear relation between total water vapour absorption and brightness temperature difference in the two wavebands is a good approximation only under these conditions. The assumption may be violated for a combination of moist tropical atmospheres and large view angles.

Methods have first been developed for the case of the sea surface. Compared with a land surface the sea surface represents a simplification due to its spatial and temporal homogeneity and, most importantly, due to the fact that the value for the emissivity of water is well known and is stable in space and time.

It has been demonstrated (Prabhakara *et al.* 1974, McMillin 1975, Deschamps and Phulpin 1980, McClain *et al.* 1983, Singh 1984, Strong and McClain 1984, McMillin and Crosby 1984, Llewellyn-Jones *et al.* 1984, McClain *et al.* 1985, among others) that the sea surface temperature can be retrieved with an accuracy better than 1 K by a linear combination of the two adjacent measurements. The linear equations to retrieve the temperature have the following general form:

$$T_s = a_0 + a_1 T_4 + a_2 T_5 \qquad (6)$$

or

$$T_s = a_0 T_4 + a_1 (T_4 - T_5) + a_2 \qquad (7)$$

where T_s represents the estimated surface temperature, while T_4 and T_5 correspond to the measured brightness temperatures in channels 4 and 5 of the AVHRR sensor. Values for a_0, a_1, and a_2 are found through regression analysis with measurements made from drifting buoys and ships, or through modelling the atmospheric transmittance for different conditions by adequate codes. McMillin and Crosby (1984), Schlüssel (1987), Fedichev (1990) and Prata (1993) give detailed theoretical developments of the physical principles on which the split-window technique is based. The high accuracy of the temperature retrieval may be attributed to several factors:

(a) The emissivity of water in the thermal infrared is well known and close to one (~ 0.99). At the same time its spectral and angular dependence in the thermal infrared between 8 and 14 μm is negligible.

(b) The difference between the temperature of the sea surface and the overlying air is usually relatively small.

(c) The temperature of the sea surface varies only slowly in space and time and can thus be retrieved with good accuracy from measurements with low spatial resolution such as those from the AVHRR.

Numerous coefficients have been published for the variables in Equations 6 and 7. While Llewellyn-Jones *et al.* (1984) published a set of coefficients which take into account regional differences as well as differences in the view angle of the sensor, several other authors published "global" coefficients which are supposed to give reasonably accurate temperatures for all possible atmospheric conditions as well as for viewing angles up to 45 degrees (*e.g.* McClain *et al.* 1985). While discussions and comparisons of several split-window algorithms may be found in Cooper and Asrar (1989), Nejedly (1986) and Schlüssel (1987), the historical evolution of the concept is described in detail by McMillin and Crosby (1984).

Witnessing the good performance of split-window algorithms over sea surfaces, several researchers have tried to transfer the method to land applications. However, an accurate retrieval of surface temperatures over land is considerably more difficult than over sea. Several factors affect the accuracy of the retrieval over land:

(a) The emissivity over land may vary considerably in space and time with absolute values ranging from about 0.91 to 0.98 in the 8 - 14 μm window. Differences in the surface type, the soil moisture content, the soil texture and surface roughness, the percentage soil covered by vegetation, as well as the phenological stage and the structure of the vegetation cover may all have significant influences.

(b) Emissivities over land may have a significant spectral dependency. Although the difference between the spectral emissivities in wavebands 4 and 5 of the AVHRR will generally be smaller than 0.01, this variation will have a significant influence on the retrieved LST (Becker 1987, Coll *et al.* 1994a, Coll *et al.* 1994b).

(c) The effective emissivity, *i.e.* the emissivity of the objects actually seen by the instrument at a particular view angle, depends on the view angle as well as on the anisotropy of the surface (Gossmann 1987, Choudhury 1989, Wan and Dozier 1989, Casselles and Sobrino 1989, Labed and Stoll 1991, Sobrino and Casselles 1991).

(d) Surface temperatures may be considerably higher over land than over sea. Thus the linearisations of the Planck law developed for the sea surface may not be adequate for the land surface temperature ranges. In addition the saturation of the AVHRR sensors at 320 K (47°C) may cause problems for the measurement of LSTs in very hot environments.

(e) LST may vary considerably within the area covered by one AVHRR pixel. This leads to the question of the meaning of a temperature measurement made over an area of at least 1.2 km^2. Similarly, the "surface" may not be clearly defined over vegetation.

(f) LST undergoes strong diurnal changes. This renders the time of measurement an important issue.

(g) Over land, air temperatures may differ considerably from the surface temperature.

(h) Due to differences in the altitude of the terrain, the atmospheric pathlength may vary considerably and linear regressions between atmospheric attenuation and brightness temperature difference established for one altitude may not be appropriate for other altitudes.

Of all the factors affecting the accuracy of the retrieval, the problems related to the unknown spectral emissivities are known to be the most important ones. In fact, no satisfactory solution to the problem has been found to date, since emissivities vary in space as well as in time and data on emissivities for natural surfaces are extremely rare. Measurements on emissivities have been made, for example, by Buettner and Kern (1965), Vincent *et al.* (1975), Taylor (1979), Sutherland (1986), Schmugge and Becker (1989) and Labed and Stoll (1991). However, it is important to note that many of the measurements have been made through integration over a broad spectral range (*e.g.* 8-15 μm). This will result in relatively low values, since this range includes certain distinct minima. Therefore, the measurements do not necessarily represent the correct values with which to analyse AVHRR data. In addition, the measurements are performed on small samples, often in laboratories, and the difficulty of assigning one representative value of emissivity to an area as seen by the instantaneous field of view (IFOV) of a sensor remains.

In order to find a solution, other methods have therefore been proposed for the correction of the atmospheric influence. These include the use of radiosonde data as input into a simplified atmospheric transmission model (Price 1983) or the use of satellite soundings, as for instance from the TIROS Operational Vertical Sounder (TOVS) onboard the NOAA satellites, for the retrieval of atmospheric temperature and water vapour profiles (Olesen and Reutter 1989). However, these methods suffer from spatial non-representativity of the data (radiosonde), inaccuracies in the measurements (TOVS), and differences in the scale of the measurements between the TOVS and the AVHRR. Consequently, the split-window technique remains the only practical solution also for land surfaces, since it is extremely simple and can be sufficiently accurate if adequate information on surface emissivities is available.

3.2. ALGORITHMS

We now present some of the more widely used split-window algorithms which have been developed for LST retrieval from AVHRR data over the last decade.

Price (1984) published a split-window algorithm for land surfaces including a term for the correction of the emissivity influence. He studied the problem over agricultural land and anticipated an accuracy of the order of ± 3 K:

$$T_s = T_4 + 3.33 \, (T_4 - T_5) * \frac{5.5 - \varepsilon_4}{4.5} + 0.75 \, T_4 \, (\varepsilon_4 - \varepsilon_5) \tag{8}$$

where:

$T_s =$	surface temperature	[K]
$T_4 =$	brightness temperature in channel 4	[K]
$T_5 =$	brightness temperature in channel 5	[K]
$\varepsilon_4 =$	emissivity coefficient for channel 4	[dimensionless]
$\varepsilon_5 =$	emissivity coefficient for channel 5	[dimensionless]

Given this equation and a surface temperature of about 300K, an error of 0.01 in the estimate of ε_4 will result in a temperature uncertainty of about 0.65 K, while an error of 0.01 in the difference between ε_4 and ε_5 implies an error of 2.25K in the estimate of the surface temperature. An assumed maximum error of 0.02 in the estimate of ε_4 and of 0.01 in the estimate of the emissivity difference, therefore, leads to uncertainties in the temperature retrieval of about 3.55 K for typical land surface temperatures.

This algorithm has been widely used and results in a relatively important correction of the atmospheric water vapour influence due to the strong weight given to the difference in brightness temperatures from channel 4 and channel 5. Even with roughly estimated emissivities, the algorithm has proven to give reasonable results, especially in tropical environments with moist atmospheres (Cooper and Asrar 1989, Vidal 1991)

In an attempt to quantify the influence of the emissivity on the surface temperature estimate, Becker (1987) studied the problem theoretically. He gave a formulation of the induced error on the retrieved surface temperature (δT) as a function of the error in the estimated difference between the spectral emissivities in AVHRR channels 4 and 5 ($\delta(\varepsilon_4 - \varepsilon_5)$), as well as the deviation of the assumed mean emissivity from the true value ($\delta\varepsilon$):

$$\delta T \cong 50 \frac{(\delta\varepsilon)}{\varepsilon^2} + 300 \frac{\delta(\varepsilon_4 - \varepsilon_5)}{\varepsilon^2} \tag{9}$$

where: $\varepsilon = (\varepsilon_4 + \varepsilon_5)/2.$

For a true emissivity of 0.96 and assumed errors of 0.02 in the mean emissivity (first term on the right hand side) and of 0.01 in the emissivity difference (second term on the right hand side), the equation gives an error of 1.09 and 3.25 K, respectively. Becker concludes that uncertainties in the mean emissivity and the difference between the spectral emissivities can result in errors for the retrieved LST of up to 2 and 7 K, respectively. However, these theoretical values account for extreme cases. In natural environments, and considering the surface variability within one AVHRR pixel, the bias will be considerably reduced.

From measurements made during the HAPEX[1] experiment, Schmugge and Becker (1989) and Schmugge et al. (1991) conclude that the last term in Equation 9, depending on the difference between the emissivities in channels 4 and 5, may be neglected for their study area, since the difference was not significant. This is an important conclusion, since this equation shows that the emissivity difference between the two wavelengths is the most critical factor. Similar conclusions may be drawn from emissivity measurements made by Vincent et al. (1975) who show that the emissivity is relatively stable in the 11-12 μm region. These results indicate that a mean emissivity value might be sufficient for a reasonably accurate retrieval over surfaces with a relatively well developed vegetation canopy. In addition, Schmugge et al. (1991) show that emissivities as measured over well-developed vegetation canopies come close to unity, thus indicating that over such surfaces temperatures corrections with a split-window algorithm can be performed rather accurately. For sparse vegetation canopies and bare soils, however, the errors may become substantial. The high emissivities measured over vegetation canopies are partly due to the effects of the vegetation structure with its numerous little cavities which are effective radiation traps and thus result in an increase of the emissivity as compared with measurements made over single leaves. Similar effects apply to any rough surface.

Grassl (1989) describes a somewhat different approach for the retrieval of LST. He proposes to derive the atmospheric error component from measurements over water bodies and to apply this atmospheric correction through an interpolation procedure over the whole image. The highest of the derived brightness temperatures in the two channels is then chosen as the surface temperature, since it will be the least influenced by the emissivity deviation from unity. The problem here is the availability of sufficiently large water bodies which assure the measurement of pure water pixels as well as the equal distribution of these waterbodies over the image. In addition, the problem of a correction for the actual emissivity deviation from unity remains.

Becker and Li (1990a) and Li and Becker (1991) propose local coefficients for the split-window algorithm as a function of the local surface emissivities in channels 4 and 5 of the AVHRR. They show theoretically that with well-estimated spectral emissivities and appropriately chosen coefficients the linearisations of the atmospheric transmission as used in the split-window approach also hold for land surfaces. However, the difficulty remains to get appropriate values for the emissivities in both channels. Their local split-window equation reads:

$$T_s = A_0 + P\left(\frac{T_4 + T_5}{2}\right) + M\left(\frac{T_4 - T_5}{2}\right) \tag{10}$$

---

[1] HAPEX: Hydrologic Atmospheric Pilot Experiment, performed in southeastern France in the framework of the International Satellite Land Surface Climatology Project (ISLSCP) in 1986. The primary objective of the experiment was to collect information for the improvement of the parameterisation of surface fluxes in atmospheric circulation models.

where
$$A_0 = 1.274$$

$$P = 1.00 + 0.15616 \frac{(1-\varepsilon)}{\varepsilon} - 0.482 \frac{\Delta\varepsilon}{\varepsilon^2}$$

$$M = 6.26 + 3.98000 \frac{(1-\varepsilon)}{\varepsilon} + 38.33 \frac{\Delta\varepsilon}{\varepsilon^2}$$

$$\varepsilon = (\varepsilon_4 + \varepsilon_5)/2 \quad , \quad \Delta\varepsilon = \varepsilon_4 - \varepsilon_5$$

and all other notations are as in Equation 8.

The given coefficients apply strictly only for NOAA 9, since the equations have been derived using the sensor response functions for the AVHRR onboard this satellite.

Their results are supported by Wan and Dozier (1989) who show theoretically that it is possible to retrieve surface temperatures for different land covers by a linear approximation through a split-window algorithm.

Prata (1993) gives a thorough discussion of the physical principles of the split-window technique over land surfaces. He derives a theoretical split-window algorithm of the form:

$$T_s = bT_4 + cT_5 + d \tag{11}$$

where:
$$b = \frac{1+\gamma}{\varepsilon_4} \left[\frac{1}{1+\gamma\, \tau_5\, \Delta\varepsilon/\varepsilon_5} \right]$$

$$c = \frac{-\gamma}{\varepsilon_5} \left[\frac{1}{1+(1+\gamma)\, \tau_4\, \Delta\varepsilon/\varepsilon_5} \right]$$

$$d = -d^* \Delta\bar{I}{\downarrow} \left(\frac{\partial B_4}{\partial T} \right)_{\bar{T}}^{-1} + \left[1-(a+b)\right] \left[\bar{T} - B_4[\bar{T}] \left(\frac{\partial B_4}{\partial T} \right)_{\bar{T}}^{-1} \right]$$

and:
$$d^* = \frac{1-\varepsilon_4 - \gamma\, \tau_5\, \Delta\varepsilon}{\varepsilon_4 + \gamma\, \tau_5\, \Delta\varepsilon}, \qquad \tau_i = \text{total atmospheric transmittance in ch.}i$$

$$\gamma = \frac{1-\tau_4}{\tau_4 - \tau_5}, \qquad \Delta\bar{I}{\downarrow} = \quad \text{constant atmospheric radiance}$$

$$\Delta\varepsilon = \varepsilon_4 - \varepsilon_5, \qquad \bar{T} = \text{mean temperature for a Taylor series}$$
$$\text{expansion of the Planck function } (T_4, T_5 \text{ or } T_a)$$

$\Delta\bar{I}\downarrow$ represents the difference in the downwelling atmospheric radiances in channels 4 and 5. For the case of the Australian continent, Prata shows that the use of a mean value of 3.6 mW m^{-2} sr^{-1} cm^{-1} for $\Delta\bar{I}\downarrow$ is acceptable. $\bar{T}$ may be represented by either T_4, T_5, or the mean atmospheric temperature (T_a) under the assumption that the different temperatures are close in magnitude.

The formula highlights the dependence of the algorithm on the spectral emissivities (ε_i) and the total atmospheric transmission (τ_i) in the two wavebands and requires that these surface and atmospheric conditions can be specified with a reasonable accuracy.

Through a long-term measurement campaign over a well documented and extensive test site in western Australia, Prata and colleagues have shown that a local split-window algorithm of the given form is applicable with good accuracy over land surfaces (Prata 1994). They measured the spatial variability as well as the temporal evolution of the surface temperature in a large wheat field and compared the empirical data with the corresponding surface temperatures retrieved from the AVHRR. They further discuss the influence of percentage vegetation cover, sensor view angle, and spatial variability of LST on the temperature retrieved from the satellite and compare the results obtained with various published split-window algorithms. Considering all error sources, Prata estimates the accuracy of a split-window algorithm over land to be of the order of ±1.5 K at best.

Coll et al. (1994a) show that atmospheric and emissivity effects can be separated and propose a split window algorithm of the form:

$$T_s = T_4 + A\,(T_4 - T_5) + B \tag{12}$$

where:
$$A = 1.0 + 0.58\,(T_4 - T_5)$$

$$B = 0.51 + 40\,(1-\varepsilon) - \beta\Delta\varepsilon$$

$$\varepsilon = (\varepsilon_4 + \varepsilon_5)/2, \quad \Delta\varepsilon = \varepsilon_4 - \varepsilon_5$$

and all other notations are as in Equation 8.

It is important to note that A is a function of the difference of the brightness temperatures in channel 4 and 5 (T_4-T_5). It therefore depends on the total atmospheric transmission in channels 4 and 5 ($\tau_i(\theta)$) which is a function of both the atmospheric water vapour content and the sensor view angle. As a consequence, A will increase with an increasing water vapour content of the atmosphere and/or increasing view angle. Coll et al. (1994a) demonstrate that this gives better results than a constant value, which assumes a linear dependence between water vapour content and atmospheric attenuation. B is a function of the mean surface emissivity in the 10.5 to 12.5 μm waveband (ε), the difference in surface emissivities in channels 4 and 5 ($\Delta\varepsilon$), and a coefficient β which decreases with increasing atmospheric water vapour content. Mean climatological values may be used for β, with β = 50 K for a tropical atmosphere, β = 75 K for a midlatitude summer atmosphere and β = 150 K for a midlatitude winter atmosphere. This parameterisation underlines that the relative importance of the emissivity effect will decrease with increasing atmospheric moisture.

Since knowledge of the mean emissivity and its spectral variation is important for an accurate retrieval, especially over sparse vegetation canopies and bare soils, methods have been proposed to estimate these values from the satellite measurements themselves. Becker and Li (1990b) and Li and Becker (1993), for example, report on a methodology for the estimation of LST and spectral emissivities from AVHRR data. This new approach relies on the additional use of channel 3 and requires the availability of night-time images for the estimation of the reflected radiation in channel 3. Besides the problem of the availability of day and night-time images and their exact coregistration, the main problem is related to the anisotropy of the reflectance in channel 3.

Casselles *et al.* (1993) propose a methodology for the estimation of a mean emissivity for each pixel from representative field measurements of selected soils and vegetation samples in the area under study. The relevant emissivity to be applied to each pixel is then retrieved as a weighted average of the two values. The corresponding weights are a function of the minimum and maximum NDVI values for the area of interest as well as the actual NDVI value of the pixel (Casselles *et al.* 1993). The underlying hypothesis is that the NDVI is linearly related to the proportion of vegetation seen by the sensor (Kerr *et al.* 1992).

More recently Coll *et al.* (1994b) proposed a method for the retrieval of the emissivity difference ($\varepsilon_4 - \varepsilon_5$) based on the brightness temperatures in channels 4 and 5. The method, however, requires a proper description of the atmosphere in order to assess the difference in the water vapour absorption of the two channels. Still, it can be useful for the assessment of the order of magnitude of the emissivity difference at selected dates when proper atmospheric data are available through measurement campaigns, for example. A major advantage is that it results in one value representing the whole pixel.

Other sources of errors when retrieving LSTs from the AVHRR are due to the large view angle of the sensor. The variation in brightness temperatures measured from a remote platform as a function of the view angle has been reported by several authors. It is mainly a function of the fractional ground cover as seen by the sensor, but also of the anisotropy of the emitted radiation field. The mean emissivity of the surface, as seen by the sensor, will thus change with the view angle. Choudhury (1989) reports on a study performed by Hatfield (1979) over wheat, which shows a change in measured brightness temperatures of up to 2 K between nadir view and a view angle of 45°, depending on the fractional ground cover of the measured surface. It is clear that the problem is much larger over fractionally covered surfaces, while over homogeneous surfaces (*e.g.* bare soil or full vegetation cover) the error is greatly reduced. Kimes and Kirchner (1983) made similar observations over row crops with variations in measured brightness temperatures of up to 6 K for viewing angles from 0° to 50°. Dozier and Warren (1982) report differences for measurements made over snow and Wan and Dozier (1989) discuss the combined influence of sensor view angle and surface slope on the "local" view angle and its impact on the land surface temperature measurement. Finally, Prata and Platt (1991) discuss similar experiences as well as the deterioration in the accuracy of the land surface temperature retrieval with increasing view angle.

The view angle will also determine the atmospheric path length from the sensor to the target. For a parallel layered medium this increase may be parameterised by the cosine of the view angle. However, for the case of the AVHRR, the influence of the Earth's curvature will notably increase the pathlength above 25 degrees and thus contribute to a rapid growth (see Figure 2). This effect will introduce a bias in the linear approximation of the split-window for large view angles (*e.g.* > 25 degrees).

Finally, Barton (1989) reports on noise in surface temperature data due to the fact that the conversion of the analog signal to digital counts depends on the on-board calibration and, therefore, may differ between the channels. For any derived parameter depending on the difference between the two measurements, this may introduce a system dependent noise of up to 0.5 K.

3.3.   DISCUSSION

The above discussion has shown that the derivation of LSTs from AVHRR measurements suffers from various inaccuracies due to the deficiencies of the sensor as well as the lack of precise information on the atmospheric state and the surface emissivity. The various problems and the magnitude of the related errors have been assessed theoretically. In order to evaluate the split-window algorithms in the light of practical applications, however, the reader will further have to consider the following issues:

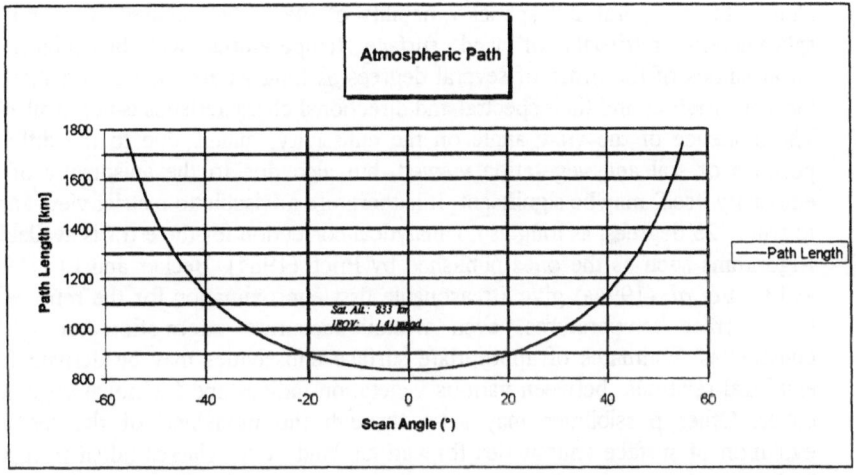

Figure 2:   Atmospheric pathlength for the NOAA AVHRR as a function of satellite view angle. Values correspond to a satellite altitude of 833 km and an IFOV of 1.41 mrad.

(a)   The acceptable error will naturally depend on the aim of the study. For the estimation of the surface energy balance or regional evapotranspiration rates, for example, highly accurate temperature estimates will be a critical component (Price 1982, Seguin and Itier 1983, Carlson and Buffum 1989). For other applications, such as the analysis of the spatial pattern of the temperature field, for instance, the absolute accuracy is not such a limiting factor and the relative accuracy between neighbouring pixels is acceptable (Vogt 1992).

(b)   Spatially consistent LST measurements such as those provided by the AVHRR, are not available from other sources, and the study of the spatial characteristics of this parameter may give important insights for the modelling of Earth-atmosphere interactions.

(c)   On regional to global scales, the accuracy of LSTs retrieved through interpolation from local point measurements will be inferior to the one achieved with the given satellite measurements.

(d)   Even a rough estimation of surface emissivities will greatly reduce the error on the LST estimate.

Based on these considerations and taking into account the specific characteristics of the problem at hand, the usefulness and applicability of the data will have to be judged case by case.

In summary, the reader should keep in mind the following points concerning AVHRR derived LSTs:

(a)   Due to the temporal and spatial variability of the surface emissivity over land, split-window retrievals of land surface temperatures will be subject to uncertainties of the order of several degrees as long as accurate information on local emissivities and their spectral and directional characteristics is not available.

(b)   The influence of the view angle on the emissivity, mainly due to the different portions of soil and vegetation viewed, but also due to the anisotropy of the emissivity itself, may be significant. It is therefore advisable to restrict view angles to within 25 degrees, as long as no analytical correction for the error is available.

(c)   Algorithms such as the ones published by Price (1984), Becker and Li (1990a) and Coll et al. (1994a) give a reasonable first approximation for the retrieval of land surface temperatures, even without accurate information on surface emissivities. Estimates of appropriate surface emissivities may be derived from empirical relations between various vegetation indices and fractional vegetation cover. Other possibilities may arise through the modelling of the temporal evolution of surface emissivities for various land cover classes taken from GIS layers.

(d)   The final absolute accuracy to be achieved under the given conditions will, to a large extent, depend on the surface cover. For a closed vegetation canopy, the accuracy may well be within 2 K, while for bare soils the uncertainty may substantially increase.

(e) The relative accuracy for measurements between neighbouring pixels is better than the absolute accuracy of the individual measurements. This assumption is based on the fact that, first, surface temperatures are usually underestimated, such that part of the absolute error will cancel out in a measurement of temperature differences. And second, over distances of the order of a few kilometres and considering the large pixel size, the expected differences in vegetation cover of the sampled areas are usually far from the extreme cases of bare soil and a full vegetation cover.

4. Interpretation

The quantitative interpretation of surface temperatures retrieved from AVHRR measurements requires further awareness of some specific problems related to the orbit of the NOAA satellites and the viewing geometry of the sensor.

One issue relates to the temporal shift in the time of overflight of the satellite. Odd-numbered NOAA satellites (and NOAA 14 as well) are launched into a sun-synchronous orbit with local solar times at nadir of about 13:30 h at satellite launch. Due to the specific characteristics of the orbit, however, the overflight drifts towards the late afternoon with a shift of about 30 minutes per year (Price 1991). Given an average life time of about four years for each satellite this results in a total shift of about two hours.

In addition, the local solar time will vary considerably across the swath. Given the swath width of more than 2500 km, the variation is of the order of about 1 h 30 min at European latitudes. For a local solar time of say 14:00 at nadir, the local solar time will be about 13:15 at the western edge of the image and about 14:45 at the eastern edge. Especially towards the end of the lifetime of a satellite, with nadir times late in the afternoon, such differences in time may result in significant apparent changes in surface temperature.

A daily coverage from a polar orbiting satellite may only be achieved through a wide swath. Any given location, therefore, will be seen from a variety of view angles on subsequent days. Consequently, the local solar time for the measurement will vary widely. For the NOAA satellites the repeat cycle for the orbit is nine days. This, in turn, translates to a nine-day cyclic pattern for the local solar time. Figure 3 depicts the situation for a series of measurements taken over Gibraltar in 1993. It demonstrates the cycle in the local solar time as a function of the view angle, as well as the general time trend due to the drift of the satellite orbit.

The analyst should further recognise that the size of the individual pixel will vary according to its position in the swath. While the instantaneous field of view (IFOV) of the sensor is about 1.1 x 1.1 km close to nadir, it increases to about 2.4 x 6.9 km at the edge of the swath (Figure 4). Due to this large change in the surface for individual IFOVs, temperatures retrieved from very different positions in the swath are not readily comparable.

Despite all these problems, however, AVHRR data do provide valuable information not available from other sources. It is up to the analyst to use the data in a prudent way. The

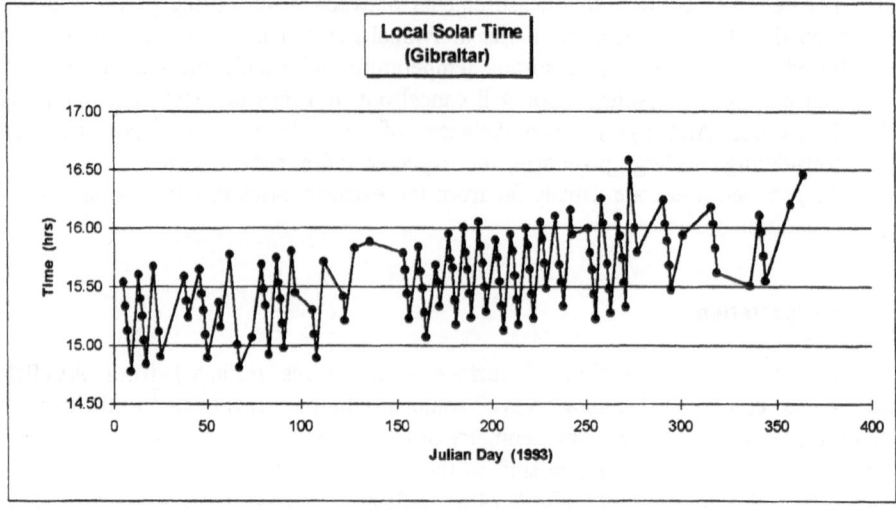

Figure 3: Local solar time for a sequence of measurements taken over Gibraltar in 1993. Irregularities in the display result from missing data or data rejected on cloudy days.

simple restriction to use only the central part of the swath (± 25° view angle), for example, will minimise most of the above mentioned problems, since atmospheric path length and pixel size will remain relatively stable (see Figures 2 and 4). At the same time, it will reduce changes in emissivities due to the percentage soil viewed by the sensor.

Such a restriction will obviously reduce the number of image acquisitions per site. However, at European latitudes and due to the converging orbits, the loss in temporal resolution will be small and largely compensated for by the noise reduction in the data.

5. Implementation

In order to implement the approaches to process AVHRR data, a number of steps will have to be followed. They include the calibration of the various channels, the conversion of the digital counts to radiances, the calculation of the corresponding brightness temperatures and, finally, the retrieval of the surface temperature taking into account atmospheric and emissivity effects.

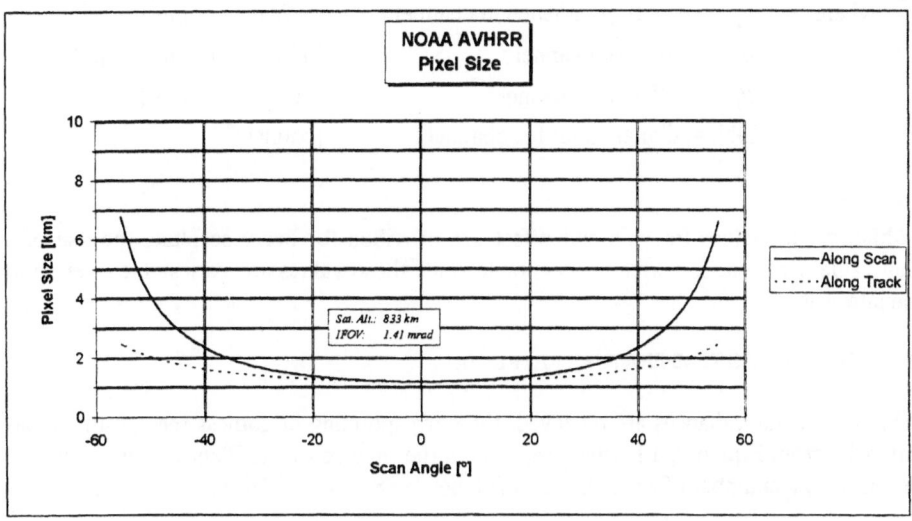

Figure 4: Pixel size of the NOAA AVHRR as a function of view angle for along scan and along track directions. Values correspond to a satellite altitude of 833 km and an IFOV of 1.41 mrad.

5.1. CALIBRATION OF CHANNELS 3,4,5

The thermal infrared sensors of the AVHRR are calibrated in-flight through the regular measurement of a blackbody of known temperature aboard the satellite and the deep space with a brightness temperature of 3 K. Both targets are measured for every scanline and the corresponding gain and offset are appended to the image data. The measurements allow for a linear calibration of the sensor and the conversion from digital counts to spectral radiances[2]:

$$L_i = \alpha_i DN_i + \beta_i \tag{13}$$

---

[2] Note that the units for equations (13) and (14) are not consistent with SI standards. However, they conform with the publications of NOAA/NESDIS and the coefficients and constants given therein.

where: L_i =   spectral radiance in channel i [mW m^{-2} sr^{-1} cm^1]

               α_i =   gain for channel i [mW m^{-2} sr^{-1} cm^1 count^{-1}]

               β_i =   offset for channel i [mW m^{-2} sr^{-1} cm^1]

               DN$_i$= digital count for channel i [count]

Although the values for gain and offset are provided for every scanline, the use of a mean value per scene is perfectly acceptable since these values are very stable over short time intervals.

5.2.   BRIGHTNESS TEMPERATURE RETRIEVAL

Once the spectral radiances are retrieved, the corresponding brightness temperatures may be derived from Equation 14, which represents the inverse of the Planck formula around the central wavenumber of each channel i (Planet 1988, Kidwell 1991):

$$T_i(L_i, v_i) = \frac{C_2 * v_i}{\ln\left(1 + \left(C_1/L_i\right) * v_i^3\right)} \tag{14}$$

where: T_i =   brightness temperature in ch. i [K]

               L_i =   spectral radiance in channel i [mW m^{-2} sr^{-1} cm^1]

               v_i =   wavenumber for channel i [cm^{-1}]

               C_1 =   constant (1.1910659 * 10^{-5})   [mW m^{-2} sr^{-1} cm^4]

               C_2 =   constant (1.438833)   [cm K]

The required wavenumbers for the different channels of the instruments on the various satellites are published in the NOAA Polar Orbiter Data Users Guide (Kidwell, 1991) and updated with the launch of a new satellite. We give the values for the satellites NOAA 6 through NOAA 12 in Table 1. More detailed information on the data availability, the data formats and sensor characteristics may be found in the chapter by Kidwell further on in this volume.

Finally, the derived brightness temperatures need to be corrected for the non-linear response of the sensor. Since only two measurements are available for in-flight calibration, the conversion necessarily is of a linear form. While the sensor of channel 3 is characterised by a linear response, the Mercury-Cadmium-Telluride (Hg-Cd-Te) sensors of channels 4 and 5 are known to have a non-linear response over the expected temperature range. The relevant correction terms have been derived through laboratory measurements before the launch of each instrument. The corresponding values are tabulated in appendix B of Planet (1988) for steps of 5K and for different blackbody

Table 1.   Wavenumbers for AVHRR channels for the temperature range 275 - 320 K (cm⁻¹). (from Kidwell 1991 and Wooster *et al.*, 1995)

Satellite	Channel 3	Channel 4	Channel 5
NOAA 6	2658.050	912.140	--.--
NOAA 7	2671.900	927.220	840.872
NOAA 8	2639.180	914.305	--.--
NOAA 9	2678.110	929.460	845.190
NOAA 10	2660.760	909.580	--.--
NOAA 11	2671.400	927.830	842.200
NOAA 12*	2639.610	921.029	837.364

* Temperature range: 270 - 310 K

temperatures. Appendix B is updated with the launch of a new satellite. The temperature of the blackbody onboard the satellite is measured in-flight through four platinum resistance thermometers and the relevant values are appended to the image data. For any given pair of target brightness temperature and internal blackbody temperature the corresponding value may be retrieved through a bi-linear interpolation of the tabulated values. The correction being of the order of 0 to ± 3.0 K for a temperature range from 200 K to 320 K (Weinreb *et al.* 1990, Rao 1993).

The brightness temperatures may then be corrected for the sensor non-linearity:

$$Tc_i = T_i + \delta T \tag{15}$$

where:
Tc_i = corrected brightness temperature in ch. *i*   [K]
T_i = brightness temperature in channel *i*   [K]
δT = correction term   [K]

It is important to realise that the correction is dependent, both on the internal blackbody temperature and the scene brightness temperature and that the correction terms for channel 4 and channel 5 may differ substantially. Note that the procedure assumes that the non-linear behaviour of the sensor in orbit is equivalent to the one established during the laboratory tests. This assumption may be violated, especially when the laboratory tests have been made long before the actual launch of the satellite (Weinreb *et al.* 1990). Taking into account the various error sources and assuming that the pre-launch non-linearity corrections remain valid after launch, Weinreb *et al.* estimate the absolute radiometric accuracy of AVHRR channels 4 and 5 to be approximately 0.55 K.

For an in-depth discussion of this method and its comparison with an equivalent correction applied to the radiance values instead of the brightness temperatures, the reader is referred to Weinreb *et al.* 1990 and Rao 1993.

5.3.   ADJUSTING FOR ATMOSPHERIC AND EMISSIVITY EFFECTS

After the brightness temperatures have been retrieved from the satellite measurements in channels 4 and 5, the surface temperature may be retrieved through appropriate algorithms as described in section 2. Any of the described split-window algorithms will give a reasonable result within the discussed limits of accuracy. The latter being a function of the various assumptions underlying a split-window approach and our knowledge on surface emissivities.

6.   Conclusions

The foregoing discussion of the various issues related to the retrieval of surface temperatures from AVHRR measurements over land has shown that the problem is well understood from the theoretical point of view. Due to inherent sensor noise and the lack of adequate data on the atmospheric state as well as the surface emissivity, however, AVHRR derived land surface temperatures are currently subject to uncertainties of the order of 2 to 4 K.

This situation may substantially improve with the availability of new instruments with improved performance characteristics as well as with new instruments for satellite soundings of atmospheric temperature and water vapour profiles. These instruments should preferably be mounted on the same platform as the infrared radiometer in order to assure contemporary measurements. However, platforms and sensors capable of such high quality measurements are only currently being designed as, for example, in the frame of the Earth Observing System (EOS) of NASA, the Meteosat Second Generation (MSG) satellite, and the ENVISAT platform of ESA.

For the time being, major improvements in the accuracy of satellite derived land surface temperatures may only be expected from the creation of relevant data banks on surface emissivities. More reliable information on the spatial, temporal and directional variability of this parameter will allow for a better evaluation of the induced error as well as for error reduction if adequate values can be provided for the region under study. At the same time, this approach highlights the general need for the integration of remote sensing and ancillary data through the use of geographical information systems (GIS). In this context the GIS will provide the spatial distribution of additional data independent from the remote measurements. This integration is especially crucial for the correct interpretation of the surface temperature, which plays a major role in the energy balance of the surface, and is closely linked to a number of atmospheric and surface parameters such as the soil moisture content.

Even under the given constraints in accuracy of the data, the usefulness of AVHRR-derived LSTs has been successfully demonstrated in many cases. Applications include the assessment of crop growing conditions, the monitoring of desertification processes, the assessment of forest fire risk, or the analysis of urban heat islands (Vidal 1989, Seguin *et al.* 1991, Roozekrans 1993, López García *et al.* 1991, Moran *et al.* 1994, Seguin *et al.* 1994, Vidal *et al.* 1994, among many others). More detailed information on the use of AVHRR derived LSTs for monitoring of crop growing conditions will be given in the paper of Seguin, this volume.

It is also important to note that the potential of satellite-derived LSTs is to be seen in regional to global scale applications. The need for adequate information on the spatial variability of the surface temperature not only over extended regions, but also within the resolution element of regional to global scale circulation models, may highlight the point. These issues are closely linked with problems of upscaling and downscaling between local measurements on the one hand and modelling exercises on the other. The analysis of satellite derived LSTs at various spatial resolutions may provide major contributions in this field.

References

Anding, D. and R. Kauth (1970): Estimation of Sea-Surface Temperatures from Space. *Remote Sensing. of Environment*, **1**, 217-220.

Barton, I.J. (1989): Digitization Effects in AVHRR and MCSST Data. *Remote Sensing of Environment*, **29**, 87-89.

Becker, F. (1987): The Impact of Spectral Emissivity on the Measurement of Land Surface Temperature from a Satellite. *International Journal of Remote Sensing*, **8**, 1509-1522.

Becker, F. and Z.L. Li (1990a): Towards a Local Split Window Over Land Surfaces. *International Journal of Remote Sensing*, **11**, no.3, 369-393.

Becker, F. and Z.L. Li (1990b): Temperature Independent Spectral Indices in Thermal Infrared Bands. *Remote Sensing of Environment*, **32**, 17-33.

Blüthgen, J. and W. Weischet (1980): Allgemeine Klimageographie. (Lehrbuch der Allgemeinen Geographie, Bd.2) (W. De Gruyter) Berlin, New York.

Buettner, K.J. and C.D. Kern (1965): The Determination of Infrared Emissivities of Terrestrial Surfaces. *Journal of Geophysical Research*, **70** (6), 1329-1337.

Carlson, T.N. and M.J. Buffum (1989): On Estimating Total Daily Evapotranspiration from Remote Surface Temperature Measurements. *Remote Sensing of Environment*, **29**, 197-207.

Casselles, V. and J.A. Sobrino (1989): Determination of Frosts in Orange Groves from NOAA-9 AVHRR Data. *Remote Sensing of Environment*, **29**, 135-146.

Casselles, V. J.A. Sobrino and E. Valor (1993): A Simple Method for Measuring and Mapping Thermal Emissivities. Proceedings of the Workshop on Thermal Remote Sensing of the Energy and Water Balance over Vegetation in Conjunction with other Sensors, 20-23 September 1993, La Londes les Maures, France, 65-68.

Choudhury, B.J. (1989): Estimating Evaporation and Carbon Assimilation Using Infrared Temperature Data: Vistas in Modeling. In: G. Asrar (ed.), Theory and Applications of Optical Remote Sensing. (Wiley) New York, 628-690.

Coll, C., V. Casselles, J.A. Sobrino and E. Valor (1994a): On the Atmospheric Dependence of the Split-Window Equation for Land Surface Temperature. *International Journal of Remote Sensing*, **15** (1), 105-122.

Coll, C. V. Casselles and T.J. Schmugge (1994b): Estimation of Land Surface Emissivity Differences in the Split-Window Channels of AVHRR. *Remote Sensing of Environment*, **47**, 1-25.

Cooper, D.I. and G. Asrar (1989): Evaluating Atmospheric Correction Models for Retrieving Surface Temperatures from the AVHRR Over a Tallgrass Prairie. *Remote Sensing of Environment*, **27**, 93-102.

Deschamps, P.Y. and T. Phulpin (1980): Atmospheric Correction of Infrared Measurements of Sea Surface Temperature Using Channels at 3.7, 11 and 12 μm. *Boundary Layer Meteorology*, **18**, 131-143.

Dozier, J. and S.G. Warren (1982): Effect of Viewing Angle on the Infrared Brightness Temperature of Snow. *Water Resources Research*, **18**, 1424-1434.

Fedichev, O.V. (1990): Theoretical Aspects of Two-Channel Infrared Remote Sensing of the Sea Surface Temperature. *Soviet Journal of Remote Sensing*, **6**.(2), 199-210.

López García, M.J., V. Casselles, J. Melía and A.J. Pérez Cueva (1991): NOAA-AVHRR Contribution to the Analysis of Urban Heat Islands. Proceedings of the 5th International Colloquium Physical Measurements and Signatures in Remote Sensing, 14-18 January 1991, Courchevel, France (ESA SP 319), 501-504.

Gossmann, H. (1984): Satelliten-Thermalbilder. Ein neues Hilfsmittel für die Umweltforschung (Fernerkundung in Raumordnung und Städtebau, Heft 16) Bundesforschungsanstalt für Landeskunde und Raumordnung, Bonn.

Gossmann, H. (1987): Thermalbilder und Oberflächentemperaturen. *Geomethodica*, 12, Basel, 117-149.

Grassl, H. (1989): Extraction of Surface Temperature from Satellite Data. in: F. Toselli (ed.), Applications of Remote Sensing to Agrometeorology (Kluwer) Dordrecht, Boston, London, 199-220.

Hatfield, J.L. (1979): Canopy Temperatures: The Usefulness and Reliability of Remote Measurements. *Agronomy Journal*, **71**, 889-892.

Kerr, Y.H., J.P. Lagouarde and J. Imbernon (1992): Accurate Land Surface Temperature Retrieval from AVHRR Data with Use of an Improved Split Window Algorithm. *Remote Sensing of Environment*, **41**, 197-209.

Kidwell, K.B. (ed.)(1991): NOAA Polar Orbiter Data Users Guide (TIROS-N, NOAA-6, NOAA-7, NOAA-8, NOAA-9, NOAA-10, NOAA-11, and NOAA-12). NOAA/NESDIS/National Climatic Data Center/ Satellite Data Services Division, July 1991, Washington, D.C.

Kidwell, K.B. (1995): NOAA AVHRR Data Availability and Formats. (this volume)

Kimes, D.S. and J.A. Kirchner (1983): Directional Radiometric Measurements of Row Crop Temperatures. *International Journal of Remote Sensing*, **4**, 299-311.

Labed, J. and M.P. Stoll (1991): Angular Variations of Land Surface Spectral Emissivity in the Thermal Infrared: Laboratory Investigations on Bare Soils. *International Journal of Remote Sensing*, **12** (11), 2299-2310.

Li, Z.L. and F. Becker (1991): Determination of Land Surface Temperature and Emissivity from AVHRR Data. Proc. 5th AVHRR Data Users' Meeting, June 25th-28th 1991, Tromsø/Norway, (EUMETSAT, EUM P09), 405-410.

Li, Z.L. and F. Becker (1993): Feasibility of Land Surface Temperature and Emissivity Determination from AVHRR Data. *Remote Sensing of Environment*, **43**, 67-85.

Llewellyn-Jones, D.T., P.J. Minnett, R.W. Saunders and A.M. Zavody (1984): Satellite Multichannel Infrared Measurement of Sea Surface Temperature of the N.E. Atlantic Ocean Using AVHRR/2. *Quarterly Journal of the Royal Meteorological Society*, **110**, 613-631.

Lorenz, D. (1973): Die radiometrische Messung der Boden- und Wasserober-flächentemperatur und ihre Anwendung insbesondere auf dem Gebiet der Meteorologie. *Zeitschrift für Geophysik*, Bd. 39, 627-701.

McClain, E.P., W.G. Pichel, C.C. Walton, Z. Ahmad and J. Sutton (1983): Multi-Channel Improvements to Satellite-Derived Global Sea Surface Temperatures.- *Advances in Space Research*, **2** (6), 43-47.

McClain, E.P., W.G. Pichel and C.C. Walton (1985): Comparative Performance of AVHRR-Based Multichannel Sea Surface Temperatures. *Journal of Geophysical Research*, **90** (C6), 11587-11601.

McMillin, L.M. (1975): Estimation of Sea Surface Temperatures from Two Infrared Window Measurements with Different Absorption. *Journal of Geophysical Research*, **80** (36), 5113-5117.

McMillin, L.M. and D.S. Crosby (1984): Theory and Validation of the Multiple Window Sea Surface Temperature Technique. *Journal of Geophysical Research*, **89** (C3), 3655-3661.

Monteith, J.L. and M.H. Unsworth (1990): Principles of Environmental Physics.- 2nd edition, (Arnold) London, New York, Melbourne, Auckland.

Moran, M.S., T.R. Clarke, Y. Inoue and A. Vidal (1994): Estimating Crop Water Deficit Using the Relation between Surface-Air Temperature and Spectral Vegetation Index. *Remote Sensing of Environment*, **49**, 246-263.

Nejedly, G. (1986): Atmospheric Corrections of NOAA-AVHRR Data. Verification of Different Methods by Ground Truth Measurements. Proc. IGARSS '86, Zürich, (ESA-SP 254) 677-682.

Olesen, F.S. and H. Reutter (1989): Determination of Brightness Temperatures of Surfaces from Satellite. Proceedings 4th AVHRR Data Users' Meeting, 5-8 September 1989, Rothenburg o.d.T./FRG (EUMETSAT, EUM P06), 263-265.

Planet, W.G. (ed.) (1988): Data Extraction and Calibration of TIROS-N/NOAA Radiometers. (NOAA Technical Memorandum NESS 107 - Rev.1) U.S. Department of Commerce, National Oceanic and Atmospheric Administration, Washington, D.C.

Prabhakara, C., G. Dalu and V.G. Kunde (1974): Estimation of Sea Surface Temperature from Remote Sensing in the 11 to 13 μm Window Region. *Journal of Geophysical Research*, **79** (33), 5039-5044.

Prata, A.J. and C.M.R. Platt (1991): Land Surface Temperature Measurements from the AVHRR. Proc. 5th AVHRR Data Users Meeting, June 25th - 28th 1991, Tromsø, Norway, (EUMETSAT, EUM P09), 433-438.

Prata, A.J. (1993): Land Surface Temperatures Derived from the Advanced Very High Resolution Radiometer and the Along-Track Scanning Radiometer 1. Theory. *Journal of Geophysical Research*, **98** (D9), 16689-16702.

Prata, A.J. (1994): Land Surface Temperatures Derived from the Advanced Very High Resolution Radiometer and the Along-Track Scanning Radiometer 2. Experimental Results and Validation of AVHRR Algorithms. *Journal of Geophysical Research*, **99**, (D6), 13025-13058.

Price, J.C. (1982): Estimation of Regional Scale Evapotranspiration through Analysis of Satellite Thermal-Infrared Data. *IEEE Transactions on Geoscience and Remote Sensing*, **GE-20**, 286-292.

Price, J.C. (1983): Estimating Surface Temperatures from Satellite Thermal Infrared Data - A Simple Formulation for the Atmospheric Effect. *Remote Sensing of Environment*, **13**, 353-361.

Price, J.C. (1984): Land Surface Temperature Measurements From the Split Window Channels of the NOAA 7 Advanced Very High Resolution Radiometer. *Journal of Geophysical Research*, **89** (D5), 7231-7237.

Price, J.C. (1991): Timing of NOAA afternoon passes. *International Journal of Remote Sensing*, **12**, 193-198.

Rao, C.R.N. (ed.)(1993): Nonlinearity Corrections for the Thermal Infrared Channels of the Advanced Very High Resolution Radiometer: Assessment and Recommen-dations.- NOAA Technical Report NESDIS 69.

Roozekrans, J.N. (1993): The Monitoring of Desertification Processes in Spain Using NOAA AVHRR Data.- Proceedings 6th European AVHRR Data Users' Meeting, 28 June - 2 July 1993, Belgirate, Italy (EUMETSAT, EUM P12), 313-322.

Schlüssel, P. (1987): Infrarotfernerkundung von Oberflächentemperaturen sowie atmosphärischen Temperatur- und Wasserdampfstrukturen. (Berichte aus dem Institut für Meereskunde an der Christian-Albrechts-Universität Kiel, Nr.161) Kiel.

Schmugge, T.J. and F. Becker (1989): Spectral Emissivity Observations in HAPEX. Proc. IGARSS'89, Vancouver, Canada, July 10th - 14th 1989, 2649-2651.

Schmugge, T.J., F. Becker and Z. L. Li (1991): Spectral Emissivity Variations Observed in Airborne Surface Temperature Measurements.- *Remote Sensing of Environment*, **35**, 95-104.

Seguin, B. and B. Itier (1983): Using Midday Surface Temperature to Estimate Daily Evaporation from Satellite Thermal Infrared Data.- *International Journal of Remote Sensing*, **4**, 371-383.

Seguin, B., J.P. Lagouarde and M. Savane (1991): The Assessment of Regional Crop Water Conditions from Meteorological Satellite Thermal Infrared Data. *Remote Sensing of Environment*, **35**, 141-148.

Seguin, B., D. Courault and M. Guérif (1994): Surface Temperature and Evapotranspiration: Application of Local Scale Methods to Regional Scales Using Satellite Data. *Remote Sensing of Environment*, **49**, 287-295.

Singh, S.M. (1984): Removal of Atmospheric Effects on a Pixel by Pixel Basis from the Thermal Infrared Data from Instruments on Satellites. The Advanced Very High Resolution Radiometer (AVHRR). *International Journal of Remote Sensing*, **5** (1), 161-183.

Sobrino, J.A. and V. Casselles (1991): A Methodology for Obtaining the Crop Temperature from NOAA-9 AVHRR Data. *International Journal of Remote Sensing*, **12** (12), 2461-2475.

Strong, A.E. and McClain, E.P. (1984): Improved Ocean Surface Temperatures From Space - Comparisons With Drifting Buoys. *Bulletin American Meteorological Society*, **65** (2), 138-142.

Sutherland, R. (1986): Broadband and Spectral Emissivities (2-18 μm) of some Natural Soils +and Vegetation. *Journal of Atmospheric and Oceanic Thermology*, **3**, 199-202.

Taylor, S.E. (1979): Measured Emissivity of Soils in the Southeast United States. *Remote Sensing of Environment*, **8**, 359-364.

Vidal, A. (1989): Estimation de l'évapotranspiration par télédétection. Application au contrôle de l'irrigation. Etudes du CEMAGREF, Série Hydraulique Agricole no. 8, Montpellier.

Vidal, A. (1991): Atmospheric and Emissivity Correction of Land Surface Temperature Measured from Satellite Using Ground Measurements or Satellite Data. *International Journal of Remote Sensing*, **12** (12), 2449-2460.

Vidal, A., F. Pinglo, H. Durand, C. Devaux-Ros and A. Maillet (1994): Evaluation of a Temporal Fire Risk Index in Mediterranean Forests from NOAA Thermal IR. *Remote Sensing of Environment*, **49**, 296-303.

Vincent, R.K., L.C. Rowan, R.E. Gillespie and C. Knapp (1975): Thermal-Infrared Spectra and Chemical Analyses of Twenty-Six Igneous Rock Samples. *Remote Sensing of Environment*, **4**, 199-209.

Vogt, J.V. (1992): Characterizing the Spatio-Temporal Variability of Surface Parameters from NOAA AVHRR Data. A Case Study for Southern Mali. (CEC-EUR Report No. EUR 14637 EN) Brussels and Luxembourg, 270 pp.

Wan, Z. and J. Dozier (1989): Land-Surface Temperature Measurement from Space: Physical Principles and Inverse Modelling. *IEEE Transactions on Geoscience and Remote Sensing*, **27**, 268-277.

Weinreb, M.P., G. Hamilton, S. Brown and R.J. Kozcor (1990): Nonlinearity Corrections in Calibration of Advanced Very High Resolution Radiometer Infrared Channels. *Journal of Geophysical Research*, **95** (C5), 7381-7388.

Wooster,M.J., Richards,T.S. and Kidwell,K. (1995): NOAA-11 AVHRR/2 - Thermal channel calibration update, *International Journal of Remote Sensing*, **16** (2), 359-363.

Acknowledgements:

The author would like to thank Michel Verstraete whose comments on the manuscript as well as clarifying discussions have greatly improved the text.

TECHNIQUES FOR GEOMETRIC CORRECTION OF NOAA-AVHRR IMAGERY

G. D'SOUZA
TREES Project
MTV Unit
Institute of Remote Sensing Applications
CEC-Joint Research Centre
I-21020 Ispra (Varese)
Italy

and

T.D.G. SANDFORD
Department of Computing
University of Bradford
Bradford BD7 1DP
United Kingdom

1. Introduction

There are an increasing number of projects that use NOAA-AVHRR data for land applications, many of which are described in subsequent chapters. All of the projects which have a mapping and monitoring component require accurate image geometric correction to ensure the quantity that they measure is an actual indicator of a land cover characteristic or its change, and not an artefact of image to image mis-registration. It is therefore increasingly important to make the geometric correction as precise as possible, and also to know the likely limits of the accuracy of the correction, in order to interpret the results in a meaningful way.

Unfortunately, the geometric correction of NOAA-AVHRR imagery is not simple. Although the NOAA orbit configuration is similar to that of other earth resources satellites, such as Landsat or SPOT (*i.e.* nearly polar, sun-synchronous orbit, period of about 110 minutes, inclination of about $99°$), the sensor view and resulting geometry and scale of the imagery is immensely different (as is the size of study area over which AVHRR data are normally used).

Any uncorrected NOAA-AVHRR image appears very distorted geometrically, due to the large swath width, the Earth curvature and the change from day to day in sensor position (among other factors). Moreover, the distortions within the image will be highly non-linear. Thus the methods commonly used for Landsat or SPOT imagery (*i.e.* ground control points selection, least-square error estimation and image warping using linear transformations to account for unknown satellite altitude or attitude variations) are not feasible for NOAA-AVHRR imagery for large areas. Even if it was possible to adequately

153

G. D'Souza et al. (eds.), Advances in the Use of NOAA AVHRR Data for Land Applications, 153–193.

identify a large number of ground control points evenly distributed across a whole NOAA-AVHRR image, a high-order polynomial of correction would be required, which would undoubtedly lead to large errors in areas of poor ground control.

Fortunately, however, a great deal of effort applied to the problem has recently led to significant advances in the understanding and solution of the geometric correction of NOAA-AVHRR imagery. If we consider Figure 1, a general view of the NOAA satellite imaging part of the Earth, the problem of relating image pixel and lines elements to Earth longitude and latitude co-ordinates, can be considered in two separate stages:

(i)   establishing the sub-satellite point (location and velocity) in relation to the Earth at the time of imaging. As the satellite orbits the Earth, the satellite is continuously scanning the Earth at a rate of 6 lines (about 6 km) per second.

(ii)   modelling the image viewing geometry as it scans 2048 pixels across track representing a swath of over 2500 km.

We shall review the history and advances made in both of these areas in sections 2 and 3 respectively. It is important to note that all the work described in this paper addresses the geometric correction of the highest resolution image data available from the NOAA-AVHRR, the so-called HRPT or LAC data. The models described herein will be equally applicable to any of the other image derived products (*e.g.* GAC) from the AVHRR sensor.

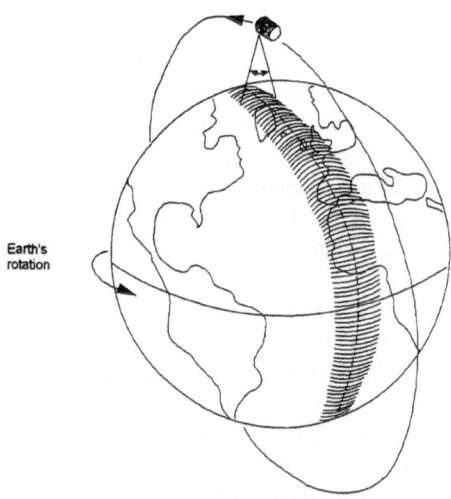

Figure 1.   Schematic illustration of the NOAA-AVHRR orbit and scanning configuration.

2. Estimating the satellite location at point of imaging

2.1 SOME BASIC ORBITAL MECHANICS

The location of the satellite (at least its position in relation to the geocentre at any one time, say at the time of imaging) can be estimated using the laws and theories of orbital mechanics. A brief review of these is therefore appropriate here.

Some of the most basic laws are those derived by Kepler from his observation of Mars around the Earth:

Kepler's Laws.

1. Each planet moves around the Sun in an ellipse with the Sun at one focus (Figure 2a).
2. The radius vector for the Sun to this planet sweeps out equal areas in equal intervals of time (Figure 2b).
3. The square of the periods of any two planets are proportional to the cubes of the semi-major axes:

$$Period = k\sqrt{a^3} \tag{1}$$

where   k   =   constant

Thus the properties of an ellipse are also important (Figure 3). The equation of an ellipse is given as:

$$r = \frac{a(1 - e^2)}{(1 + e\cos v)} \tag{2}$$

where   r   =   radius
          a   =   semi-major axis
          e   =   eccentricity (r/d), For an ellipse $0 < e < 1$
          v   =   true anomaly

The True Anomaly, v, is the most basic variable to describe the orbit. However there are other, more mathematically useful, anomalies that can be calculated:

The Eccentric Anomaly, E, is defined such that

$$r\cos v = a(\cos E - e) \tag{3}$$

$$r = a(1 - e\cos E) \tag{4}$$

This simplifies the equation of the ellipse such that it now becomes

$$r = a(1 - e\cos E) \tag{5}$$

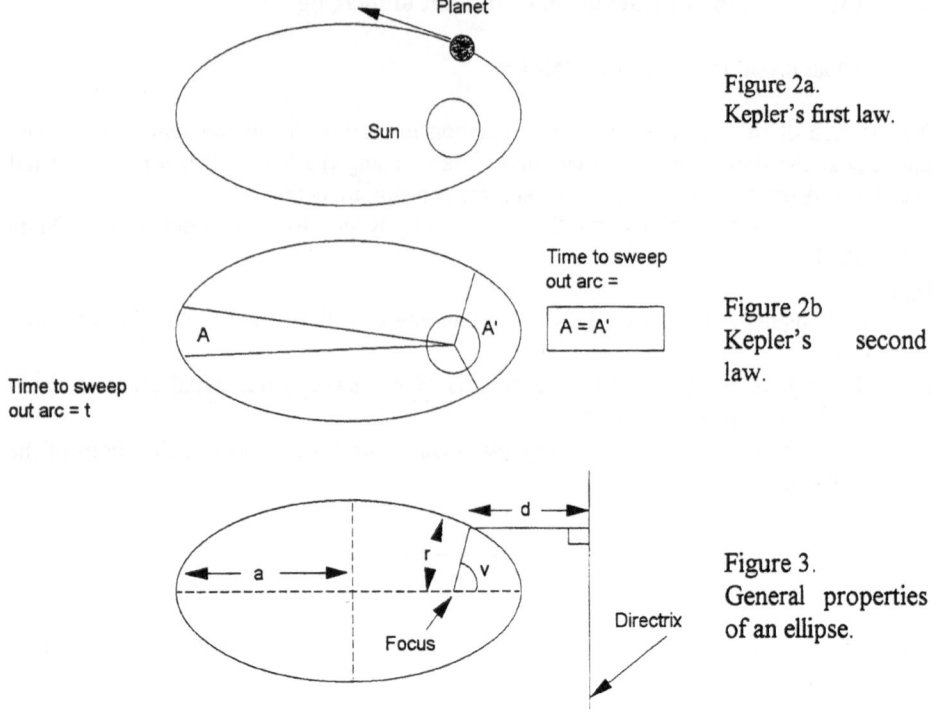

Figure 2a.
Kepler's first law.

Time to sweep out arc =

$A = A'$

Figure 2b
Kepler's second law.

Time to sweep out arc = t

Figure 3.
General properties of an ellipse.

The equation for velocity, V, then becomes:

$$V = \sqrt{\frac{\mu}{a(1-e^2)} \frac{(1+e\cos E)}{(1-e\cos E)}} \qquad (6)$$

The eccentric anomaly is used to simplify some of the more complex models of motion, (e.g. the elliptical model, see section 2.2.3), but is not really useful otherwise.

The Mean Anomaly, M: The period T of an orbit is the time taken for the satellite to complete one cycle of its orbit from perigee to perigee. According to Kepler's 3rd law and Newton's law of gravity:

$$T = 2\Pi \sqrt{\frac{a^3}{\mu}} \qquad (7)$$

Therefore, if we consider an imaging satellite orbiting with the same period but at a constant orbital speed, the angular rate of motion would be:

$$n = \sqrt{\frac{\mu}{a^3}} \qquad (8)$$

where n is the mean motion.

Then the mean anomaly, M, is the angle between the position of the satellite and the perigee, *i.e.* at a general time, t:

$$M = n(t - t_0) \qquad (9)$$

where t_0 is the time that the satellite last passed the perigee.

The mean anomaly and the eccentric anomaly are related through Kepler's equation:

$$M = E - e \sin E \qquad (10)$$

Of the three different anomalies, v is the best for positioning the satellite in space and M is the best for relating position to a time co-ordinate. They are interchangeable (though some not analytically), so there is no real preference.

Therefore, to fix the satellite on its elliptical orbit, we need at least three parameters:

semi-major axis, a
eccentricity, *e*
an anomaly, v, E or M

First, however, the ellipse has to be fixed in space. Kepler's laws state that the plane of the ellipse is fixed in space (*i.e.* NOT fixed relative to the Earth). However, if we assume that the stars are also fixed in space (which is a reasonable assumption), than we can reference all points and planes to them, given a reference plane and a fixed direction in that reference plane. The plane used by astronomers for this purpose is the Equatorial Plane, which is the projection of the Earth's equator on the star background (which is assumed fixed in space). The direction used by astronomers is the First Point of Aries, (also known as the vernal equinox), which lies on the equatorial plane at some place in the firmament.

Then, the Greenwich Hour Angle, GHA, is the angle between the first point of Aries and the Greenwich Meridian (the meridian of zero longitude) (see Figure 4). This changes at a rate of about $360°$ per day and equations relating the Greenwich hour angle to a chosen time can normally be found in astronomical almanacs.

The Orbital Plane and the equatorial plane both go through the Earth's centre (the Geocentre). The orbital plane and the perigee position are fixed by the following parameters, which are known as Keplerian elements since this is the type of orbit that they describe (Figure 5):

Ω = Right Ascension of Ascending Node
ω = argument of perigee
i = inclination

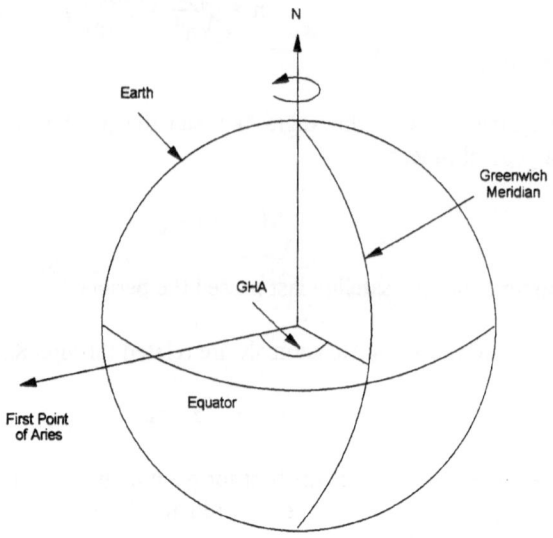

Figure 4. The Greenwich Hour Angle

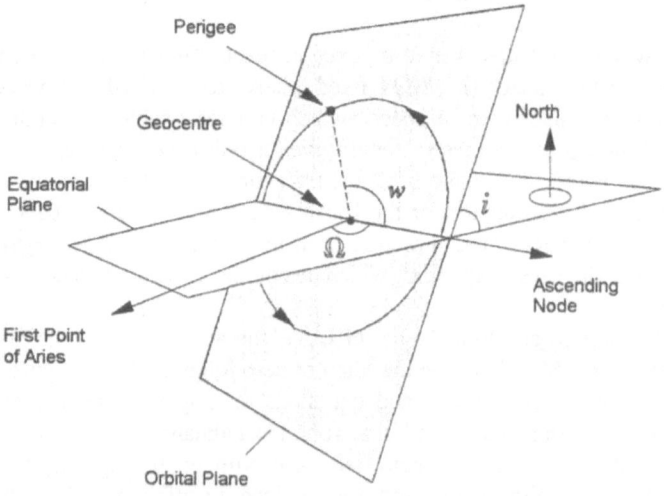

Figure 5. The Keplerian elements and orbital plane.

If there are no perturbations, the satellite would continue in its orbital plane *ad infinitum*. However, this is generally not the case, at least for low-Earth polar orbiting satellites. The main cause of perturbations is the asphericity of the Earth, *i.e.* its oblateness and irregularity. Due to the dynamic geology and its rotation about its axis, the Earth is more an ellipsoid than a sphere, *e.g.* the WGD72 model provides figures of 6378.153 km for the semi-major axis, and 6356.750 for the minor axis- a difference of more than 21 km.

The effect of all perturbations is that if we launch a satellite in a circular ($e = 0$), polar ($i = 90°$) orbit from the equator, the orbit will not remain circular but will tend to "hug" the surface a little, getting closer to the geocentre at the poles. Moreover, this effect does not stay constant but tends to have a long-term effect on the orbit, increasing the deformation as time goes on.

Other perturbations (of lesser importance) are due to the gravitation effects of the Sun and moon, atmospheric drag, solar radiation pressure (though this should be almost negligible for small, heavy satellites), and finally, possible exertion of forces originating from the satellite itself (*e.g.* attitude control boosters, orbit correction systems, *etc.*).

In general, the semi-major axis and eccentricity will vary approximately linearly with time due to atmospheric drag, the inclination remains relatively constant, the argument of perigee and the right ascension of ascending node vary linearly in time due to the Earth's oblateness, and they (as well as the mean anomaly) have a small quadratic variation in time due to the indirect effects of the changing semi-major axis (Rosborough *et al.* 1994).

It is the initial calculation and maintenance of the semi-major axis and the inclination of the NOAA satellites that controls their sun-synchronicity. If the rate of change of the right ascension of ascending node of the satellite is set to match that of the Sun, perfect sun-synchronicity will be achieved. However, the maintenance of the satellite's orbital parameters becomes more difficult with time, and the overhead pass time of the NOAA satellites (especially the afternoon overpass ones) have been observed to drift considerably over the lifetime of the satellite (Price, 1991). Morning overpass satellites tend to accelerate toward earlier Equatorial crossing times, and afternoon overpass satellites accelerate toward later crossing times, with the latter showing bigger amounts of drift (up to 2.5 hours over a 4 year period) (see Price 1991, and chapter by Kidwell). This change in overpass time can have serious implications for some analyses and some applications (for example see chapters by Seguin, Verstraete and Flasse, and Grégoire).

Because of the perturbations and subsequent variations of the satellite's orbit then, for accurate estimation of the satellite location and its velocity, very up-to-date accurate ephemeris (orbital element) information is a prerequisite. There are a few organisations and institutions that regularly monitor these parameters for a number of satellites (*e.g.* NORAD, various national and international observatories, *etc.*) and as NOAA data users needs for this information have grown, it has been becoming more readily available. However, although the different agencies provide all the necessary Keplerian elements, they are often generated by different models, so care must be taken in their use and especially in their propagation through time (see section 2.3).

The main sources of publicly available orbital elements for the NOAA satellite series are the TBUS messages broadcast daily by NOAA on the Global Telecommunication System

(GTS), or available on the NOAA Electronic Bulletin Board, and the NASA/NORAD "two-line" prediction bulletins (also known as "Two-Line Element" (TLE) data sets) which are available on a large number of electronic bulletin boards, over the Internet and even published weekly in some quality newspapers. Marsouin and Brunel (1987) also used data from the ARGOS system but these data are in general not as widely available. There is some hope that future NOAA satellites will themselves broadcast the latest TBUS messages as part of their TIP information (Hughes, *pers comm*).

2.2. NOAA SATELLITE MODELS

2.2.1. *General.* There are several satellite orbital models that may be applied to estimate the location of the NOAA satellite at any given time. These range in complexity, sophistication, level of ancillary data requirement and also in the level of attainable accuracy. Again, it is the use and number of NOAA-data applications that have led to an increased effort and recent improvement in the understanding and solution of the problem of instantaneous satellite position estimation.

In the following section, the type and theory of the various models as they developed through time to their current level of sophistication is reviewed. A comparison of their relative levels of accuracy is then provided in section 2.3 below.

2.2.2. *Simple Circular Model.* The simplest satellite model to describe and use is that which assumes a circular orbit around a spherical Earth. Assuming some recent orbital elements are available, the true anomaly can be found at the time of interest (*i.e.* at the start of imaging) by:

$$v = M + n(t - t_0) \tag{11}$$

This is the simplest version of the circular formula. For a more complex one that accounts for short-term periodic variations, see (26).

Then the angular position from the Equator-crossing point is :

$$\theta = v + \omega \tag{12}$$

The distance to the satellite is then given by:

$$r_s = h_s + R_E \tag{13}$$

where h_s =   nominal height of the satellite
and R_E is the radius of the Earth

The inherent assumption in this simple circular orbit model is that the satellite radiometer scans along a great circle that passes directly beneath the satellite at right angles to the satellite direction motion. Thus spherical trigonometry is valid (see section 3.1) and the (static Earth) latitude and longitude of the sub-satellite point can therefore be given simply by:

$$\phi' = \sin^{-1}(\sin i \sin \theta) \tag{14}$$

$$\lambda' = \cos^{-1}\left(\frac{\cos \theta}{\cos \phi}\right) \tag{15}$$

where ϕ' and λ' are the latitudinal and longitudinal displacement from the right ascension of the ascending node of the viewpoint in a non-rotating Earth model (Ho and Asem, 1986). [It is important to note that care must be taken to select the correct quadrant for the inverse sine and cosine functions in these equations].

During time t, the Earth would obviously have rotated a certain amount. This rotational effect will only effect the longitudinal not the latitude calculation, so an adjustment has to be made according to:

$$\lambda = (\lambda' - \Lambda t) + \lambda_0 \tag{16}$$

where λ_0 =   longitude at equator crossing
             Λ =   Earth angular velocity

In general, estimates of the satellite location with this method will be poor especially if only nominal values are used (see section 2.3), and most users have added refinements to improve the estimation. For example, Ho and Asem (1986) used the simple circular model equations as the basis for their technique for mapping NOAA images of France, but added various improvements, including the use of ground control points (GCPs) to improve the estimates of satellite height and inclination.

2.2.3. *Simple Elliptical Models.* The assumption of a circular orbit is clearly simplistic. Indeed Brush (1985) showed that the circular orbit assumption, against a more realistic elliptical orbit, could lead to along-track errors in the determination of the satellite location of over 6 seconds (36 lines). An improvement on the simple circular orbit model may be made by assuming a more realistic elliptical orbit, though this does become more mathematically complex. Determining the position of the satellite in such an orbit requires the solution of Kepler's equations. First, the mean anomaly at time t must be found

$$M_t = M_0 + n(t - t_0) \tag{17}$$

[Note that the time of interest, t, can normally be read off the data transmitted as part of the HRPT (or LAC) data stream. Every scan line of data collected by the NOAA-AVHRR system has associated with it a time (though this has been found to drift significantly with real time)].

Then, by iteration, the eccentric anomaly E can be found by solution of Kepler's equation:

$$M = E - e\sin E \tag{18}$$

and the true anomaly is given by:

$$= 2\tan^{-1}\left(\sqrt{\frac{(1+e)}{(1-e)}}\tan(\frac{E}{2})\right) \tag{19}$$

Finally the latitude and longitude can be determined as per the simple circular case.

In this case, the distance from the centre of the Earth to the satellite can be found by:

$$r_s = a(1 - e\cos E) \tag{20}$$

2.2.4. *Simple Models plus daily changes in elements.* A further refinement that can be made to the simple models is to make corrections to account for the changing in time of the argument of perigee and the right ascension of the ascending node, both of which vary with time due mainly to the oblateness of the Earth.

Suitable first order corrections for the NOAA satellites are (see Roy, 1978, Ho and Asem, 1986, Price, 1991) for the change in the argument of perigee:

$$\dot{\omega} = 1.5 J_2 R_E^2 \frac{\sqrt{\mu}}{a^{3.5}} (2 - \frac{5}{2}\sin^2 i) \quad rad / s \tag{21}$$

and for the change in the right ascension of the ascending node:

$$\dot{\Omega} = -1.5 J_2 R_E^2 \frac{\sqrt{\mu}}{a^{3.5}}\cos i \quad rad / s \tag{22}$$

2.2.5. *Complex Special Models using mean elements.* As mentioned before, the satellite's orbit is perturbed by a number of factors, notably the oblateness of the Earth, so it cannot be modelled very accurately by Kepler's laws alone. Indeed, it is the Earth's gravitational field that has by far the greatest short-term effect on the orbit of an Earth's satellite. Virtually all other perturbations will only effect the long-term behaviour of the satellite. As most orbital dynamists are ONLY interested in the long-term perturbations (as it is these that allow studies of the geopotential, air density at high altitudes, *etc.*), in their calculations, the short-period effects are often averaged out and mean elements are supplied.

These mean elements do not necessarily hold for any particular time (not even for the epoch time) and their use in circular or elliptical models has therefore led to substantial errors. In particular, the argument of perigee and hence the mean anomaly (which are crucial to the calculation of the satellite position in simple models) can differ significantly to the actual, instantaneous (not mean) element values.

For the most accurate estimate of the satellite location, an estimate of the true (instantaneous) elements, or their rate of change, must be made. However, both TBUS and TLE data sets contain mean elements. The difference between these mean elements and the actual instantaneous elements at any one time depends on how the mean elements were calculated.

Kloster (1989) proposed an empirically-derived approximation to a special model that gives deviations from the circular orbit, and claimed that this worked well for satellites such as NOAAs which are in nearly circular orbits.

$$\delta r = 4.6 + 1.5\cos(2v) - 7.9\sin v - ae\cos(v - \omega) \text{ km.} \tag{23}$$

$$\delta v = 0.006\sin(2v) + 0.125(1 - \cos v) + 2e\frac{180}{\Pi}(\sin(v - \omega) + \sin\omega)\text{deg} \tag{24}$$

Other perturbation models, however, take _mean_ elements as their input then add in the perturbations. Note that since the mean value will depend on the orbital theory used, it is very important to derive true elements with the same theory as that by which the mean ones were calculated, or accuracy will be lost (Sandford 1992b).

Hoots and Roehrich (1980) describe three models for propagation of NORAD TLE data sets for low-Earth orbiting satellites, and the results from the most appropriate of these, SGP4, are compared with the results from other methods in section 2.3. This model takes TLE mean elements as its input.

Although the use of these models represents a significant increase in complexity, one of the major advantages is that, as well as the satellite location (in three-dimensional space), the satellite's velocity is also provided, and this can then be propagated further in the orbit more accurately.

2.2.6. *Complex "Osculating Elements" Models.* The true (or instantaneous) elements are also known as osculating elements and they are constantly changing. One of the most common methods of obtaining the osculating elements from mean elements or *vice versa* (*i.e.* to add in, or remove the dominant short-term effects) is by modified use of the Brouwer (1959) formulations. These (long and complex) equations are used in the calculation of TBUS (mean) elements so, if this is the type of ephemeris data available, the Brouwer (1959) equations could be used to determine the short-term period variations and subsequent propagation of the orbit (Rosborough *et al.* 1994).

This method is used in the so-called BROLYD algorithm (Nagle, 1986) which is quite widely used, *e.g.* by ESA in the generation of SHARP format images (ESA, 1988), by the MARS project (Sharman *et al.* 1992) and in the NOAA-AVHRR navigation software developed at the University of Colorado (*Rosborough et al.* 1994). It is generally reported as the most appropriate method for NOAA satellites (when TBUS data are available).

2.3.   INTERCOMPARISON OF METHODS.

In this section, the differences in satellite location as calculated by the different methods described in section 2.2 are compared. TBUS and TLE data for the same day (19 April 1992) were obtained and sub-satellite locations (latitude and longitude) at two-minute intervals starting at 15.44Z for a whole orbit were calculated using the following combination of models and elements:

(a)   BROLYD model using TBUS elements
(b)   Circular model using TBUS elements
(c)   Elliptical model using TBUS elements
(d)   SGP4 model using TLE elements
(e)   Kloster algorithm using TBUS elements.

Only a relative comparison is possible. It is impossible to obtain latitude and longitude control points for a whole orbit at two-minute intervals and, although the TBUS parts I-III sometimes contain sub-satellite latitude and longitude values at this frequency for "reference orbits", these have been found to be very inaccurate (Sandford, 1992b). Thus for the comparison, BROLYD+TBUS results were adopted as reference, and the relative differences of the other methods compared with its results.

The distance between the sub-satellite position as predicted by the BROLYD+TBUS combination and by the model under test was found using the approximation:

$$\text{distance} = 111\sqrt{\delta\theta^2 + (\delta\lambda\cos\phi)^2} \text{ km.} \qquad (25)$$

The height of the satellite above the ground was also calculated at each point and, again the difference between the value for the model under test and that for the BROLYD+TBUS combination was calculated.

To illustrate the various effects, graphs of distance error against time and height error against time have been produced. Latitude is also shown on each graph to depict any relationship between error and orbital position. Note that the graphs have been plotted on different vertical scales to show detail more clearly. Also, note that the results quoted here differ slightly from those in Sandford (1992b) due to modifications to the method by which the sub-satellite latitude and longitude are calculated from satellite position and the use of a slightly different implementation of the BROLYD model.

Figure 6 shows the difference between predicted position from the BROLYD model and the predicted position for the chosen model for the circular and elliptical models. Similarly, Figure 7 shows the difference for the predicted satellite height. The errors are considerable- up to 20 km in distance and 15 km in height for the circular model, and 15 km and 10 km respectively for the elliptical model.

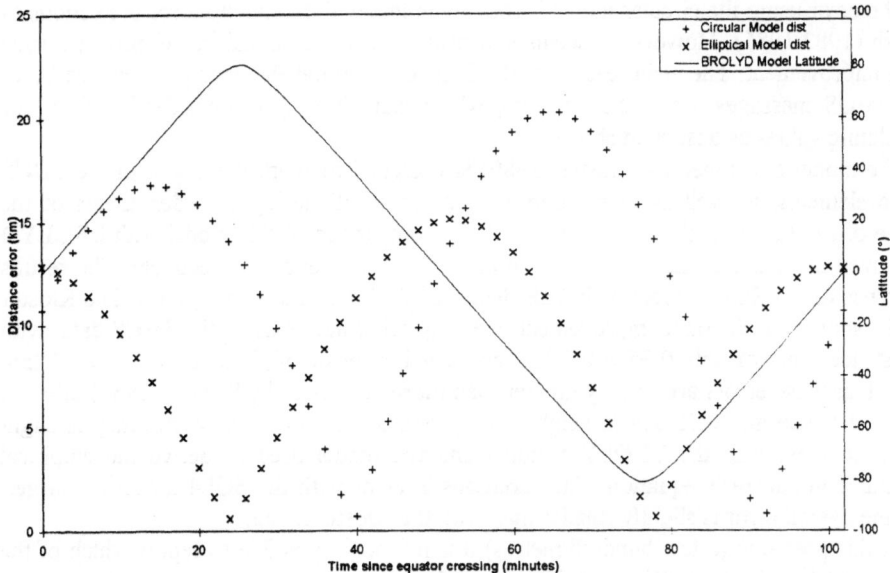

Figure 6. Simple orbital model results relative to the BROLYD model results-distance.

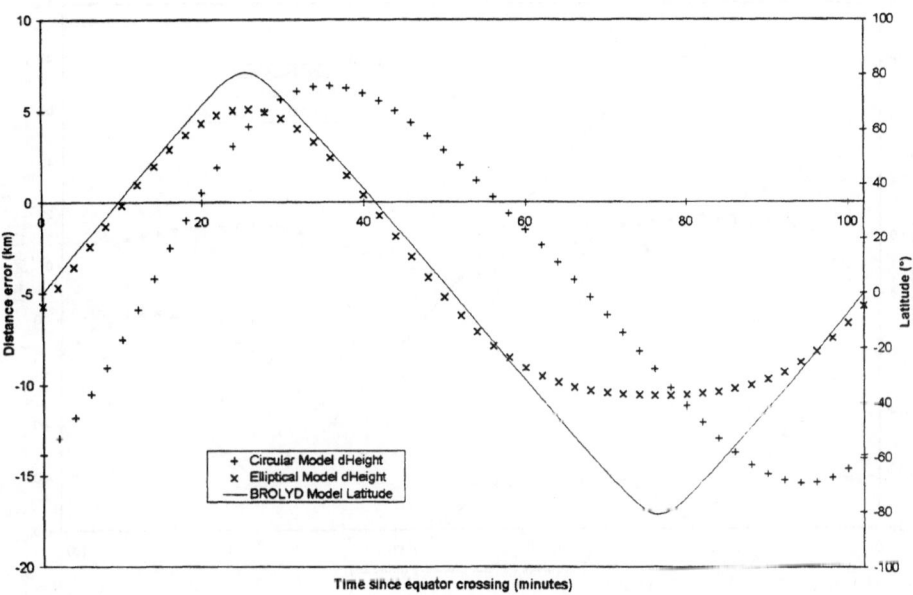

Figure 7. Simple orbital model results relative to the BROLYD model results -height.

The error using the circular model is roughly in line with the timing error predictions of Brush (1982,1985). However it seems surprising that the elliptical model provides such little improvement. The main reason for this is that the orbital elements given in part IV of the TBUS messages are mean elements, which can differ quite considerably from the osculating values as described above.

The model developed by Kloster (1989) is designed to work directly with the TBUS mean elements, as well as taking into account some of the higher order effects of the shape of the Earth on the orbit. Results of the comparison of this model with BROLYD are given in Figure 8 and 9 for position and height differences respectively. The results from use of the SGP4 model with TLE data are plotted on the same graphs. The Kloster model shows considerable improvement over any other model using the TBUS data, with a distance error of only 0.95 ± 0.1 km, and a height error of about only 0.4 ± 0.1 km. Note that these errors are slightly greater than those suggested by Kloster (1989) of ≈ 0.4 km in distance and 0.15 km in height. The probable cause of this discrepancy is slight differences between the BROLYD model and the model used to derive the empirical constants in Kloster's equation. The variations in error with the SGP4 model are larger, but the overall error is slightly smaller than with the Kloster model.

At this precision (a few hundred metres) it is no longer possible to report which of the three models (BROLYD/Kloster/SGP4) is "right". In any case, however, the divergences between the three over the time period of a day with the use of up-to-date orbital elements are equivalent to sub-pixel errors in navigating AVHRR images.

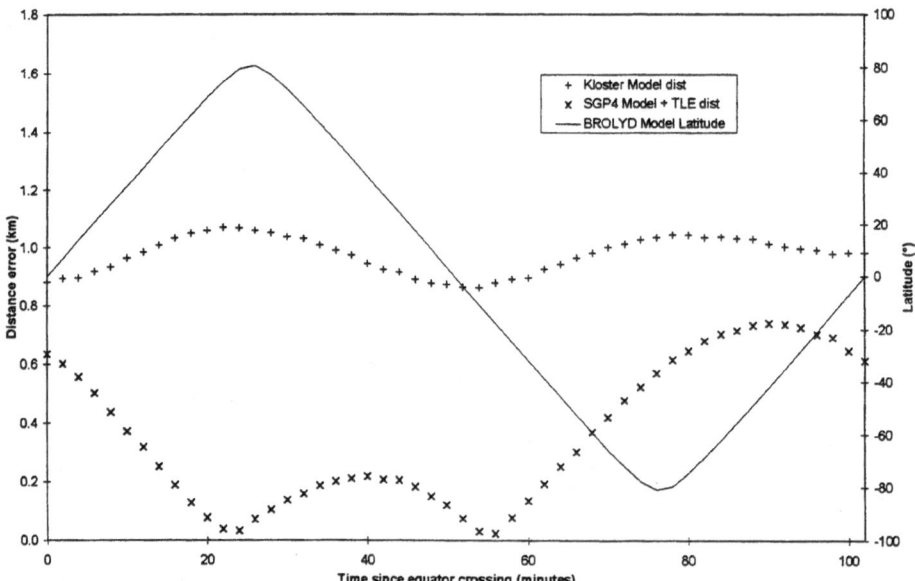

Figure 8. Kloster and SGP4 orbital model results relative to the BROLYD model results- distance

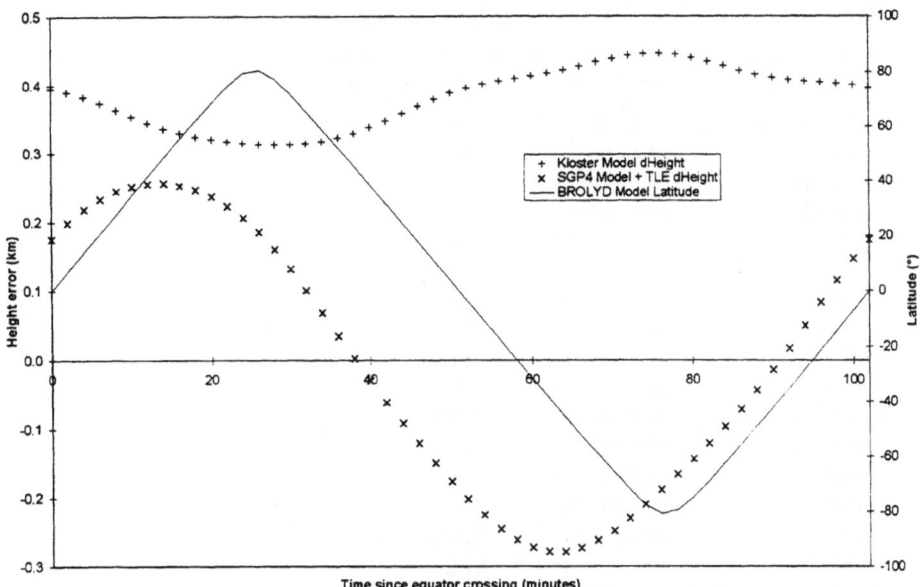

Figure 9. Kloster and SGP4 orbital model results relative to the BROLYD model
results- height

2.4.   PROPAGATION OF ERRORS WITH TIME.

All of the comparisons shown above have been made from the same orbit on the same day
for NOAA-11 using TBUS or TLE elements from the same day. (These are by convention
generated for a moment at which the satellite is crossing the Equator in a northwards
direction (the "ascending node")). However, timely orbital element data are not always
available, and users of NOAA AVHRR data often have to perform geometric corrections
with orbital elements of a few days (and sometimes weeks!) old.

Kloster (1989) and Nagle (1986) carried out some sensitivity tests, and another test
reported by Sandford (1992b) involved the assessment of errors resulting from the use of
older ephemeris data. TBUS and TLE data for NOAA 11 were collected for eight dates at
approximately one week intervals over March and April 1992. The epoch dates and times,
plus ephemeris type for each set of data are given in Table 1 below.

These dates and times provide 16 points at which the latitude of the satellite should be
zero. The actual position at each of these times was calculated for each of the eight sets of
TLE or TBUS data, as appropriate for the model under test, and the results are provided
below. The three models compared were:

(a)   the BROLYD model
(b)   the Kloster model
(c)   the SGP4 model

Table 1. Ephemeris Data Used for Long Term Investigation

Epoch Date	Epoch Time	Ephemeris Type
92/03/07	00:35:58.052	TBUS
92/03/07	22:41:54.597	TLE
92/03/14	00:53:31.533	TBUS
92/03/14	09:23:29.905	TLE
92/03/22	00:58:55.427	TBUS
92/03/22	19:40:51.583	TLE
92/03/28	01:28:23.409	TBUS
92/03/28	20:10:19.457	TLE
92/04/05	01:33:36.921	TBUS
92/04/05	10:03:34.949	TLE
92/04/10	00:33:06.271	TBUS
92/04/10	19:15:01.802	TLE
92/04/19	00:26:06.613	TBUS
92/04/19	10:38:03.562	TLE
92/04/26	00:43:16.103	TBUS
92/04/26	09:13:13.088	TLE

TBUS data were used with the BROLYD and Kloster models, and TLE data were used with the SGP4 model. The circular and elliptical models were not used as the short term results already indicated that their locational accuracy was poor, even with the use of recent ephemeris data.

The results for each tested model were plotted as latitude error against time from epoch. Figures 10-12 show the results for the BROLYD, Kloster and SGP4 models respectively. The data have been plotted on different vertical scales to show more detail. Both BROLYD and Kloster models show a quadratic relationship between error and time, and a fitted curve is shown on each graph. The equation for the curve is :

$$y = -0.0025x^2 - 0.0033x - 0.0091 \qquad \text{for the BROLYD model, and}$$
$$y = -0.0023x^2 - 0.0019x + 0.0093 \qquad \text{for the Kloster model}$$

The SGP4 model is about twice as good, with peak error of about only 3.5° at 60 days from the nearest available ephemeris data, compared with 6° for the other two models. The principal difference between the SGP4 model and the other two, and the main reason for the locational improvement in time, is that SGP4 takes account of atmospheric drag.

Rather than a single quadratic relationship, the SGP4 results show a family of curves, one for each set of ephemeris. The reason for this is that the atmospheric drag parameter varies considerably over a quite short time period, and the "instantaneous" value provided in the two line elements will only be valid for short to medium term predictions.

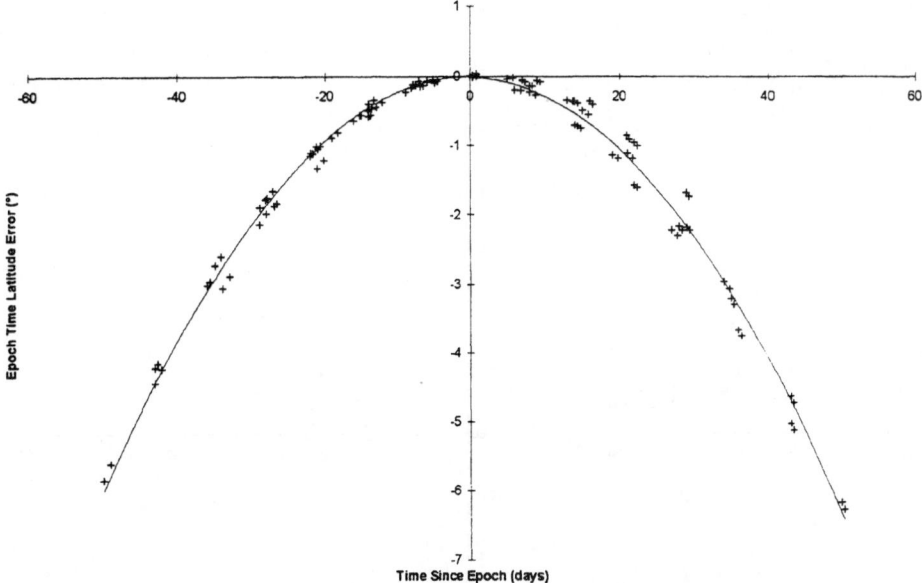

Figure 10. Latitude error against time from epoch- BROLYD+TBUS model..

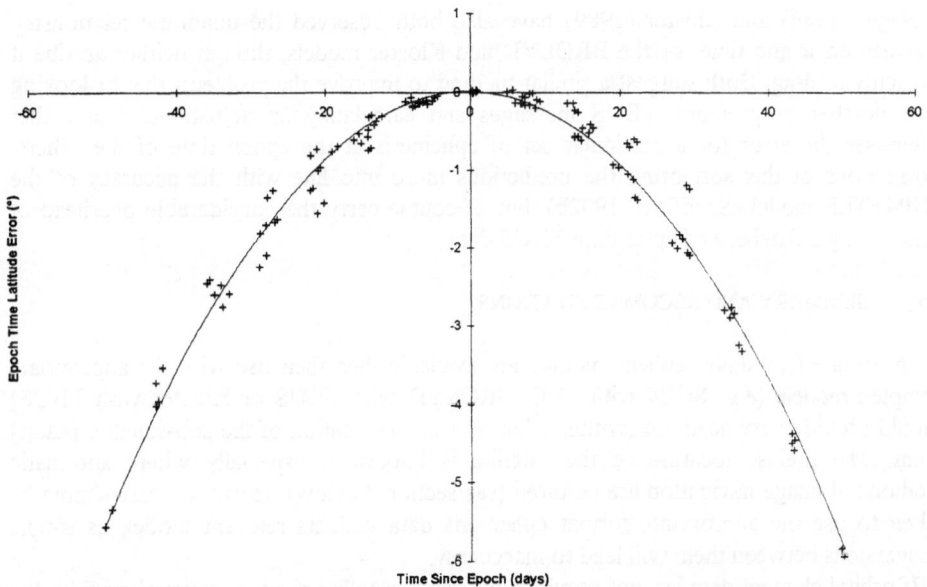

Figure 11. Latitude error against time from epoch- Kloster+TBUS model.

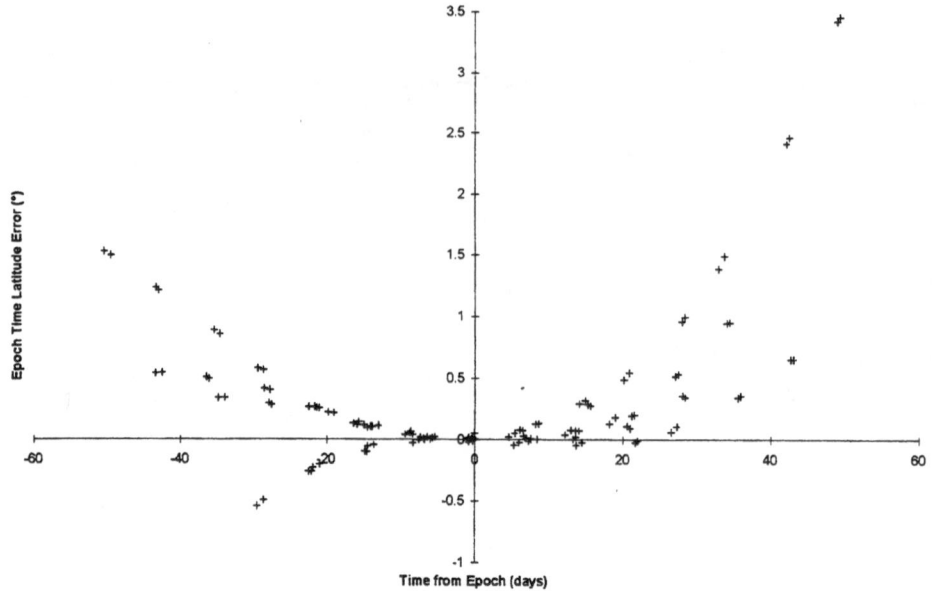

Figure 12. Latitude error against time from epoch- SGP4+TLE model.

Nagle (1986) and Kloster (1989) have also both observed the quadratic relationship between error and time for the BROLYD and Kloster models, though neither ascribe it explicitly to drag. Both suggest a similar method to improve the problem- that of looking at a number of previous TBUS messages and calculating an adjustment factor that minimises the error for a particular set of ephemeris at the epoch time of the others. Corrections of this sort bring the predictions more into line with the accuracy of the SGP4+TLE model (Sandford, 1992b), but of course carry the considerable overhead of maintaining a database of up to date TBUS data.

2.5. SUMMARY AND RECOMMENDATIONS.

If up-to-date (say daily) ephemeris data are available then their use with the appropriate complex models (*e.g.* SGP4 with TLE, BROLYD with TBUS or Kloster with TBUS) should provide very accurate (within 1 km or 1 pixel) location of the sub-satellite (nadir) point. The precise location of the satellite is important, especially where automatic methods of image navigation are required (see section 3 below). However, care should be taken to use the appropriate format ephemeris data with its relevant model, as simple conversions between them will lead to inaccuracy.

If orbital element data are not regularly available, satellite position estimations with the propagation of the SGP4+TLE model will still give relatively good (± 1° or about 100 pixels) for a month either side of the date of the TLE data. Most of this error will be in the along-track direction. The SGP4+TLE model is thus the recommended method when orbital element data are not regularly available, though similar accuracy could also be obtained with BROLYD+TBUS or Kloster+TBUS methods as long as atmospheric drag is also accounted for.

Where daily ephemeris information is not available, sub-satellite location estimation will be poor and the image navigation methodology will have to allow for a corrective shift in the along-track direction to account for the possible movement of the image relative to the predicted position. Note that even with up-to-date ephemeris data and the use of complex models, this shift methodology may still be necessary. This is because the clock on board the satellite (which labels each image with a time-stamp) has been known to drift significantly with respect to time (Marsouin and Brunel 1991). Thus the time used to estimate the satellite position (if read off directly from the data stream) may contain some error, and therefore lead to inaccuracy. More recently, NOAA have started to publish regular information on the expected drift of the satellite clock, and the date and amount of the corrections made to it, and the information is also included in the TBUS messages. In the future, it may be possible to use these data to avoid the necessity of a "shift" methodology (*Rosborough et al.* 1994).

The circular and elliptical models will produce only crude estimates of the satellite location, and will therefore only be useful where general location of the satellite location is required (as for example for antenna pointing to the satellite). Moreover, simple circular models only yield estimates for the sub-satellite latitude and longitude- the height is normally assumed to be the nominal value (or estimated crudely from GCPs). The height of the satellite is crucial to the next phase of the geometric correction procedure- the modelling of the image geometry, as changes in the satellite altitude above the geoid will directly affect the scale of the imagery (amongst other things), and the other distinct advantage of the complex models is that they also provide a good estimate of this parameter.

2.6. CO-ORDINATE CONVERSION.

Keplerian to Earth-Fixed Conversion. The complex orbital models yield the satellite position and velocity in three dimensions in Keplerian element form. Although simple conversion to geodetic or geocentric co-ordinates are possible, Rosborough *et al.* (1994) provide more precise formulae for the conversions, from Keplerian element to Cartesian Co-ordinates (the so-called Earth-Fixed system which has the x-axis intersecting the Greenwich meridian and the z-axis along the Earth spin vector).

As a first step, the true anomaly is obtained from the mean anomaly. Although equation (11) provided us with a simple formula, the relationship is actually transcendental, so a series expansion could be used (Smart, 1953):

$$v = M + (2e - \frac{1}{4}e^3)\sin M + (\frac{5}{4}e^2 - \frac{11}{24}e^4)\sin 2M + \frac{13}{12}e^3\sin 3M + \frac{103}{96}e^4\sin 4M + .. \qquad (26)$$

The distance to the satellite as a function of the true anomaly is the given by the equation of an ellipse :

$$r_s = \frac{a(1-e^2)}{(1+e\cos v)} \tag{2}$$

The Earth-fixed cartesian position component of the satellite are then:

$$x_s = r_s\left[\cos\Omega\cos(\omega+v) - \sin\Omega\sin(\omega+v)\cos i\right] \tag{27}$$

$$y_s = r_s\left[\sin\Omega\cos(\omega+v) + \cos\Omega\sin(\omega+v)\cos i\right] \tag{28}$$

$$z_s = r_s\sin i\sin(\omega+v) \tag{29}$$

and the inertial velocity of the satellite, expressed in the same Earth-fixed co-ordinate system is:

$$\dot{x}_s = \frac{x_s l e}{r_s p}\sin v - \frac{1}{r_s}\left[\cos\Omega\sin(\omega+v) + \sin\Omega\cos(\omega+v)\cos i\right] \tag{30}$$

$$\dot{y}_s = \frac{y_s l e}{r_s p}\sin v - \frac{1}{r_s}\left[\sin\Omega\sin(\omega+v) + \cos\Omega\cos(\omega+v)\cos i\right] \tag{31}$$

$$\dot{z}_s = \frac{z_s l e}{r_s p}\sin v + \frac{1}{r_s}\sin i\cos(\omega+v) \tag{32}$$

where the parameters p and l are :

$$p = a(1-e^2) \tag{33}$$

$$l = \sqrt{\mu p} \tag{34}$$

Earth-Fixed to Geodetic Co-ordinates. The Earth-fixed co-ordinates can be related to the geodetic co-ordinates (which specify the geodetic latitude (ϕ^*), longitude (λ) and height above the reference spheroid (h) of a point) by the following equations:

$$x = (N+h)\cos\phi^*\cos\lambda \tag{35}$$
$$y = (N+h)\cos\phi^*\sin\lambda \tag{36}$$

$$z = [(1-f)^2 N+h)] \sin \phi \; ^*$$ (37)

where f is the flattening of the Earth's reference spheroid and N is given by:

$$N = \frac{a_e}{(1 - e_e^2 \sin^2 \Phi^*)^{\frac{1}{2}}}$$ (38)

where a_e is the semi-major axis of the reference ellipsoid and e_e is related to the flattening by the expression:

$$e_e^2 = 1 - (1-f)^2$$ (39)

Given x, y and z the computation of the geodetic latitude and height are not as straightforward. The longitude can easily be calculated from:

$$\lambda = \tan^{-1} \frac{y}{x}$$ (40)

but the closed form solutions for height and latitude are much more involved and a numerical solution is normally implemented (Borkowski 1989). If the point lies on the reference spheroid (h = 0) then the conversion between cartesian and geodetic co-ordinates is greatly simplified as spherical trigonometry is valid. This then is normally the assumption for the surface being imaged. However, the general (non-zero elevations) are still needed since it is necessary to compute the geodetic latitude of the satellite which may be necessary in the computation of the spacecraft altitude (see section 3.6).

2.7.   INSTANTANEOUS TRACK ORIENTATION.

In celestial mechanics, the ground-track orientation of the satellite is given by the angle a, where (Figure 13):

$$\cos \alpha = \frac{\cos \Delta \Omega - \cos \Pi \cos(\varpi + v)}{\sin \Pi \sin(\varpi + v)}$$ (41)

However, this is only true in inertial co-ordinates. In Earth-fixed co-ordinates, the velocity vector must be adjusted to compensate for the velocity of the sub-satellite point due to the Earth's spin (Figure 13). Then the ground-track orientation is given by:

$$\tan \beta = \frac{V \sin \alpha - R_E \omega_E \cos \psi}{V \cos \alpha}$$ (42)

where ω_E = 360 ° /day

174

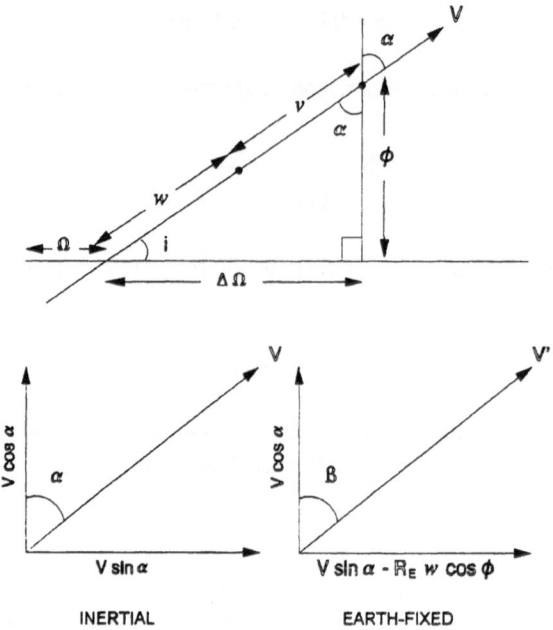

Figure 13. Instantaneous track orientation.

Representations of the Ground Track. If we attempt to plot the ground track on a map frame, the track is not necessarily a straight line. This is because of the Earth's rotation (Figure 14) and because the type of map projection may not project a plane as a straight line anyway, *e.g.* a Mercator projection (as illustrated in Figure 15).

If $\Delta\lambda$ is the angle that the Earth has rotated through during one orbital period T:

$$\Delta\lambda = \varpi_E T \tag{43}$$

then we can adjust the period to gain the repeat ground-track that we desire, *e.g.* if we require the satellite to make m orbits in n days, we solve:

$$m\omega_E T = 360\,° \tag{44}$$

for *T*, using the fact that :

$$T = 2\Pi\sqrt{\frac{a^3}{\mu}} \tag{6}$$

Such a repeating ground-track is called "m:n resonance" (Sowter, 1993).

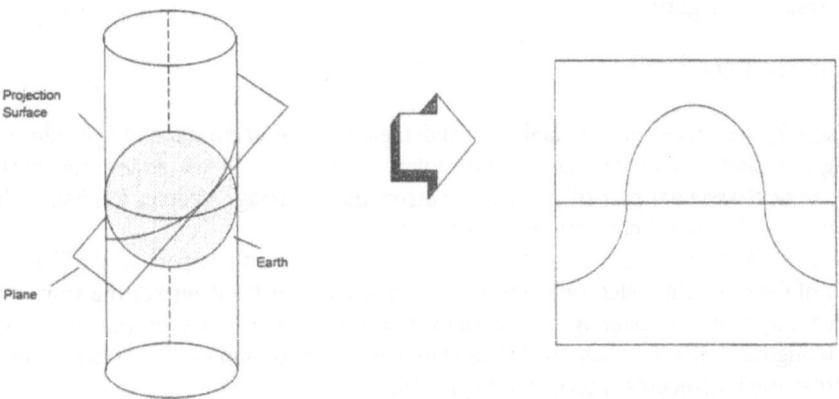

Figure 14. Map representation of ground track.

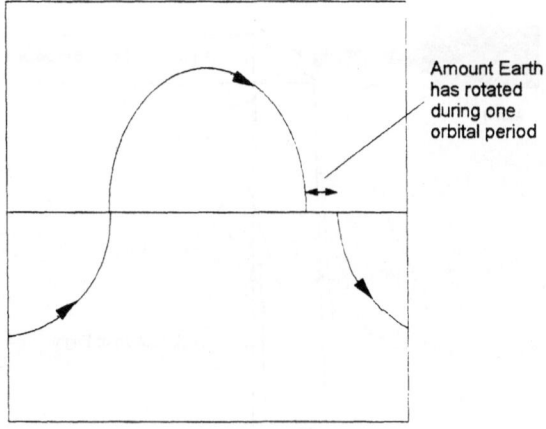

Figure 15. Earth rotation effect.

3. Image Navigation.

3.1 IMAGE GEOMETRY.

In section 2, we reviewed the methods (and their relative accuracy and reliability) for locating the satellite and the sub-satellite point. In this section we review methods to model the next essential part of geometric correction, the image geometry. First, a short revision of the AVHRR behaviour is appropriate.

The AVHRR scanner uses a rotating mirror fixed at 45° with respect to the Earth and the axis of the observing telescope. The telescope is aimed at the centre of the scan mirror in such a way that the resulting scan pattern describes a great circle on the surface of a non-rotating Earth perpendicular to the satellite's direction of motion (Emery *et al.* 1989). The across-track geometry is shown in Figure 16.

If we assume that the pixel is formed by a view of the ground from a fixed instantaneous field of view (IFOV), then (see Figure 17):

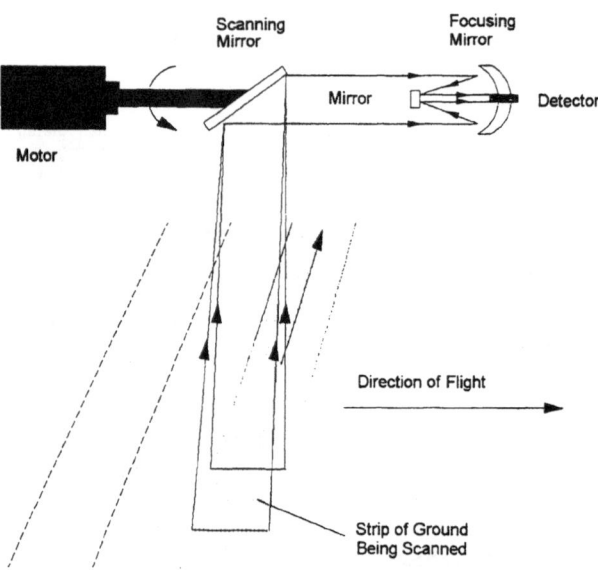

Figure 16. NOAA-AVHRR scanning system.

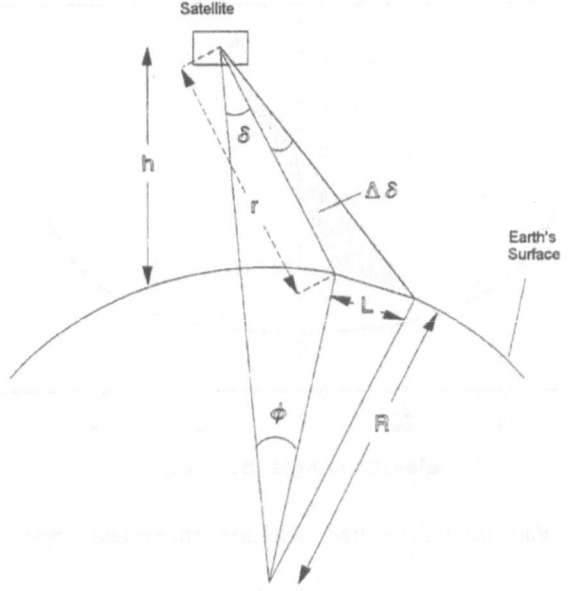

Figure 17. NOAA-AVHRR across-track scanning geometry.

$$L = \left\{ \frac{(R_E + h_S)\cos\delta}{\sqrt{1 - (1 + \frac{h}{R})^2 \sin^2\delta}} - R \right\} \Delta\delta \qquad (45)$$

Of later importance:

$$\psi = sin^{-1}\left[\left(1 + \frac{h}{R}\right)sin\delta\right] - \delta \qquad (46)$$

$$r = (R_E + h_S)\cos\delta - R_E \sqrt{1 - (1 + \frac{h_S}{R_E})^2 \sin^2\delta}\,^r \qquad (47)$$

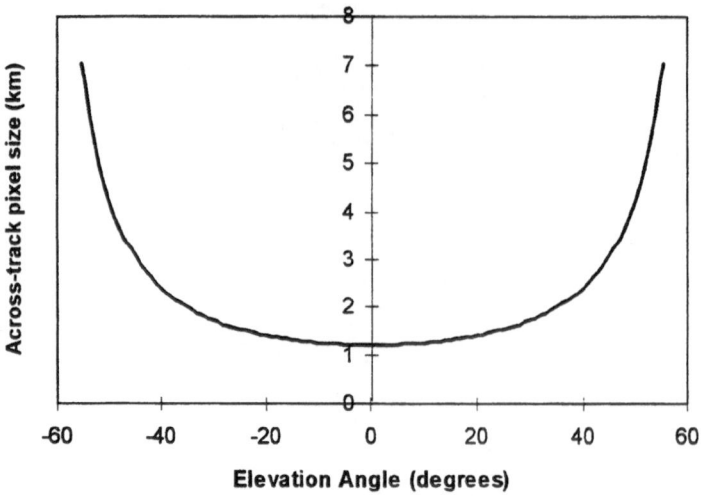

Figure 18. Variation of across-track pixel size with elevation angle

If we plot the value of "L" the across-track pixel size across the scene from edge to edge we see (Figure 18) that the across-track pixel size increases to nearly 7 km (in the across-track direction) at the AVHRR edge of swath (55.4°). This is important to note, and users of NOAA-AVHRR data must be aware of the changing pixel size (*i.e.* spatial resolution) with off-nadir location. Some users exclude the very edge of swath parts of NOAA-AVHRR imagery (say the end 256 pixels). Brush (1985) proposes a "lineariser" to replicate pixels at the edge of swath to yield a uniform scale across the swath. In all cases, it is apparent that because of this effect, any ground-control pointing procedure must take place on undistorted imagery, and root mean square errors when quoted are only valid for uncorrected data, in terms of pixel displacements (not kilometre accuracies).

The pixel size may also change in the along-track direction with changes in the satellite altitude, but these are normally very small during the time taken to acquire an image (say 15 minutes).

Because of the variations in pixel size in the across-track and (to a lesser degree) in the along-track direction, regular features on the ground will have a distorted representation on the image (Figure 19). Lines of latitude and longitude for instance, will appear sigmoidal in the image with the greatest distortion furthest away from the nadir track. Note that the lines of latitude will be further distorted closer to the poles (Figure 20).

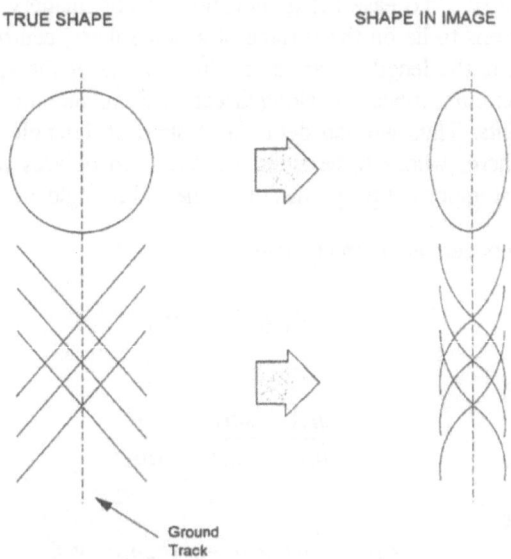

Figure 19. Distortion in imagery.

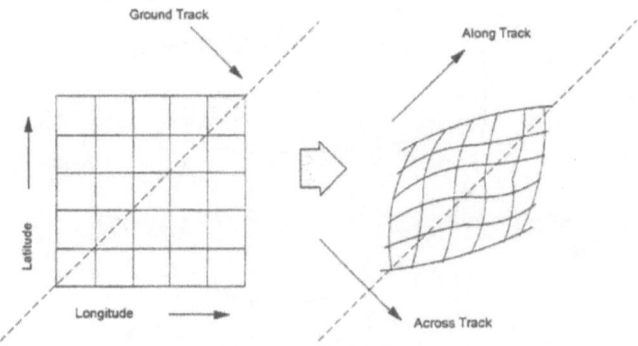

Figure 20. Latitude and longitude lines as represented in NOAA AVHRR imagery.

Spherical Trigonometry. To ease the geolocation of our imagery, we can translate all our angles and positions to lie on the surface of a unit sphere, centred on the geocentre. On such a unit sphere, the length of arc along the surface of the sphere is equal to the angle which the arc circumscribes (*i.e.* along Great Circles), and any two great circles will intersect at two points. Thus we can define a "Spherical Triangle" as a 3-sided figure drawn on a unit sphere, whose three sides are made up of arcs of three great circles (Figure 21). The most important properties of a spheroid triangle are:

(i) the relationship between the internal angles:

$$A + B + C \leq 180° \tag{48}$$

(ii) the sine rule:

$$\frac{\sin A}{\sin a} = \frac{\sin B}{\sin b} = \frac{\sin C}{\sin c} \tag{49}$$

(iii) the cosine rule:

$$\cos a = \cos b \cos c + \sin b \sin c \cos A \tag{50}$$

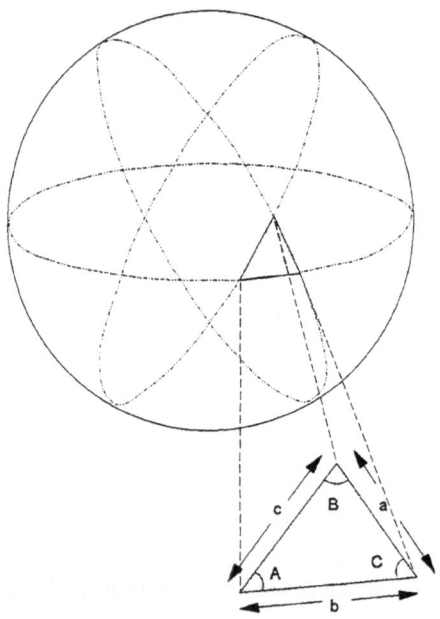

Figure 21. The spherical triangle and spherical trigonometry.

3.2   DIRECT IMAGE NAVIGATION.

Given the satellite position and an image acquired by the satellite at that time, there are two functions a user may want to image:
(i)   given a particular pixel and line co-ordinate, calculate the latitude and longitude to which it corresponds (a process referred to as Direct Navigation or Direct Referencing);
(ii)   given a particular latitude and longitude, calculate the corresponding pixel and line co-ordinates in the image (a process referred to as Indirect Navigation or Inverse Referencing).

The first of these two functions is mathematically easier, but it is the second which is required for image remapping or resampling to other (more useful) projections. We shall consider direct navigation first, referring to the scan geometry defined in the model as specified by Ho and Asem (1986) and Emery *et al.* (1989) and shown in Figure 22. This (simple) model assumes a spherical Earth. Emery *et al.* (1989) justify the use of this assumption by claiming that if the centre location of the image is located as precisely as possible (perhaps using a complex orbital model plus a shift to account for timing errors), then the application of a locally spherical model would provide sufficiently accurate results (at least over a relatively small area).

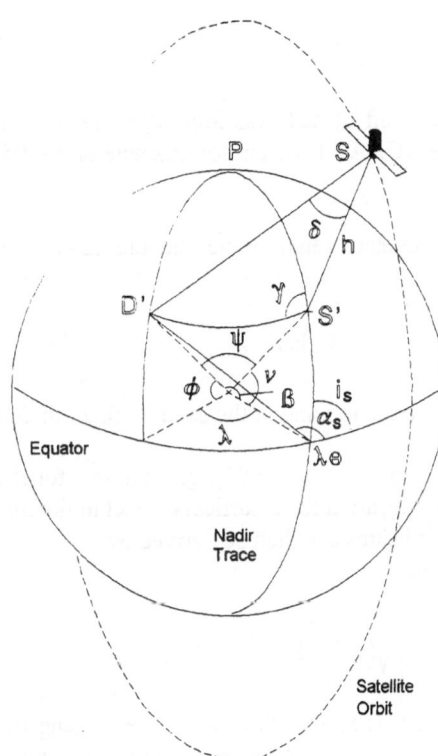

Figure 22.
NOAA   satellite image geometry (as used in Emery *et al.* 1989, and Ho and Asem, 1986).

From Figure 22 (a static spherical Earth), four principal trigonometric equations can be deduced:

$$\cos\beta = \cos\psi\cos\theta + \sin\psi\sin\theta\cos\gamma \tag{51}$$

$$\sin(i-j) = \frac{\sin\psi\sin\gamma}{\sin\beta} \tag{52}$$

$$\sin\Phi_D{}' = \sin\beta\sin j \tag{53}$$

$$\cos\lambda_D{}' = \frac{\cos\beta}{\cos\Phi_D{}'} \tag{54}$$

For the point D' being imaged we can write the relationship between the scan mirror angle δ and the geocentric angle ψ from nadir as:

$$\psi = \sin^{-1}\left\{\left[(R_E + h_s)/R_E\right] - \sin\delta\right\} - \delta \tag{46}$$

where the off-nadir viewing angle is defined by:

$$\delta = p\delta_i \tag{55}$$

where δ may be defined as negative to the left of nadir (as looking in the satellite flight direction) and positive otherwise. For the NOAA-11 system for example $\delta i = 0.955 \times 10^{-3}$ radians.

There is also a relationship between the satellite zenith angle and the off-nadir viewing angle (see triangle SOD' in Figure 22)

$$z = \sin^{-1}\left[((R_E + h_s)/R_E)\sin\delta\right] = \psi + \delta \tag{56}$$

where ψ is negative to the left of the nadir trace looking in the satellite flight direction and positive otherwise.

According to Ho and Asem (1986) and Emery et al. (1989), given a time for the centre line of the image (t), a pixel (p) and line (l) location for a particular pixel in the image, the corresponding latitude and longitude co-ordinates can then be derived by:

(i)   calculating θ (the angular span) from

$$\theta = t\left[\sqrt{(\mu/r^3)}\right] \tag{57}$$

(ii)   then calculating δ and ψ by equations (55) and (56) successively, using the radius as that of (an assumed spherical) Earth plus the nominal height of the satellite.

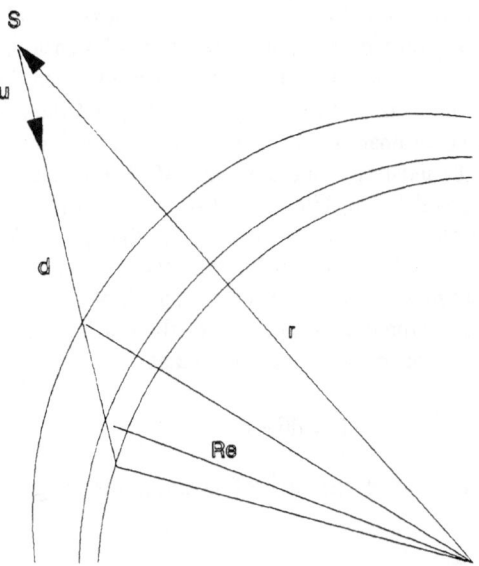

Figure 23. The geometry used by Rosborough *et al.* 1989, to estimate point of intersection of scanner view with Earth's surface (shown in 2D only).

(iii)   then determine β and φ from equations (51) and (52) successively
(iv)   finally, θ_D' and λ_D' can be calculated by equations (53) and (54) respectively.

To account for the rotation of the Earth during time t, the geographic longitude then has to be corrected by (Ho and Asem, 1986):

$$\lambda_D = (\lambda_D{}' - \dot{\Lambda}_n t) + \lambda_0 \tag{58}$$

Although mathematically relatively simple, the assumptions of a circular orbit and spherical Earth plus the use of the nominal satellite height (see equation (56) for example) should not be expected to produce very high accuracy. Both Ho and Asem (1986) and Emery *et al.* (1989) propose the use of one or more GCPs from which better estimates of satellite height (and inclination angle) could be made. Even with these modifications, however, the method is still not sufficiently accurate for large areas (Brush, 1988).

The use of timely satellite ephemeris data with complex orbital element models are useful in that they already provide good estimates of satellite height and velocity (from which inclination can be determined well). The use of this calculated (rather than assumed) satellite height plus corrections for the Earth's oblateness provides much better results. Rosborough *et al.* (1994) provide a sophisticated three-dimensional vector-based method for locating the latitude and longitude of a particular point once the 3-D location of the satellite is known (from the results of the complex model) and the look direction of its sensor is calculated (from co-ordinate conversion). Their major improvement on the technique described above can be illustrated in the triangle SOD' for Figure 22.

They do not assume a spherical Earth, but assert that if the latitude at the point of intersection is unknown, then so too is the Earth radius at that point. With respect to Figure 23, they state that at the point of intersection (D') the vector $\bar{r}$ from the geocentre to the satellite is known as is the unit vector $\bar{u}$ (the look vector of the sensor). However, neither the magnitude of the look vector (d) nor the Earth radius at this point of intersection are known. Using vector formulae they show that :

$$\vec{R} = d\bar{u} + \bar{r} \tag{59}$$

and, using spherical trigonometry and the law of cosines, the magnitude d can be expressed by:

$$d = -\bar{r}.\bar{u} - \left[R^2 - r^2 + (\bar{r}.\bar{u})^2\right]^{1/2} \tag{60}$$

An iterative procedure is proposed: initiating d from equation (60) by setting R to the spherical Earth radius. With this value for the magnitude for the look vector, the vector from the geocentre to the surface point can then be found from equation (59). This will provide us with an improved estimate for the Earth radius, and the procedure can be repeated until some desired level of convergence is obtained (Rosborough *et al.* 1994).

Given the satellite location and direction of look vector, Puccinelli (1976), also using a vector-based approach, proposed the location of the point D' by determining the intersection of the look vector with the equation that defines the surface of an ellipsoid:

$$\left(\frac{x^2 + y^2}{a^2} + \frac{z^2}{b^2}\right) = 1 \tag{61}$$

These two latter, relatively sophisticated, approaches both yield much improved results over the simpler techniques, especially over large areas. Another big advantage of these approaches is that they both offer the possibility of making attitude corrections to the calculations if these can be measured or determined in some way (see section 3.5).

3.3   INDIRECT IMAGE NAVIGATION.

In indirect referencing or navigation, given the latitude and longitude of a particular point, we try to determine the corresponding line and pixel location in the image. This is the transformation required for the remapping (interpolation/resampling) of images into mapped projections (*i.e.* we would like to compute the pixel and line location that corresponds to each map location in our chosen (output) projection). While the direct navigation could be solved analytically, the indirect navigation requires a much more complicated iterative procedure that is normally accomplished by numerical techniques. The iteration is due to the fact that the time, t, from the right ascension of the ascending node of the satellite orbit to the time when the relevant pixel is imaged, is not known and must therefore be derived (Emery *et al.* 1989).

With respect to Figure 22 again, we first assume that the time at which the line of interest is imaged is the same time as that when the first line was imaged. Then (Ho and Asem, 1986):

$$t = t_f + t_0 \tag{62}$$

From t, we can then calculate the static Earth longitude (λ_D') from the given geographic longitude (λ_D) using:

$$\lambda_D' = (\lambda_D - \dot{\Lambda}_n t) + \lambda_0 \tag{63}$$

(NB: This equation ignores the motion of the right ascension of ascending node which can contribute an error of a few kilometres in longitude over large orbital segments due to across track skews).

Using equation (54) we can then calculate β, and then φ, ψ and θ by equations (53), (52) and (51) successively. A new time estimate can then be made from equation (57). If the difference between this new time estimate and the first time is larger than some predetermined threshold (0.01 seconds say), then we can make the new time estimate the first plus half the difference and repeat the whole process, until some desired level of convergence for t is reached. Finally, using equation (56) δ can be calculated and finally the pixel number p can be derived from equation (55) and the line number would be (t-t$_f$)x line rate +1 (Ho and Asem, 1986; Emery *et al.* 1989). Again, because of the inherent assumptions, large errors are inevitable, though improvements may be obtained if GCPs are used or if better estimates for satellite height and location are made from complex orbital models.

In the 3-D vector-based system of Rosborough *et al.*(1994) (see Figure 22) the vector $\bar{R}$ to the specified latitude and longitude is not known, neither is d, nor $\bar{r}$ (at the time of image line acquisition). An initial guess can be made for the satellite location (based on either an approximate latitude and longitude estimate of the sub-satellite point or an approximate time from the equatorial crossing). From this, the satellite position, velocity vector and the look vector of the radiometer are all calculated and rotated into the so-called spacecraft-fixed co-ordinate system. The resulting look vector will not (in all likelihood) lie in the scanning plane (across-track) direction. Then the projection of this look vector into the satellite pitch plane (which corresponds very nearly with the orbital plane) will relate to the amount that the satellite is either ahead or behind of where it should be. The satellite location can then be corrected by this amount and the procedure can be repeated until the calculated look vector does lie in the scanning plane, at which point the desired line location of the given geodetic co-ordinates is found (Rosborough *et al.* 1994).

Whichever method is adopted, the indirect navigation method is clearly much more computationally demanding than the direct method. This is unfortunate because the image mapping procedure requires it for every pixel in the output image. One way of accelerating the process is to run the <u>direct</u> navigation transformation for a grid of points and create a look-up table of latitude and longitude references or "tie -points" for example for every 40 pixels and every 40 lines. Then, for any given latitude and longitude, instead of running the indirect navigation algorithm, the corresponding pixel and line locations can be obtained by interpolation (for example, bilinear interpolation) from the look-up table of tie-points. Such a method will obviously contain some generalisation, but can yield considerable savings in computing time. In many cases, the loss of accuracy by interpolation has been shown to be minimal, especially when the grid size of the look-up table is chosen carefully (Rosborough *et al.* 1994). Obviously, interpolation over coarser grids and at more off-nadir parts of the images will result in more generalisation and poorer geometric fidelity.

Vector overlays, to plot over the images in the satellite projection, and to estimate the shift required for example, can be generated by "threading through" linear vector databases of coastlines, rivers, *etc.* Incidentally, solar and satellite azimuth and zenith angles can also be calculated for this grid of "tie-points" once their latitude and longitude locations have been determined.

3.4   APPLICATION EXAMPLES

Both NOAA (in their Level 1b format data) and ESA (in their SHARP format) data, supply a grid of "tie-points" together with the image data. The NOAA Level 1b format contains tie-points (latitude and longitude control points) corresponding to every 40 pixels for each line (see Kidwell, 1992). These "tie-points" are believed to have been calculated using the elliptical orbit/spherical Earth and mean elements model with recent ephemeris data (Emery *et al.* 1989). As mentioned previously, the assumptions used in this process are normally valid only for small areas of interest, and as a result, several users have found the tie-points provided in the NOAA data to be too inaccurate, especially for use over larger areas- displacement errors of over 6 pixels have been found on occasions. However, NOAA are currently improving their method of generating tie points to include in Level 1b data (see chapter by Kidwell).

The ESA-SHARP format imagery contains tie-points at a spacing of every 32 pixels and every 16 lines. These are generated within the SHARK software (Fusco *et al.* 1989) which is used at the ESA network of receiving stations. The tie-points are generated using the complex BROLYD model with recent TBUS data (which are obtained frequently from the NOAA bulletin board). However, partly to overcome the problems with satellite clock drift and partly to compensate for older TBUS data, there is also some operator interaction in the generation of the imagery and its relevant ancillary data. A coastline overlay is generated with SHARP 4-minute images, and the operator estimates the shift required to fit the coastline to the image acquired (more details can be found in the chapter by Arino). The along-track shifts are assumed to be time related and the across-track shifts are assumed to be results of satellite-roll. The accuracy of the resulting tie-points is generally good for the 4-minute scenes (1440 lines). However, in many cases, some operator error is evident, particularly in images where the operator has fitted the overlay to some part of the image (say where there is only a small amount of coastline visible, *e.g.* West Africa) with a resulting inaccuracy at different parts of the image. The assumption that all across-track errors are due to satellite-roll itself can also be a cause of inaccuracy. The coastlines, country borders and latitude/longitude graticule generated in the SHARP process are also supplied with the ESA-SHARP format imagery (as bit overlays in the higher bits of the SHARP format 16-bit words), thus users of the data can judge for themselves whether the fit made (and thus also the fit of the tie-points themselves) are adequate for their areas of interest. Another useful aspect of the SHARP-format image, is that the TBUS information (the latest available at the ESA receiving station) is also provided. Thus users may, should they prefer to do so, run their own orbital model calculations without recourse to a satellite ephemeris source.

Intercomparison of different geometric correction methods is difficult. Most reports provide only overall root mean square (r.m.s) errors with little indication of location and number of control points used in the derivation of transformations. This is often a poor indicator of the quality of image geometric correction, especially if the r.m.s. error cited is derived from fitting transformations to the same set of control points. A much better measure of geometric correction quality is obtained if the r.m.s. error is calculated from an independent set of evenly distributed points. However, as this is often impractical, comparison of geo-corrected imagery with ancillary data such as digital databases, or examination of image to image registration provide other methods of independent evaluation of goodness of fits. The examination of movie loops of geo-corrected imagery, and the calculation of maximum value composites also provide useful qualitative indicators of image to image matching.

In an initial evaluation of the most suitable geometric correction method for the TREES project (see chapter by Malingreau *et al.*), three methods were evaluated: one based on the grid of points supplied in the NOAA Level 1b data; one based on the software and ephemeris database provided by the University of Colorado (and described in Rosborough *et al.* 1994); and one based on the use of a BROLYD+TBUS model with an additional interactive stage. The TREES project requires geometric correction of images covering very large areas, so the tests were performed on areas typically of 1500 kilometres square. The accuracy of geometric correction was judges by the overlaying of digital vector database information on the geometrically-corrected imagery (Achard and D'Souza, 1994).

The NOAA model was found to work poorly over the large areas. There appeared to be non-uniform shifts across the images (of the sort that would be obtained with erroneous estimates of satellite height), although mostly along-track correspondence was relatively good. The Colorado software provided a significant improvement over the use of the NOAA-supplied estimates, but also suffered from slight displacements in the across-track direction. The Colorado software does include methods to estimate and compensate for satellite attitude drifts which should provide additional precision, but these were not implemented in the tests (see section 3.5). The best correspondence between the geometrically-corrected images and the vector overlays (generated from the CIA World Data Base II (Anderson *et al.* 1973)) was provided by the third method tested: based on the use of the BROLYD model plus an operator-determined shift to account for satellite time and roll errors. At the general scale, image to map correspondence was very good at off-nadir as well as at sub-satellite parts of the image. At some parts of the images, the image to map correspondence was so good, that errors in the digital data-base became apparent (*e.g.* the location and shape depiction of some islands and other coastal regions). In general, the quality of fit over all the images was found to be within about 2-3 pixels everywhere. In the TREES project, a second-stage interactive process is used for better image to image correspondence. Where the quality of digital map data is good, and there is sufficient coastline information, image to map (GCP) matching is carried out. However, where the digital data are poor and/or there are few useable coastline features, a master image is created and image to image matching is carried out. Note that this latter option is only possible once indirect navigation (image remapping and resampling) has been made.

A more refined and automated geometric correction methodology is provided by the JRC-MARS (Monitoring Agriculture by Remote Sensing) project, within its SPACE software (Software for Processing of AVHRR data for the Communities of Europe). This software is designed to calibrate and geometrically-correct daily (historical and in real-time) NOAA-AVHRR imagery for Europe, and to do so in a nearly automatic manner requiring little human interaction (see chapter by Vossen). The software takes ESA SHARP Level-1 data as input, and from the TBUS data supplied with these images, uses the BROLYD model to generate latitude and longitude tie-points at a spacing of every 22 pixels and lines This grid-spacing has been optimised to yield less than 0.1 pixel interpolation errors at European latitudes. Using these tie-points, vector overlays (corresponding to small map chips) are deformed to the satellite projection. Over 600 map chips for Europe and the north coast of Africa have been defined (generated from WDB II) and these are stored and accessed from a chip library data-base. The use and performance of each of the map chips is carefully monitored so that new ones can be added or poor ones deleted as necessary. The SPACE software excludes the most off-nadir 128 pixels on each side of AVHRR images, and where there are cloud-free parts of coastline within the area considered, an automatic matching of the map chips with the land-sea image boundary is made (by binary map correlation). Next, a three-stage error-modelling phase screens out unlikely image-map chip matches, and from the remaining data, an average shift in the along-track and across-track directions is estimated. Along-track errors are assumed to be due to satellite time clock errors, so the equator crossing time is adjusted accordingly. Across-track errors are assumed to be due to errors in the calculated equator crossing longitude and the right ascension of ascending node as provided by the TBUS is therefore adjusted to compensate. These corrections are then saved for later use and the whole grid of tie-points is recalculated. For all the chip-matches, residual errors are then calculated and a first-order polynomial of correction is calculated and the image remapped (entailing only one resampling step). The calculation of the first-order polynomial correction is assumed to compensate for any satellite attitude and/or other unaccounted minor errors. Note that all the image to map comparison is made in the raw satellite projection, so only one resampling step will be required, and the root mean square errors are valid in terms of pixel distances. The shape of the area of interest (Europe) and the large amount of coastline everywhere, facilitates the image-map correlation (although cloud cover and inaccuracies in the digital database can cause problems) (Sharman *et al.* 1992).

Within the Monitoring Tropical Vegetation (MTV) Project, another AVHRR processing chain exists for the automatic processing of a large number of images of West Africa acquired at the ESA-Maspalomas ground receiving station (Malingreau and Belward, 1995). This area presents another challenge, as the amount of coastline available for geometric correction is much more limited. Thus, this software uses the ESA-generated tie-points for a first fit, followed by a more precise fitting based on a number of *image* chips (of about 50 lines by 50 pixels). These have been defined from a historical set of NOAA-AVHRR imagery of the area for image to image correlation. From the results of the correlation, a polynomial of correction is estimated and applied to the whole image. Again, only one resampling step is carried out. Image to map correlation would have provided very biased results for this area, since there is much land not near coast. Image to image correlation however is also complex. For example, the season for which the image is acquired must also be considered, since the land cover changes rapidly in this part of the world, and hydrological features can also vary. The image chip database for correlation must therefore also be multi-seasonal ideally.

The GEOCOMP system (Robertson *et al.* 1992) which provides an automated, high throughput system for generating high precision AVHRR composite images, includes an even more sophisticated geometric correction module. It also uses correlation of image to map chips, and image to image chips, but some of its image chips are synthesised from spatially-degraded Landsat MSS and MOS-1 MESSR images. An overall r.m.s. error of better than 800 metres is often claimed with the use of 10 or more control points. The GEOCOMP system also incorporates the use of a coarse-scale digital terrain model.

The geometric correction process used for the Global 1 km Land Cover Project (see chapter by Belward), aims for positional accuracy of 1000m or less (Eidenshink and Faundeen 1995). It incorporates a BROLYD+TBUS model, control point matching and correction for terrain elevation. The control point matching is based is based on image to map correlation using map chips derived from the Digital Chart of the World (DCW) and World Vector Shoreline (WVS) databases, and the digital elevation data used for correction of the terrain effects is ETOPO5. No quantitative evaluation of the effect of the use of such coarse spatial resolution topographic data is yet available, though Eidenshink and Faundeen (1995) claim that registration errors of up to 12 km can occur for extremely off-nadir pixels in areas of high relief.

Note that in nearly all land-based applications of NOAA-AVHRR data nearest neighbour resampling is carried out. While preserving the digital values measured by the sensor itself, this method itself carries an inherent pixel locational error of half a pixel.

3.4   SATELLITE ATTITUDE ESTIMATION.

Even if the most sophisticated methodology is used for automatic renavigation of the imagery there may still be some residual locational errors. Some of these may originate from deviations from normal in satellite attitude. In fact relatively large displacement errors may occur for even quite small variations in satellite roll, pitch and yaw. Pitch errors provide a very similar displacement effect as timing errors, and satellite roll errors are very similar to errors that would occur with errors in the equatorial crossing longitude. Satellite yaw errors, however have a more rotational effect, and are much more difficult to compensate for.

Rosborough *et al.* (1994) provide a method of calculating the satellite attitude from a set of ground control points and also a way of improving the direct or indirect referencing once this is known. Puccinelli (1976) also provides methods for calculating the intersection points of the sensor view with the Earth's surface given the satellite's attitude (though note that Puccinelli uses a different and somewhat unconventional definition of the axes of roll, pitch and yaw). Both these corrections provide small improvements to the geometric correction of imagery, though they can also introduce errors if care is not taken in the careful identification of GCPs. In effect, the attitude corrections proposed by Rosborough *et al.* (1994) and by Puccinelli (1976) have a similar effect of correction as do the polynomials of correction as described above.

4.   Conclusions.

Any type of NOAA-AVHRR geometric correction used will to some extent be a trade-off between: desired levels of accuracy, the area of application, the nature of application, the required speed of correction and the facilities and supporting ancillary data sets available.

For small study areas, or for applications where precise geometric correction is not necessary, simple models may be satisfactory, but recent advances in the sophistication of complex orbital models, sophisticated image geometry models and increased availability of timely satellite ephemeris information have proved that it is also becoming increasingly possible to geometrically correct large NOAA AVHRR images with reasonable accuracy (to about 2-3 pixel accuracy throughout the image, though this accuracy may decrease significantly where very edge of images are used). It is unlikely that this level of fit can currently be improved either by satellite location model or image geometry model improvements alone. Therefore, for applications requiring better geometric correction fits, some form of image to map or image to image correction is necessary. The exact method employed will to some extent depend on the area under consideration and the required use of the data.

Indeed, it now appears that the limiting factors are not the *models* of satellite position and image geometry, but in the amount and reliability of ancillary information. For instance, an along-track shift between predicted and observed positions may be due to satellite time clock error, an error in equatorial longitude crossing time, an error in satellite pitch or satellite yaw or a combination of any of these. Similarly, an across-track error may be due to a combination of all or any of satellite roll, satellite yaw and right ascension of ascending node error. Until more information can be obtained for any of these, for example detailed satellite time clock errors, or predictive and validated models of satellite equator crossing times and longitude, or satellite attitude reports at the time of imaging, it is likely that (in the foreseeable future at least) an "all-encompassing" polynomial transformation, derived by either manual or automatic ground control pointing, will still have to be applied to attain image registration to the nearest pixel or sub-pixel.

5.   Acknowledgements

The authors would like to thank Vivienne Coleman for her splendid assistance in drawing many of the figures for this text

6.   Bibliography and References.

Achard, F. and D'Souza, G., 1994. Collection and Pre-Processing of NOAA-AVHRR 1 km resolution data for tropical forest resource assessment. TREES Series A: Technical Document No. 2, EU-JRC+ ESA, Report EUR 16055 EN, 58pp. Available from JRC.

Anderson, D.E., Angel, J.L. and Gorny, A.J., 1973. World Data Bank II: Content, Structure and Application, Office of Geographic and Cartographic Research, Central Intelligence Agency, Washington,D.C.

Borkowski, K.M.,1989. Accurate algorithms to transform geocentric to geofetic coordinates, *Bulletin Geodesique*, **63**, 50-56.

Brouwer, D., 1959. Solution to the problem of artificial satellite theory without drag, *Astronomical Journal*, **64**, 378-397.

Brunel, P. and Marsouin, A., 1987. An operational method using Argos orbital elements for navigation of AVHRR imagery, *Int. Journal of Remote Sensing*, **8**, 569-578.

Brush, R.J.H., 1982. A real-time retrieval system for images from polar orbiting satellites, PhD Thesis, University of Dundee, UK, 243pp.

Brush, R.J.H., 1985. A method for real-time navigation of AVHRR imagery, *IEEE Trans on Geoscience and Remote Sensing* , **GE-23, 6**, 876-887.

Brush, R.J.H., 1987. The navigation of AVHRR imagery, *Int. Journal of Remote Sensing*,9,1491-1502.

Eidenshink, J.C. and Faundeen, J.L., 1995. The 1 km AVHRR global land data set: the first stages in implementation, *Int. Journal of Remote Sensing*,17,77-98.

Emery, W.J. and Ikeda, M., 1984. A comparison of geometric correction methods of AVHRR imagery, *Canadian Journal of Remote Sensing*,10,46-56.

Emery, W.J., Brown, J. and Nowak, Z.P., 1989. AVHRR image navigation: summary and review, *Photogramm,. Eng. and Remote Sensing*, 55, 1175-1183.

ESA 1988. SHARP-1. Technical Specification of CCT format. ESA-Earthnet Program Office. Feb 1988.

Forrest, R.B.,1981. Simulation of orbital image-sensor geometry. *Photogramm,. Eng. and Remote Sensing*, 47, 1187

Fusco, L., Muirhead, K. and Tobiss, G., 1989: Earthnet's Coordinated Scheme for AVHRR Data. *International Journal of Remote Sensing*, 10, 625-636.

Ho, D. and Asem, A., 1986. NOAA AVHRR image referencing ,*Int. Journal of Remote Sensing*, 77,895-904.

Hoots, F.R. and Roehrich, R.L.,1980. Models for propogation of NORAD element sets. Project Space-Track Report No. 3. Aerospace Defence Command, Peterson AFB. CO.

Kidwell, K., 1991. NOAA polar orbiter data users guide. National Environmental Satellite Data, and Information Service, National Climate Data Center, Satellite Data Services Division, July 1991.

Kloster, E., 1989. Using TBUS orbital elements for AVHRR image gridding, *nt. Journal of Remote Sensing*, 10, 653-659.

Kloster, E. and Farrelly,B.A., 1984. Mapping the Fram strait ice edge using AVHRR imagery and NOAA satellite orbital data. In Procs of IGARSS'84 Symp, Strasbourg 27-30 Aug 1984 (ESA SP 215) 369-372.

Legeckis, R. and Pritchard, J., 1976. Algorithm for correcting the VHRR imagery for geometric distortions due to the Earth curvature, Earth rotation and spacecraft roll attitude errors. NOAA Tech Memorandum NESS 77, NESS, Washingtion, DC, USA.

Marsouin, A. and Brunel, P., 1991. Navigation of AVHRR images using ARGOS or TBUS bulletins, *Int. Journal of Remote Sensing*,12, 1575-1592.

Marsouin, A. and Brunel, P., 1992. Systematic navigation errors on NOAA-12 AVHRR images, *Int. Journal of Remote Sensing*,14, 171-176.

Nagle, F.W.,1986. A description of prediction errors associated with the TBUS-4 navigation message and a corrective procedure. NOAA Technical Memorandum NESDIS 16, edited by National Technical Information Service (NTIS), USDoC, US.

Price, J.C., 1991. Timing of NOAA afternoon passes, *Int. Journal of Remote Sensing*, 12, 193-199.

Puccinelli, E.F.,1976. Ground location of satellite scanner imagery,*Photogramm,. Eng. and Remote Sensing*, 42, 537-543.

Robertson, B., Erickson, A., Friedel, J., Guindon, B., Fisher, T., Brown, R., Teillet, P.M., D'Orio, M. Cihlar, J. and Sanz, A., 1992. GEOCOMP: A NOAA AVHRR data geocoding and compositing system.

Rosborough, G.W., Baldwin, D.G. and Emery, W.J., 1994. Precise AVHRR image navigation. Paper accompanying software , *IEEE Trans Geoscience and Remote Sensing*, 32, 644-657.

Roy, A.E., 1978. Orbital Motion. Adam Hilger Ltd. Bristol, UK

Sandford, T.D.G.,1992a. A review of image navigation methods for NOAA satellites,. Research Report CS 28 92. Bradford University, Dept of Computing, 18pp.

Sandford, T.D.G.,1992b. A comparison of orbital prediction algorithms for NOAA satellites,. Research Report CS 29 92. Bradford University, Dept of Computing, 18pp.

Sharman, M., LeLerre, A., Barnes, I., and Bierlaire, P., 1992. Software for Processing AVHRR data for the Communities of Europe (SPACE): algorithms, benchmarks and standards. JRC-Ispra, Italy.

Smart, W.M., 1953. *Celestial Mechanics*, John Wiley & Sons Inc., New York.

Sowter, A., 1993. Orbital mechanics for Remote Sensing, Eurocourse Notes, Jan 1993, 78pp.

CLOUD DETECTION USING AVHRR DATA

K.T. KRIEBEL
Deutsche Forschungsanstalt für Luft- und Raumfahrt,
Institut für Physik der Atmosphäre,
Postfach 11 16,
D-82230 Wessling,
Oberpfaffenhofen,
Germany

ABSTRACT. The split-window facility offered by the AVHRR instrument is used to good purpose in an advanced cloud detection and analysis package called APOLLO. This is based on threshold tests to distinguish between cloud-free, fully-cloudy and partially-cloudy pixels over land and sea surfaces. If necessary, a further test to distinguish between cloud, snow and ice is applied. The final result is a cloud mask which enables users to use only those pixels which are most appropriate for their desired application, *i.e.* fully-cloudy pixels for cloud products, cloud-free pixels for surface products and both partially-cloudy and fully-cloudy pixels for cloud cover amount estimation. In this chapter we present the design of the algorithms included in the scheme, and also discuss validation of the derived products which has been performed in several case studies and shows reasonable results.

1. Introduction

One of the most important procedures in the use of AVHRR data for land applications is to ensure that physical parameters or derived measurements (*e.g.* NDVI, surface temperature, *etc.*), are representative of the land surface itself, and not contaminated by cloud. For example, a small or thin cloud within the IFOV of a land surface pixel may significantly reduce its calculated surface temperature or vegetation index. Thus, for many of the land applications described in this book, some scheme for identifying and excluding pixels which are partially or fully-cloudy from the analysis have been adopted. On the other hand, some information on cloud properties themselves is frequently required for land applications, *e.g.* to estimate rain rate or amount of sunshine. Cloud properties related to these applications (*e.g.* cloud top temperature, or height) can be derived from fully-cloudy pixels. Thus, it is often important to distinguish three types of pixels: cloud-free pixels, partially-cloudy pixels and fully-cloudy pixels.

Cloud detection in satellite data relies essentially on the amount of spectral information given for each pixel. One broad channel in the visible region (like the one on the Meteosat sensor for example), would allow the satisfactory detection of bright clouds over the dark ocean (except in areas of sunglint) but not over bright arid land surfaces. One channel in the 10 to 12 μm region would allow the satisfactory detection of high clouds because they are colder than the surface, but would not be able to distinguish low clouds from land surfaces. Thus, a combination of both spectral channels (visible/near-infrared and thermal

195

G. D'Souza et al. (eds.), Advances in the Use of NOAA AVHRR Data for Land Applications, 195–210.

infrared) is frequently used with geostationary satellites such as Meteosat and GOES. The so-called bispectral technique examines a two-dimensional scattergram of reflectance versus temperature. From this scattergram, specific regions can be identified and labelled as surface, low cloud or high cloud, according to their position in the two-dimensional feature space. However, there remains a certain amount of uncertainty, often leading to a considerable amount of error in the products derived.

The AVHRR sensor was the first one to offer the possibility of two split-windows. That is, the visible/near infrared region is split into two channels, one below 0.7 μm and the other one above; and the thermal infrared region is also split into two channels, one around 11 μm and the other around 12 μm. Additionally, a fifth channel at 3.7 μm is included in the AVHRR system. In this spectral region, reflected solar and emitted thermal radiance are of similar magnitude, resulting in interesting but often difficult-to-interpret signals.

Such enhanced spectral information as offered by the AVHRR over previous sensors offers the possibility of checking each pixel separately according to its spectral properties, *i.e.* to apply a set of threshold tests to each pixel to determine the amount of cloud contained within it. The difficulty is in the setting of the thresholds for particular regions and for particular seasons, even when physical quantities such as radiances, reflectances and equivalent brightness temperatures to define the thresholds are used (instead of raw digital counts). A sophisticated set of threshold tests has been developed by Saunders and Kriebel (1988) in a package called APOLLO ('AVHRR Processing scheme Over cLouds, Land and Ocean') which is applied quasi-operationally at DLR, CMS Lannion in France (Derrien, *pers. comm.*), and at several other operational receiving stations around the world. The NOAA has also developed a sophisticated set of tests for cloud, collectively called CLAVR (Stowe *et al.*, 1991), and these have been adopted as the standard set of cloud detection tests for the Global 1 km Land Cover Project (see chapter by Belward). Together, the APOLLO and CLAVR methods form the most widely-used set of cloud detection algorithms used world-wide.

In this chapter, the different threshold tests used in the APOLLO scheme are described in some detail (Saunders and Kriebel, 1988; Kriebel *et al.*, 1989; Gesell, 1989) and some validation results are also included.

2. Threshold Tests

Figure 1 shows the mean spectral transmittance of a typical atmosphere together with the spectral ranges of the five AVHRR channels. Also indicated on the diagram are the main absorbing atmospheric gases in the different spectral channels.

The basic idea of cloud detection is based on the apparent contrast in the measured radiances of cloud-free and cloudy pixels. Clouds are bright in channels 1 and 2, *i.e.* they have high reflectance. In channels 4 and 5 they emit radiance according to their temperature (normally the temperatures of the viewed surfaces of cloud is very cold compared with sea or land surfaces) and their emissivity values (which for most clouds are close to unity). With the exception of low and thin clouds, clouds normally contrast

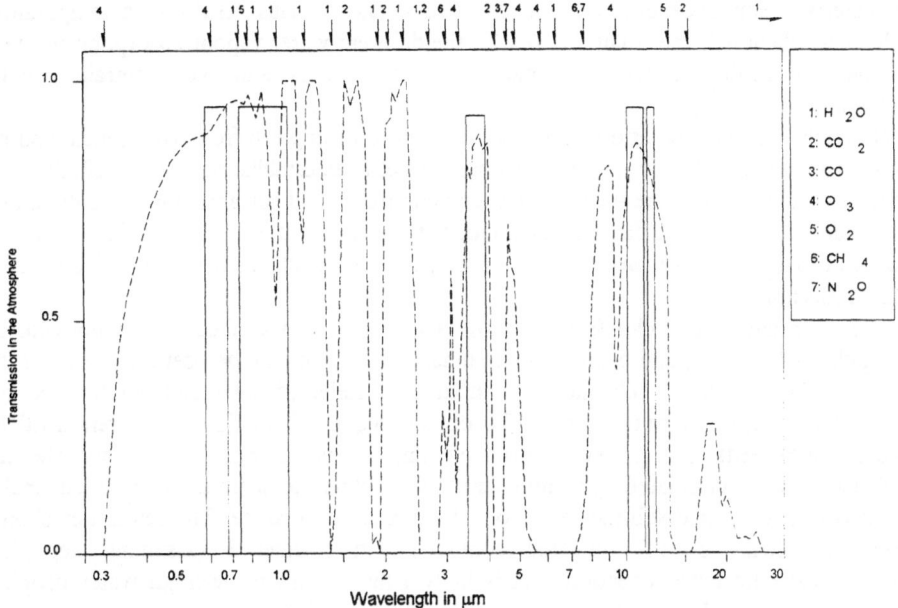

Figure 1. Mean transmittance of a typical atmosphere, spectral ranges of the five AVHRR channels, and the main absorbing atmospheric gases.

strongly to the relatively darker and warmer land and sea surfaces. The Earth's surface may be considered under four broad categories: ocean, vegetated land, arid land, and snow/ice.

- Ocean surfaces are dark in channels 1 and 2 and warm in channels 4 and 5 with an emissivity of 0.99.
- Vegetated land is dark in channel 1, bright in channel 2 and also usually quite warm with an emissivity around 0.96.
- Arid land is bright in channels 1 and 2 and also warm with emissivities usually less than 0.96 due to the influence of the reststrahlen bands in minerals and quartz sand.
- Snow and ice are normally very bright in channels 1 and 2 and relatively warm (in comparison to cloud temperatures) in channels 4 and 5 with an emissivity of around 0.99.

From this generalisation, it follows that during daytime, cold thick clouds can be easily distinguished from all kinds of other surfaces except snow and ice. However, warm thick clouds are difficult to detect over arid land and snow/ice, and thin transparent clouds are difficult to distinguish everywhere. Thin clouds themselves may be determined from their characteristic emittance difference in channels 4 and 5. This results in a temperature difference which is higher than that due to the different water vapour absorption in the 2 channels. In particular, thin ice clouds, *i.e.* cirrus clouds, and even contrails can be identified.

Normally, unvegetated arid land is very difficult to distinguish from low clouds, and no satisfactory universally-applicable procedure has yet been established. One possibility may be to take into account the usually high temperature of arid surfaces during daytime and to decide that a bright surface is not a cloud if the temperature is higher than about 15 degrees C. Another possibility may be to use channel 3 reflectance together with channel 1 and 2 reflectance.

The discrimination of clouds from snow and ice-covered surfaces is not possible by using channels 1, 2, 4, and 5 only, but channel 3 at 3.7 μm can be useful in deriving the phase of the medium. In channel 3, a mixture of reflected solar and emitted thermal radiance is measured during daytime. To obtain unique information, a separation of the two radiance parts is required. In principle this is easy because the emitted thermal radiance could be calculated by using channel 4 brightness temperature and the channel 3 sensitivity function and subtract this from the measured radiance. The remainder should then theoretically be the reflected component in channel 3 (assuming negligible bidirectional reflectance variations). It is known that at this wavelength water droplets reflect much more than ice particles (cirrus) and both reflect more than a snow covered surface. This behaviour is quite similar to that at 1.6 μm wavelength, so it will be equally applicable with ATSR bands.

The problem with this separation is the unknown emissivity in channel 3. This could be overcome by using measurements at night of the same surface to determine the emissivity, provided that it is the same during day and night. However, this is unlikely to be the case, partly due to changes in humidity. A simple approach to replace emittance by reflectance was proposed by Ruff and Gruber (1983). If transmittance is set to zero, emittance (= absorptance) is 1 - reflectance. The relation describing the measured radiance can now be solved for reflectance leaving temperature as the only remaining unknown. This is the same as that obtained from channel 4 provided that channel 4 emissivity is sufficiently well known, which is usually the case. Setting transmittance to zero is acceptable for thick clouds and all kinds of cloud-free surfaces but obviously not acceptable in the case of thin clouds. However, no more acceptable technique is currently known to derive 3.7 μm reflectance.

Based on the preceding discussion, five threshold tests have been defined to determine whether each pixel is cloud-free or not (Figure 2). The philosophy is that a pixel is labelled as cloud-free only if none of the tests finds cloud contamination.

Two tests use simple brightness (*i.e.* reflectance) and temperature thresholds. Two other tests use the split-window information supplied from AVHRR: the ratio of channel 2 to channel 1; and channel 4 minus channel 5 brightness temperature. The other test is a "spatial homogeneity test" performed with channel 4 brightness temperature data. These are described in some detail below. Separate tests have to be made according to whether night-time or daytime images are being analysed, and whether the pixel being processed is over land or sea.

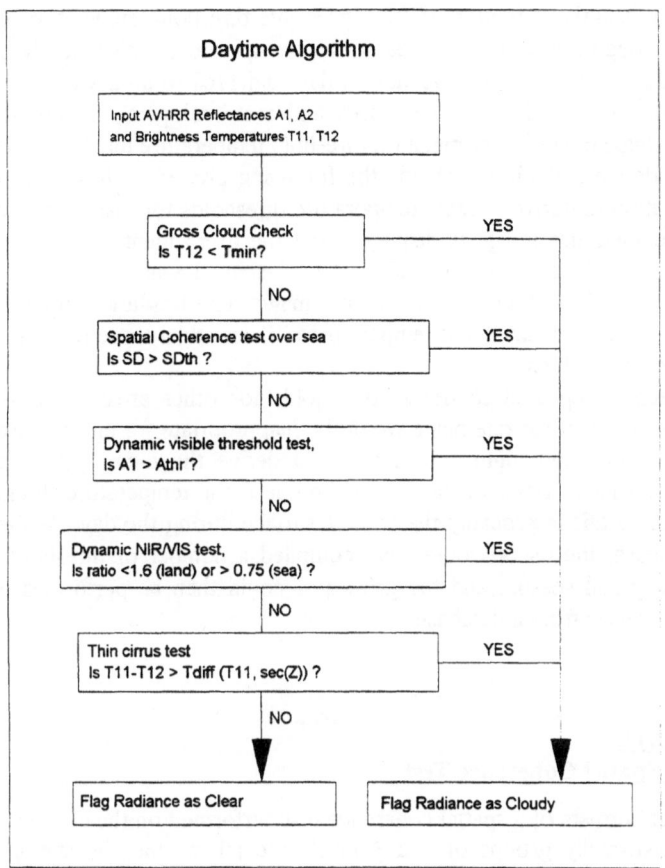

Figure 2. Daytime cloud detection algorithm implemented in APOLLO.

Test 1:
Daytime or Night-time over Land or Water
Gross Cloud Check (Gross temperature test):

The first test, known as the gross cloud check, is a simple thermal infrared threshold test using the brightness temperature calculated from channel 5 (or channel 4 if it is not available), as a check on cloud contamination. The channel 5 brightness temperature is used in preference to others because clouds are thought to have a greater optical depth at these wavelengths. If the brightness temperature is below a certain threshold temperature, the pixel is flagged as being cloud-contaminated.

The key point is in the definition of the appropriate threshold temperature. Over the sea it is relatively straightforward as the sea surface temperature varies slowly in space and time. Over land, however, the large day-to-day and area-to-area variability in surface temperatures due to the different land uses and meteorological conditions makes the definition of a single overall discrimination threshold temperature much more difficult

In order to derive suitable thresholds the following procedure has been used in new areas of application to derive suitable temperature thresholds for that area. A channel 4 or 5 brightness temperature image is displayed and the users identify visually (and draws polygons around or places the cursor on) land and sea areas which they see were likely to be the coldest, yet cloud-free surfaces in the image. The brightness temperatures over these areas are then determined and temperatures 2K less than these measured values are adopted as threshold values.

This approach is repeated to derive thresholds for other areas, as it is likely that threshold values derived for one place would be inappropriate for another area. It is also repeated for daytime and night-time images, to derive the most appropriate, suitable thresholds for particular times of day. Over the coast, the temperature threshold for the sea can be used as this is probably the coldest surface during the day. At the end of the "calibration" phase, the users would have compiled a number of thresholds relating to area, time of day and season, and the gross test could then be performed by accessing these threshold values from a database.

Test 2:
Daytime (Water):
Temperature Spatial Coherence Test

The second test consists of a spatial coherence test performed on the channel 4 brightness temperatures. Normally groups of 3 x 3 pixels are taken, and the average channel 4 brightness temperature and variance of those pixels which have not been labelled as cloudy by the previous test is calculated. Over cloud-free water surfaces, the variance will of course be very low, so if the variance exceeds a low value, the pixel will be labelled as cloudy (and will, therefore, fail this test). However, the test is not applicable for land pixels during daytime, nor for coast pixels, where the contrast between land and water pixels will be too great.

Test 3:
Daytime
Dynamic Visible Threshold Test:

In older implementations of APOLLO, single reflectance thresholds were set for determination of cloudy pixels. *e.g.* any pixels over 0.3-0.15 % reflectance were classified as cloudy ones. However, because of marked variations within images, more refined methods that determined variable thresholds based on surface/cloud histograms were developed. These techniques came to be known as dynamic reflectance threshold tests, and they use the reflectances calculated for channel 1 and 2. Surface and cloud populations can normally be distinguished in reflectance histograms of about 32 x 32 pixel arrays, allowing a suitable threshold to be set at a relatively high reflectance. All pixels with reflectances above this threshold may then be assumed to be cloud-contaminated. Identifying a cloud-free peak in this way and then setting a threshold value removes uncertainties due to the variation in calibration and bi-directional reflectance variations due to different solar and satellite angles.

The following taken from Saunders (1988) describes how the dynamic reflectance threshold technique works:

For the chosen 32 x 32 box, "First the number of radiances that make up the histogram peak value and the corresponding radiance I_{pk} are determined and if the peak is significant (*i.e.* contains more than 0.5% of the total population) the process continues. The lower I_1 and the upper I_2 limits of the histogram are then determined. If the low radiance end I_1 of the histogram is within a reflectance m (0.025 for water, 0.035 for land) of the histogram peak and the peak radiance is less than over a typical cloudy scene I_{max} (0.1 for water and 0.15 for land) then the peak is assumed cloud-free and the visible threshold T is set at a reflectance n (0.012 for water, 0.017 for land) above I_{pk}".

Having obtained the appropriate threshold value in this manner, every pixel within 32 x 32 boxes is assessed, unless it is flagged as having sunglint or low sun angles. It is deemed cloudy if its reflectance exceeds that of the determined threshold. Over sea, channel 2 reflectances should be used as they are less sensitive to aerosol and molecular effects, and absorbed more by water. Over land, however, channel 1 reflectances should be used since the reflectances of land surfaces in this channel are much lower (in general) than channel 2, which should increase the contrast between land and cloud. Over coastal areas, identification of a cloud-free peak is more difficult, so a fixed reflectance could be used in this case (say 15%).

The arrays may be calculated from adjacent or moving filters (*i.e.* overlapping 32 x 32 areas) to avoid possible discontinuities, though this would of course be much more computationally demanding. Clearly, this test cannot be used with night-time images, since the reflectances in these data will be zero.

Test 4:
Daytime Test:
Dynamic VIS/NIR ratio test

The fourth test makes use of the ratio of channel 2 to channel 1 reflectances. This ratio (Ch2/Ch1) should be close to 1.0 for clouds, as the reflectance of clouds only decreases slightly at near-infrared wavelengths, and anisotropic effects are similar in both channels and hence should cancel each other out. Over cloud-free water, however, enhanced backscattering at the shorter wavelengths due to molecular and aerosol scattering causes visible reflectances to be often twice those in the near-infrared (outside sunglint), giving values of about 0.5 for the ratio. Over land with healthy green vegetation, the reflectance increases markedly at the near-infrared wavelengths compared with the shorter ones. Even over desert or during the winter when the vegetation is dormant, the reflectance is higher at longer wavelengths (except over snow and ice), ensuring that the ratio value is always greater than 1.0 for the land surfaces.

Thus, thresholds could be specified that classify a pixel as cloud-free only if the ratio is less than 1.6 (over land) or greater than 0.75 (over sea). Again, this test will only work with daytime imagery, and only for areas not affected by sunglint or not at extremely low sun angles.

Test 5:
Daytime and Night-time:
Split-window difference: (Thin Cirrus Test)

The fifth test proposed makes use of the temperature difference between calibrated channels 4 and 5, so it can only be applied to sensors where both channels are present. The temperature difference can be used to detect most types of clouds including semi-transparent cirrus ones because of the different emissivities of cloud at the two wavelengths. The major possible exception of detectable cloud by this test is that of uniform low cloud. For most cloud-free pixels, the maximum temperature difference between channels 4 and 5 which would be expected to be due to the temperature dependence of the Planck function being only about one degree K. However, it has been shown that brightness temperature differences of up to six degrees K (for channel 4 minus channel 5 brightness temperatures) can be obtained over cloud due to the optical properties of thin cloud being very different at the two wavelengths. For clear sky radiances the differences are often less than 1K but the value will vary with total column water amount and the satellite zenith angle.

Saunders and Kriebel (1988) used a set of pre-computed clear sky AVHRR brightness temperatures to show the expected dependence of the temperature difference on the cosine of the satellite zenith angle and the channel 4 brightness temperature. Computed temperature difference values ($T_{11}-T_{12}$) were plotted against channel 4 brightness temperatures (T_{11}) and the inverse of the cosine of the satellite zenith angle ($\sec\vartheta$) for 117 different tropical and mid-latitude maritime atmospheres selected from a data set assembled by NOAA. For values of T_{11} between 260 and 310K the maximum computed clear sky $T_{11}-T_{12}$ values were noted for these atmospheres for a range of $\sec\vartheta$ values, and threshold values at least 0.25 K higher than any of the computed clear sky $T_{11}-T_{12}$ values were compiled in a look-up table An extract of this look-up table is given below, and the whole table is provided in the Saunders and Kriebel (1988). All pixels with a $T_{11}-T_{12}>$ (threshold value) are then identified as cloud-contaminated, and fail this test.

Although the derivation of the look-up table thresholds is made from tropical and mid-latitude maritime atmospheres, the tests have nevertheless been shown to be effective in detecting thin cirrus cloud and the edges of thicker cloud too, in both daytime and night-time imagery, in a range of other locations. Obviously this test cannot be applied to those data where only one of the NOAA thermal channels (4) is available (for example, NOAA 10).

Table 1. Temperature thresholds in degrees K for the (T_{11}-T_{12}) cloud detection.

T_{11}(K)	secϑ values				
	1.0	1.25	1.50	1.75	2.0
260	0.55	0.60	0.65	0.90	1.10
270	0.58	0.63	0.81	1.03	1.13
280	1.30	1.61	1.88	2.14	2.30
290	3.06	3.72	3.95	4.27	4.73
300	5.77	6.92	7.00	7.42	8.43
310	9.41	10.74	11.03	11.60	13.39
320	14.01	16.02	16.06	16.80	19.61

Figure 3 shows the algorithm used to separate partially-cloudy pixels from fully-cloudy ones. Cloud-filled pixels are determined by applying the ratio test (channel 2 to channel 1 reflectance) and the spatial coherence test a second time but only to those pixels which are not totally cloud-free, i.e. to those pixels which have failed at least one of the previous five tests described. The ratio test looks for a peak around 0.9 in the histogram and flags all pixels within a small distance from this peak as fully-cloudy. The spatial coherence test then uses a relaxed threshold to distinguish homogeneous clouds from broken cloud fields over both ocean and land surfaces. If snow and ice may occur, a further test is applied which is based on the 3.7 μm reflectance, but combined with other channels to distinguish clouds from snow (Gesell, 1989). The result is stored in a cloud mask which is a 32 bit integer file. Herewith the user can preselect any meaningful combination of test results to obtain the appropriate subgroups of pixels. For night-time images, the ratio test cannot be used, and so a combination of the spatial coherence test and the channel 3, channel 4 and channel 5 differences are used (see Figure 3).

For the detection of partially-cloudy pixels in night-time imagery, a simpler scheme to the one shown in Figure 2 has to be applied, and this is summarised in Figure 4. Channels 1 and 2 reflectances are of course zero, and channel 3 measures emitted radiance only which can be converted reliably to brightness temperatures, just as for channels 4 and 5. As well as the spatial coherence and gross cloud checks, differences between the three spectral channels are also used to detect all possible cloud types (see Saunders and Kriebel, 1988 and Figure 4). If the channel 4 minus channel 3 brightness temperature difference is more than 1K, low cloud or fog can be assumed. If the channel 3 minus channel 5 brightness temperature difference is more than 1.5K, thin clouds or a partially filled field-of view can be assumed. The spatial coherence test is applied over ocean and land with a relaxed threshold of 1K. The same test is also used to determine fully-cloudy pixels. This means that if the variance is below 1K, the test does not flag the pixel as cloudy, but if another test finds the pixel to be cloudy the spatial coherence test flags the

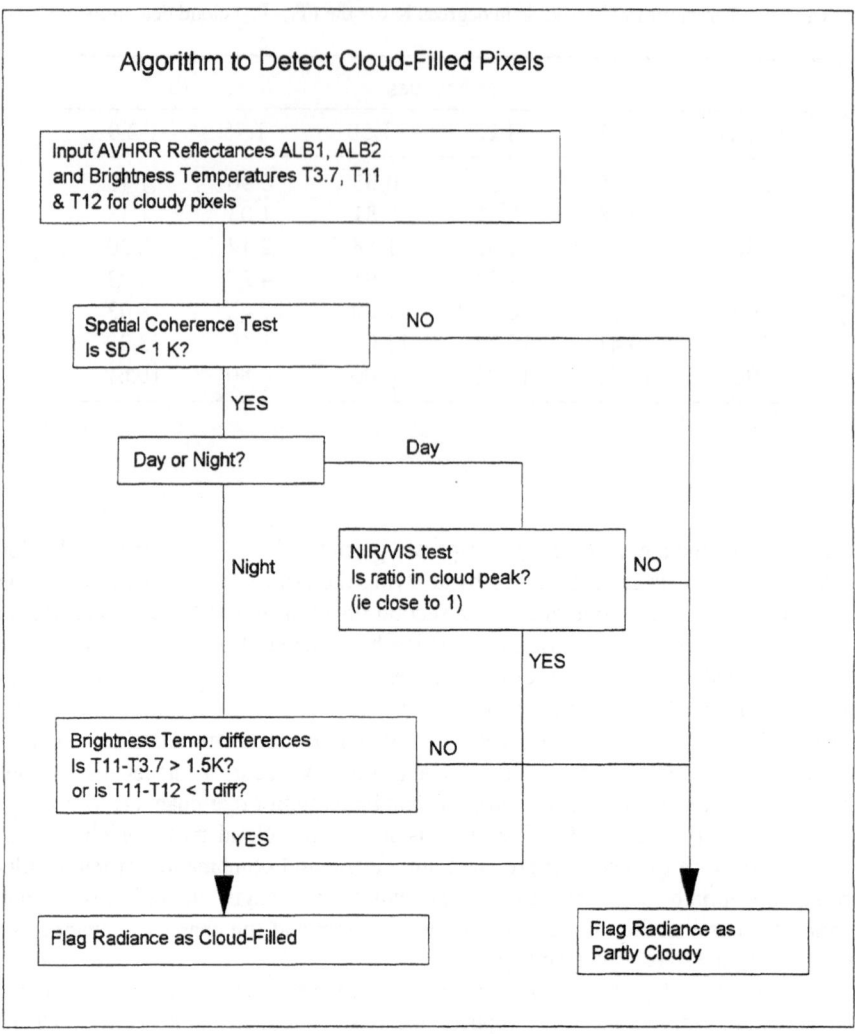

Figure 3. Algorithm for the determination of fully-cloudy from partially-cloudy pixels.

pixel as fully-cloudy. Fully-cloudy pixels are also assumed if either channel 4 minus channel 5 temperature difference is below 1K or channel 4 minus channel 3 temperature difference is larger than 1.5K.

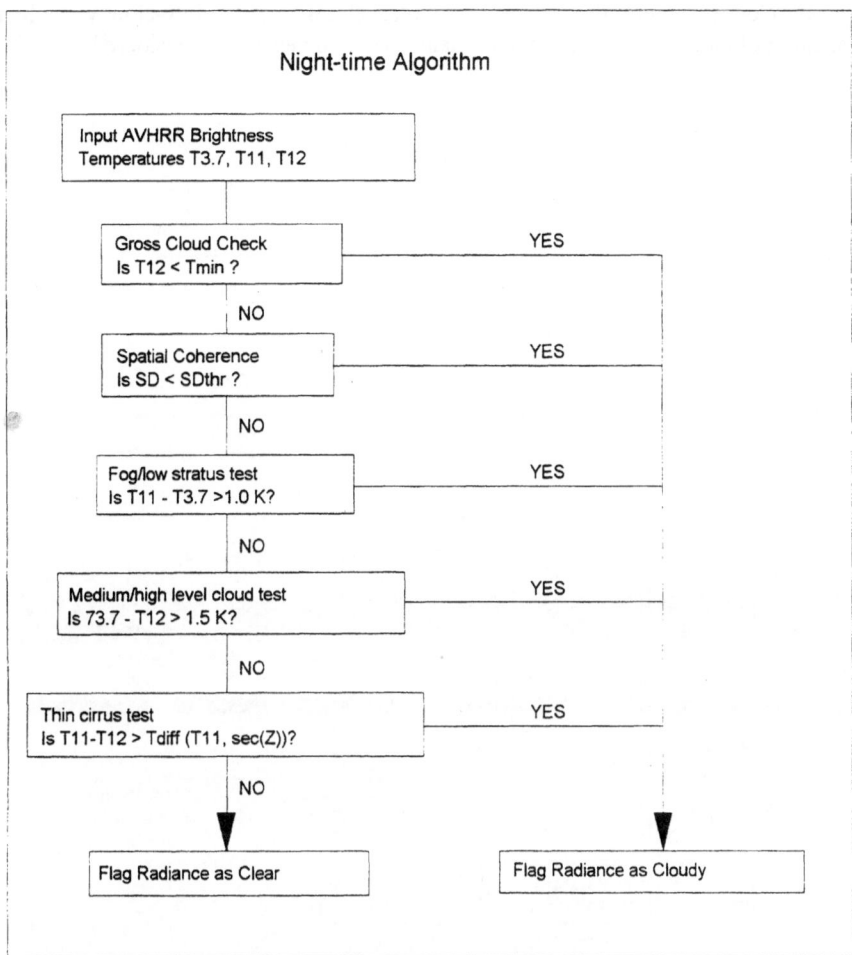

Figure 4. Night-time cloud detection algorithm.

3. Cloud Products

In Figure 5, the whole APOLLO scheme is shown. At DFVLR, a standard input routine, the SHARK package developed by ESA/ESRIN is used. This package reads HRPT data and produces calibrated and georeferenced (navigated) image files (see chapter by Arino) which in turn are used by APOLLO. The scheme shown in Figure 5 has been described from the start of the cloud mask detection. This is considered as the standard procedure which every pixel should undergo because all algorithms that derive information on surfaces or clouds require that a pixel is either cloud-free or fully-cloudy. Only the determination of cloud cover needs the partially-cloudy pixels to be considered as well.

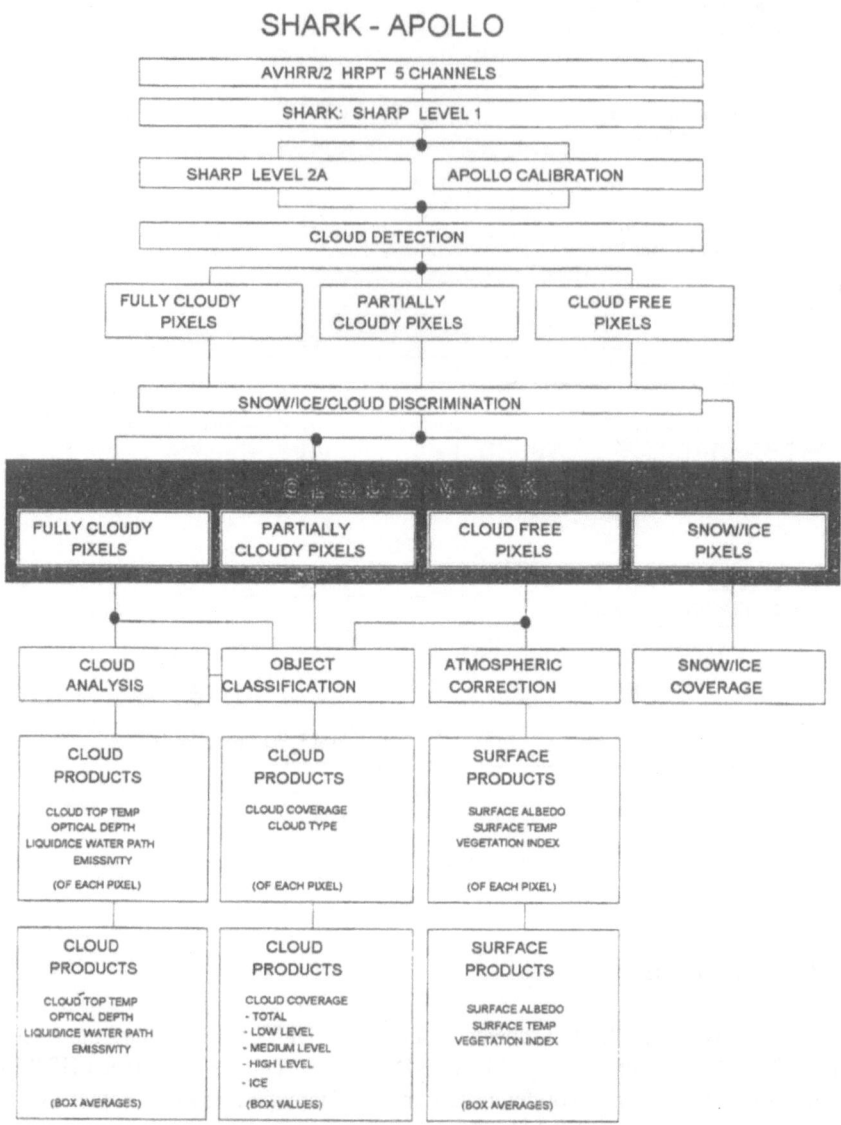

Figure 5. Flow chart of the SHARK/APOLLO package implemented at DFVLR.

Within APOLLO, only a few application algorithms have been included, but several other sophisticated application algorithms can be incorporated easily. Surface products such as albedo, NDVI and SST can also be derived. In APOLLO the main emphasis to date has been on the detection of cloud and the identification of various cloud properties. First, all fully-cloudy pixels are classified into 4 groups called low, medium and high clouds according to their channel 4 cloud top temperature, and thin clouds according to a low albedo together with a higher temperature than that of thick clouds. Cloud cover of partially-cloudy pixels is determined by using a linear interpolation between their cloud-free and their fully-cloudy parts. The latter information is taken from the nearest neighbours which are cloud-free or fully-cloudy, thus relying on the assumption of horizontal homogeneity. Cloud classification of the partially-cloudy pixels is made by assigning them to that type which occurs most frequently in a preset neighbourhood.

Channel 1 reflectance is used to obtain cloud reflectance (see Kriebel et al., 1989). In the case of water clouds, a parameterisation by Stephens (1978) is applied to obtain cloud optical depth, liquid water path and emissivity. With ice clouds, optical depth is obtained by applying the same scheme together with a correction factor to account for different reflectance of ice clouds compared with water clouds of the same optical depth (Platt et al., 1980). Ice water path and emissivity are derived from cloud reflectance by using parameterised relations given by Starr and Cox (1985). Cloud top temperature is derived from channel 4 temperature according to a scheme developed by Saunders (1988).

4.   Validation of Cloud Products

When the accuracy of the derived cloud products is determined by independent measurements, error estimates due to the parameterisation schemes used yield a root mean square error in the order of 30%. Validation would require measurements of the same quantity in the same volume of air at the same time. This can only be achieved approximately. One of the main reasons to use channel 1 reflectance to obtain optical depth is the absence of water vapour absorption in its spectral sensitivity region. Discounting the absorption in the cloud condensation nuclei, the photons entering the cloud from above are either reflected or transmitted. Usually, cloud reflectance is composed of about 60% of the incoming irradiance and about 40% is diffusely transmitted. Because both quantities are of comparable size, one can assume that the photons which leave the cloud into the upward direction have interacted with cloud droplets in all cloud layers, not only in the top layers. This means that the upward radiance carries information on the optical density from all layers of the cloud but the information content is not evenly distributed. Nevertheless, it does not seem too unrealistic to assume that the liquid water path derived from channel 1 cloud reflectance is connected to the vertical integral over the cloud liquid water content as, for example, derived from in-situ airborne measurements.

Another problem is the difference in the measured volume. The AVHRR data comprise about 1 km^2 in the horizontal plane or even more. The aircraft data can only give one or more slant paths through the cloud from cloud top to cloud bottom or vice versa. Even if this is repeated as quickly as possible, the data never comprise the whole km^2 but just a few slant lines within this km^2. An assumption on cloud homogeneity is unavoidable. Cloud homogeneity can be checked by a horizontal flight through the considered cloud. The limiting factor of such in-situ measurements is the time difference between satellite overpass and airborne measurement. How long is a cloud constant in time? This certainly depends on cloud type, stratiform clouds changing less quickly than convective clouds. Due to such considerations, the validation is restricted to stratiform clouds and the time difference is aimed to be less than 15 minutes. Because it is very difficult to conduct an aircraft equipment at the correct cloud at the correct time, the cases where a useful validation could be made are rare. Presently, we have just three validations of the liquid water path in stratiform clouds, three more have just been recorded and are not yet evaluated. There are also three validations of the optical depth of thin cirrus clouds by means of an airborne lidar. Two have been made with an upward looking lidar which requires the optical depth below the aircraft to be obtained from other sources. The third has been performed by means of a downward looking lidar from above the cirrus. The derivation of the optical depth from lidar backscatter measurements requires some calibration of the lidar. This has been described by Ruppersberg et al. (1992).

The results are shown in Table 2 and Figure 6. Table 2 summarises the three validation results of the stratiform cloud cases. The first two were obtained with the Johnson-Williams hot-wire probe and the third with PMS probes. The different sign of the difference between airborne and satellite data seems to reflect the different sensitivity of the two instrument types for small cloud droplets in particular and, more generally, the accuracy limit of such airborne measurements which is in the order of 30%. Therefore, the agreement is quite acceptable, but the statistics are by far insufficient. Figure 6 shows a comparison of optical depth data obtained from a downward looking airborne lidar with simultaneous satellite derived data. The vertical extension of the lidar data band reflects the lidar error estimate. The satellite derived data are based on the processing of fully-cloudy pixels. Whenever a pixel is not fully-cloudy, the optical depth is set to zero, as can be seen in Figure 6 from 13:33 to 13:34 UTC. The agreement is also quite good.

Table 2. Validation results of the liquid water path in gm^{-2}. The aircraft data are higher than the satellite data in the case of the Johnson-Williams probe, but lower in the case of the PMS probe.

Aircraft data	Case 1. 19.1.87 Johnson-Williams 155	Case 2. 19.1.87 Johnson-Williams 103	Case 3. 24.5.88 PMS (FSSP/OAP) 8.4
APOLLO	95	69	14.6
ΔLWP LWP$_{aircraft}$	0.39	0.33	-0.74
ΔLWP LWP$_{aircraft}$	-0.63	-0.49	0.42

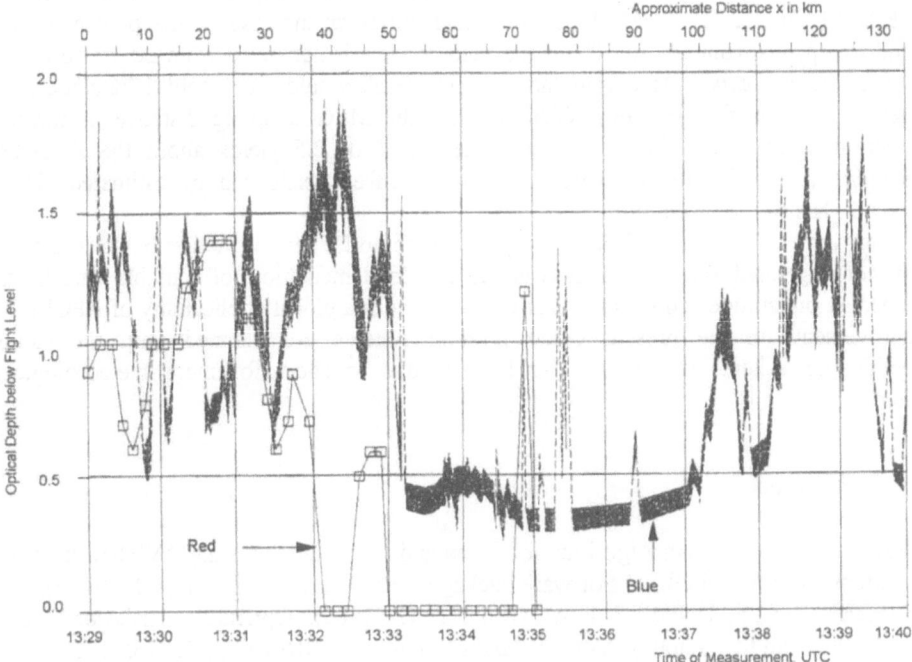

Figure 6. Comparison of optical depths derived from satellite data with APOLLO (marked as red and indicated by —□—) with those derived from airborne lidar data (marked as blue and indicated by ----). The vertical width of the lidar data indicated the range of uncertainty. Zero optical depth in the satellite data is due to partially-cloudy pixels.

5.   Outlook

So far it has been demonstrated that AVHRR data are well suited to allow for a good quality cloud detection which may form the basis for exclusion of cloudy pixels for analysis of land applications, or alternatively may form the basis for cloud classification and cloud analysis. The latter yield results with uncertainties in the 30% range which seems to be quite acceptable considering all the simplifications and assumptions which have been necessary. Future satellite systems which are designed to measure surface and cloud properties should provide at least the spectral information of AVHRR and a spatial resolution comparable with that of the AVHRR. The future Meteosat system, Meteosat Second Generation (MSG), is designed with a sampling distance (spatial resolution) of 3 km. This seems to be at the lower limit of spatial resolution which still allows for the above mentioned philosophy to neglect partially-cloudy pixels and to derive products from cloud-free and fully-cloudy pixels only. The same is true for AVHRR GAC data which are

1 by 4 km² averages expanded to the GAC pixel size. This is supported by a recent study of METEO-France for EUMETSAT which identifies an increase of the percentage of cloud-free pixels from 5% to 10% if the field-of-view is narrowed from 20 km to 8 km (Derrien, *pers. comm.*). An extrapolation of this result would give about 20% cloud-free pixels with 3 km field-of-view which is about the MSG sampling distance. A further reduction of the field-of-view by the same factor of 2.5 yields about the AVHRR resolution and a fraction of 30% to 40% cloud-free pixels can be estimated. This, however, has still to be confirmed.

Improvements in the cloud detection package comprise the distinction between clouds and non-vegetated (arid) land as well as the implementation of variable thresholds depending on latitude and season. This would result in a global applicability of APOLLO. Improvements in the present cloud product scheme are planned with the cloud classification scheme which will also be extended to allow for more meteorological criteria.

6. References

GESELL, G., (1989). An Algorithm for Snow and Ice Detection using AVHRR data: An Extension to the APOLLO Software Package. *Int. J. Remote Sensing*, **10**, 897-905.

KRIEBEL, K.T., SAUNDERS, R.W., GESELL, G. (1989). Optical Properties of Clouds Derived from Fully-cloudy AVHRR Pixels. *Beitr. Phys. Atmosph.*, **62**, 165-171.

PLATT, C.M.R., REYNOLDS, D.W., ABSHIRE, N.L., (1980). Satellite and Lidar Observations of the Albedo, Emittance and Optical Depth of Cirrus Compared to Model Calculations. *Mon. Wea. Rev.*, **108**, 195-204.

RUPPERSBERG, G.H., RENGER, W., (1992). Shadow technique for improved inversion of lidar data to cirrus and contrail optical depth. ESA-IRS, Technical Translation 1263, Frascati, 20 pp.

RUFF, I., GRUBER, A., (1983). Multispectral Identification of Clouds and Earth Surfaces Using AVHRR Radiometer Data. Preprints 5. Conf. Atm. Radiation, October 31 - November 4, 1983, Baltimore, Md., AMS, Ed., Boston, Mass., USA.

SAUNDERS, R.W., (1988). Cloud Top Temperature/Height: A High Resolution Image Product from AVHRR. *Meteorol. Magazine*, **117**, 211-221.

SAUNDERS, R.W., KRIEBEL, K.T., (1988). An Improved Method for Detecting Clear Sky and Cloudy Radiances from AVHRR Data. *Int. J. Remote Sensing*, **9**, No.1, 123-150.

STARR, D.O'C., COX, S.K., (1985). Cirrus Clouds, Part I: A Cirrus Cloud Model. *J. Atmos. Sci.*, **42**, 2663-2681.

STEPHENS, G.L., (1978). Radiation Profiles in Extended Water Clouds II: Parameterization Schemes. *J. Atmos. Sci.*, **35**, 2123-2132.

STOWE, L.L., McCLAIN, E.P., CAREY, R., PELLEGRINO, P., GUTMAN, G.G., DAVIS, P., LONG, C., HART, S., (1991). Global Distribution of Cloud Cover Derived from NOAA/AVHRR Operational Satellite Data. *Adv. Space Res.*, **11**, (3)51-(3)54.

RECENT ADVANCES IN ALGORITHM DEVELOPMENT TO EXTRACT INFORMATION FROM AVHRR DATA

M. M. VERSTRAETE and S. FLASSE[‡]
The European Commission's Joint Research Centre,
Institute for Remote Sensing Applications,
I - 21020 Ispra (VA),
Italy.

ABSTRACT. Recent advances in the use of remote sensing data to monitor terrestrial environments are presented and the rationale for vegetation indices and the issues related to their proper interpretation are reviewed. A method to evaluate these vegetation indices is proposed. Physical models describing the bidirectional reflectance of natural surfaces are introduced and the role and importance of inversion methods are underlined. An empirical reflectance model is used in the analysis of actual AVHRR data, and the results are discussed. The paper ends with a discussion of the needs and priorities for further research and applications in this area.

1.   Introduction

Satellite remote sensing technologies can provide large amounts of data on the state and evolution of the terrestrial environments. The usefulness of these radiative data, collected in space to address concrete problems at the surface of the Earth entirely, depends on our capability to interpret them and to extract useful information on the processes of interest. In this paper, we will discuss the interpretability of the data in the context of remote sensing in the optical domain, although similar issues arise in other spectral bands.

Various methods have been proposed over the past few decades to analyse optical remote sensing data, and these are clearly designed for specific applications. For instance, mapping requires a high spatial resolution and can be done on the basis of a single image. Other applications, including all those which can be classified as event detection, require frequent data sets at a spatial resolution which must be adequate for the event to be identified, and are based on differences between successive measured values. Another large class of applications is related to the monitoring of one or more aspects of vegetation cover and, here, the identification of the presence, amount and possibly other properties of plant canopies is of direct interest. In this case, vegetation indices are often used.

A complementary approach to investigate terrestrial surfaces consists of developing mathematical models that describe explicitly the transfer of radiation in the atmosphere and its interaction with the Earth's surface. To the extent these models are based on

---

[‡] Current address: Natural Resources Institute, Central Avenue, Chatham Maritime, Kent ME4 4TB, United Kingdom.

G. D'Souza et al. (eds.), Advances in the Use of NOAA AVHRR Data for Land Applications, 211–229.

fundamental physical theories, they can help explain the nature of the processes responsible for the measurements, and thereby provide a detailed and reliable insight into the nature and structure of the surface.

Recent advances in the understanding of the transfer of radiation in the atmosphere and the surface environments have permitted new tools to be developed. Specifically, new vegetation indices have been proposed to address the limitations of existing indices, and improved reflectance models are now available to investigate more directly the physical processes that condition the measured signals. Some of these advanced methods are discussed below, and the opportunities arising from the new satellite sensors now under development are outlined.

2.   Vegetation indices

Vegetation indices constitute a subset of a more general class of spectral indices, which can be designed for any number of applications. In this paper, we will focus on the quantitative estimation of the properties of the vegetation cover over terrestrial areas, using remote sensing data acquired in the solar spectral range. Many other types of targets and spectral bands could be considered, but these fall outside the scope of this discussion.

Green living vegetation grows through the assimilation of atmospheric carbon dioxide in leaves - a biochemical process known as photosynthesis. This process relies on specific pigments, in particular chlorophyll, to absorb solar radiation. This radiative energy is then used to synthesize organic molecules which form the basis of all plant and animal life. However, not all solar radiation is useful for this process. In particular, the energy associated with electromagnetic waves whose wavelength exceeds about 700 nm is not sufficient to drive biochemical reactions. The structure of plant leaves is well adapted to this physical constraint: chlorophyll molecules absorb very well in the visible wavelengths, especially in the blue and red, but the bulk of the radiation beyond this threshold of 700 nm is simply scattered by the cell walls and does not contribute to the heating of the plant (*e.g.*, Gates, 1980). The spectral reflectance profile of a typical leaf is shown in Figure 1, together with representative spectra for a few other natural surfaces, such as soils, water, snow, and clouds. Since only live green plants exhibit a strong spectral gradient around 700 nm, this feature (signature) can be exploited to detect the presence of vegetation. Such an approach requires only two spectral measurements, on either side of this threshold.

The spaceborne sensor most often used to repetitively monitor vegetation over large regions is the Advanced Very High Resolution Radiometer (AVHRR). Various versions of these instruments have been flown on the US National Atmospheric and Oceanic Administration (NOAA) series of platforms over the last ten years, so that long time series of data are available, at least at the low spatial resolution known as Global Area Coverage (GAC). Table 1 summarizes the main characteristics of this sensor. It can be seen that the first two channels sample the spectral reflectance on both sides of the 700 nm threshold mentioned above, and this feature provides a unique opportunity to investigate vegetation processes, since the latter exhibits a much stronger reflectance gradient at that point than

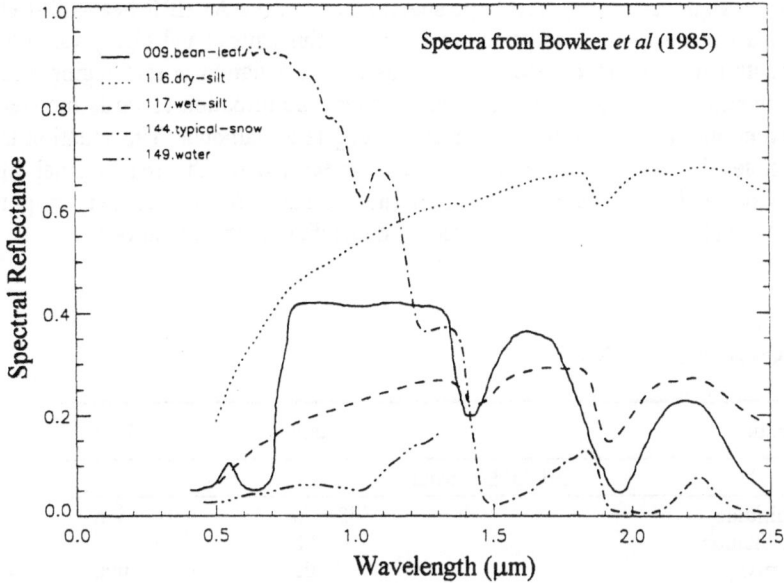

Figure 1. Spectral profiles of typical natural surfaces.

other natural surfaces. The strong spectral differences between the reflectance of green living vegetation and all other surfaces, combined with the technological opportunity to acquire spectral data on either side of this threshold resulted in the creation of simple mathematical expressions known as 'vegetation indices'. These empirical formulae aim enhance this spectral contrast, and are designed so that higher index values correspond to a higher probability that the target being observed contains green live vegetation.

Various indices have been proposed in the literature, starting with simple expressions such as $\rho_{nir} - \rho_{red}$ or ρ_{nir}/ρ_{red} (Pearson and Miller, 1972). The most widely used index, however, is the Normalized Difference Vegetation Index (Rouse *et al.* 1974), defined as:

$$\text{NDVI} = \frac{\rho_{nir} - \rho_{red}}{\rho_{nir} + \rho_{red}} \tag{1}$$

where: ρ_{nir} and ρ_{red} are the observed reflectances in the red and near-infrared channels, respectively.

Of course, other surface types may exhibit somewhat similar spectral responses. It can be seen from Figure 1 that a wet soil has a larger spectral reflectance gradient in that region than a dry one. A change in soil moisture may therefore be mistaken for an increase in vegetation (*e.g.* Huete *et al.* 1985). Similarly, the atmosphere, whose effects are always present in space observations, differentially influences the spectral measurements in complex ways (*eg.*, Kaufman, 1989). The large spectral band of the AVHRR near-infrared channel includes a water vapour absorption band, so that this natural and highly variable atmospheric constituent tends to reduce the transmission of radiation in that region and the reflectance measured in space. Hence, an atmospheric perturbation towards moister conditions could be mistaken for a decrease in green vegetation amount. The situation is even more complex in the case of aerosols: these affect mostly the red channel of AVHRR, but may result in a net attenuation of the measured reflectance over bright surfaces or in a net enhancement over dark surfaces, depending on the conditions.

Table 1. Overview of AVHRR

Characteristic	Value	Unit
Orbital Elements		
Nominal altitude	833-870	km
Orbital inclination	98.8	°
Orbital period	102	min
Equatorial crossing time (odd)	2:30(D)-14:30(A)	hr (LT)
Equatorial crossing time (even)	7:30(D)-19:30(A)	hr (LT)
Repeat period	9.2	day
Revisit capability	1	day
Instrument geometry		
Total scan angle	0 to ±55.4°	
View zenith angle	0 to ±68.9°	
Swath width	2700	km
Time range on scan line at Eq.	1:37	hr:min
Pixels per scan line	2048	
Instantaneous field of view	1.39-1.51	mrad
Ground resolution (nadir)	1.1 x 1.1	km
Ground resolution (edge)	2.4 x 6.9	km
Radiometric resolution	10	bit
Spectral definition		
Band 1	0.58-0.68	μm
Band 2	0.72-1.10	μm
Band 3	3.55-3.93	μm
Band 4	10.3-11.3	μm
Band 5	11.5-12.5	μm
Data products		
HRPT and LAC nadir resolution	1.1 x 1.1	km
GAC nadir resolution	3.3 x 5.5	km
GVI nadir resolution	12 x 30	km

Sources: Kidwell (1986), Vogt (1992).

Another source of perturbation on the values of vegetation indices is the dependence of the individual spectral reflectance measurements on the particular geometry of illumination and observation at the time of the observation (*eg.*, Goward *et al.* 1991; Pinty *et al.* 1993; Cihlar *et al.* 1994). This topic will be discussed further below.

A number of indices have been proposed in the literature to address some of these issues, the main objective being to manipulate the mathematical expression in such a way that the resulting index remains sensitive to the presence of vegetation, but becomes less affected by the undesirable perturbing factors. These include the Perpendicular Vegetation Index (PVI) of Richardson and Wiegand (1977); the Soil-Adjusted Vegetation Index (SAVI) of Huete (1988); the Transformed Soil-Adjusted Vegetation Index (TSAVI) of Baret *et al.* (1989, 1991); the Atmospherically Resistant Vegetation Index (ARVI) of Kaufman and Tanré (1992); and the Modified Soil-Adjusted Vegetation Index (MSAVI) of Qi *et al.* (1994), among others.

Recently, Pinty and Verstraete (1992a) proposed the Global Environment Monitoring Index (GEMI), defined as follows:

$$GEMI = \eta \, (1 - 0.25\eta) - \frac{\rho_{red} - 0.125}{1 - \rho_{red}} \qquad (2)$$

where:

$$\eta = \frac{2(\rho_{nir}^2 - \rho_{red}^2) + 1.5\rho_{nir} + 0.5\rho_{red}}{\rho_{nir} + \rho_{red} + 0.5} \qquad (3)$$

This non-linear index was designed specifically to be less sensitive than other indices to atmospheric perturbations and to soil effects (over dark to medium soils), and to be very sensitive to bright surfaces and clouds in the visible region.

A comparison of GEMI with NDVI on actual AVHRR data confirms the theoretical performance of GEMI (Flasse and Verstraete, 1994). Figure 2 shows the NDVI (panel a) and the GEMI (panel b) computed from the same AVHRR-GAC data of Africa, acquired on 16 May 1989. This figure provides an overall feel for the differences and similarities between the two indices. It can be seen that both indices agree on the location of dense vegetation. However, they do differ significantly in more xeric environments, and GEMI appears to provide more detailed information where the vegetation cover is limited, as in the Sahel. GEMI also maintains a higher dynamic range in desertic regions such as the Sahara desert than NDVI. The NDVI distinguishes better than GEMI between land masses and water bodies. Finally, the two indices behave quite differently in terms of clouds: while they may be hard to distinguish from soils with the NDVI, they stand out very clearly in the GEMI image. These results are discussed in more detail in Flasse and Verstraete (1994).

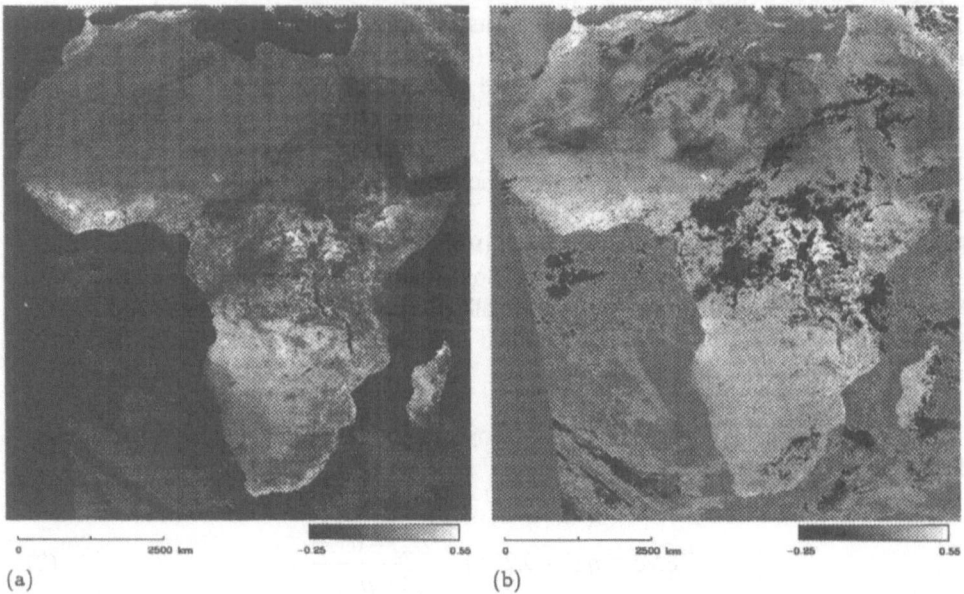

Figure 2. Mosaic of AVHRR-GAC data from three consecutive orbits for Africa on May 16, 1989. NDVI is shown in panel (a) and GEMI in panel (b). The scale of grey values has been selected to cover an equal range of index values in both panels.

The existence of multiple indices suggests that an objective method should be designed to assess the advantages, drawbacks, and limits of applicability of these indices, so that they can be used appropriately. This evaluation must be made with a specific purpose in mind. For example, we can assess the capacity of a vegetation index to serve as predictor for a biophysical quantity such as the vegetation fractional cover (σ) *i.e.* the proportion of the ground covered by plants. Hopefully, the value of the index varies when this fractional cover increases: we want the index to be sensitive to the signal of interest. However, we also require it to be insensitive to the undesirable perturbations, which introduce the equivalent of noise in the index value. The importance of this noise can be estimated by computing the value of the index at a given fractional cover for a range of (unknown) soil or atmospheric conditions. This will yield a maximum and a minimum index value corresponding to the selected fractional cover. These values bracket the uncertainty in the index value with respect to these perturbing factors. Such a computation can be repeated for a set of fractional covers, and the two curves of the maximum and minimum index values can be drawn as a function of the parameter of interest. Ideally, one would like these two curves to be as close to each other as possible.

For example, Figure 3 shows how simulated values of NDVI and GEMI vary as a function of the fractional vegetation cover, when the underlying soil background is allowed to vary from dark to medium bright soils. It is obvious that GEMI is much less affected than NDVI by changes in soil brightness (which is a strong function of soil wetness, and hence recent precipitation events) over this specific range of soils.

A measure of the noise introduced in the index by the perturbation is provided by the area between these curves, and this leads to an objective signal-to-noise criterion to evaluate an index (Leprieur *et al.* 1994):

$$S/N = \frac{VI(\sigma_M) - VI(\sigma_m)}{\int_{\sigma_m}^{\sigma_M} [\max(VI(\sigma_i)) - \min(VI(\sigma_i))]d\sigma} \quad (4)$$

where: σ_m and σ represent the minimum and maximum fractional vegetation covers for which data are available, respectively, and where the average is taken over the set of soil or atmospheric conditions considered. In practice, this integral may be approximated by a finite sum if the index values have been computed for a discrete number of vegetation covers. The values of the signal-to-noise ratio corresponding to the situation depicted in Figure 3, over the full range of fractional cover, are 8.58 for GEMI and 2.86 for NDVI.

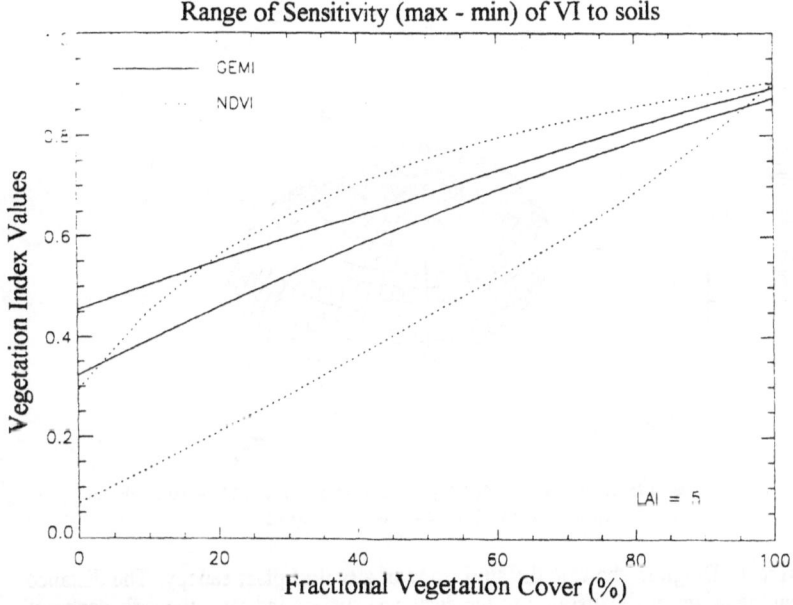

Figure 3. Variability of NDVI and GEMI as a function of the fractional vegetation cover when the soil type is unknown.

3.   Bidirectional reflectance models

All remote sensing observations made with a small instantaneous field of view instrument are bidirectional measurements, because most surfaces are anisotropic and the measurements are therefore specific to the particular geometry of illumination and observation. Figure 4 shows a typical bidirectional reflectance distribution function for a vegetated surface.

As a result, it can be seen that the same surface, simultaneously observed from different directions, yields different measurements. Directional effects hinder the direct interpretation of the observations, but also provide, in principle, an opportunity to retrieve information on the structure of the surface (Verstraete and Pinty, 1992).

Bidirectional reflectance models are mathematical equations that describe how the reflectance of the medium under consideration depends on the intrinsic properties of that medium but also on the geometry of illumination and observation. In other words, these models describe the anisotropy of the reflectance of the observed medium in terms of a set of parameters specifying how this medium interacts with the incoming radiation, for the

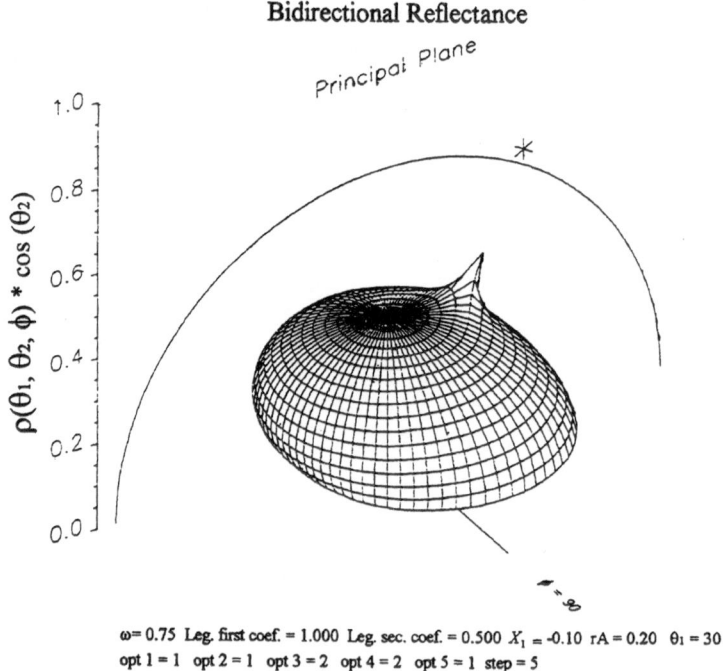

$\omega = 0.75$ Leg. first coef. $= 1.000$ Leg. sec. coef. $= 0.500$ $X_1 = -0.10$ rA $= 0.20$ $\theta_1 = 30$
opt $1 = 1$ opt $2 = 1$ opt $3 = 2$ opt $4 = 2$ opt $5 = 1$ step $= 5$

Figure 4. Diagram showing the anisotropy of a typical plant canopy. The distance from the origin of the diagram to the displayed surface indicates the reflectance of the canopy in the given direction.

particular geometry of illumination and observation. As far as satellite remote sensing is concerned, each observation provides a measurement of such bidirectional reflectance. The inversion of such bidirectional reflectance models against observations permits, in principle, the retrieval of quantitative information on the optical and geometrical properties of this medium.

If the objective is to describe the reflectance field in great detail and with high accuracy, complex models including a large number of parameters must be used. However, if the goal is to use remote sensing data to characterize the medium, then the model can only contain a very small number of parameters, because the inversion procedure cannot identify a single solution when there are too many parameters (typically more than 3 to 6). In this section, we focus exclusively on these simple invertible models. Further information on bidirectional reflectance models can be found in the literature (*e.g.* Pinty and Verstraete, 1992b; Myneni and Ross, 1991; Goel, 1988).

Two categories of invertible models can be distinguished, depending on whether they are physically-based, *i.e.* describe the relevant processes in terms of measurable physical quantities, or whether they are empirical, in which case there is no constraint on the form of the model equation. In this latter case, however, the model parameters have no particular meaning (i.e. cannot be measured). This distinction is made for various reasons: the physically-based models are the only ones which can (1) provide a clear understanding of the processes at work, (2) yield information on the actual optical and structural properties of the medium, and (3) be explicitly validated. The process of model validation is discussed at length in Pinty and Verstraete (1992b). Empirical models are also useful, but only in specific applications, such as the normalization of the data with respect to angular effects, the computation of surface albedo, or the provision of a lower boundary condition for atmospheric models, for example.

The basic mechanism of 'inverting' a reflectance model against a set of remote sensing observations consists of adjusting the values of the model parameters until this model can effectively 'explain', *i.e.* reconstruct, the variance in the observations. The degree to which the model 'fits' the measurements is evaluated by computing the 'distance' between the model and the observations, for instance with a least square estimator, and the problem of finding the optimal values of the model parameters (*i.e.* those that best describe the observations) is reduced to that of locating the minimum of that function. Standard optimization procedures exists to perform that task, although rather strict computational conditions must be met if the parameters are to be retrieved with a certain degree of accuracy. Details on the inversion have been provided elsewhere (*e.g.* Verstraete *et al.* 1994), but the essence of the procedure can be summarized as follows.

A bidirectional reflectance distribution function (BRDF) model capable of describing remote sensing measurements may be expressed by an equation of the form:

$$z = f(x_1, x_2, \ldots x_n; y_1, y_2, \ldots, y_m)$$

(5)

where the reflectance of the surface z is expressed as a function of n independent variables x_i defining the conditions of observations, and m state variables y_j. Usually, $m \gg 1$. When remote sensing is used to characterize the surface, the independent variables x_i correspond to the spatial, temporal, spectral and directional conditions of the observation: they are precisely known. The dependent variables y_j, however, are the unknowns to be estimated.

In principle, if we knew the values of x_i and y_j, we could predict the corresponding observed value. In practice, we take one measurement $\hat{z}$ under conditions specified by x_i and would like to retrieve a 'best' estimate of the state variables y_j. Whenever $m > 1$, Equation (5) cannot be inverted analytically because we have only one equation and more than one parameter to estimate. We can take multiple observations, however, and form a system of M simultaneous equations:

$$z_1 = f_1 (x_{11}, x_{12}, ..., x_{1n}; y_{11}, y_{12}, ..., y_{1m})$$
$$z_2 = f_2 (x_{21}, x_{22}, ..., x_{2n}; y_{21}, y_{22}, ..., y_{2m})$$
$$.$$
$$.$$
$$\quad (6)$$
$$.$$
$$z_M = f_M (x_{M1}, x_{M2}, ..., x_{Mn}; y_{M1}, y_{M2}, ..., y_{Mm})$$

but we now have M equations and M times m unknowns. This set of equations can be simplified and solved if the following conditions are verified: (1) the system does not change significantly between measurements (*i.e.* the values of the state variables y_j are unchanged for all measurements), (2) the functional form of the equation (f) does not change from one observation to the next, (3) observations taken for various values of the independent variables x_i result in significantly different values of the variable of the state z, and (4) more observations are taken for various conditions x_i than there are parameters y_j to retrieve ($M > m$). Optionally, one or more of the parameters y_j can be specified on the basis of other sources of information.

When these conditions are met, the system of equation above can be rewritten:

$$z_1 = f (x_{11}, x_{12}, ..., x_{1n}; y_1, y_2, ..., y_m)$$
$$z_2 = f (x_{21}, x_{22}, ..., x_{2n}; y_1, y_2, ..., y_m)$$
$$.$$
$$.$$
$$\quad (7)$$
$$.$$
$$z_M = f (x_{M1}, x_{M2}, ..., x_{Mn}; y_1, y_2, ..., y_m)$$

The characterization of the state of the system by estimating the 'best' values of the state variables y_j is achieved by numerically varying the values of these variables in the model until the difference between the predicted (z) and observed ($\hat{z}$) values is minimized. Mathematically, the problem is to minimize a figure of merit function such as:

$$\delta^2 = \sum_{k=1}^{M} \left[\hat{z}_k - f\left(x_{k1}, x_{k2}, \ldots, x_{kn}; y_1, y_2, \ldots, y_m\right) \right]^2 \tag{8}$$

Standard numerical algorithms are available to minimize this expression, and produce the optimal values of the parameters y_j that account for the observed variability of the measurements $\hat{z}_k$. For this approach to be useful, the numerical algorithms must optimize globally (be insensitive to local minima) and the minimum of the function must be well defined for each one of the parameters y_j (Renders and Flasse, 1995). In addition, the sensitivity of the inversion algorithm to initial conditions and the influence of the noise in the data on the retrieved values must be fully documented.

When this inversion can be done with a physical model, the parameters have a well defined meaning and can often be measured in the field. This means that the accuracy of the inversion and the validity of the model can be controlled by independently measuring these quantities *in situ*. The inversion of an empirical model proceeds exactly in the same way, except that the model parameters are not physical quantities, and cannot be measured in the field. These models cannot therefore be validated in this way: they can only be verified by comparing the reflectances predicted by the model to other reflectance data for the same target. In a way, empirical models may predict the right reflectance value for the "wrong" reason.

The spectral directional hemispherical reflectance of a surface, often known as the spectral albedo of that surface, is a parameter of great importance in a number of applications, and especially for climate models. This surface property represents the proportion of the incoming radiation which is reflected in all directions, when the illumination is in the form of a collimated beam. This quantity is normally obtained by integrating the bidirectional reflectance over all view angles, keeping the illumination direction fixed. Both physical and empirical models can be used to estimate this quantity. In practice, the model is first inverted against reflectance data to characterize the anisotropy of the surface, and the same model, driven with the parameters just retrieved, is then used to estimate the albedo, either by generating reflectances in a large number of directions and summing, by analytical or numerical integration.

4.   Inversion of BRDF models against AVHRR data

The theoretical developments outlined above are only intended to provide an overview of the problems associated with the inversion of a BRDF model against observations. Many issues have been ignored for the sake of brevity or clarity, but are discussed further in the literature (*e.g.* Verstraete *et al.* 1994). Nevertheless, the actual implementation of this approach with satellite data poses its own set of issues.

First of all, some of the conditions required for the inversion are not quite met. For instance, since the AVHRR instrument acquires only one data point (and hence one particular illumination and view geometry) for each location at a time, it is necessary to accumulate data over time to gather a large enough data set. When the presence of clouds is taken into account, this implies that the collection of 5 to 10 observations of a given site may require a couple of weeks or more. If this occurs at a time of rapid changes in the vegetation cover, the system under observation is not constant during the period of measurement. Another general problem is that the BRDF models currently available may not be able to represent the full complexity of the surface, especially if it is highly heterogeneous spatially.

These difficulties can be partially addressed by reducing to the strict minimum the period over which observations are gathered (but this requires that a very small number of parameters be retrieved), or by assigning the values of some of the parameters *a priori*, so that only a subset of the model parameters need to be retrieved. Alternatively, simple empirical models can be inverted against the data because they more often require the smallest number of parameters, and can be applied to complex surfaces. The feasibility of this approach when using AVHRR data was discussed in Flasse *et al.* (1993), where it was shown that the values of the parameters retrieved by inversion of a semi-empirical model were capable of simulating the observed reflectances, and also to predict relatively well other measurements under different geometries of illumination and observation. As a result, this approach can be used to compute reasonably accurate spectral albedos (directional-hemispherical reflectance) by integrating the empirical model over the whole hemisphere of observation for a fixed illumination, and to yield spectral data sets decontaminated of the directional effects. Such data sets can be considered as 'normalised' with respect to the geometry of illumination and observation, and could serve to estimate vegetation indices freed from these effects.

One way to illustrate this approach consists in comparing NDVI values estimated on the basis of surface albedos with those based on the original bidirectional reflectances (Flasse, 1993). A one-year-long data set of AVHRR data was first assembled for the Konza Prairie in the US. The semi-empirical surface model of Rahman *et al.* (1993) was inverted on successive periods of 15 days; this resulted in the production of time series of model parameters. Based on these values, the time evolution of the spectral albedo of the surface was evaluated, and these values served to assess the NDVI. Figure 5 exhibits the values of the vegetation index based on the computed albedos ($NDVI_{ALB}$) and on the measured reflectances ($NDVI_{MVC}$), composited over the same period (15 days) used for the inversion of the model (Holben, 1986). Clearly these two indices are not equivalent: $NDVI_{ALB}$ corresponds to an integrated value over a 15-day period, while $NDVI_{MVC}$ results from one daily value supposed to be representative for that period.

The $NDVI_{MVC}$ trace presents plateaus, indicating that a particular NDVI value remained the maximum value over successive compositing periods. On the other hand, the $NDVI_{ALB}$ trace shows more variability and probably represents better the actual variations of albedo during the corresponding periods of time. Another interesting feature of this Figure is that $NDVI_{ALB}$ is generally lower than $NDVI_{MVC}$. This is to be expected, since the Maximum Value Composite (MVC) selects the largest possible NDVI values, which may be significantly affected by the anisotropy of the surface. Since NDVI values decrease with the solar zenith angle and since albedos are computed with a fixed solar zenith angle for the entire time series, the total amplitude of the $NDVI_{MVC}$ trace may also be partly due to

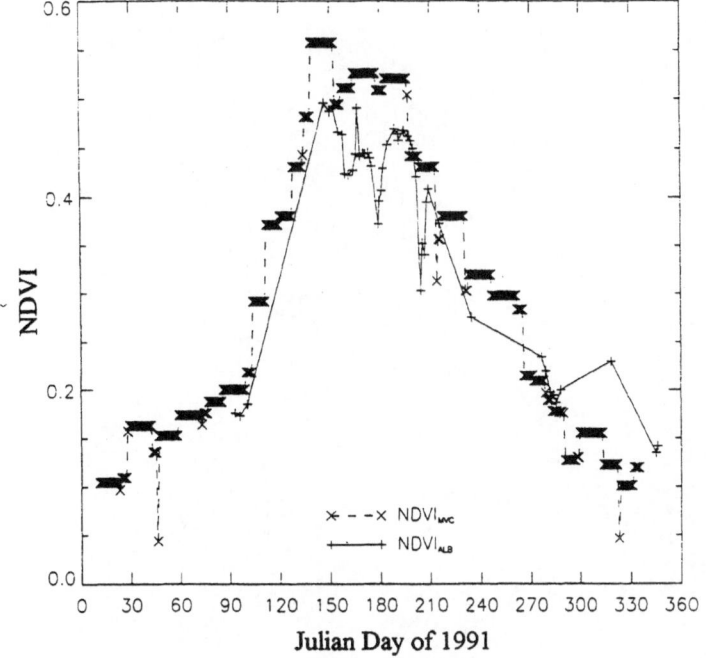

Figure 5. NDVI time series over a Konza prairie. (i) NDVI$_{ALB}$ (+-+) is computed from albedos, and (ii) NDVI$_{MVC}$ (x-x) is computed with the MVC technique, from bidirectional reflectances.}

the seasonal evolution of the position of the Sun. This comparison underlines the importance of directional effects on the interpretation of AVHRR data.

To the extent that the parameters retrieved by inverting an empirical model against the observations have no precise physical meaning, they are not intrinsically useful. They only allow the reconstruction of bidirectional reflectances or the computation of the albedo, as indicated above. Nevertheless, it appears that these empirical parameters do not acquire arbitrary values either. In fact, when their values are displayed as a function of space, they exhibit recognizable features. These parameters may therefore prove useful in some applications, such as land cover classification.

To explore further this opportunity, we have applied the inversion approach over a time series of 20 AVHRR-GAC images of November 1987, over Africa. These continental images were obtained by stitching successive NOAA orbits, as explained in Belward et al. (1992). Figure 6 shows one of the images used in the time series. On that image, bidirectional effects are very much visible at the edges of each swath. It is worth noting that while directional effects are sometimes undesirable, in this case they constitute the very source of the information being sought. Figure 7 presents maps of the three empirical parameters retrieved by inversion of the semi-empirical bidirectional reflectance model

Figure 6. AVHRR-GAC composite image for the African continent for the 17th November 1987. Channel 2 Top-Of-Atmosphere reflectance is shown. The bidirectional effect is clearly evident on each of the three successive orbits.

(Rahman *et al.* 1993) against the AVHRR-GAC time series for channel 1. The surface parameter k appears to take most of the bidirectional effects into account (the inclination of stripes corresponds to that of the orbits), whereas the surface parameters ρ_0 and Θ appear to represent spatial features. The proper interpretation of these spatial features will require further investigation, bearing in mind the theoretical difficulties described earlier. Nevertheless, such empirical parameters may turn out to be useful in specific applications, such as land cover assessments.

Figure 8 represents the spectral albedo values for AVHRR channels 1 and 2, for the same region of Africa, computed by integrating the bidirectional reflectance model, driven by the retrieved parameters of Figure 7, over the whole upper hemisphere. This approach provides non-directional spectral albedo values which represent the integration of information over short time series. The variations observed in such variables, from one

Figure 7. Maps of the parameters ρ (top), Θ (middle) and k (bottom) of the empirical bidirectional reflectance model, retrieved by inversion (for channel 1 only).

short time series to another, are representative of actual changes of the observed media (surface and atmosphere) rather than of bidirectional effects.

The applications presented above constitute initial attempts to demonstrate the feasibility of operationally inverting bidirectional reflectance models against satellite data, and AVHRR data more specifically, at continental scale. These examples stress the potential of the approach.

CH1 (albedo) [0.0, 0.3]

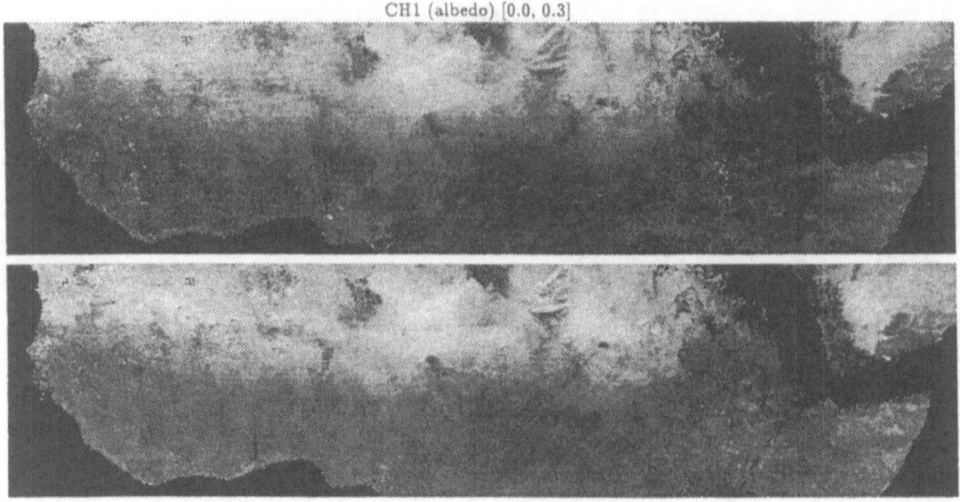

Figure 8. Maps of albedos computed by the integration over the whole hemisphere of the bidirectional reflectance models driven by the surface parameters of Figure 7, but for channel 1 (top) and 2 (bottom).

5.   Discussion and conclusion

The need for appropriate tools for the exploitation of remote sensing data will be enhanced with the arrival of new instruments. Both the European Space Agency (ESA) and the U.S. National Aeronautic and Space Administration (NASA) are currently building a new generation of Earth Observation instruments, which will fly by the end of this decade. Other advanced instruments will precede and follow. The general characteristics of these new sensors include a better spatial resolution (*e.g.* the Medium Resolution Imaging Spectrometer, MERIS), an improved radiometric resolution and on-board calibration mechanisms (all new sensors mentioned here), a much higher spectral resolution (*e.g.* the Moderate Resolution Imaging Spectrometer, MODIS, as well as MERIS), and in some case a much improved directional coverage (*e.g.* Multi-angle Imaging Spectro-Radiometer, MISR), while maintaining a global and almost daily coverage. Overviews of the design characteristics and specifications of these instruments can be found in NASA (1993) and ESA (1993). The immediate consequence of these developments will be a much larger rate of data acquisition, and the need for automatic data processing algorithms. In the meantime, AVHRR data will continue to be used to document the state and evolution of terrestrial surfaces over large areas, while high resolution instruments such as SPOT and Landsat will be exploited for detailed intensive studies over relatively small regions.

Since all imaging sensors observe the Earth's surface with a small aperture instrument, and as these surfaces are anisotropic, the measurements are specific to the particular geometry of illumination and observation. Bidirectional reflectance models can be used to either exploit this property, or to remove the variance due to the geometry from the remote sensing data. Examples of this approach were given above, and its limitations were discussed. Similarly, all observations acquired by space-borne instruments are subject to atmospheric effects, since the solar radiation must traverse the atmosphere twice before entering the sensor. Furthermore, since data are accumulated over significant time periods, it will become even more important to design and implement tools that can take advantage of the dynamic evolution of the targets to gain knowledge about the surface. To simulate the processes of data acquisition, such models will have to take into account the orbit of the platform and the characteristics of the instruments. Accounting for both directional and atmospheric effects is therefore an inescapable requirement for all applications.

End users of satellite remote sensing data are unlikely to be very interested in scientific and technical issues such as orbital models, illumination geometry, radiometric calibration, directional and atmospheric effects, *etc.* On the other hand, the solution to these questions is not part of the mandate of national and international space agencies either. It is therefore likely that highly specialized research groups will develop the scientific know-how and technical competence to address these issues, and evaluate these new approaches in pilot studies. In parallel, operational groups will take the charge of acquiring the data from the space agencies and deliver useful products to the users based on the latest algorithms provided by the research groups.

The scientific problems discussed in this paper will remain recurrent issues in the field of remote sensing for the foreseeable future. It is likely that vegetation indices will continue to proliferate, in response to the requirements of specific applications and progress in the understanding of the physical mechanisms that control these measurements. Similarly, better BRDF models and improved inversion procedures will appear. The emergence of ever faster computers (in particular due to parallel processors), may permit the inversion of surface-atmosphere reflectance models on an operational basis. End users should, however, strive to maintain a basic understanding of the radiation transfer issues and of the practical processing steps that have been applied, in order to avoid over-interpreting the data they use. It is hoped that this chapter may contribute to this objective.

Acknowledgements

The authors are grateful to H. Cambridge for the support he provided in processing the AVHRR-GAC data used for this work, and to Giles D'Souza for his thorough editing of the final version of the manuscript.

References

Baret, F., G. Guyot and D. J. Major (1989), TSAVI: A vegetation index which minimizes soil brightness effects on LAI and APAR estimation, *12th Canadian Symposium on Remote Sensing and IGARSS '89*, 10-14 July 1989, Vancouver, Canada, 1355-1358.

Baret, F. and G. Guyot (1991), Potential and limits of vegetation indices for LAI and APAR assessment, *Remote Sensing of Environment*, **35**, 161-173.

Belward, A. S., J. V. Vogt, A. Falk-Langemann, S. Saradeth and H. Cambridge (1992, Peparation of AVHRR GAC data sets for global change studies, *in Proceedings of the Central Syposium of the 'International Space Year' Conference*, 30 March-4 April 1992, Munich, Germany, Volume EAS SP-341, 19-23.

Cihlar, J., D. Manak, and N. Voisin (1994), AVHRR bidirectional reflectance effects and compositing, *Remote Sensing of Environment*, **48**, 77-88.

ESA (1993), *ENVISAT*, Special Issue of *ESA Bulletin*, No. 76, ESA Publications Division, Noordwijk, 7-60.

Flasse, S. P. (1993), *Extracting Quantitative Information from Satellite Data: Empirical and Physical Approaches*, Institute for Remote Sensing Applications, Joint Research Centre, Commission of the European Communities, Ispra, Italy, EUR 15409 EN, 199pp.

Flasse, S. and M. M. Verstraete (1994), Monitoring the environment with vegetation indices: Comparison of NDVI and GEMI using AVHRR data over Africa, in *Vegetation, Modelling and Climatic Change Effects*, Edited by F. Veroustraete and R. Ceulemans, SPB Academic Publishing, The Hague, 107-135.

Flasse, S. P., M. M. Verstraete and D. J. Meyer (1993), Inverting a bidirectional reflectance model to remove directional effects in AVHRR data in *Proceedings of the 6th European Data Users' Meeting*, Belgirate, Italy 29th June-2nd July, 1993, 79-86.

Gates, D. M. (1980), *Biophysical Ecology*, Springer-Verlag, Berlin, 611 pp.

Goel, N. S. (1988), Models of vegetation canopy reflectance and their use in estimation of biophysical parameters from reflectance data, In *Remote Sensing Reviews*, **4**, 1-222.

Goward, S., B. Markham, D. Dye, W. Dulaney, and J. Yang (1991), Normalized difference vegetation index measurements from AVHRR, *Remote Sensing of Environment*, **35**, 257-277.

Holben, B. (1986), Characteristics of maximum-value composite images from temporal AVHRR data, *International Journal of Remote Sensing*, 7, 1417-1434.

Huete, A. R., R. D. Jackson, and D. F. Post (1985), Spectral response of a plant canopy with different soil background, *Remote Sensing of Environment*, **17**, 37-53.

Huete, A. R. (1988), A soil-adjusted vegetation index (SAVI), *Remote Sensing of Environment*, **2**, 295-309.

Kaufman, Y. J. (1989), The atmospheric effets on remote sensing and its correction, In Asrar, G. Editor, *Theory and Application of Optical Remote Sensing*, Wiley Series in Remote Sensing, 336-428, John Wiley and Sons, New York.

Kaufman, Y. J. and D. Tanré (1992), Atmospherically resistant vegetation index (ARVI) for EOS-MODIS, *IEEE Transactions on Geoscience and Remote Sensing*, **30**, 261-270.

Leprieur, C., M. M. Verstraete, and B. Pinty (1994), Evaluation of the performance of various vegetation indices to retrieve vegetation cover from AVHRR data, *Remote Sensing Reviews*, **10**, 265-284.

Myneni, R. B. and J. Ross (1991), *Photon-Vegetation Interactions. Applications in Optical Remote Sensing and Plant Ecology*, Springer-Verlag, Berlin, 565p.

NASA (1993), *1993 EOS Reference Handbook*, NASA Earth Science Support Office, Washington, DC 20024, 145 pp.

Pearson, R. L. and L. D. Miller (1972), Remote mapping of standing crop biomass for estimation of the productivity of the short-grass prairie, Pawnee National Grasslands, Colorado, in, *Proceedings of the 8th International Symposium on Remote Sensing of Environment*, ERIM, Ann Arbor, MI, 1357-1381.

Pinty, B. and M. M. Verstraete, (1992a), GEMI: A non-linear index to monitor global vegetation from satellites, *Vegetatio*, **101**, 15-20.

Pinty, B. and M. M. Verstraete, (1992b), On the design and validation of bidirectional reflectance and albedo models, *Remote Sensing of Environment*, **41**, 155-167.

Pinty, B., C. Leprieur, and M. Verstraete (1993), Towards a quantitative interpretation of vegetation indices. Part 1: Biophysical canopy properties and classical indices, *Remote Sensing Reviews*, **7**, 127-150.

Qi, J., A. Chehbouni, A. R. Huete, Y. H. Kerr, and S. Sorooshian (1994), A modified soil adjusted vegetation index (MSAVI), *Remote Sensing of Environment*, **48**, 119-126.

Rahman, H., B. Pinty, and M. M. Verstraete (1993), Coupled surface-atmosphere reflectance (CSAR) model. 2. Semi-empirical surface model usable with NOAA Advanced Very High Resolution Radiometer data, *Journal of Geophysical Research*, **98**, 20,791-20,801.

Renders, J.-M. and S. P. Flasse (1995), Hybrid methods using Genetic Algorithms for global optimization, *IEEE Transactions on Systems, Man, and Cybernetics*, in print.

Richardson, A. J. and C. L. Wiegand (1977), Distinguishing vegetation from soil background information, *Photogrammetric Engineering and Remote Sensing*, **43**, 1541-1552.

Rouse, J. W., R. H. Haas, J. A. Schell, D. W. Deering and J. C. Harlan (1974), Monitoring the vernal advancement of retrogradation of natural vegetation, Final report, Greenbelt, MD, National Aeronautics and Space Administration (NASA), Goddard Space Flight Center (GSFC).

Verstraete, M. M. and B. Pinty (1992), Extracting surface properties from satellite data in the visible and near-infrared wavelengths, in *TERRA-1: Understanding the Terrestrial Environment, The Role of Earth Observations from Space*, Edited by P. M. Mather, Taylor and Francis, London, 203-209.

Verstraete, M. M., B. Pinty, and R. Myneni (1994), Understanding the biosphere from space: Strategies to exploit remote sensing data, in *Proceedings of the 6th ISPRS International Symposium on Physical Measurements and Signatures in Remote Sensing*, Val d'Isère, France, 17-21 January 1994, CNES, 993-1004.

DERIVATION OF GEOPHYSICAL PARAMETERS FROM AVHRR DATA

Gérard DEDIEU

LERTS,
Unité mixte CNES-CNRS,
18 avenue Edouard Belin,
31055 Toulouse Cedex,
France.

ABSTRACT. In this chapter we present recent research developments in the use of data derived from the NOAA-AVHRR together with various types of models and ancillary data. The synergistic use of models and data is a promising way for the future use of remotely-sensed data. First, it provides a quantitative retrieval of geophysical parameters, such as surface albedo, which are closely related to satellite measurements. Second, this approach is attractive since coupling models and data may allow the estimation of model parameters (e.g. leaf area index) which are not directly linked to radiance measurements *per se*.

In this chapter we present two examples of studies which illustrate the combined use of satellite data and models to retrieve geophysical parameters. The objective of the first study is the assessment of surface albedo from AVHRR directional measurements in the shortwave channels. The aim of the second study is to estimate vegetational Net Primary Productivity at the global scale. In addition, these studies require measurements of the highest possible precision, and their consideration will illustrate the current state-of-the-art in data processing capabilities.

In their respective domains, these studies are at the forefront of current remote sensing applications, and the results presented here are preliminary. Further work is needed, particularly regarding validation. However, we think that these two studies represent significant examples of the new trends in the use of satellite measurements and deserve to be presented in the framework of this book.

1. Surface albedo from AVHRR data

1.1 INTRODUCTION

This section is based on the work of F. Cabot and G. Dedieu, and is detailed in several related publications (Cabot and Dedieu 1993, Cabot *et al*, 1993a and 1993b).

Solar albedo of land surface is a key parameter required by general circulation models (GCMs) to compute surface energy budgets. Satellite measurements in the solar spectrum appear to be the only tools which may provide its estimate with the necessary global coverage and required frequency. Not only is the total albedo for the whole solar spectrum needed, but also the estimate of surface albedo in several spectral domains of this spectrum. Since absorbing and scattering processes in the atmosphere are spectrally

231

G. D'Souza et al. (eds.), Advances in the Use of NOAA AVHRR Data for Land Applications, 231–263.
© 1996 *ECSC, EEC, EAEC, Brussels and Luxembourg.*

dependent, the radiative effect in the atmosphere of the solar energy reflected by the surface will depend on the spectral properties of the surface.

Satellites provide bidirectional reflectance measurements of a surface-atmosphere system in several wavelength bands, with a high temporal frequency, and a reasonable spatial resolution. However, since natural surfaces are seldom lambertian, bidirectional reflectance is not equivalent to albedo. Relative variations between hemispherical reflectance derived from nadir measurement and real albedo are high and may reach 150%, depending on the solar zenith angle and characteristics of the surface (Eaton and Dirmhirn 1979, Kimes and Sellers 1985).

Another problem to consider is that in order to monitor vegetation or land cover, many studies need measurements of surface reflectance over long periods of time. But, since each of the measurements taken from a satellite on various dates is acquired under different geometric conditions, surface anisotropy introduces additional signal perturbations which combine with atmospheric absorption and scattering. The observed variation of reflectance caused by surface directional effects can be rather large, sometimes more than one order of magnitude (Gutman *et al*, 1989, Roujean *et al* 1992). Thus, there is a great need to normalise reflectance to a fixed reference, *e.g.* nadir or hemispheric reflectance.

Therefore, it appears necessary to develop a method which is able to take into account the anisotropy of the surface, in order to compute accurate hemispherical reflectance as well as to get normalised sets of data. Several problems arise when one wants to derive surface directional properties with current satellite data. First, only a narrow range of illumination and observation angles can be obtained from current polar or geostationary orbits, and sampling is not sufficient to retrieve hemispherical reflectance through a simple numerical integration of the measurements (Eaton and Dirmhirn 1979, Kimes and Sellers 1985, Kimes and Holben 1992). Second, surface and atmospheric characteristics may change between two consecutive observations. Changes in surface structural and optical properties lead to changes in directional behaviour while different atmospheric conditions decrease the accuracy of surface reflectance retrieval.

In this study, we used both satellite measurements and an analytic bidirectional reflectance model to derive surface albedo. In short, our approach lies in the fitting of model parameters using a set of NOAA-AVHRR measurements. Then, in order to check the ability of models and fitted parameters to describe directional behaviour, we used the model to predict the reflectance that a different satellite would measure in a different plane. The predicted signal is then compared with actual measurements of this different sensor, in this case, the Meteosat visible channel, VIS.

A number of bidirectional models have been developed following different methods and different aims (e.g. Ross 1981, Rahman *et al* 1993, Hapke 1981). In this study, we used the model developed by Rahman *et al* (1993). Results obtained with the whole set of the above-mentioned models, as well as further details about the processing, can be found in Cabot and Dedieu (1993).

The model of Rahman *et al* (1993) has been used together with measurements acquired by the NOAA-AVHRR and Meteosat VIS sensors over western Africa. These meteorological satellites have been used as they provide radiance measurements with a high temporal repeatability, one of the best angular sampling frequencies currently available, and with a ground resolution suited to the retrieval of surface parameters on a global scale. First, we present the results obtained over a single pixel located in an arid area of Niger. Then, we describe the results obtained for a larger region corresponding to the HAPEX-Sahel experiment area.

1.2. DATA

Our test site is a zone covering one square degree, East of Niamey. Its bounds are from 2°E to 3°E and 13-14°N (Figure 1). This area corresponds to the square degree of the HAPEX-SAHEL experiment. Sahelian vegetation in this region exhibits a well-marked cycle driven by the rain regime: this feature would be valuable as a benchmark of the method described hereafter over seasonally vegetated areas.

1.2.1 *Satellite Data.* The NOAA-11 AVHRR data set we used is composed of 167 full-resolution AVHRR images (HRPT) acquired almost daily, between 11:30 and 15:30 UTC, over a six-month period from May to October 1991. These images have been acquired and pre-processed by the AGRHYMET ground receiving station in Niamey and include visible (channel 1: 0.58-0.68 µm), near infrared (channel 2: 0.73-1.1 µm), and both thermal infrared channels (channel 4:10.33-11.25 µm and 5:11.40-12.38 µm). Pre-processing consists of thermal channel calibration and geometric registration to a Plate Carrée projection with a 1.1 km ground resolution. This image data set covers the whole area where the HAPEX-SAHEL experiment took place in 1992.

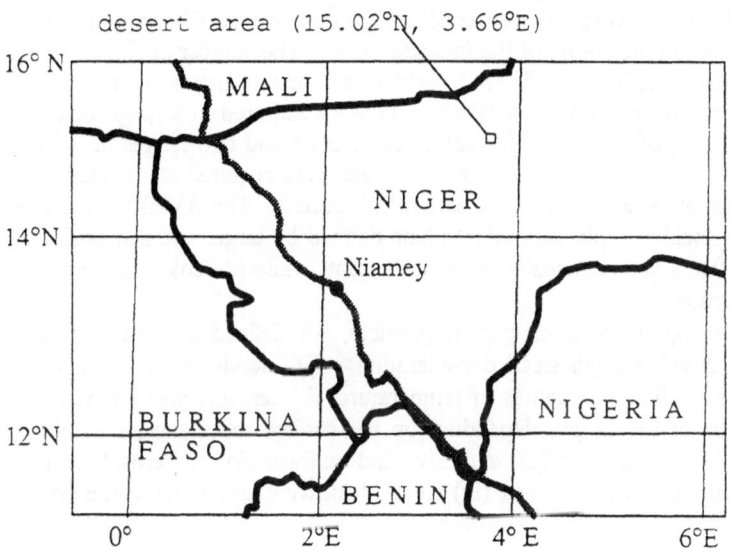

Figure 1. Study Area, with the location of the pixel used to explain the approach over bare soils.

Meteosat is a meteorological geostationary satellite. Its sensors provide measurements of the radiance leaving the atmosphere every 30 minutes in three spectral bands: (i) shortwave or visible (0.35-1.1 μm), (ii) thermal infrared (10.5-12.5 μm) and (ii) the so-called water vapour channel (5.7-7.1 μm). The resolution at nadir is 2.5 by 5 km for the visible channel, and 5 by 5 km for the other channels. The Meteosat-4 data used were derived from the full-resolution data by resampling to 30 km once every three hours. This product is called Meteosat B2 and is processed and delivered by the European Space Operations Centre of the European Space Agency (ESA/ESOC) to serve the International Satellite Cloud Climatology Project (ISCCP).

1.2.2. *Meteorological data.* In order to correct satellite measurements for atmospheric perturbations, we derived atmospheric water vapour content from quasi-daily radio soundings performed by the Meteorological Service of Niger at Niamey Airport. No *in situ* measurements were available for ozone and aerosol optical depth. Mean monthly ozone content is taken from the London *et al* (1976) climatology. We assumed a continental aerosol model (McClatchley 1971) and set aerosol optical thickness to a constant value of 0.2 at 550 nm. Since we selected only clear sky measurements through the filter process described below, we assumed that value to be statistically representative of the turbid atmosphere (Faizoun *et al* 1993). We used these data to achieve atmospheric correction using the SMAC method (Rahman and Dedieu 1993) based on the 5S radiative transfer code (Tanre *et al* 1990).

1.2.3. *Geometric configuration.* As already mentioned in the introduction, it is nearly impossible to get a complete sampling of measurements in viewing angle, illumination angle and relative azimuth (Figure 2). Most of the sampling is constrained by latitudinal position of the observed site as well as local time observation. Observation angles depend on the orbit and on the design of the imaging device. The number of similar satellites with different orbital plane is not sufficient to achieve a complete angular sampling.

For our test site, NOAA-11 AVHRR data were acquired in a range of 0° to 60° for view zenith angle, of 10° to 60° for solar zenith angle and two ranges of 0-30° and 120-240° for relative azimuth, whereas Meteosat data were acquired at 19° view zenith angle and in a range of 0-180° for relative azimuth (Figure 3). The AVHRR measurements lie mostly in the incidence plane (vertical plane defined by target and sun positions), where directional effects are presumably most important, while Meteosat data are acquired in different directions.

In order to get as clear images as possible, we defined a filtering process of the reflectances based on physical considerations: (i) clouds produce high values of reflectances as well as low values of temperature; (ii) aerosols may increase or decrease the reflectance of the target, depending on the surface reflectance, but often induce a decrease of the temperature; (iii) extensive and uniform clouds, aerosols and haze often lead to a decrease of the contrast; (iv) cloud shadowing and partial cloud cover increase the contrast.

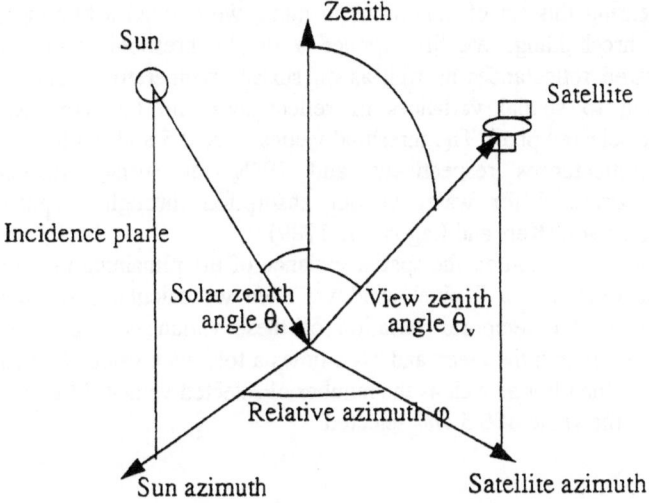

Figure 2. Geometric configurations and angle-notations used in this chapter.

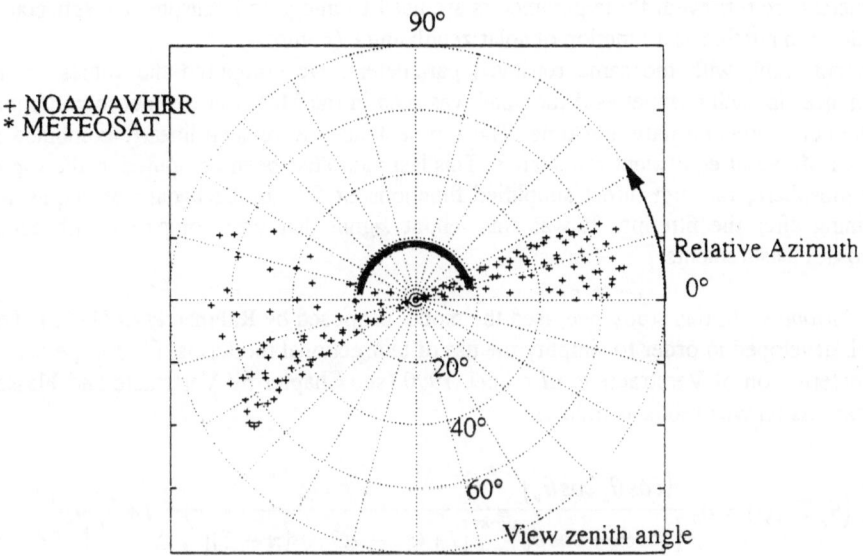

Figure 3. Angular sampling of NOAA-AVHRR(+) and Meteosat (*) acquisitions.

1.2.4. *Filtering*. Bearing this set of conditions in mind, we defined a filtering process based on two-level thresholding. We first applied a simple thresholding on values of visible and near infrared reflectances as well as on surface temperature, followed by a thresholding regarding to spatial variances of reflectances and temperatures in the neighbourhood of the selected pixel. The threshold values were 0.5 and 0.6 for the visible and near infrared reflectances respectively, and 293K for surface temperatures. Temperatures were corrected for water vapour absorption through a split-window formula adapted for bare soil (Kerr and Lagoaurde 1989).

After this first filter, we computed the spatial variance of the remaining measurements in the three channels over a 7 x 7 pixel window. Then we calculated the mean and standard deviation, s, of the temporal variation of these variances. We retained the measurements which fell within the mean and +/- s times a tolerance value. This tolerance value rules the strain of the filter as well as the number of selected values. After some tests over a bare soil target, the value of 0.5 was selected.

1.3.   METHOD.

The method is based on the fit of the parameters of a bidirectional reflectance model against surface bidirectional reflectances derived from AVHRR measurements. Surface reflectances are extracted from the images through a filtering process aimed at eliminating atmospherically contaminated values, whether from aerosols or clouds. Once model parameters are retrieved, these parameters are used to integrate hemispherical reflectance (albedo) computable as a function of solar zenith angle (Figure 4).

Concurrently, with the same retrieved parameters, we computed the values of the reflectances in each channel as if the pixel was seen in the Meteosat geometric conditions for the corresponding date and time. These reflectances were then linearly combined to obtain a Meteosat equivalent reflectance. This last value has been computed at the top of the atmosphere, through direct simplified functions of 5S, for each date of acquisition remaining after the filtering. It was this rebuilt signal that was compared with actual Meteosat measurements.

2.3.1. *Models*. In this study, we used the model proposed by Rahman *et al* (1993). This model, developed in order to simplify the use of bidirectional models is, for a large part, a parameterisation of Verstraete *et al* model, 1990 (see Chapter by Verstraete and Flasse). The expression we used is as follows:

$$\rho(\theta_s, \theta_v, \varphi) = \rho_0 \frac{(cos\,\theta_s\,cos\,\theta_v)^{k-1}}{(cos\,\theta_s + cos\,\theta_v)^{1-k}} \frac{1-\Theta^2}{[1+\Theta^2 - 2\Theta\,cos(\pi - \xi)]^{3/2}} \left(1 + \frac{1-\rho_0}{1+G}\right) \quad (1)$$

The parameters to be retrieved by regression are ρ_0 which represents a mean level for the reflectances, and k and Θ which account for the anisotropy of the surface. Θ ranges from -1 (backward scattering) to +1 (forward scattering).

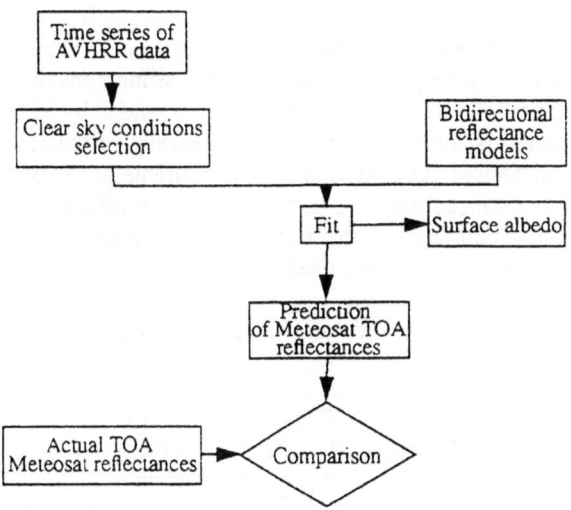

Figure 4. Flow diagram of the processing procedures used in this study.

In this equation Θ_s and Θ_v are solar and viewing zenith angles respectively, and ξ is the phase angle defined by:

$$cos\xi = cos\theta_s\ cos\theta_v + sin\theta_s\ sin\ \theta_v\ cos\varphi \qquad (2)$$

where ϕ is the relative azimuth.

G is given by

$$G = \sqrt{tan^2\ \theta_s + tan^2\ \theta_v - 2\ tan\theta_s\ tan\theta_v\ cos\ \varphi} \qquad (3)$$

This analytical expression of the bidirectional reflectance is based upon physical and geometrical considerations on the transfer of radiation through a porous medium. Though developed for horizontally homogeneous vegetative canopies, it is fairly applicable over semi-infinite media composed of uniformly-distributed scatterers such as bare soil.

1.3.2. *Inversion scheme: hypothesis and limits.* Most of the bidirectional reflectance models are based on the physics of the interactions of light with ground and vegetation.

However, solving the equations of these interactions analytically requires a set of assumptions involving semi-infinite media, homogeneity, and linear combinations of the effects produced by different media (soil, canopy). These assumptions are seldom valid, especially when working with the size of ground resolution provided by NOAA-AVHRR.

The bidirectional model is fitted against atmospherically corrected bidirectional reflectances derived from filtered AVHRR radiance measurements. This means that, at this stage, we assumed that atmospheric and surface directional effects can be accounted for separately, without the coupling of the two processes. Pinty *et al.* (1989) have emphasised the importance of the algorithm used to fit the models and recommended the accuracy of the E04JAF routine from the Numerical Algorithm Group library, which implements a quasi-Newton algorithm. We also used this routine to minimise the root mean square error defined below:

$$rmse = \sqrt{\frac{\sum_{i=1}^{N}\left[\rho_{mes_i} - \rho_{mod}\left(\theta_{v_i}, \theta_{s_i}, \varphi_i, \bar{p}\right)\right]^2}{N-1}} \qquad (4)$$

where subscripts *mes* and *mod* are for measurements and model respectively. N is the number of samples used and vector p contains the complete set of parameters of the model.

The model has been fitted against visible and near infrared reflectances separately, using the whole set of "clear sky" reflectances. We did not set any constraints on parameters, an approach which perhaps can be improved in the future. Then we compared the data with the reconstituted signal.

1.4. RESULTS AND DISCUSSION

1.4.1 *Fit of the bidirectional reflectance model for a single pixel.* For the sake of clarity, we first present the results obtained for a single pixel in the northern part of Niger (Figure 1), in a desert area where we assumed no temporal change in surface characteristics.

The aim of the fitting process is to retrieve a set of parameters which makes the directional behaviour of the model close to the observations. In order to evaluate the quality of the retrieval of the parameters we used two criteria: the root mean square error and the correlation coefficient.

For this desert area, we assume that surface properties remain constant during the period considered. Besides an angular sampling as complete as possible, the least square inversion requires a minimum number of measurements, a condition which is fulfilled by using the whole set of measurements over the period.

The results of the fits are given in Table 1. The r.m.s. errors of the fit, in terms of the surface reflectance, are 0.0246 (less than 8.25% of the mean) and 0.0244 (6.8% of the mean) in the visible and near infrared wavelengths, respectively. The correlation

Table 1: Results of the fit of directional model against NOAA AVHRR data

	Visible Channel				Near Infrared Channel		
Parameters		Correlation	r.m.s.e	Parameters		Correlation	r.m.s.e
ρ_{sol}	0.284			ρ_{sol}	0.376		
k	0.866	0.630	0.0246	k	0.857	0.783	0.0224
Θ	0.039			Θ	0.033		

coefficients we obtained are small: less than 0.63 in the visible channel and 0.78 in the near infrared channel.

We assumed that the temporal variations of the signal over our arid test site could be attributed mainly to differences in observation geometry. The way these variations can be simulated by the models is presented in Figure 5. One must keep in mind that some residual atmospheric effects may still influence the observations. Assuming that variations of the directional reflectances are mainly due to varying angular conditions when observing an anisotropic surface, the maximum observed variation is of the order of 32% in the visible and 30% in the near infrared.

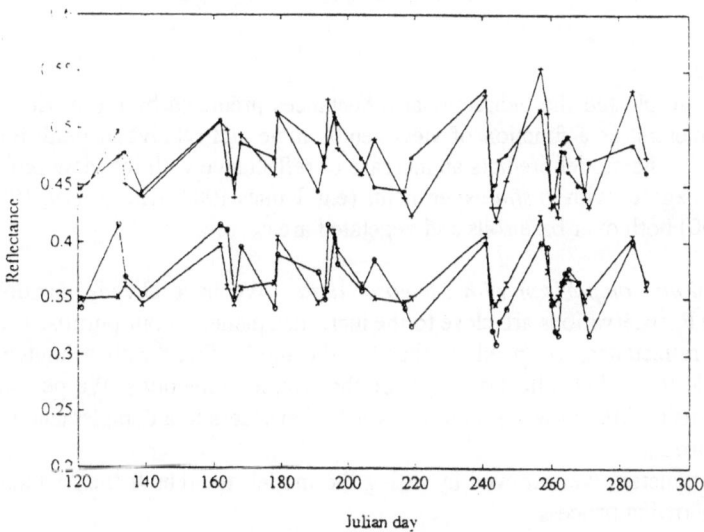

Figure 5. Temporal variations of original and fitted signal (-o- measured visible channel, -x- modelled visible channel, -+- measured near infrared channel, -*- modelled near infrared channel).

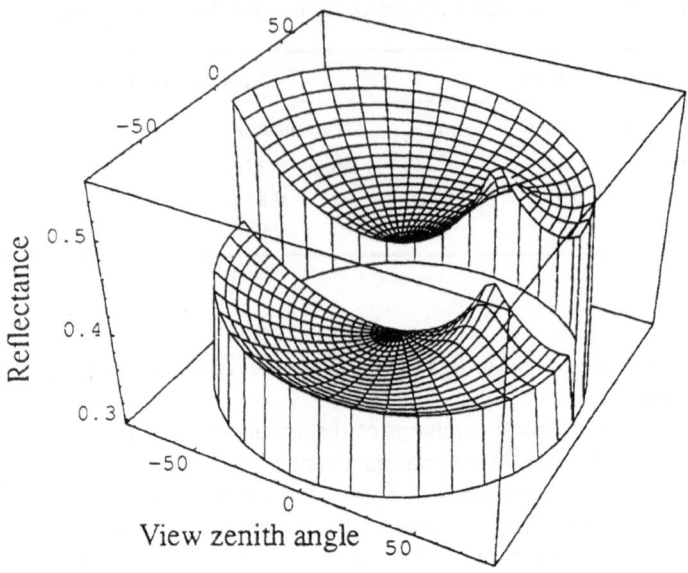

Figure 6. Bidirectional reflectances of the surface as predicted by the model of Rahman *et al*, (1993) with parameters fitted by using AVHRR data over a desert area. Foreground: visible channel; background: near infrared channel.

In Figure 6 we plotted the bidirectional reflectances predicted by the model, with the retrieved parameters, as a function of view zenith angle and relative azimuth for a solar incidence of 45°. The model predicts an increase of reflectance with the view zenith angle, a feature which agrees with *in situ* experiments (e.g. Kimes 1983, Kimes *et al*, 1985, Irons and Smith 1990) both over bare soils and vegetated areas.

1.4.2. *Validation: comparison with Meteosat data.* We have already mentioned that NOAA-AVHRR observations are close to the incidence plane. As our purpose is to derive hemispherical reflectance, we need to check if the model fitted with a limited angular sampling is able to predict reflectance outside the sampled directions. We performed this verification by using Meteosat observations which give access to a complementary angular sampling (Figure 2).

With the parameters we retrieved by fitting the model, we rebuilt the Meteosat signal through the following process:

• Simulation of surface reflectances for AVHRR visible and near infrared channels, using a model with retrieved parameters but for geometric configuration of Meteosat for each day selected through filtering

- Simulation of Meteosat surface reflectance, *i.e.* broadband reflectance (0.35-1.1 μm). This spectral simulation is based on a linear combination of AVHRR channels 1 and 2, weighted by the relative spectral response of the two satellites (Arino *et al.* 1991).

$$\rho_M = 0.484 \, \rho_{VIS} + 0.516 \, \rho_{NIR} \qquad (5)$$

where ρ_M is the Meteosat reflectance, ρ_{VIS} and ρ_{NIR} are the NOAA-AVHRR visible and near infrared reflectances respectively.

- Simulation of the reflectance that Meteosat would measure at the top of the atmosphere (TOA). Atmospheric characteristics (gaseous and aerosol contents) are the same used to correct AVHRR measurements. TOA reflectances are predicted for two times a day (11:30 and 14:30).
- Simulation of the digital counts that the Meteosat visible sensor would supply. Digital counts are also predicted for 11:30 and 14:30.

Top-of-atmosphere (TOA) reflectances and digital counts were then simulated for each clear sky day and compared with actual Meteosat TOA reflectances and counts.

Figures 7 and 8 compare simulated and measured TOA digital counts and reflectances respectively. The actual measurements are quite well predicted, and small variations of reflectances between 11:30 and 14:30 are reproduced. This means that the spectral combination works well and that the three channels are correctly intercalibrated. The only large difference between simulations and measurements (day 208 at 14:30) may be attributed to a change in atmospheric conditions (probably sub-pixel-sized, unresolved clouds) during the time lag between NOAA and Meteosat acquisitions. The amplitude observed in the plot of digital count is due to the change of solar zenith angle from 11:30 to 14:30, an effect which is corrected for when working with reflectances, since a normalisation by the cosine of the solar zenith angle is applied. Statistics of comparisons are shown in Table 2.

Table 2. Results of the comparison of model predicted counts (DC) and Top of Atmosphere reflectances with genuine Meteosat measurements.

Correlation Coefficient	Root Mean Square Error DC)	Root Mean Square Error (TOA reflectance)
0.94	4.06	0.014

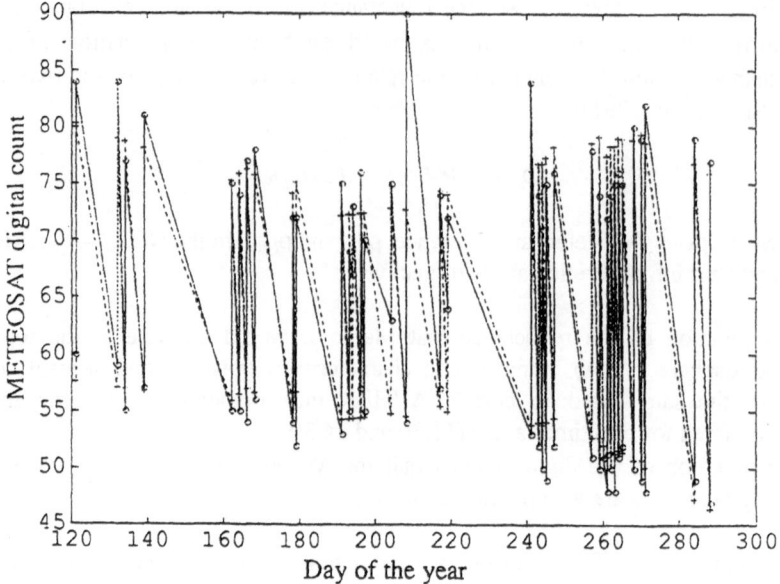

Figure 7. Comparison between original Meteosat signal (8-bit digital counts,- and simulation (--) with the model of Rahman *et al* (1993).

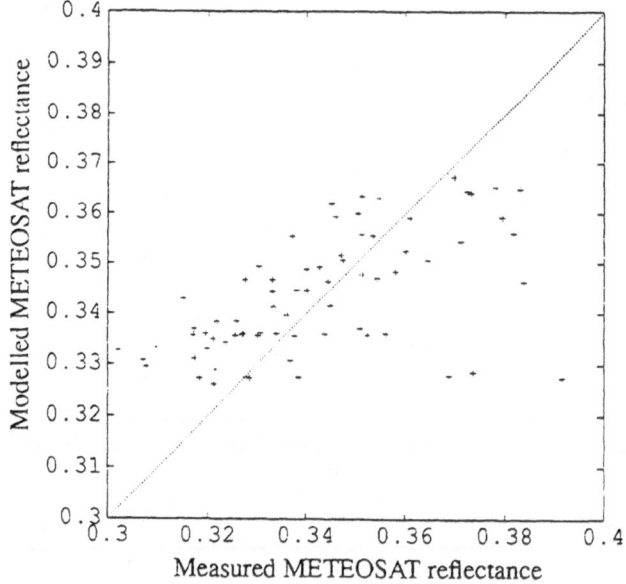

Figure 8. Comparison between TOA reflectances measured by Meteosat and simulation with the model of Rahman *et al* (1993).

The root mean square error is about 6% (*i.e.* 0.014), which is slightly better than the r.m.s.e. obtained when fitting the models against AVHRR data (Table 1), but more surprisingly the correlation coefficients are larger when comparing actual and simulated Meteosat measurements. The increase in the correlation coefficient is partly due to the larger variation of solar zenith angle. Another part of the explanation could be the use of the same atmospheric conditions to compute the atmospheric transmission both downward and upward. If errors in surface reflectances due to atmospheric effects behave with a zero mean noise, the fits of the model will give a good estimate of model parameters, while r.m.s.e. and correlation coefficients will reflect the level of noise. As the radiative transfer model we used is objective, Meteosat reflectances predicted at TOA are unaffected by uncertainties about atmospheric characteristics. This holds only if uncertainty about atmospheric effects are small, does not infer too much on the retrieval of model parameters, and does not lead to systematic bias. Therefore, if these conditions are satisfied, the comparison of predicted and measured reflectances at TOA would work better than at the surface.

In our study, despite a rather large variation of zenith and azimuth angles, phase angles range only from 0-100° for NOAA-AVHRR and 0-45° for Meteosat: phase angles of the data sets used for training and validation lie in a similar domain. Thus, as the bidirectional model strongly depends on the scattering phase function, the comparison of reflectances predicted by the models with reflectances derived from Meteosat measurements does not constitute a complete validation. However, bidirectional effects are greatest in the incidence plane, while directional behaviour is nearly lambertian when the distance to this plane increases. Since the model has been designed to account for this behaviour, one can expect that it will be fairly well reproduced. This issue is far less limiting if one is concerned with the normalisation of remotely sensed data to a common reference. In this case, the model is fitted with data and used to predict reflectance in a reference direction close to the ones of the sample.

1.4.3. *Albedo.* Direct albedo, $A(\Theta_s)$, for a given solar zenith angle, Θ_s, is computed by numerically integrating the fitted model in all viewing directions in a hemisphere. For the sun at nadir, spectral albedos we derived are 0.354 and 0.459 in visible and near infrared bands respectively. Albedo as a function of solar zenith angle is plotted in Figure 9: albedo in both spectral bands increases when solar zenith angle increases, which is in agreement with ground measurements (Kimes 1983, Kimes *et al*, 1985, Irons and Smith 1990).

1.5.   APPLICATION OF THE METHOD TO THE HAPEX-SAHEL ZONE

The method described above has been applied to a large number of pixels corresponding to the whole HAPEX-Sahel site, using the data set described in section 2.2 (1991 data). As large parts of the area are covered by vegetation which exhibits seasonal variations, we cannot assume that surface properties are constant as we had for the desert area. Therefore, we used here, a running temporal window whose size is determined by the number of "clear sky" measurements available and the number of parameters to be retrieved.

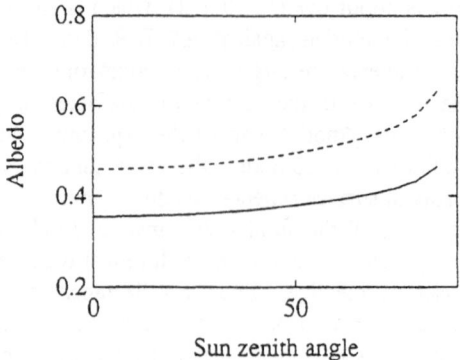

Figure 9. Albedo as a function of solar zenith angle.

In order to illustrate what can be obtained for areas covered by sparse vegetation, we provide here some results for the site of Bani Zoumbou. For this site, the statistics of the results of the fit are presented in Table 3.

For this site, we computed surface albedo in both channels. Spectral surface albedos can then be used to compute Normalised Difference Vegetation Index (NDVI) for a given solar zenith angle (Figure 10), in order to normalise NDVI to a fixed geometric reference. It should be noted that the behaviour of the NDVI computed from albedos is somewhat different from what could be expected of Maximum Value Composite technique (Holben 1986).

The method described above has also been applied to the whole HAPEX Site, on a pixel by pixel basis. The actual image of bidirectional reflectances for a given day has been compared with the bidirectional reflectance image predicted by the model. The model was fitted against observations acquired during a ten-day period. It can be seen on Figures 11 and 12 that the model works well, even if it seems to lower the contrasts. The reason for this decrease of contrast is not clear and will be investigated. The image of surface albedo for the HAPEX-Sahel region is shown in Figure 13. Figures 14 to 16 present the mapping of the three parameters of the model of Rahman *et al* (1993). Further analysis is required before firmer conclusions can be drawn, like the use of these parameter maps as a new source of information. However, the interesting feature is that none of the three parameters seems noisy, as could have been the case if either the model or its calibration did not perform sufficiently well.

Table 3. Results of the bidirectional model fit for Bani Zoumbou (sparse vegetation)

Visible		Near Infrared	
r.m.s.e	correlation	r.m.s.e.	correlation
0.009	0.889	0.010	0.917

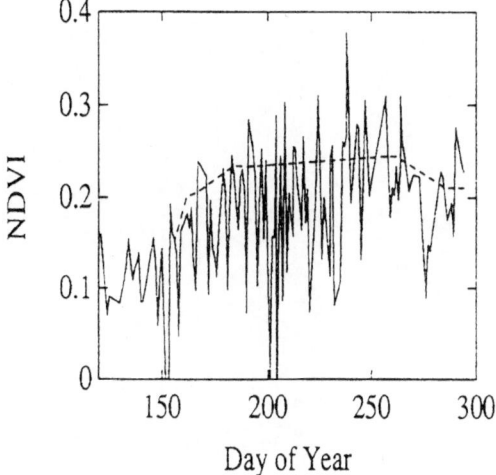

Figure 10. Temporal evolution of the NDVI.

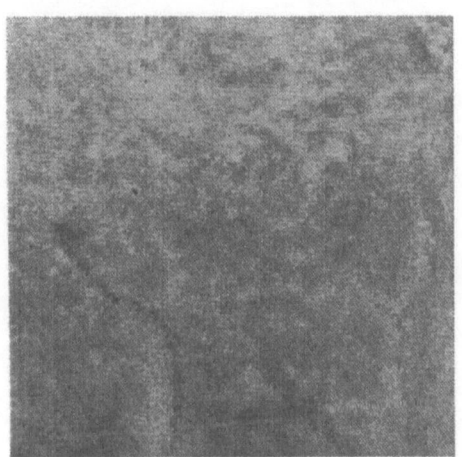

Figure 11. Actual bidirectional reflectance image over the HAPEX-Sahel site.

Figure 12. Model predicted bidirectional over the Hapex-Sahel site reflectance image.

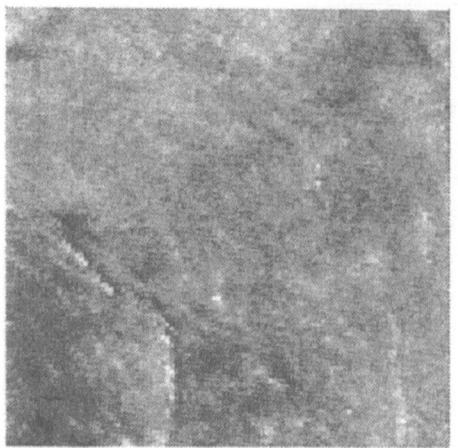

Figure 13. Direct albedo derived from angular integration of the bidirectional model

Figure 14. Mapping of the ρ_0 parameter of Rahman *et al*'s (1993) model

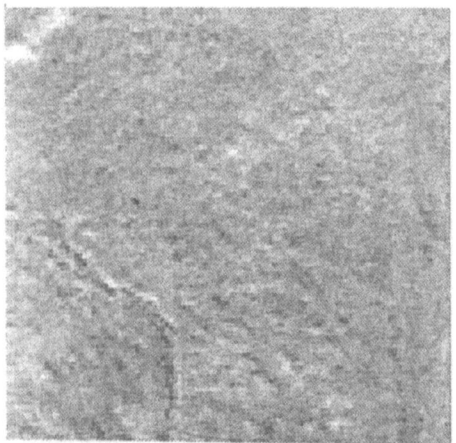

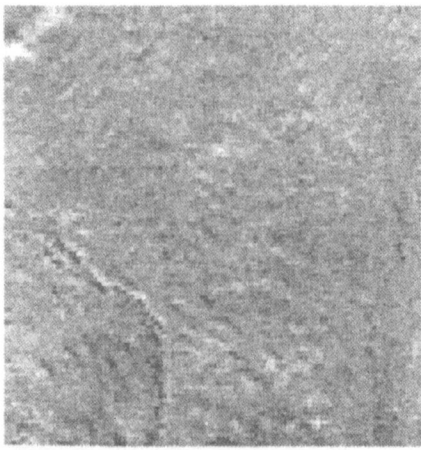

Figure 15. Mapping of the k parameter of Rahman *et al*'s (1993) model

Figure 16. Mapping of the Θ parameter of Rahman *et al*'s (1993) model

1.6.   CONCLUSION ON COUPLING BIDIRECTIONAL MODELS AND SATELLITE DATA

The general purpose of this study was to analyse the ability of directional models to supplement the limited angular sampling provided by Earth observing sensors in order to normalise satellite data or to estimate a surface albedo. Over bare soil, radiances measured by spaceborne sensors can be successfully used to fit surface reflectance bidirectional models.

The most attractive reason for using bidirectional models is the assumption that the underlying physical concepts would permit the prediction of directional behaviour in any direction, even if the set of angles available for parameter adjustment is limited, and eventually to retrieve surface parameters such as Leaf Area Index or canopy structure. In order to check this assumption, we predicted the reflectances at the top of atmosphere that the Meteosat sensor would measure and compared these predicted reflectances with the reflectances actually measured. This comparison gave very satisfactory results, with r.m.s. errors of only 0.014. However, while NOAA-AVHRR and Meteosat viewing geometries are quite different, the range of phase angles is rather close. Thus, this comparison is only a partial validation of our initial assumption. In addition, it appears that the normalisation of multitemporal data sets to a common geometric reference is feasible. A difficulty is the application of our approach over vegetated areas, for global scale studies, because of the temporal evolution of the vegetation which makes the filtering and fitting operations more difficult.

2.   Net primary productivity at the global scale from AVHRR

Another important physical parameter that scientists would like to determine from a combination of remotely-sensed data and models, particularly for carbon cycle studies, is the amount of productivity for particular areas. Gross Primary Productivity, *GPP*, is the total amount of carbon fixed by its plants in photosynthesis. Some of this fixed carbon is used to provide the energy required by plant growth or for maintenance of existing living tissues : the total amount used in this way is referred to as autotrophic respiration (R_a). Net Primary Productivity[*], *NPP*, is the difference between *GPP* and R_a *i.e.* net carbon fixation.

When a plant or some of its components (*e.g.* leaves, fine roots, *etc.*) die, organic matter falls on the soil surface (surface litter) and is eventually dispersed through the soil profile. This dead organic matter is decomposed by soil micro-organisms. Decay of soil organic matter and surface litter releases gaseous carbon compounds (*e.g.* CO_2, CH_4). This emission is referred to as heterotrophic respiration (R_h). Net ecosystem production, *NEP*, is the difference between *NPP* and R_h on an annual basis. Generally, annual *NPP*

---

[*]   Productivity differs from production. Net primary productivity is most commonly measured as dry organic matter synthesised per unit area time (g.m^{-2}. year^{-1}). Net production is expressed as dry organic matter synthesised per unit time (t.year^{-1}). Biomass is the dry matter of living organisms present at a given time per unit area (g.m^{-2}). (see Whittaker *et al.*, 1975).

and R_h are approximately balanced. A positive *NEP* corresponds to soil formation and/or increase of living biomass (stem, roots).

When dealing with the global carbon cycle, the main quantity to assess is the net ecosystem production, since *NEP* corresponds to the amount of carbon stored in soil and vegetation. Therefore, this fixed carbon does not contribute to the greenhouse effect. Unfortunately, there are few estimates of *NEP* and its estimates at the global scale are difficult, particularly if the climate is changing rapidly. From the modelling viewpoint, most of current approaches calculate *NPP* and R_h separately, and then estimate *NEP* as the difference between the two.

Characteristic time scales of processes occurring in vegetation and soils range from seconds to centuries, as shown in Table 4.

2.1.   NET PRIMARY PRODUCTIVITY MODELLING : A SHORT BIBLIOGRAPHY

We will consider here models aimed at estimating geographic patterns of *NPP* at the global scale. These models integrate experimental and theoretical knowledge in a set of equations designed to compute *NPP* as a function of varying environmental conditions.

These models use spatially referenced information on climate, soils, vegetation, *etc*. The spatial reference is grid-cell based, with a given geographic projection and resolution (generally 0.5° x 0.5° or 1° x 1°). For every grid-cell, parameters and forcing variables must be specified.

Model parameters are considered constants valid everywhere, or only for a particular ecosystem. For example, maps of vegetation types often have attributes like the percentage of tree cover or maximum *LAI*. Forcing variables correspond to variables which may vary both in time and space, such as precipitation and temperature, and which influence the results. Even for a steady state vegetation distribution, *NPP* depends on precipitation, temperature, and so on. Therefore, in addition to the spatial dimension, models also have a temporal dimension which is the time step used for running

Table 4. Characteristic time scales of processes occurring in vegetation and soils.

Time scale	Examples of processes
0 ... 1 day	photosynthesis, respiration, transpiration.
1 day...1 month	*NPP*, phenology (budburst, flowering, ...), water balance.
1 year	reserve formation, stem biomass increase, *NPP*, *NEP*, litterfall.
1 year - 5 years	short turnover soil carbon, longevity of conifer leaves (1 to 2 years).
5 years - 100 years	long turnover soil carbon (~30 years), corresponding to a high lignin content, turnover of forest carbon (~20 years).
100 years - 1500 years	stabilised soil carbon, charcoal.

(integrating) the model. Apart from parameters and forcing variables, models also include state variables which describe the state of the system (*i.e.* the ecosystem) at every time step : soil moisture, leaf surface, biomass. A global model for *NPP* estimate is made of a set of equations, parameter maps and files of variables (*e.g.* to describe weather conditions).

In this section we first present some examples of models aimed at estimating global *NPP*. We distinguish between three types of models : statistical, parametric and mechanistic. This categorisation of each is sometimes arbitrary, since typological boundaries between models are often fuzzy. We describe each model schematically before presenting some results and discussing their respective strengths and weaknesses.

2.1.1. *Statistical models.* Statistical models applicable at the global scale are based on the assumption that, more than vegetation type, climatic factors are driving *NPP*. This assumption connects to empirical schemes used to predict geographic patterns of potential vegetation by relating physiognomic vegetation types to annual precipitation (or potential evapotranspiration), *PET* and growing-season temperature (*e.g.* Holdridge, 1947). Statistical models are built by using regression between experimental measurements of *NPP* and climate variables such as precipitation or temperature.

The prototype of statistical models is the Miami model (Lieth, 1975). Lieth used more than 1,000 measurements of annual *NPP* to relate *NPP* with precipitation (*P*) and temperature (*T*) :

$$NPP(T) = 3000/[1 + exp(1.315 - 0.119T)] \qquad (6)$$

and

$$NPP(P) = 3000 [1 - exp(-000664\, P0)] \qquad (7)$$

where the *NPP* unit is grams of photosynthesised dry matter[†] per m^2 and per year, *T* is the mean annual temperature (°C), *P* is the annual precipitation (mm). *NPP* is estimated by using both of the above equations and the final estimate is the minimum of the two.

This model assumes that climate is the major driving factor of *NPP* (*e.g.* Woodward, 1987). However, it is based on a drastic simplification of primary productivity processes. Indeed, actual evapotranspiration (*AET*) is a better indicator of water resources available to plants than precipitation and temperature. Therefore, Lieth (1975) proposed an improved version of his model, called the Montreal model, where *NPP* is a function of *AET* (in mm) :

$$NPP(AET) = 3000 [1 - exp(-0009695(AET-20))] \qquad (8)$$

Several authors adapted the Miami or Montreal models in order to estimate *NPP* and carbon fluxes at the monthly time scale (*e.g.* Box, 1988, Esser, 1991).

[†] To estimate *NPP* in grams of carbon, a factor of 0.45g of carbon per gram of dry matter can be used (Atjay *et al.*, 1979). Energy content of dry biomass is of about 17.8 Joule per gram.

2.1.2. *Parametric models.* Whereas statistical models empirically relate the variable of interest (*NPP*) to factors (*P, T, etc.*) which directly influence it, parametric models try to describe the main processes occurring within a plant or an ecosystem. These models provide a description of the processes at a macroscopic level, based on a generalisation and simplification of ecophysiological knowledge.

We present here an example of a parametric model which is based on the use of satellite measurements (Ruimy *et al.* 1993). The approach is very similar to the one used by Heimann and Keeling (1989), and both are based on the work of Monteith (1972, 1977) on crops. Schematically, at each time step and for each grid-cell, the following variables are passed to the model :

- photosynthetically active radiation reaching the top of the canopy (*PAR*). *PAR* corresponds to visible solar radiation, *i.e.* wavelengths ranging from 0.4 to 0.7 μm.
- fraction of *PAR* which is absorbed by the canopy (*APAR = Absorbed PAR*). An absorption coefficient, ε_i, is defined as the ratio *APAR/PAR*.
- photosynthetic efficiency, ε_b, which is the amount of organic dry matter photosynthesised per unit of absorbed *PAR*. The ε_b coefficient is in grams of dry matter per megajoule (1 MJ - 10^6 J) of *APAR* ($g_{(drymatter)}$/MJ).

NPP for a given period is the result of the integral over time :

$$NPP_{g(dry\ matter)} = \int_{t_0}^{t} e_b(t)\varepsilon_i PAR(t)dt \qquad (9)$$

The use of this model at the global scale requires that every variable is defined at each time step for every grid-cell. Ruimy *et al.* (1993) ran this model with a weekly time step and 1° x 1° resolution. *PAR* absorption coefficient, ε_i, is derived from remotely-sensed measurements in visible and near infrared spectral intervals. Indeed, both experimental (*e.g.* Asrar *et al.* 1984) and theoretical studies (Sellers, 1985, 1987) showed that the PAR absorption coefficient, ε_i, can be estimated from various combinations of visible and near infrared reflectances as measured by satellite sensors. (1993) *et al.* used the following relationship :

$$\varepsilon_i = -0.025 + 1.25\ NDVI \qquad (10)$$

where NDVI is the so-called normalised difference vegetation index :

$$NDVI = \frac{\rho_{pir} - \rho_{vis}}{\rho_{pir} + \rho_{vis}} \qquad (11)$$

where ρ_{vis} and ρ_{pir} are visible and near infrared reflectances, respectively.

Photosynthetically active radiation, *PAR*, is derived from global solar radiation, R_g, which is more frequently measured than *PAR*, while:

$$PAR = \varepsilon_c\, R_g \qquad (12)$$

Ruimy *et al.* (1993) set ε_c at *0.48*, while R_g was obtained from calculations by a climate model. In the near future, R_g estimates will also be derived from satellite based algorithms (*e.g.* Dedieu *et al.* 1987, Whitlock *et al.* 1990, Whitlock *et al.* 1993).

For each of the 10 vegetation types taken into account, the values of the photosynthetic efficiency, ε_b, were derived from published results. On average, for an annual vegetation cycle, ε_b varies from 0.5 to 3 g.MJ^{-1}. One of the difficulties encountered is the scarcity of experimental measurements of ε_b. In addition, most of the published results are only valid for above-ground biomass and some empirical relationship has to be used to take into account production of below-ground biomass. Finally, a variety of definitions are used by the various authors for ε_b. Some of them used intercepted *PAR* instead of absorbed *PAR*.

2.1.3. *Mechanistic models.* Mechanistic models integrate the information derived from empirical and theoretical studies to describe vital processes underlying plant growth and survival. These models account for fundamental processes such as photosynthesis, respiration, water use, assimilate allocation, and their dependence on environmental conditions (meteorology, soil type, nutriments). Their objective is to explain and predict the functioning of vegetation.

We will consider here the *Terrestrial Ecosystem Model* (*TEM*) proposed by Raich *et al.* (1991) (see also Melillo *et al.* 1993). *TEM* is an *Ecosystem/tissue model* (Ågren *et al.* 1991) based on a rather sophisticated modelling of vegetation functions, connected to a nutrient-cycling model. We present this model, since, as far as we know, it is the only mechanistic model conceived for, and applied at, the global scale and already published.

The *TEM* contains five state variables and three compartments described in the following figure :

TEM contains only one compartment for living vegetation therefore it does not distinguish between the various tissues (leaves, stems, roots). Much attention has been paid to the nitrogen cycle. *TEM* operates with a monthly time step. Monthly net primary productivity, *NPP*, is the difference between gross primary productivity, *GPP*, and autotrophic respiration R_a :

$$NPP_t = GPP_t - R_{at} \qquad (13)$$

252

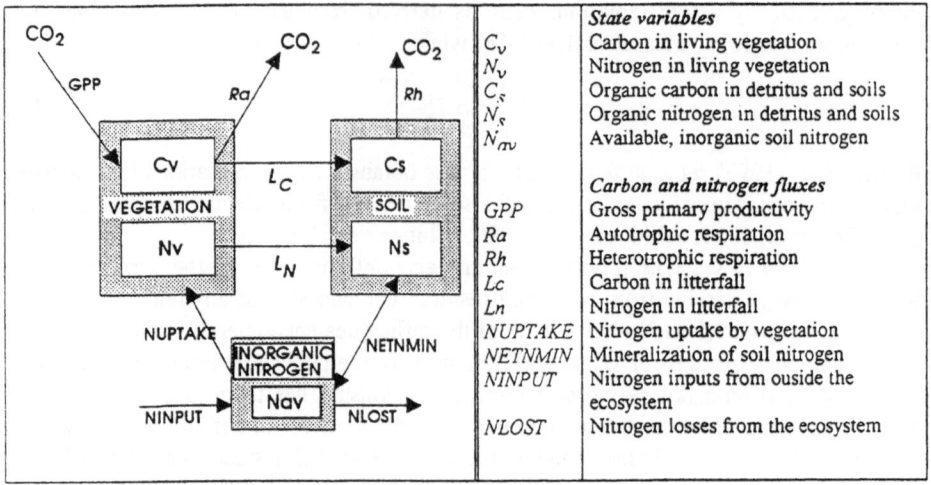

Figure 17. The Terrestrial Ecosystem Model (TEM, from Raich *et al.* 1991).

In its current version, *TEM* assumes a potential vegetation and annual balance of nitrogen and carbon pools. Annual net ecosystem productivity is zero, in other words, *GPP* is balanced by autotrophic and heterotrophic respiration. Litterfall, L_c, is balanced annually by the heterotrophic respiration. In the same way, nitrogen uptake by vegetation is equal on an annual basis to nitrogen in litterfall and to net nitrogen mineralisation (*NETNMIN*). Net nitrogen mineralisation is the amount of inorganic nitrogen produced during decomposition less that immobilised by decomposer organisms.

Nitrogen inputs include nitrogen provided by precipitation, dry deposition and nitrogen fixed by soil micro-organisms. Nitrogen losses (*NLOST*) include gaseous losses of nitrogenous compounds, losses of soluble inorganic and organic nitrogenous compounds.

TEM computes the gross primary productivity as a function of irradiance of photosynthetically active radiation, *PAR*, atmospheric CO_2 concentration, moisture availability, mean air temperature, photosynthetic capacity, and, indirectly, nitrogen availability :

$$GPP = \left(C_{max}\right) \frac{PAR}{k_i + PAR} \frac{C_i}{k_c + C_i} (TEMP)\left(A_c\right)(KLEAF) \qquad (14)$$

C_{max} is the maximum rate of carbon assimilation by the entire plant canopy under optimal environmental conditions (g.m^{-2}.month^{-1}). k_i is the irradiance of *PAR* at which the assimilation is equal to $0.5C_{max}$. C_i is the concentration of CO_2 inside leaves and k_c is the value of C_i at which the assimilation is equal to 0.5 C_{max}. *TEMP*, A_c and *KLEAF* are unitless coefficients expressing the influence of air temperature, relative nutrient availability and plant phenology, respectively. These coefficients are estimated using empirical parameterisations, not detailed here.

Autotrophic respiration, R_a, is the sum of growth respiration, R_g, and maintenance respiration, R_m. Maintenance respiration is a direct function of plant biomass, C_v, and increases with air temperature, T, with a Q_{10} of 2.0 :

$$R_m = K_r(C_v)exp(0.0693T) \tag{15}$$

where K_r is the respiration rate of the vegetation per unit of biomass carbon at 0 °C (g.g^{-1}.month^{-1}).

Growth respiration is estimated to be 20% of the difference (if > 0) between *GPP* and R_m.

2.2. SOME RESULTS

Net primary productivity estimated by some models for the whole of the terrestrial biosphere and a few vegetation types are summarised in the following Table 5:

Table 5. Net primary productivity estimated by some models for the whole of the terrestrial biosphere and a few vegetation types.

Model	① Gt(C)	② g(C).m^{-2}.y^{-1}	③ g(C).m^{-2}.y^{-1}	④ g(C).m^{-2}.y^{-1}	⑤ g(C).m^{-2}.y^{-1}
Statistical models					
Lieth, 1975	53	1000	400	400	50
Box, 1988	68				
Esser, 1991	??				
Fung[1] *et al.* 1987	47	800	300	600	100
Parametric models (satellite based)					
Heimann & Keeling, 1989	56				
Ruimy *et al.* 1993	60	450	500	300	100
Mechanistic models					
Melillo *et al.* 1993	53	1098	393	238	120

Annual net primary productivity of Earth vegetation ① and mean *NPP* for : ② Evergreen tropical forest, ③ Savanna, ④ Boreal forest, ⑤ Tundra.

[1] *Strictly speaking, the approach of Fung et al. quoted here is not model-based but uses satellite measurements to derive seasonal evolution of NPP from annual in-situ estimates.*

The results must be analysed with caution. Indeed :

- some of the authors are dealing with potential vegetation (Melillo *et al.* 1993 and Box, 1988) while others are considering actual vegetation (Ruimy *et al.* 1993 and Heimann and Keeling, 1989);
- *NPP* is estimated for a mean present climate (Box, Lieth, Melillo *et al.* Fung) or for a specific year (1986 for Ruimy *et al.* 1978-1979 for Heimann and Keeling but with 1983-1984 satellite data);
- vegetation types and their geographic extent depend on the authors.

The most noticeable difference in *NPP* estimates occurs for tropical forest between the results of Ruimy *et al.* (1993) and the other assessments. The lack of experimental measurements for tropical forest (but also for the other vegetation types) does not allow us to conclude on this difference. According to some measurements (Saldarriaga and Luxmoore, 1991), *NPP* of undisturbed moist tropical forest is weak since maintenance respiration, linked to biomass, is high. But the discussion is still open.

2.3. DISCUSSION

We have described some examples of models aimed at estimating *NPP* on a global scale. Some models go further by including decomposition processes and the nitrogen cycle.

2.3.1. *Which type of model?*

Statistical models are easy to implement and they probably have the capacity to provide rather accurate *NPP* estimates if they are established with a large number of measurements. However, they were established for a set of environmental conditions which characterise the present, assuming a nearly steady state. They might fail to provide accurate estimates for different or continuously evolving conditions like that of a climate transition. The relevance of modelling is not only to provide quantitative results but also to be a tool for improving our understanding of processes. From this standpoint, the heuristic capability of statistical models is weak.

The *parametric model* we described above (Ruimy *et al.* 1993, close to the model of Heimann and Keeling, 1989) is based on high temporal resolution satellite measurements which serve as inputs to a simplified and macroscopic model. This type of model needs to know a large number of photosynthetic efficiencies for each vegetation type, which raises several issues such as measurement consistency. In addition, these models have no predictive capacity, since input variables are derived from observations. However, these parametric models have the advantage of being based on remotely-sensed measurements which provide relevant and global information on vegetation cycles, taking into account seasonal and interannual variability.

Mechanistic models generalise, at scales of about 100 x 100 km, processes which have been studied at the scale of plant component or whole plant. Mechanistic models are rather complex, and require data which are generally not available over long time periods and for every ecosystem. The generally adopted solution is to calibrate the model (see below). Mechanistic models are prognostic models which, in principle, can be interactively coupled with climate model (*e.g.* Melillo *et al.*). These constitute a general framework to structure our current understanding of biosphere processes and provide guidelines to define experiments by suggesting a new functioning hypothesis.

2.3.2. *Model calibration.* Calibration of a model is the process by which all or part of the parameter values are adjusted by regression or iterative techniques so that the results of the model fit with the available observations.

Calibrating statistical models is relatively easy, since (for example) knowing precipitation, temperature and *NPP* for a large set of data is sufficient to determine the four parameters of the Miami model.

Mechanistic models are based on a set of parameters derived either from experiments or from calibration : it is difficult, even for a single vegetation type, to measure all the needed parameters throughout the life span of the plant. However, the number of field experiments will probably increase in the future.

The *TEM* calibration uses observations acquired in a limited number of sites. These observations consist of the amount of carbon (C) and nitrogen (N) present in vegetation and soils, C and N fluxes, *NPP*, and climate data. Nine of the model parameters which are controlling C and N fluxes (Figure 17) are adjusted so that the model reproduces observations. The other parameters are derived from published information. The set of parameters for a given type of vegetation is then used for any grid-cell where the same vegetation type is present.

A different approach for calibration has been used by Heimann and Keeling (1989). They introduced, into an atmospheric transport model, observed and predicted CO_2 fluxes exchanged between the various sources and sinks (biosphere, oceans, fossil fuels). They did not calibrate the *NPP* sub-model, but instead adjusted the rate of heterotrophic respiration so that simulated atmospheric CO_2 concentration reproduced concentrations measured by ground stations. This is a very global calibration applied to only one parameter.

2.3.3. *The validation issue.* A first remark is that none of the *NPP* models have been validated at the global scale. Only the approach of Heimann and Keeling (1989) has been validated globally as described above, but the resolution of this type of validation based on the use of atmospheric transport models is weak and some error-compensating effects may exist.

Validations are generally performed with local measurements of *NPP*, or carbon fluxes. The model is then calibrated with a sub-set of measurements and validated with the rest of the available measurements (different sites and/or different environmental conditions). But we have here the opposite problem compared with the ones which arise when using atmospheric transport models : validation applies to only small areas, and a number of measurements are required to increase its reliability.

Calibration and validation issues are closely linked since both depend on the availability of consistent observations at the global scale. Some possible solutions will be described below and particularly the potential role of satellite observations as a new source of global, high resolution information.

2.4. A POSSIBLE SOLUTION FOR THE CALIBRATION OF VEGETATION PROCESS MODEL

We previously mentioned the difficulty of calibrating/validating mechanistic models at the global scale. The available tools either provide a detailed but local validation (field experiments) or a very global one which integrates the whole carbon fluxes (atmospheric transport models). Satellite observations may help to resolve the dilemma, since they provide consistent, global and high spatial and temporal resolution data. However, satellite sensors only measure radiances exiting at the top of the atmosphere, and most of surface parameter retrievals are indirect.

Apart from well-known applications like dynamic mapping of vegetation, remotely-sensed measurements may provide two different kinds of information :

- *Parameterisation of some phenological stages* in connection with environmental conditions. For example, time series of vegetation indices could be analysed together with meteorological data in order to parameterise the beginning of the vegetation cycle over large areas.
- *Parameter retrieval* : remotely-sensed measurements may be used to invert a sensitive parameter[‡] of the model which is then used to force the model. This is what is done by Ruimy *et al.* (1993) and Heimann and Keeling (1989) when they retrieve *PAR* absorption coefficient. A slightly different approach retrieves leaf area index *LAI*, from reflectance measurements, and injects it into the model (Running *et al.* 1989, Band *et al.* 1993). These injections of *LAI* or *APAR* are very constraining forcings, since photosynthesis strongly depends on them. Some attempts also exist to derive canopy stomatal resistance from thermal infrared measurements (Taconet *et al.* 1986). All these approaches are diagnostic in essence. The retrieval of model parameters through inversion schemes has several drawbacks, such as low frequency because of cloud cover and weak accuracy due to the separate use of each observation.

Satellite-derived data may also provide other possible benefits. For validation first, a canopy radiative transfer model can be coupled to a mechanistic model to predict *LAI*. These coupled models can predict the reflectances a satellite instrument would measure. Temporal evolution and the level of canopy reflectances depend mainly on vegetation state, ground cover and *LAI*, all parameters which also strongly influence *NPP*.

[‡] In the rest of the text we will generally use the word parameter for both parameters and variables.

Comparison, both in time and space, of observed and predicted reflectances constitutes a tool for validating vegetation functioning models (*e.g.* Moulin and Fischer, 1993). This validation is mainly applicable for vegetation types which exhibit marked seasonal cycles, but is probably not suited for evergreen forests.

This kind of direct approach could be extended to several spectral domains (see Bouman, 1991, for RADAR wavelengths). By coupling a vegetation functioning model and a land surface flux model (SVAT's) predicting surface temperatures or thermal fluxes (in the infrared or microwave wavelengths), both measurable from satellite, could be considered.

We may notice, first, that canopy reflectances and temperatures as observed from space are closely connected to two fundamental processes of vegetation functioning : absorption of solar energy and transpiration. Simulating these reflectances and temperatures accurately put strong constraints on functioning models, at least for most of the vegetation types. We can, therefore, look for methods which could benefit from these constraints to adjust model parameters and variables. Such methods already exist in the different context of meteorology and are referred to as data assimilation methods.

Parameters which can be derived directly from satellite measurements are not necessarily the most pertinent for the model. A better approach seems to add to vegetation models the ability to predict remotely-sensed measurements. These measurements will be used to periodically constrain the model which evolve the rest of the time under the influence of external (meteorology) or internal (carbon uptake, phenology) factors. The main advantage of this approach is first to allow the adjustment of parameters which are not directly observable from satellites and second to respect the internal dynamics to the model and the knowledge of processes as described by the equations.

In the field of meteorology, the question is usually to define the set of initial conditions from which an AGCM will be run for weather forecasting. All the available information on the difference between predicted and actual state is contained in the difference between predicted and actual observations (Talagrand, 1987). The role of data assimilation is to reduce these differences through optimisation techniques, leading to an improvement of initial conditions. A similar approach is now being tested (*e.g.* Moulin and Fischer, 1993) to initialise parameters of vegetation process models.

Remotely-sensed data assimilation in vegetation models seems to be a promising approach. However, it is worthwhile to notice that satellite observations are only measurements of electromagnetic radiation and therefore we need to couple functioning and radiative transfer models.

Satellite observations in the solar spectrum depends on *LAI* and ground cover. If we intend to relate observations to model predicted reflectances, functioning models must be able to describe, specifically, plant components needed for such prediction. Lumping the whole biomass into one "black box" is not appropriate. In addition, remote sensing measurements are instantaneous, and it will probably be necessary to decrease the time step of models working at the monthly scale.

Several functioning models are being developed which include several compartments : *e.g.* FOREST-BGC (Running and Coughlan, 1988) and the model proposed by Janecek *et al.* (1989). These two models were initially designed for forests and are now being extended to every biome. LERTS is also developing a model (Kergoat and Dedieu, 1993) which, from the beginning, accounts for possible satellite data assimilation. This model includes a number of compartments. This number may be changed according to vegetation type or for specific purposes. Note that more attention must be paid to assimilate allocation when the number of compartments is increased.

2.5 CONCLUSION

We have briefly presented some of the current research topics which deal with vegetation and climate interactions. The focus was on the modelling of vegetation functioning at the global scale for relatively short time periods, typically from a few days to a few years. Net primary productivity is the main feature of biosphere and we presented various methods for its estimation at the global scale. We did not try to synthesise the large number of existing models designed for local scale applications.

We stood intentionally within the objective of predicting ecosystem functioning in response to climate changes. This standpoint explains the large part devoted to modelling and to model strengths and weaknesses. Clearly, ecosystem modelling is just beginning and large uncertainties remain. However, despite their drawbacks, models are essential tools which will favour the vital development of a multi-disciplinary dialogue by providing a general framework for field experiments.

We did not discuss some important questions, such as the links which may exist between vegetation functioning and biome models. The coupling of these two types of models will probably be one of the major areas of research in the coming years.

References and Bibliography for the "Surface albedo" topic

Arino, O., Dedieu, G., and Deschamps, P.Y., 1991: Determination of land surface spectral reflectances using Meteosat and NOAA/AVHRR shortwave channel data. *International Journal of Remote Sensing*, **13**, 2263-2287.

Cabot, F., and Dedieu, G., 1993: Surface albedo from space : coupling bidirectional models and remotely sensed measurements. Submitted to *Journal of Geophysical Research*.

Cabot, F., Dedieu, G., and Maisongrande, P., 1993a : Surface albedo from space over HAPEX SAHEL sites. To be published in the *Proceedings of the 6th AVHRR Data User's Meeting*, EUMETSAT-JRC, Belgirate, Italy, 28th June-2nd July, 1993.

Cabot, F., Maisongrande, P., and Dedieu, G., 1993b : Monitoring the AVHRR calibration. To be published in the proceedings of the *International Geoscience and Remote Sensing Symposium* (IGARSS'93), Tokyo, Japan, August 18-21, 1993.

Deering, D.W, Eck, T.F., and Otterman, J., 1990 : Bidirectional reflectances of selected desert surfaces and their three-parameter soil characterisation. *Agricultural and Forest Meteorology*, **52**, 71-93.

Eaton, F.D., and Dirmhirn, I., 1979: Reflected irradiance indicatrices of natural surfaces and their effect on albedo. *Applied Optics*, **18**, 994-1008.

Faizoun, C.A., and Dedieu, G., 1993 : Atmospheric effects on NOAA/AVHRR shortwave measurements: Sensitivity study and use of atmospheric climatology to correct AVHRR time series. To be published in *Proceedings of 6th AVHRR Data User's Meeting*, EUMETSAT-JRC, Belgirate, Italy, 28th June-2nd July, 1993.

Goward, S.N., Markham, B., Dye, D.G., Dulaney, W., and Yang, J., 1991 : Normalized difference vegetation index measurements from the advanced very high resolution radiometer. *Remote Sensing of Environment*, **35**, 257-277.

Gutman, G.G., Ohring, G., Tarpley, D., and Ambroziak, R., 1989 : Albedo of the U.S. Great Plains as determined from NOAA-9 AVHRR Data. *Journal of Climate*, **2**, 608-617.

Gutman, G.G., 1991 : Vegetation indices from AVHRR: an update and future prospects. *Remote Sensing of Environment*, **35**, 121-136.

Hapke, B., 1981 : Bidirectional reflectance spectroscopy 1. Theory. *Journal of Geophysical Research*, **86**, 3039-3054.

Holben, B.N., 1986 : Characteristics of maximum-value composite images from temporal AVHRR data. *International Journal of Remote Sensing*, **7**, 1417-1434.

Irons, J.R., and Smith, J.A., 1990 : Soil surface roughness characterization from light scattering observations. *Proceedings of IGARSS'90*, Washington, USA, 20-24 May, 1990, 1007-1010, 1990.

Kerr, Y., and Lagouarde, J.P., 1989 : On the derivation of land surface temperature from AVHRR data. *Proceedings of the 4th AVHRR data users' meeting*. Rothenburg, FRG, 157-160.

Kimes, D.S., and Sellers, P., 1985 : Inferring hemispherical reflectance of the Earth's surface for global energy budgets from remotely sensed nadir or directional radiance values. *Remote Sensing of Environment*, **18**, 205-223.

Kimes, D.S., and Holben, B.N., 1992 : Extracting spectral albedo from NOAA-9 AVHRR multiple view data using an atmospheric correction procedure and an expert system. *International Journal of Remote Sensing*, **13**, 275-289.

Kimes, D.S., 1983 : Dynamics of directional reflectance factor distribution for vegetation canopies. *Applied Optics*, **22**, 1354-1372.

Kimes, D.S., Newcomb, W., Tucker, C.J., Zonneveld, I.S., Van Wijngaarden, W., De Leeuw, J., and Epema, G.F., 1985 : Directional reflectance factor distributions for cover types of Northern Africa. *Remote Sensing of Environment*, **18**, 1-19.

London, J., Bojkov, R.D., Oltmans, S., and Kelley, J.I., 1976 : Atlas of the global distribution of total ozone July 1957 - June 1967, NCAR/TN-113+STR.

McClatchey, R.A., Fenn, R.W., Selby, J.E.A., Garing, J.S., and Volz, F.E., 1971 : Optical properties of the atmosphere. (AFCLR-71-0279) Air Force Cambridge Research Lab., Bedford, Mas.

Pinty, B., Verstræte, M.M., and Dickinson, R.E., 1989: A physical model for predicting bidirectional reflectances over bare soil. *Remote Sensing of Environment*, **27**, 273-288.

Rahman, H., and Dedieu. G., 1993 : A simple method for the atmospheric correction of satellite measurements in the solar spectrum. Submitted to *International Journal of Remote Sensing.*

Rahman, H., Pinty, B., and Verstræte, M.M., 1993 : A coupled surface-atmosphere reflectance (CSAR) model part 1: model description and inversion on synthetic data, Submitted to *Remote Sensing of Environment.*

Ross, J.K., 1981 : The radiation regime and architecture of plant stands, 1981 : Dr W. Junk Publishers, Boston.

Roujean, J.L., Leroy, M., and Deschamps, P.Y., 1992 : A bidirectional reflectance model of the Earth's surface for the correction of remote sensing data. *Journal of Geophysical Research,* **97,** (20) 455-468.

Roujean, J.L., Leroy, M., and Deschamps, P.Y., and Podaire, A., 1992 : Evidence of surface bidirectional effects from a NOAA/AVHRR multitemporal data set. *International Journal of Remote Sensing,* **13,** 685-698.

Shibayama, M., and Wiegand, C.L., 1985 : View azimuth and zenith, and solar angle effects on wheat canopy reflectance. *Remote Sensing of Environment,* **18,** 91-103.

Tanré, D., Deroo, C., Duhaut, P., Herman, M., Morcrette, J.J., Perbos, J., and Deschamps, P.Y., 1990 : Description of a computer code to simulate the satellite signal in the solar spectrum: the 5S code. *International Journal of Remote Sensing,* **11,** 659-668.

Verstræte, M.M., Pinty, B., and Dickinson, R.E., 1990 : A physical model of the bidirectional reflectance of vegetation canopies, 1. Theory. *Journal of Geophysical Research.* **95.** (11) 755-765.

References and Bibliography for the "net primary production" topic

Ågren, G.J., McMurtie, R.E., Parton, W.J., Pastor, J., and Shugart, H.H., 1991 : State-of-the-art of models of production-decomposition linkages in conifer and grassland ecosystems. *Ecological Applications,* **1,** 118-138.

Asrar, G., Fuchs, M., Kanemasu, E.T., and Hatfield, J.L., 1984 : Estimating absorbed photosynthetic radiation and leaf area index from spectral reflectance in wheat. *Agronomy Journal,* **76,** 300-306.

Band, L.E.., Patterson, P., Nemani, R., and Running, S.W. 1993 : Forest ecosystem processes at the watershed scale: incorporating hillslope hydrology. *Agricultural and Forest Meteorology,* **63,** 93-126.

Bascatowe, R., Adams, J.A., Keeling, C.D., Moss, D.J., Whorf, T.P., and Wong, C.S. 1980 : Response of atmospheric carbon dioxide to the weak 1975 El Nino. *Science,* **210,** 66-68.

Bolin, B. 1986 : How much CO_2 will remain in the atmosphere - The carbon cycle and projections for the future. In *greenhouse effect, climate change and ecosystems, SCOPE 29,* Bolin, Doos, Warrick and Jäger (Eds), John Wiley & Sons, pp 93-155.

Bonan, G.B. 1991 : Atmosphere-Biosphere exchange of carbon dioxide in boreal forest. *Journal of Geophysical Research,* **96,** D4, 7301-7312.

Bouman, B.A.M. 1991 : The linking if crop growth models and multi-sensor remote sensing data. *Proceedings of the 5th int. Coll. on Physical Measurements and Signatures in Remote Sensing.* Courchevel, France, 14-18 January 1991, ESA SP-319, 583-588.

Box, E. 1988 : Estimating the seasonal carbon source-sink geography of a natural, steady-state terrestrial biosphere. *Journal of Applied Meteorology*, **27**, 1109-1123.

Dedieu, G., Deschamps, P.Y., and Kerr, Y.H. 1987 : Satellite estimation of solar irradiance at the surface of the earth and of surface albedo using a physical model applied to Meteosat data. *Journal of Climate and Applied Meteorology*, **26**, 79-87.

Esser, G. 1991 : Osnabrück Biosphere Model : structure, construction, results. In : *Modern Ecology : basic and applied aspects.* Esser, G. and Overdieck (Eds), Elsevier, Amsterdam, London, New York, Tokyo, pp 679-709.

Fung, I.Y., Prentice, K., Matthews, E., Lerner, J., and Russel, G. 1983 : Three dimensional tracer model study of the atmospheric CO_2: response to seasonal exchanges with the terrestrial biosphere. *Journal of Geophysical Research*, **88**, 1281-1294.

Fung, I.Y., Tucker, C.J., and Prentice, K.C. 1987 : Application of Advanced Very High Resolution Radiometer to study atmospheric-biosphere exchange of CO_2. *Journal of Geophysical Research*, **92**, 2999-3015.

Heimann, M., and Keeling, C.D. 1989 : A three-dimensional model of atmospheric CO_2 transport based on observed winds. 2. Model description and simulated tracer experiments. In : D.H. Peterson (Ed) : *Aspects of climate variability in the Pacific and the Western Americas. Geophysical Monograph* 55, pp 237-274.

Holdridge, L.R. 1947 : Determination of world formations from simple climatic data. *Science*, **105**, 367-368.

Janecek, A., Benderoth, G., Lüdeke, M.K.B., Kindermann, J., and Kohlmaier, G.H. 1989 : Model of the seasonal and perennial carbon dynamics in deciduous-type forests controlled by climatic variables. *Ecological Modelling*, **49**, 101-124.

Keeling, C.D., Piper, S.C., and Heimann, M. 1989 : A three-dimensional model of atmospheric CO_2 transport based on observed winds : 4. Mean annual gradients and interannual variations. In : D.H. Peterson (Ed) : *Aspects of climate variability in the Pacific and the Western Americas. Geophysical Monograph* 55, pp 305-363.

Kergoat, L., and Dedieu, G. 1993 : A Generic Model for Global Carbon Study and Satellite Data Assimilation. To be submitted for the proceedings of the *4th International CO_2 Conference*, Carqueiranne, France - September 13-17, 1993.

Lieth, H. 1975 : Primary production of the major vegetation units of the world. In : *Primary Productivity of the Biosphere*, Lieth, H., and Whittaker, R.H., (Eds), Berlin-Heidelberg-New York, Springer Verlag, pp 237-263.

Martin, P. 1993 : Vegetation responses and feedbacks to climate : a review of models and processes. *Climate Dynamics*, **8**, 201-210.

Melillo, J.M., McGuire, A.D., Kicklighter, D.W., Moore, B., Vorosmarty, C.J., and Schloss, A.L. 1993 : Global climate change and terrestrial net primary production. *Nature*, **363**, 234-240.

Monteith, J.L. 1972 : Solar radiation and productivity in tropical ecosystems. *Journal of Applied Ecology*, **9**, 277-294.

Monteith, J.L. 1977 : Climate and the efficiency of crop production in Britain. *Royal Society of London, Philosophical Transaction, Series B*, **281**, 277-294.

Moulin, S., and Fischer, A. 1993 : Simulation of the temporal variations of NOAA/AVHRR reflectances. Coupling of functional model and satellite data. To appear in the proceedings of the *6th AVHRR Data Users' Meeting*, EUMETSAT-JRC, Belgirate, Italy, 28th June - 2nd July 1993.

Porter, J.R., Bragg, P.L., Rayner, J.H., Weir, A.H., and Landsberg, J.J. 1982 : The ARC Winter Wheat Model - principles and progress. *British Plant Growth Regulator Group, Monograph 7, 'Opportunities for manipulation of cereal productivity'*, (ed. A.F. Hawkins and B Jeffeoat).

Prentice, I.C., Cramer, W., Harrison, S.P., Leemans, R., Monserud, R.A., and Soloman, A.M. 1992 : A global biome model based on plant physiology and dominance, soil properties and climate. *Journal of Biogeography*, **19**, 117-134.

Raich, J.W., Rasterrer, E.B, Melillo, J.M., Kicklighter, D.W., Steudler, P.A., Peterson, B.J., Grace, A.L., Moore, B., and Vorosmarty C.J. 1991 : Potential net primary productivity in South America: application of a global model. *Ecological Applications*, **1**, 399,429.

Ruimy, A., Dedieu, G., and Saugier, B. 1993 : Methodology for the estimation of terrestrial net primary production from remotely sensed data. Submitted to *Journal of Geophysical Research*.

Running, S.W., and Coughlan, J.C. 1988 : A general model of forest ecosystem processes for regional applications. I. Hydrologic balance, canopy gas exchange and primary production processes. *Ecological Modelling*, **42**, 125-154.

Running, S.W., and Gower, S.T. 1991 : A general model of forest ecosystem processes for regional applications. II. Dynamic C allocation and nitrogen budgets. *Tree Physiology*, **9**, 147-160.

Running, S.W., Nemani, R.R., Peterson, D.L., Band, L.E., Potts, D.F, and Oierce, L.L. 1989 : Mapping regional forest evapotranspiration and photosynthesis by coupling satellite data with ecosystem simulation. *Ecology*, **70**, 1090-1101.

Saldarriaga, J.G., and Luxmoore, R.J. 1991 : Solar energy conversion efficiencies during succession of a tropical rain forest in Amazonia. *Journal of Tropical Ecology*, **7**, 233-242.

Sellers, P.J. 1985 : Canopy reflectance, photosynthesis and transpiration. *International Journal of Remote Sensing*, **6**, 1335-1372.

Sellers, P.J., 1987 : Canopy reflectance, photosynthesis and transpiration. II. The role of biophysics in the linearity of their interdependence. *Remote Sensing of Environment*, **21**, 143-183.

Sellers, P.J., Mintz, Y., Sud, Y.C., and Dalcher, A. 1986 : A simple biosphere model (SiB) for use within General Circulation Models. *Journal of Atmospheric Sciences*, **42**, 505-531.

Spitters, C.J.T., van Keulen, H., and van Kraalingen, D.W.G. 1989 : A simple and universal crop growth simulator: SUCROS87, in *Simulation and systems management in crop protection*, R. Rabbinge, S.A. Ward and H.H. van Laar (Eds), Simulation Monographs 32, PUDOC, Wageningen.

Taconet, O., Bernard, R., and Vidal-Modjar, D. 1986 : Evapotranspiration over an agricultural region using a surface flux/temperature model based on NOAA/AVHRR data. *Journal of Climate and Applied Meteorology*, **25**, 284-307.

Talagrand, O. 1987 : Assimilation des observations météorologiques. In *Climatologie et Observations Spatiales*. Summer School of CNES 1986. CEPADUES - Editions.

Tans, P.T., Fung, I.Y., and Takahashi T. 1990 : Observational constraints on the global atmospheric CO_2 budget. *Science*, **247**, 1431-1438.

Whitlock, C.H., Charlock, T., Staylor, W.F, Pinker, R.T., Laslo, I., DiPasquale, Ritchey. 1993 : WCRPSRB SW Data Product Description - Version 1.1 NASA TM 107747.

Whitlock, C.H., Staylor, W.F., Darnell, W.L., Chou, M.D., Dedieu, G., Deschamps, P.Y., Ellis, J., Gautier, C., Frouin, R., Pinker, R.T., Laslo, I., Rossow, W.B., and Tarpley, D. 1990 : Comparison of surface radiation budget satellite algorithms for downwelled shortwave irradiance with Wisconsin FIRE/SRB surface-truth data. *Proceedings of the Seventh Conference on Atmospheric Radiation, July 23-27, San Francisco, California. Published by the American Meteorological Society, Boston, Mass.* pp 237-242.

Woodward F.I. 1987 : *Climate and Plant Distribution*. Cambridge University Press.

LAND COVER MAPPING FOR GLOBAL CHANGE RESEARCH

G. SAINT
Centre National d'Etudes Spatiales,
18 avenue Edouard Belin,
31055 TOULOUSE CEDEX,
FRANCE.

ABSTRACT. Global change research, its cause and effect, and its relationship to human activity relies on the physical understanding of the Earth's climate and its interactions with oceans and land. Land cover is a significant boundary condition for all climatic processes. Therefore, knowledge, detection and characterisation of land cover and its change is important for the understanding and prediction of overall global change. Global land cover characterisation is now evolving from classical methods which were based on well-defined, well-documented and often ground-based categories, to a system based more on remote, repetitive and continuous mapping of surface parameters through the use of adapted remote sensing systems. New techniques are developing for surface parameter estimation and these should be particularly well suited to near future observations systems.

1.   Introduction.

Studies on "Global Change" are being cited more and more in several large programs, often as justifications of projects dealing with the observation of the Earth, with the understanding of the processes that cause and effect climate change, and with the forecasting of eventual global climate evolution. Beside the natural evolution which has shaped our environment, the industrial era has brought new stresses on the components of the entire Earth system. The processes which induced the "natural" changes are now greatly affected by modifications of some new driving forces and/or boundary conditions. While the impact of the economic activities has remained small or undetected until the last two centuries, continued developments can now have a significant impact on many aspects of our environment which can lead to dramatic and sometimes disastrous changes. Among the factors which can induce global environmental changes, land cover is certainly one of the most easily perceived. The most common definition for land cover refers to the description of the land surface physiognomy in general. Land use, referring to the way land cover is exploited for human benefit, can induce much change of land cover at various scales. A detailed description of the land surface will therefore have to associate both land use and land cover descriptions to allow studies of mechanisms by which changes can and do occur.

A detailed description will normally lead to a classification in which the definition of each category, and of each boundary between each category, will largely depend on the purpose of that classification. The large diversity of classifications that currently exist

G. D'Souza et al. (eds.), Advances in the Use of NOAA AVHRR Data for Land Applications, 265–278.
© 1996 *ECSC, EEC, EAEC, Brussels and Luxembourg.*

becomes a very important problem when comparing two different descriptions either at the same time or at different times for different places. Land cover changes can only be assessed when a strict homogeneity between classifications can be assured. One way to overcome the difficulty is to base the description of land cover on physical parameters which can characterise the different mechanisms by which land cover is established, or by which it is changing. This approach is certainly more suited to quantitative studies where interactions between the physical, chemical and biological processes have to be understood. It is also more easily related to collections of physical measurements of properties of the surface and remotely-sensed data are generally included at first in that physical approach. However, when changes have to predicted, or when scenarios have to be established to assess the effects of particular social or political decisions, the relationship between human impact on land cover and the various physical parameters has to be determined, which is generally very difficult to establish. Thus, the qualitative description of land cover, allowing a more direct description or understanding of the physiognomy of the surface, still has to be relied on for characterising land cover distribution. Any mapping of land cover will then have to provide (i) a strict and complete description of the taxonomy used (including a strict definition of limits between categories), and (ii) a way of bridging the gap between the qualitative description adopted and a *physical* characterisation of the categories, including both physical parameters and underlying mechanisms which will induce the dynamics of each categories.

The aim of many projects, which are related to the International Geosphere Biosphere Program (IGBP), is based on the need to establish and express such relationships. Remote sensing, due to its physical and global capabilities is one of the tools which has to be developed and utilised for successful studies in this area. While the overall target is global understanding, the objectives can only be attained through multidisciplinary and multiscale approaches : both the qualitative description and the physical understanding of land cover categories, and of their processes, have to be made at local and/or regional scales. Local descriptions will generally be related to basic "pure" components of a given category, while regional descriptions will have to integrate the natural heterogeneity up to the level of ecosystems. At this scale, some generic description of many processes has to be linked and the type and level of interaction determined. In general, AVHRR-like wide field observations will be related to high spatial resolution measurements, and the complementarity between the high frequency and global capability of the former, together with the capability of the latter to describe the "pure" component, will have to be fully exploited.

2.   Typical Examples of Land Cover Description.

2.1.   REGIONAL MAPPING OF LAND COVER

Many different maps of land cover exist at different scales, having been made for different purposes. Vegetation maps for local areas, for example, have been compiled at very detailed scales to provide information on parameters as diverse as vegetation canopy extent, structure or floristic composition. In general, the extension of detailed local to regional-scale maps to continental or global-scales often incorporates a gross simplification of their taxonomic legend, in order to maintain a certain amount of homogeneity at the coarser scale. The resulting simplified maps then become less useful for ecological studies.

An example of a good regional/continental vegetation map is the one of Australia (Carnahan 1990). This presents an extensive classification by patterns of structural forms and arrangements of plant height and spacing. The description is well suited to large area characterisation and uses 10 categories : closed forest; open forest; woodland; open woodland; scrub and heath; shrubland; open shrubland; grassland and sedgeland; herbland; pasture and cropping. It is derived from a more detailed classification which combines the structural form of the top and most significant lower strata of the vegetation cover, together with the upper stratum coverage and the family of dominant species. The vegetation code associated with an "homogeneous" area is then composed of five symbols:

1. the floristic code of the upper stratum chosen from 22 codes,
2. the growth form code of the upper stratum chosen from eight principal forms (tall trees, medium trees, low trees, tall shrubs, low shrubs, hummock grasses, tussocky or tufted grasses and graminoids, other herbaceous plants)
3. the cover code of the upper strata corresponding to four ranges of coverage (less than 10%, 10-30%, 30-70%, greater than 70%),
4. the floristic code of the lower stratum,
5. the growth form code of the lower stratum.

Obviously, the categories defined for floristic codes and growth form codes are suited to the Australian vegetation types and could not be extended to areas outside the continent. However, the descriptions and examples associated with the major areas of Australia allows quite detailed comparison for different areas at different times. For example, the detailed characterisation of each cover type (while still subject to some interpretation in many cases), allows a study of changes which may have occurred during the last 200 years. Changes and conversion from each category of land cover to another could then be synthesised, to show, for example, the impact of human action and the difference resulting between Aboriginal and European land use practices (Table 1).

Table 1. Estimate of change in vegetation type for Australia from the 1780s to 1980s (from Carnahan 1990).

Areas in '000 km^2	from : "Natural" vegetation (1780s)								
to : Present vegetation (1980s)	1	2	3	4	5	6	7	8	Total gain
1 = Forest		15							15
2 = Woodland	75							20	95
3 = Open woodland	100	260						40	400
4 = Tall shrubland									
5 = Tall open shrubland				280					280
6 = Low shrubland				25					25
7 = Low open shrubland						300			300
8 = Grassland/Pasture	140	320		155	30	15			710
Total natural vegetation	690	1570	1650	1230	1370	380	100	500	
Total present vegetation	390	1070	2000	770	1620	90	400	1150	

2.2.   GLOBAL LAND COVER DATA SETS

The above cartography adapted to the vegetation types of Australia shows some consistency over the entire continent as well as over two centuries. This is made possible by the exhaustive work of gathering and documenting ground truth to ensure a clear understanding of each category definition, and of the limits between these categories. In the case of global research, several data sets have been collated, either in digital or analog form. These global data sets have been assembled for various purposes, for example, boundary conditions for General Circulation Models (GCMs), delineation of homogenous areas for global ecosystems modelling, or they were obtained from climatologic data regionalisation. Some of the more widely-used global data sets available today are:

- Matthews (1983) and Henderson-Sellers (1986) used for GCM boundary conditions, in which land cover categories are classified according to surface roughness, albedo, stomatal resistance and any other vegetation parameter that is important for the interaction with the atmosphere.

- Matthews (1983) and Olson (1983) which are more related to studies of global primary production and can provide information on the vegetation types to which parameters synthesising (or driving) the production of biomass can be associated, *e.g.* carbon content, efficiency coefficient, *etc.*
- Holdridge map categories which are more related to a regionalisation of the major climatological variables: temperature, precipitation and evapotranspiration. These maps are more related to potential vegetation than actual vegetation.

While being the best data sets available today, all of the above suffer from a lack of consistency and cannot therefore be used together to compare or associate parameters from different descriptions. For example, Figure 1 (taken from Townshend 1991) shows the discrepancies between some of the cartographies. In general differences are related to: (i) the definition of each category, (ii) the subsequent delineation of these defined categories derived from many individual sources, and then (iii) the generalisation or adaptation to a global description suitable for global scale studies. Within each cartography, even parameters which should be relatively constant and consistent may show great variations, from one season to the other or from place to place. For example, albedo for desert areas can be found to vary from between 0.2 to 0.5. The seasonal variations of vegetation can also show strong heterogeneity within the same class (for savanna for example).

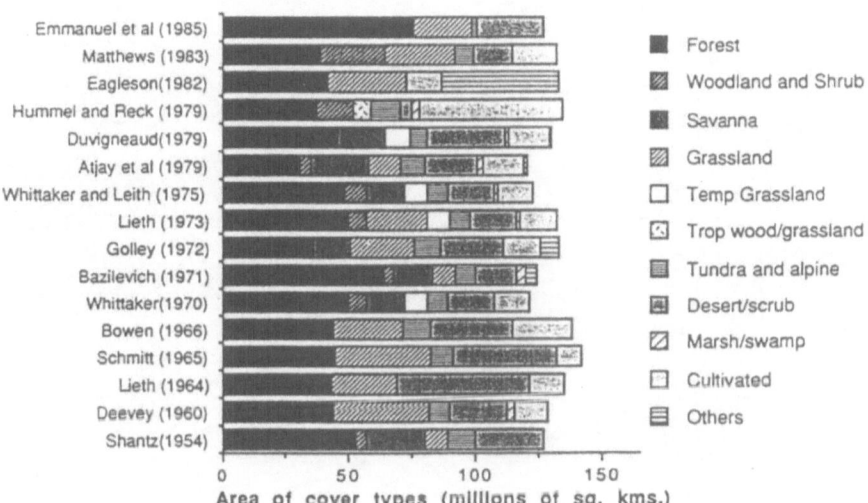

Figure 1. Variations in estimates of global land cover classes based on collations of collations of cartographic sources (from Townshend *et al.* 1991).

2.3. TOWARDS MAPPING OF PHYSICAL PARAMETERS

Because of the limitations of the existing global scale data sets, there has to be further methodological development in the collection and collation of new land cover data sets, which will lead to a definition or classification that will be more suited to the type of modelling associated with global change studies. Definition of the main physical parameters necessary for global modelling exercises is therefore in progress (bearing in mind that some studies might not easily integrate physical descriptions, *e.g.* studies on the mechanisms driving changes in land use). The resulting list of parameters must be based on the processes which are inter-related in the evolution of both the physical climate system and the land surfaces. Figure 2 shows all of the links between the main processes which are important to global change research. It is important to note that "land cover" by itself is not apparent in that diagram. While land use is one of the main factors that could induce changes in terrestrial ecosystems, land cover has to be considered as the product of several parts of the system : terrestrial ecosystems, terrestrial energy/moisture and soils.

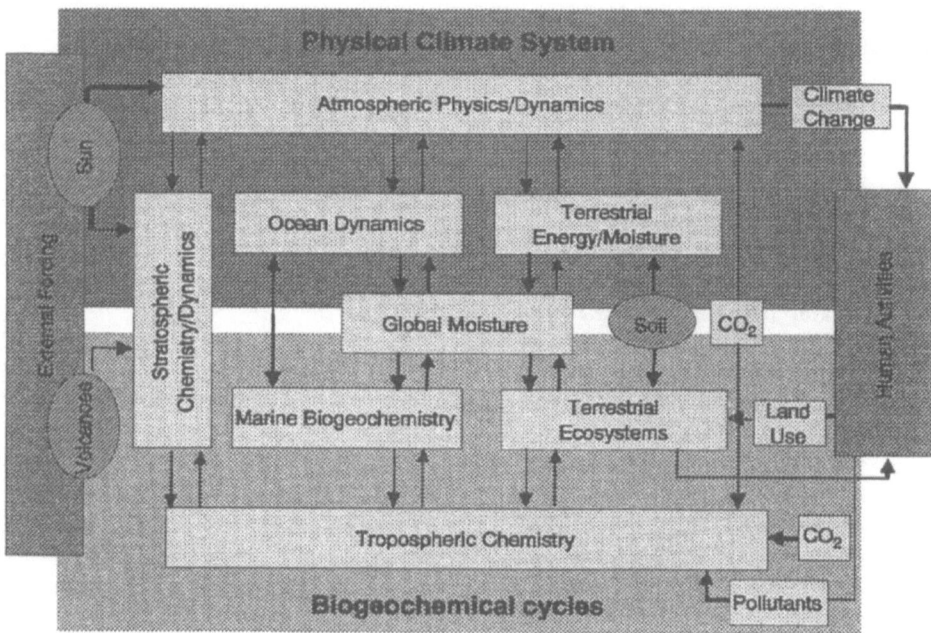

Figure 2. Showing the relationship between the main processes important for global change research.

To answer questions about, or to derive simulation models for, each of the components presented above, various data sets will have to be assembled. As is currently the case with existing ones, some will have to be adapted to climatic studies, providing boundary conditions for parameters which are important to soil-vegetation-atmosphere exchanges and their evolution through the seasons. Others will have to be adapted to the study of fluxes of carbon or gases between surface and atmosphere. Data sets allowing regional-to-global scale modelling of the biosphere, with all its processes from local to ecosystem levels, will also have to be developed. These should ensure correct understanding of the response of ecosystems to, and their interaction with, climate change.

Together with a list of parameters that must be mapped, the precision and accuracy which are needed for quantitative use must also be provided. For example, the impact of surface albedo on climate is now better understood through the use of existing general circulation models (GCMs), and it can be estimated that regional variations of albedo of some percent can affect the trend both locally and for areas a certain distance away. The precision is particularly important when changes must be studied. Some absolute quantities might not be known accurately, but their changes may be more accessible. For example, the absolute stock of carbon in some ecosystems will most certainly be very difficult to determine while its *variations* might be more readily obtained from fluxes between vegetation and atmosphere.

3.   Remote Sensing and Land Cover Characteristics

As previously described, land cover mapping is often used to delineate categories of land areas which have the same characteristics for some purpose or that at least have a narrow range of values for individual characteristics. A more direct way to estimate the same characteristics might be more suitable for global mapping purposes. If some parameters can be computed from physical measurements, their value might be more accurate than simply making hypotheses based on their possible value for some particular land cover category. Furthermore, the spatial description that could be obtained from complete gridded measurements (as provided by remotely-sensed data) will allow a more complete analysis of the transition areas which are always the most important for a study of changes.

Among the many current projects which are dedicated to the mapping of land cover, approaches which aim at deriving physical land *characteristics* rather than land *cover* categories are better suited to global change studies. For other needs where land cover maps are still needed, the usual techniques of classifying several sets of parameters can be used (see next section). To follow the former approach, the most useful way to define the required measurements (to determine the right physical parameters) is to first establish relationships between the processes and desired surface parameters, and then to derive the relationships between these parameters and what measurements are possible. The capability of each measurement for the determination of useful information must be assessed as well as the conditions under which measurements have to be made. As for any spatially and time variant process, the sampling strategy has to be carefully defined in order to address the right scale.

As usual, complementary measurements can be made to infer accurate land characteristics. The association of remotely-sensed data with ground data can also be a way of deriving valid quantities. In this case, even while they cannot be related alone to the information which is needed, their spatial distribution can be used to design the measurement plan for ground observations. For example, when land cover maps are needed for primary production estimates, the conversion efficiency used in the Monteith-like approach cannot be obtained directly from remotely-sensed data but areas which display similar temporal-radiometric behaviour will most likely correspond to homogeneous areas of conversion efficiency.

3.1. GENERAL STRATEGY FOR THE SELECTION OF REMOTELY-SENSED MEASUREMENTS

The scale of any map is related to the spatial variability of the phenomena that is mapped. When dealing with dynamics, the temporal variations also have to be taken into account to define a better way to access the information. For land surfaces, the processes that are to be studied can be associated with spatial and temporal scales in a simple manner which will simplify the design of observation systems. Three major categories of processes can be associated to specific spatial and temporal scales as given in Table 2 below.

The parameters or variables of each of these processes generally depend on each other, and relationships between the different spatial and temporal scales have to be introduced in the models or analysis. For example, growth and production are functions of the amount of energy and water which can be exchanged between land surface and atmosphere, whilst at longer time scales ecosystem dynamics are strongly related to the capability of each to develop under some environmental conditions.

The same separation can also be used for the type of properties which can be associated to remotely-sensed data:

Table 2. Three major categories of land surface processes and their spatio-temporal scales.

Processes	Spatial Scale	Temporal Scale
Exchanges surface - atmosphere	A few mm to metres	A few seconds to days
Growth - Production	Metres to few tens of metres	Days to weeks
Ecosystems dynamics	Hundreds of metres to kilometres	Months to some years

- exchange processes are related to radiative properties or fluxes characteristics which can often be inferred directly from remotely-sensed data,
- growth processes are related to the physical and biological properties which are generally not related directly to remotely-sensed data but have to be inferred using either empirical or biophysical models,
- ecosystems dynamics can generally be related to the identification of land cover which is mostly done using temporal features.

3.2.   PROCESSES AND PARAMETERS

The most important parameters for each of the above processes can be presented as follows :

3.2.1. *Exchanges.* Water and carbon fluxes between land surface and atmosphere strongly depend on the amount of solar energy available, the albedo which controls the amount of energy retained by the surface, rainfall, air and vegetation temperatures and wind. Resistances (mainly aerodynamic and stomatal) are the control for water exchanges while carbon fluxes depend strongly on atmospheric CO_2 concentration and stomatal resistance. The exchanges between vegetation and soil depend heavily on rooting volume, transfer coefficients from soil to roots, thermal inertia and temperature and humidity profiles.

3.2.2. *Growth and production.* The most important parameters which govern growth and production (and therefore the seasonal dynamics of land surfaces) are: the absorbed photosynthetically active radiation (PAR) for which the structure of the canopy, including vegetation coverage of the soil, is an important factor; the water content of the vegetation, which can limit production, and the conversion efficiency which is strongly dependant both on vegetation species and phenological stage; and the capability of the soil to provide the nutrients, which is dependent on the nutrient contents and also on the nature of the soil which can be associated to porosity, conductance, *etc.*

3.2.3. *Ecosystems dynamics.* Generally, ecosystem dynamics are characterised by changes of its species composition and of its structure. This is already well described in the classical physiognomic cartography, except that the time evolution is, in general, not easy to map.

3.3.   PARAMETERS AND REMOTELY-SENSED MEASUREMENTS.

As is usual in remote sensing, each set of measurements can be associated with different properties of the surface, the interpretation skill being to use the most effective hypothesis to derive surface characteristics from remotely-sensed data.

Each measurement has to be defined by its spectral characteristics (wavelength, bandwidth), spatial resolution, time/date of acquisition and possibly other properties for new systems (geometric conditions, polarisation, *etc.*). Complete descriptions of the type of information which can be extracted from the existing systems are given in previous chapters (especially those by Dedieu and Verstraete and Flasse). Due to the large variety of surface characteristics which have to be derived for land cover mapping, it is obvious that complementary systems have to be used to provide information for each process, at each of the necessary space and time scales.

AVHRR-like measurements, providing global coverage with high temporal repetitivity, are well suited to the monitoring of seasonal and long-term changes. However, the lack of high spatial resolution has to be supplemented by less frequent observations that will give information on the heterogeneity of the land surface. Thus, spatial and temporal information can then be mixed to provide overall vegetation characteristics.

3.4.   EXISTING INITIATIVES FOR LAND COVER MAPPING

Following the above rationale, IGBP has defined a Land Cover Change Project which aims at providing a global land cover map and information on changes at a resolution of 1 km which is consistent with the AVHRR spatial resolution (IGBP 1990, Townshend 1992). Land cover classification should focus on a limited number of categories including evergreen forest, deciduous forest, shrub, grassland, desert, snow, ice, tundra, cropland and urban areas. To derive these categories from remotely-sensed data, while other research studies are providing methods to extract information from AVHRR data, the basic problems of delineating boundaries of categories and of defining higher levels of classification still remain, together with the problem of associating surface characteristics and categories. The project is designed around selected pilot study sites on which methodologies to extract both information on characteristics, and the capability to define categories, will be tested. The test sites are selected in connection with the different Core Projects of IGBP, mainly the Global Change and Terrestrial Ecosystems (GCTE), the Biological Aspects of the Hydrological Cycle (BAHC) and the International Global Atmospheric Chemistry (IGAC) projects. After the methodological developments and validation in each of the sites, synthetic methods to derive a global map of changes will be developed.

To overcome the difficulties of defining categories and their attributes for ongoing studies on global models, in parallel "simple" surface characteristics such as vegetation index, albedo, solar radiation fluxes, evapotranspiration and surface temperature will be derived from existing data and particularly from AVHRR multi-temporal data sets.

The Global 1 km archive derived from all AVHRR acquisitions, by a network of about 30 stations, will be assembled and pre-processed by USGS, NASA, NOAA and ESA (see later chapters by Arino and by Belward). The specifications of the products that are suitable for the extraction of land cover information have been designed by IGBP (Townshend 1992) together with methods for corrections of AVHRR data to improve the quality of the data.

4.   Methods for Extraction of Land Cover Characteristics and Categories

Among the numerous methods which have been developed to extract information on land surface from remotely-sensed data, many of them are based on classification, either automatic (unsupervised) or trained on specific objects (supervised). These methods are now well known for any type of object and will not be detailed here, except for special aspects which are related to the temporal evolution and to the global extension of the data sets. As described above, the principal aim of global research is to provide models of the components of the entire Earth system to establish the possible scenarios related to future evolutions. The models which have to be established are based on physical and biological properties and these are associated to processes which are related to the nature of the land surface. Some methodological concepts on information extraction will then also be presented.

4.1.   CLASSIFICATIONS FOR LAND COVER CATEGORIES

The use of AVHRR-like information for global land cover mapping is related first to the need to reduce the amount of data to process. The handling of data corresponding to a wall to wall coverage of continents is still quite difficult, especially when time-series have to be used, e.g. one measurement every week or every ten days for each 1 km pixel. Second, the spatial integration due to the 1 km resolution can filter out high spatial frequency variations which are not fundamental for global studies, once local heterogeneity has been integrated into regional models using some selected high resolution data. Third, AVHRR offers the capability of providing high repetitivity which could lead to a useful monitoring of seasonal changes. As presented in other chapters, the original daily information can be filtered to get rid of the effects of atmosphere (bad atmospheric conditions due to water vapour or aerosol contents or clouds) and to dampen the modifications to the useful signal related to directional effects because of the changing acquisition geometry. In general, after filtering, for most regions the resulting information is weekly measurements with the exception of some areas, especially around the Equator where cloud cover is present for long periods of time, when composites of longer time periods have to be made. A weekly frequency is quite compatible with the rate of change of major vegetation types and the growth/production processes could be adequately characterised at that frequency.

This new dimension of analysis has to be taken into account in classification methods. The *spectral* signature is replaced by the *temporal* profile which is supposed to be a characteristic of the land cover type. Typical methods for classification can be adapted to that temporal feature and provide classes of temporal signatures which can then be associated on some specific sites to classical land cover types. The main problem is to determine the best correspondence between classes of temporal signatures and expected land cover types.

To reduce the volume of data which has to be processed, compression techniques to retain only the most significant temporal features for each pixel can be applied. These may be based on, for example, Fourier filtering where each component can be associated with particular periods of variations (Viovy 1990). Generally, the first five or six components give classification results in a small number of classes coherent with global scale studies. However, all these analyses are based on the data themselves, without introduction of any other information, either on the causality of temporal evolution or on the possible underlying mechanisms.

4.2. PROCESS-RELATED METHODS FOR LAND COVER MAPPING

To introduce *a priori* knowledge about processes that could help either land cover identification or land surface characteristics extraction, new methods are now being developed, ranging from semi-empirical methods, to integration into biophysical models (Viovy 1990, Fischer 1992). Some examples are given here.

4.2.1. *Introduction of causality.*

As presented above, knowledge about the processes which drive growth, about exchanges at different temporal scales and about land surface state at a particular time can provide a good prediction of its state for another later time. One way of introducing that prediction capability is to model vegetation radiometric evolution using the statistical Markov model (Viovy 1991). Each state of the vegetation development can be associated with some range of values for remotely-sensed data. Transitions between states can be statistically characterised. The probability that a particular vegetation cover changes from one state to another can then be given *a priori*, either from ecological knowledge or from statistical analysis on a certain data set. The process consists of iteratively comparing the prediction made from past time to the actual measurement, then updating the knowledge of present state and predicting the next state or set of measurements that should be obtained through the observation system. This method can be used for two purposes:

- to detect anomalous change in one particular land cover, anomaly that could be due to environment conditions or human actions,
- for classification purposes related to the classical land cover mapping. At each time an observation is obtained or for each time series, the best Markov model can be determined and this should correspond to the most probable land cover category which fits with the series of measurements.

4.2.2. *Introduction of ecological knowledge on the processes.*

As in the previous method, prediction can be achieved at each time, not only using some statistical knowledge about the processes but also information about the impact of environmental conditions. For example, in semi-arid regions, the water availability is the main factor driving vegetation seasonal growth. Knowing land cover type, for each possibility of rainfall episodes, some prediction of the most probable response of vegetation to that rainfall can be established. The interpretation technique is then not only related to recognition of a radiometric temporal profile but also to the way the vegetation type responds to the environmental conditions, producing a different characteristic radiometric profile. The identification can be made using either classical classification techniques or Markov models.

This method is more suited to the general problem of detecting significant changes : changes always occur from year to year especially on natural and cultivated vegetation where water availability can be the limiting factor to growth. It is essential that radiometric changes which are due to environment changes and their impact on land cover characteristics are not detected as possible land cover changes, then *a priori* knowledge on the way land cover characteristics respond to environmental condition variations can be used to filter out these anomalies which are not due to actual land cover changes.

4.2.3. *Integration of remotely-sensed data with models of growth.* To concentrate on the possibility of using environmental conditions and to rely mostly on process models, new methodologies introducing extensive models of vegetation growth as observed by remote sensing instruments are being developed. These are aimed at using remotely-sensed data as indications of correct predictions of biophysical models. They should reduce uncertainties related to observational problems such as the effects of atmosphere or directional properties on remotely-sensed measurements. In cases of complex mixing of land cover types within the resolution cell of AVHRR-like images, the spatial integration of different radiometric responses can very much confuse the interpretation. Here, the use of some spatial description of heterogeneity obtained from a reduced number of high spatial resolution acquisitions can greatly complement the *a priori* knowledge of time evolution of the land surfaces if they were driven solely by growth processes and environmental conditions. The frequent measurements of AVHRR-like systems can then be used for a better estimation of the most important parameters which can represent the surface evolution. For crops for example, the entire seasonal evolution depends strongly on both environment conditions and planting date. When the planting date can be refined iteratively as more and more measurements are acquired, the synchronisation of plant model with environment conditions will be improved gradually, leading first to a better evaluation of the type or proportion of crops and second to some prediction of its future seasonal evolution.

For global changes studies, the advantage of such techniques is that they can provide ways of accessing parameters that cannot be directly measured but only inferred from the model. For example, carbon assimilation depends first on the exchanges and growth processes (especially on the capability of vegetation to convert photosynthetic energy to biomass), and second on energy and water availability throughout the season. While the above technique allows the matching of the internal processes to environment conditions, the model will allow the derivation of the correct value for conversion efficiency at each date, taking into account the estimated phenological stage. The final carbon flux will then be determined from the best values of the different relevant parameters.

5. Conclusion

The different problems for land cover mapping have been outlined. If the definitions of land cover characteristics are to be matched with the needs of global change research, classical categories are not always the best way to characterise the land surface.

Parameters related to the processes which have to be understood for global modelling need to be accessed through remote sensing data and possibly complementary ground measurements. Also, new techniques for information extraction are now being developed and should be adapted to existing data sets.

At the same time, some new observational systems are being designed to better match the spatial and temporal scales which must be addressed. In a few years both operational (VEGETATION/SPOT4) and experimental (EOS) systems will together provide much better data sets which will allow efficient wall to wall mapping of both land cover types and land cover characteristics.

6. References and related bibliography

Carnahan, J., 1990. *Atlas of Australian Resources, VEGETATION.*, Australian Surveying and Land Information Group.

Fischer, A., 1992. *Suivi de la croissance des cultures en zone hétérogène au moyen d'informations satellitaires. Complémentarité avec les modèles de croissance.* Thesis Université Paul Sabatier Toulouse, January 1992

IGBP, 1990. (International Geosphere Biosphere Program). *The Initial Core Projects.* IGBP Report No. 12.

Townshend, J., 1992. (ed) Improved Global Data for Land Applications : a proposal for a new high resolution data set. IGBP Report No 20.

Townshend, J., Justice, C., Li, Gurney, and MacManus, W.C.L., 1991. Global land cover classification by remote sensing : present capabilities and future possibilities. *Remote Sensing of Environment,* **35**, 243-256

Turner, B.L., Moss, R.H, and Skole, D.L., 1993. (eds) *Relating Land Use and Global Land Cover Change.* HDP Report No. 5

Viovy, N., 1990. *Etude spatiale de la biosphère terrestre : intégration de modèles écologiques et de télédétection.* Thesis Université Paul Sabatier Toulouse, November 1990.

Viovy, N, and Saint, G., 1991. Intégration d econnaissances qualitatives sur les modèles de fonctionnement pour le suivi de la végatation à partir de séries temporelles d'images de télédétection. 5th Coll. on Spectral Signatures Courchevel France January 1991

NOAA-AVHRR BASED TROPICAL FOREST MAPPING FOR SOUTH-EAST ASIA, VALIDATED AND CALIBRATED WITH HIGHER SPATIAL RESOLUTION IMAGERY

J.P. MALINGREAU, F. ACHARD, C. ESTREGUIL, H.J. STIBIG and G. D'SOUZA

Commission of the European Union,
Joint Research Centre,
Institute of Remote Sensing Applications,
21020 Ispra (Varese),
Italy.

ABSTRACT. The AVHRR instrument represents a major tool in the development of global approaches to the mapping and monitoring of vegetation. The case of tropical forests forms a particularly important challenge as (i) tropical deforestation represents a very real, urgent and contentious environmental issue, (ii) tropical ecosystems are complex, widely distributed and exhibit specific seasonal and man-induced dynamics, (iii) they tend to occur in areas of relatively unfavourable atmospheric conditions for the optical tools available for operational use.

Following widespread concern about the loss in extent of tropical forest and its likely impact on biodiversity loss, climate, biochemical cycles and economic development, and following considerable debate and controversy about the actual rates of deforestation in different areas, the CEU-ESA TREES project was established to derive a set of procedures and methods to improve global analyses of forest cover extent and condition. One of its central objectives was the compilation of a comprehensive pan-tropical forest map, at a nominal 1 km resolution, based on the analysis of NOAA-AVHRR imagery, depicting tropical forest extent, density and character (seasonality).

NOAA-AVHRR images are obtained, pre-processed and classified according to the adopted legend, and related to other information in a dedicated information system. The large-area results obtained at the 1 km resolution are then verified by comparison with ancillary data (e.g. conventional forest maps) and calibrated by reference to classifications of forest type and area derived from higher spatial resolution Landsat TM, MSS, SPOT and ERS-1 SAR imagery.

The processes carried out in the TREES project provide a valuable example to any other projects aimed at the assessment of vegetation at continental or global scales. They are therefore presented here, with specific examples drawn from the application of the TREES methodology to application in continental and insular South-East Asia.

1.   Introduction

1.1   THE TREES PROJECT

In an earlier chapter, Saint highlighted the importance of general land cover mapping in global change research. Tropical deforestation is one of the most significant forms of this type of change. as it has been identified as an important component of the global carbon

G. D'Souza et al. (eds.), Advances in the Use of NOAA AVHRR Data for Land Applications, 279–309.

cycle (Houghton *et al.* 1987) and it is known to effect regional climate and hydrology (Dickinson and Henderson-Sellers 1988, Salati *et al.* 1990). Moreover, changes in tropical forest extent and condition will have a large impact at regional and local scales, severely effecting local population habitats and biodiversity, as well as local and regional economies. Following widespread concern about the loss in tropical forest extent and its likely impact on biodiversity loss, climate, biochemical cycles and economic development, and also in response to the considerable debate and controversy about the exact rates of deforestation in different areas (*e.g.* Myers 1989, WRI 1989), the TREES project was established in 1990 as a joint activity by the Communities of the European Union and the European Space Agency. Its overall aim was to derive a set of procedures and methods to improve global analyses of forest cover extent and condition. Its specific objectives, include:

- the compilation of a comprehensive pan-tropical forest map at about 1:1 million scale (or 1 km resolution)
- the study of seasonality effects in tropical forests as evidenced in relatively long time-series of satellite-derived information
- the modelling of deforestation processes and dynamics
- the investigation of the usefulness of new sensors for tropical forest mapping and monitoring, in particular the potential of the Synthetic Aperture Radar (SAR) and Along-Track Scanning Radiometer (ATSR) flown onboard the ERS-1
- the development of an operational, continuous monitoring system, aimed at pinpointing the most active deforestation areas (hot-spots) which should then become the focus of more detailed and intensive observation.

In this presentation, we will focus on the first of these specific objectives- the compilation of a comprehensive pan-tropical forest map. Because of the large area to be covered and the requirement for highly repetitive observations (to overcome persistent cloud-cover problems and to document vegetation seasonality variations), only data from the NOAA-AVHRR sensor can provide a suitable blend of the spectral, spatial, radiometric and temporal resolutions required for this application. Furthermore, the operational footing of the AVHRR instruments on the NOAA series of satellites was felt to guarantee a desirable level of continuity for any operational procedures that may be recommended for subsequent phases of the project.

In the TREES project, NOAA-AVHRR data are obtained, pre-processed and classified according to a specified legend (described below), and related to other ancillary information in a specially-designed information system, called the Tropical Forest Information System (TFIS) (D'Souza *et al.*, 1993). The large-area results derived at the 1 km resolution are then verified by comparison with ancillary data (*e.g.* conventional forest maps, *etc.*) and the measurements thus obtained are calibrated by reference to forest classifications derived from higher spatial resolution Landsat TM, MSS, SPOT and ERS-1 SAR imagery. Obviously the description of the various stages of NOAA-AVHRR data collection, pre-processing and analysis, as well as the processes of verification with other datasets, some of which are derived from imagery of a higher spatial resolution, will provide a useful illustration of a large-scale detailed land-cover mapping exercise carried out primarily with the use of NOAA-AVHRR imagery.

Detailed examples provided in this chapter will be drawn from the TREES methodology as applied over tropical South-East Asia including the countries of Thailand, Myanmar, Malaysia, Indonesia and Papua New Guinea. This region of the world is very diverse in terms of climate, topography, population distribution and forest type, which ranges from the highly seasonal sub-tropical forests of continental South-East Asia (*e.g.* Myanmar, South China) to the equatorial wet types of insular South-East Asia (*e.g.* Sumatra, Kalimantan). It is also an area that has witnessed, in recent times, considerable and diverse transformation of a large part of its forest area into a complex mosaic of degraded and mixed formations. The major agents of this change have been shifting cultivation, logging and clearing for plantation estates, food-crop agriculture and transmigration programs. The whole region therefore provides a good test for the use of NOAA-AVHRR data for the application at hand. In the following sections, we illustrate the range of methodologies that have been developed in order to utilise the full information content of the data to map all these varying types of forests occurring in such different climates and terrain.

1.2 LEGEND CONSIDERATIONS

1.2.1 Satellite-based analyses. In any mapping exercise, the determination of a suitable legend is one of the first and most important considerations. In this case, the determination of the classification legend is both a key and a controversial issue in the AVHRR analysis. Traditional methods of forest classification based on detailed large-scale, local observations of canopy structure and floristic composition are clearly inappropriate to work based on remotely-sensed imagery, especially when large areas at relatively low spatial resolution are considered. However, it is widely accepted that satellite imagery offers the only feasible solution for global-scale mapping of vegetation, so satellite image-based vegetation classification schemes can only be founded on those classes which could be reliably distinguished, given the capabilities and shortcomings of the observation systems at hand.

To ensure widespread acceptability, usage and compatibility with conventional applications then, these classes should be chosen so that they satisfy as many as possible of the information needs of foresters, ecologists, climate modellers, and other interested scientists. Two ways of achieving this may be envisaged. One may be to go up the hierarchy of classes to conduct the analysis at the highest levels of generalisation. Such is the case in recent global forest assessments (WRI 1989, Myers 1989), and of the TREES project approach: to carry out a crude (forest/non-forest) classification of land cover at global levels. There is a growing demand for such a global data set interpreted in basic terms which would be a highly valuable as a base for an overall assessment of deforestation processes. Despite the apparent simplicity of the classification legend, many conceptual difficulties remain. Among them, that of semantics. The very concepts which are at the core of the debate such as "*deforestation*" and "*forest degradation*" lack universal definition and application. When is a tropical forest canopy closed or open, when is it pristine or degraded? Obviously the term deforestation means different things to different people. Selective logging for the forester is a form of sustainable management which leaves the forest cover in a state allowing regeneration, but for the conservationist this very same activity will constitute *"deforestation"*. A climate modeller will not be concerned about the type of forest or its inherent biodiversity *per se*, but by changes in albedo, canopy roughness, radiation balance and other physical parameters. On the other hand, the ecologist's interest will range from a species count to the net primary productivity of the ecosystem. Consequently, a common denominator must be found if a variety of users are to be satisfied, at least at the more generalised level of interest. Starting with the classical inventory of needs does not seem to be a promising way to define such a common denominator and there is room for new thinking on the role of remote sensing data as a unifying driver.

A possibly novel approach to satellite-based forest classification relates to the more intensive use of data derived from space sensors. At the outset, it is important to remember that such data are of a simply radiative nature only. That is, any vegetation-related information which can be extracted from the data is always a function of the radiative property of the canopy. A corollary is that any biospheric element or process which does not induce a measurable change in the radiative characteristic of the canopy cannot be detected! Ideally space observations must be interpreted and analysed with models which can, in a systematic and quantitative fashion, trace and simulate links between a set of radiometric measurements and the state of a plant community. A range of models can be identified with different degrees of sophistication. Empirical models attempt to establish statistical links between remotely sensed data and observations on the ground or some proxy of desired information (*e.g.* vegetation indices and green biomass through statistical regression). Physical models, in contrast, link the sensor signal directly to a surface parameter, but biospheric models attempt to relate a specific surface parameter to a plant process (or its by-product). Examples of the latter may be seen in the attempts to link net primary productivity, vegetation indices and the absorption of photosynthetically-active radiation. Currently, such approaches are still under development and cannot be applied in an operational perspective to all kinds of vegetation (see earlier chapter by Dedieu).

From the perspective of global mapping of vegetation, then, if we want to extract new information from satellite data sets, we must ask how radiation data can be exploited not only to identify objects but also to provide continuous physical measurements of vegetation without explicit reference to a pre-set classification. Two examples illustrate this point:

- it has been shown that spectral contrasts, whether related to albedo or to brightness temperature can lead to a spatial separation between evergreen tropical forest cover and surrounding agricultural land. Within the forest canopy itself, different levels of degradation should translate into a continuum of radiometric changes which can only be identified as fields of varying albedo, surface temperature and the like. These differences may be, in the first instance, simply thresholded to suit a binary forest/non-forest classification. At a second level, however, thresholds of forest degradation must be specifically defined, since the radiometric information is non-discrete (see examples of this in section 3). New methods of image classification, such as mixture modelling or fuzzy classification, which present a continuous, rather than discrete, classification product may also assist the improved exploitation, presentation and acceptance of remotely-sensed results;

- forest seasonality is a characteristic explicitly used in tropical forest classifications (*e.g.* tropical deciduous forest, semi-deciduous forests, dry seasonal forests, monsoon forests, *etc.*), but while such classification is binary, in the sense that a forest is considered as seasonal or not, it avoids the question of the intensity or even the facultativeness of the seasonality process (Woodward 1986). Time series of satellite-derived vegetation index data are now available over a period of ten years or so, and these can be used as an indication of the seasonality of the photosynthetic process (Justice *et al.* 1985, Tucker *et al.* 1985). Again, the data provided by satellite sensors show a continuum of signal amplitudes from the "evergreen" to the fully seasonal pattern. Such forest canopy related information is highly relevant for a series of analyses (Malingreau 1990): seasonal flushes affect albedo and evapo-transpiration, and inter-annual comparisons provide indications of the climate impacts on the tropical forest vegetation; drought and stress related to seasonal effects are also an indication of the forest susceptibility to fire, one of the most common agent of land clearing in the tropics.

Some of procedures intrinsic to these concepts cannot yet be implemented- they are still in their infancy in some cases, and dependent on more research and reprocessing of data. Wherever possible, however, they were borne in mind in the derivation of a suitable globally-applicable TREES-project legend, and in the classification of the AVHRR imagery according to this legend.

1.2.2 TREES Project global-scale legend. After detailed reviews and preliminary examinations of spatial, spectral and temporal characteristics of NOAA-AVHRR imagery over tropical forest areas, it was felt that a meaningful, appropriate and useful legend for the pan-tropical map should be based essentially on two parameters that were felt to be determinable with a high degree of reliability for all tropical areas: the amount of forest cover within 1 k^{m2} pixels, and the phenological character of the forest area overall. Based on these two criteria, the following five classes, on which to base the global TREES legend, were identified:

1. Dense evergreen forest (pixels containing 70% or more of forest cover, which is mainly of an evergreen nature)
2. Fragmented evergreen forest (pixels containing between 40 and 70% forest cover, which is mainly of an evergreen nature)
3. Dense seasonal forest (pixels containing 70% or more forest cover, which is mainly of a seasonal/deciduous nature)
4. Fragmented seasonal forest (pixels containing between 40 and 70% or more forest cover, which is mainly of a seasonal/deciduous nature)

5. Non-forest, characterised by pixels with less than 40% forest cover.

As will be shown in the following sections, for some areas, under certain conditions, more detailed classes can sometimes be derived (*e.g.* mixed forest/savanna in Africa, mixed deciduous or semi-evergreen and dry dipterocarp forests in continental South-East Asia, *etc.*) but these are normally aggregated into a set of the above five classes in order to provide consistency at the global level of abstraction.

2. AVHRR data acquisition, screening and pre-processing

2.1 COLLECTION NETWORK

Apart from the ESA-Earthnet African network covering the African region (with receiving stations at Maspalomas and Niamey, and more recently at Nairobi and La Réunion), there was no systematic effort in the collection of a pan-tropical AVHRR data set at the time of the start of the TREES project. The situation has improved considerably since the start of the Global AVHRR 1 km project (see chapters by Arino and by Belward later in this book), but initially the TREES project had to rely on specific arrangements with individual station operators in order to secure the required amounts of data, which demanded considerable time and effort. In accordance with other AVHRR-based land-cover studies, priority for image acquisition was placed on afternoon passes (normally from the odd-numbered satellites) when illumination and thermal contrast conditions are more suitable for discrimination of different land surfaces from each other.

Much of the data collection, screening and pre-processing was carried out on a per-continent basis, and the current number of images (full passes) of good quality data that has been obtained by the TREES project, and their source, is summarised by continent below.

Region	Receiving Stations	Number of images
South-East Asia	Bangkok, Townsville, S. China, NOAA-LAC,	280
Africa	Maspalomas, Niamey, Nairobi	150
Central and South America	Baton Rouge, Cachoeira Paulista , NOAA-LAC	100

More specifically for the areas described in this chapter, over continental and insular South-East Asia 128 full High Resolution Picture Transmission (HRPT) tracks from 12 orbits over 191 days (during the dry season November-May 1991) were provided by the Bangkok receiving station. For Indonesia and Malaysia, 80 images were provided, some from the same receiving station, and others from the NOAA-Large Area Coverage (LAC) archives in Washington. A further 130 images were provided from the Townsville HRPT receiving station which covered Irian Jaya and Papua New Guinea.

2.2   SELECTION CRITERIA

Given the large amount of satellite imagery available and their highly variable quality, one of the first procedures to be organised was an intensive screening process. This involves the display and visual examination of false colour composites of every available acquisition in order to select and retain for further analysis, only those images which present acceptable characteristics in terms of cloud cover, haze/smoke, angular position in the path, image integrity, *etc.* Priority of acquisition was placed on close-to-nadir passes in order to avoid excessive bi-directional reflectance distortions and strong atmospheric attenuation. Because of the high variability from area to area in the tropics, however, criteria related to image quality in terms of cloud/haze has to be kept flexible and thresholds of acceptability are thus regionally dependent. For example, around the Equator experience has shown that the screening must try to maximise the selection of bits of "clear" data in order to assemble image mosaics with all the retained portions. In areas with a more pronounced dry season, several acceptable-quality full passes may usually be obtained during the dry season and these can be more easily combined into a set of multi-temporal image composites in which the cloud and atmospheric distortions may be minimised (see section 3.2.2).

For the selection, processing and subsequent classification of the AVHRR imagery, the land mass is considered in a convenient number of "*image windows*". These windows have been selected after some consideration of the prevailing ecological and climatological conditions which ultimately control the type of existing vegetation and likely acquisition of good quality imagery. These factors will have an important impact on the type of methodology that can be adopted for forest classification in any one area. For example, areas with a distinct dry season will contain more deciduous forests than areas characterised by constantly wet conditions. There will also be a greater likelihood of obtaining cloud-free imagery over the former (at least during the dry season), which may therefore facilitate more convenient multi-temporal image acquisition and classification. In such areas, the acquisition of imagery ranging from the beginning to the end of the dry season should be feasible and will allow optimum discrimination of forest from non-forest areas, as well as discrimination of forest areas which are predominantly evergreen from those that are predominantly deciduous (Stibig *et al.* 1993). However, in constantly wet areas, no preferred time period for image acquisition may be defined and multi-temporal classification may consequently be much more difficult or even unfeasible. In this case, cloud-free parts of single images (obtained at any time in the year) have to be classified independently and then aggregated to form the final classification (as in, for example, Gastellu-Etchegory *et al.* 1993). The division of the tropical land mass into a number of image windows may therefore be considered as a first, relatively crude stratification of the tropics for the purposes of AVHRR-based forest mapping.

The image windows selected for the TREES project are shown in Figure 1.

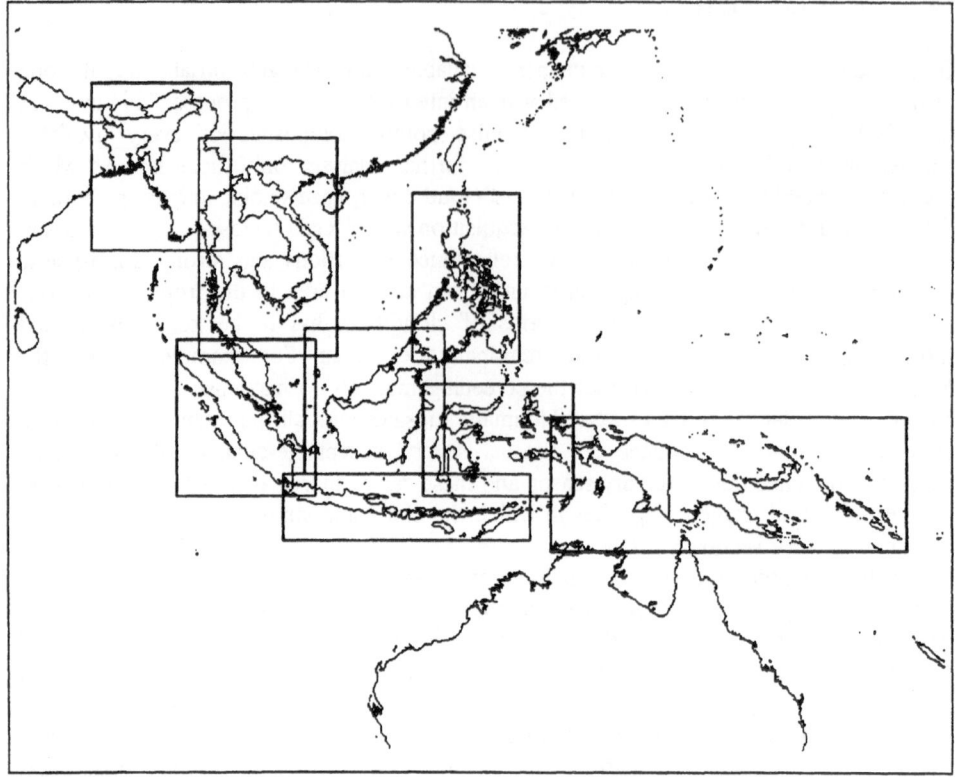

Figure 1. TREES project image windows for South East Asia.

In summary, for each of the defined image windows, all available images were screened in order:
- to retain data with least cloud cover, atmospheric moisture and other contamination
- to obtain a representative sample of data over the dry season for seasonal formations
- to select images as close as possible to nadir conditions as pixels closer to nadir have resolutions closer to the nominal and least affected by atmospheric path length and bi-directional reflectance effects.

2.3   PRE-PROCESSING

Although Horning and Nelson (1992) in their work over Madagascar reported that only the first two AVHRR channels were useful for tropical forest area determination, most other studies had demonstrated the utility of Channel 3 (Malingreau *et al.* 1990, D'Souza and Malingreau 1994), and Channel 4 has also proved useful in cloud and haze masking. So the TREES project chose to process the first four AVHRR channels for further analysis. The latest available visible channel calibration coefficients were used to calculate Channel 1 and 2 top-of-atmosphere reflectances and on-board derived calibration coefficients were used to convert Channel 3 digital counts to radiances and Channel 4 counts to brightness temperatures (according to procedures outline in preceding chapters of this book). However, atmospheric correction was not applied because of the difficulty with the application of accepted procedures, and the paucity or unreliability of ancillary atmospheric data in tropical regions.

Standard cloud detection procedures were only successful in masking large areas of dense cloud, so more detailed cloud and haze-masking was effected for each image by the use of a simple clustering classification based on all four channels. Geometric correction was effected by the use of complex orbital model (assuming sheroidal Earth and elliptical satellite orbit) for a first-stage correction, followed by an interactive ground control point (GCP) selection phase and second-stage correction (as outlined in an earlier chapter by D'Souza and Sandford). The GCPs were selected from the digital version of the World Data Bank (WDB) II coastline layer, which was also used to generate a land/sea mask. In general, root mean square errors were kept to below 1.5 pixel for over 20 GCPs for each image correction. The images were geo-matched to a Plate Carrée projection with a spatial resolution of 0.01 degrees in latitude and longitude.

For South-East Asia, five image windows were defined and these are listed below (see also Figure 1 above):

1.	Myanmar	15.0-30.0°N	87.0-100.0°E
2.	Thailand	5.5-25.0°N	97.0-110.0°E
3.	Sumatra/Peninsular Malaysia	7.0°S-7.0°N	95.0-108.0°E
4.	Borneo	5.0°S-8.0°N	107.0-120.0°E
5.	Irian Jaya/Papua New Guinea	12.0°S-0.0°N	130.0-150.0°E

3. AVHRR data analysis

3.1 IMAGE CLASSIFICATION METHODOLOGIES

3.1.1 *Background.* Good quality AVHRR images contain a lot of useful information. Well made false colour composites, especially those based on Channels 3, 2 and 1 assigned to the red, green and blue colour channels respectively, are very informative to the trained interpreter- *i.e.* someone familiar with the colour renditions and scale of imagery, *etc.* Various gradations of forest density and condition can often be distinguished, and there are a lot of other indicators related to forest condition that can be seen, such as smoke plumes from active fires, fire scars, the manifestation of roads or tracks, *etc.* Indeed, visual interpretation has often proved to be one of the most effective forms of information extraction from AVHRR imagery. However, it can also be considered very time-consuming over large areas, somewhat subjective and the results are often interpreter-dependent and error-prone especially in areas of fragmented forest..

Although image classification is a considerable generalisation of the information content inherent in satellite-derived multispectral imagery, it does produce easily-understood thematic layers of information that are more acceptable and easily understood in the wider user community. For example, statistics such as the percentage of forested areas within certain watersheds or countries may be provided from a classified, thematic layer. We have therefore endeavoured to produce such a thematic map, but have also borne in mind that the unclassified images themselves have a great deal of other inherent information. The TREES project is considering ways of accessing and visualising images (or sequences of images) at the same time as the classified result, in order to present the full information content to potential users. However, for the purposes of this presentation we will concentrate on the methodologies used in the derivation of classification results and derived statistics.

The diversity of conditions in terms of topography, type and condition of forest, weather, availability of imagery, *etc.*, for the different image windows has necessitated considerable experimentation with different classification methodologies, based essentially on spectral, spatial and temporal discriminators, and these are discussed more fully below.

3.1.2 *Spectral discriminators.* The examination of spatial-spectral transects across known forest/non-forest transitions of various gradients (Malingreau *et al.* 1989, Achard and Blasco 1990, Estreguil *et al.* 1992) showed that, of the most useful derived information from AVHRR sensors, the Normalised Difference Vegetation Index (NDVI) and Channel 3 radiances were the most informative- behaving in a markedly different manner for forest as opposed to non-forest areas. This effect can be seen in Figure 2a- a radiometric transect taken from some processed images in eastern Thailand. The transect crosses areas of dense evergreen, fragmented evergreen and seasonal forests as well as areas of plantations and other non-forest land cover. Both NDVI and Channel 3 radiances show good contrasts for the different cover types, with high NDVI and very low Channel 3 values for dense forest areas- as expected for the cool, multi-storey, high green biomass vegetation, and comparatively lower NDVI and higher Channel 3 radiances for non-forest cover types. The relative magnitude of the signals in the two channels depends on the exact nature of the cover type- generally the value of the NDVI is directly proportional to the amount of tree canopy and green vegetational cover, and the Channel 3 radiance is inversely proportional to them. Thus as the transect crosses from dense forest to more degraded forest, then to agriculture and more grassland-dominated areas, the Channel 3 radiance values increase and NDVI values decrease. The NDVI behaviour reflects the presence and/or absence of green phytomass matter, while the Channel 3 radiance behaviour is thought to be mainly due to both the emissivity temperature and the reflectance contrast between a green cover (cool, dark) and dry vegetation or bare soil cover (hot, bright).

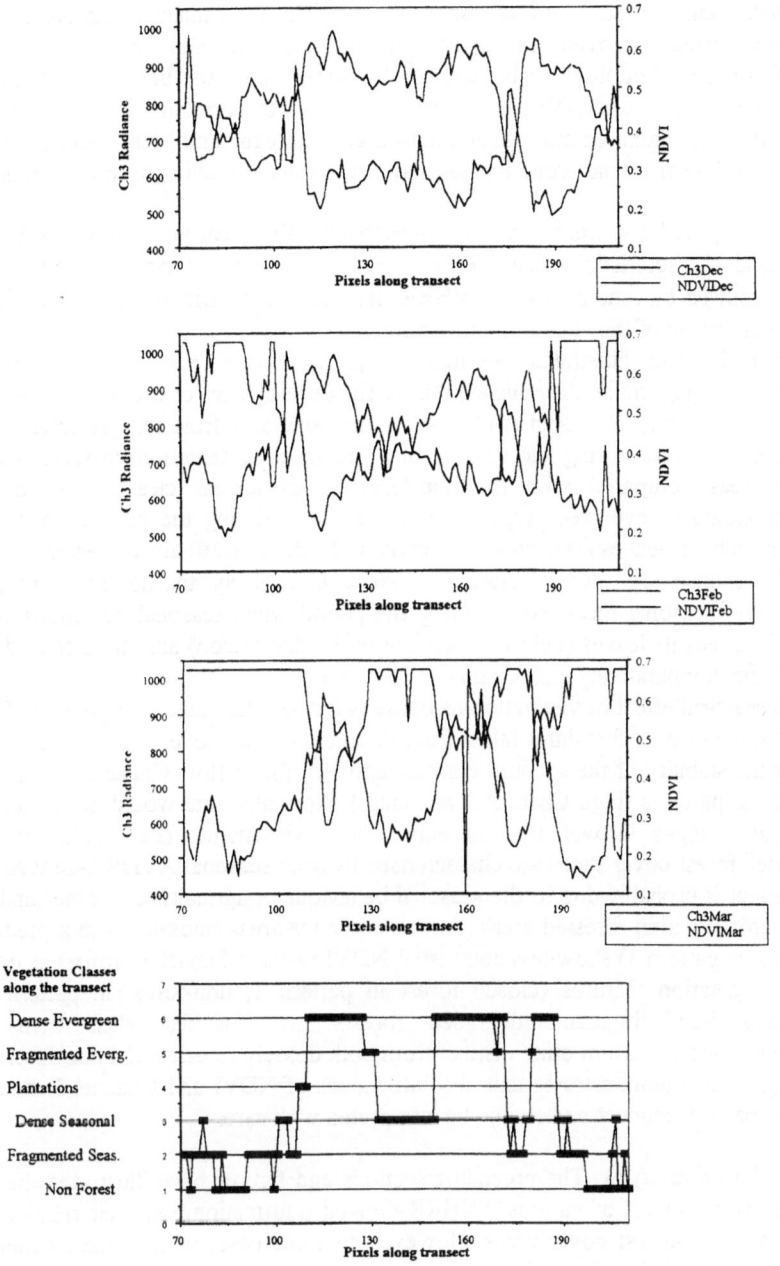

Figure 2a. Radiometric transect in Eastern Thailand

Transect analyses such as these, showed that Channel 3 radiances and NDVI would be the most useful derived information to provide maximum discrimination in any classification methodology to be adopted. In some cases, the behaviour of the ratio of Channel 3 radiance to NDVI has been shown to be well correlated to transitions between forest, densely vegetated mosaics and non-forest (Estreguil and Malingreau 1995), though *absolute* values from particular images may not always be applicable to other images.

3.1.3 *Temporal discriminators*. The seasonality of tropical vegetation is a fundamental ecological characteristic which can be used as an important classification criteria, *i.e.* when captured by time series of AVHRR data, seasonal variations are powerful keys to the identification of the vegetation formations.

Figure 2a also illustrates the multi-temporal spectral responses of seasonal and evergreen vegetation as they change during the development of the dry season in eastern Thailand. The spatial transect is plotted for the same area from images obtained at three different key dates during the dry season. The spectral/ temporal characteristics of the forest areas compared with the non-forest areas appear clearly: the forest areas corresponding to evergreen types or seasonal types during the period when most live vegetation bears leaves (December and early in the dry season) are characterised by a high NDVI response and a low Channel 3 radiance. Conversely, the non-forest pixels or the pixels over seasonal forest types during the period when seasonal deciduous vegetation normally sheds its leaves (February and late in the dry season) are characterised by a low NDVI and comparatively high Channel 3 response.

The seasonal effect of vegetation response is further illustrated in Figure 2b. This figure shows samples over five dates taken from three broad land cover types in Sumatra. It too shows the stability of the spectral contrast between forest (low Channel 3 radiances) and non-forest patterns (high Channel 3 radiances). Normally, one would not expect marked seasonal changes in wet tropical equatorial environments (Laumonier 1992). Yet, degraded forest cover types are characterised by such seasonal AVHRR-derived patterns. This effect is probably due to the seasonal behaviour of agriculture or other under-storey crops grown within forested areas. Also disturbance areas (mosaics with a predominance of forest in pattern 3) show less contrasted NDVI *versus* Channel 3 profiles as opposed to stable vegetation features (closed forest in pattern 1, non-forest in pattern 2). This illustrates how fragmented/degraded forests may manifest their own peculiar, spectral/temporal pattern quite distinct from both densely forested and non-forested areas. The figure also demonstrates that absolute values of NDVI and Channel 3 radiances are highly image dependent and should be interpreted with care.

3.1.4 *Visual analysis*. The preceding sections and figures have illustrated the potential discriminatory effect of various AVHRR-derived information and their relationship with forest and non-forest cover types. However, in some cases where image quality is very poor, such detailed radiometric analysis is not possible, and visual interpretation becomes necessary. This relies on an interpreter's skill in separating features while making allowances for internal atmospheric or illumination variations, and the remarkable human eye-brain capacity for recognising minute differences in tone and spatial pattern. In cases where visual interpretation is necessary, it is normally based on a colour composite made up of AVHRR Channels 3, 2 and 1 assigned to the red, green and blue colour channels respectively, although in some cases individual channels are also used (Malingreau *et al*. 1989). The delineation of the major vegetation formations are made with the assistance of existing vegetation maps, high spatial resolution images and fieldwork reports as available. Once confirmed, the boundaries are digitised and incorporated into the information system. These boundaries are then overlaid on multi-temporal sequences of AVHRR images to confirm the delineation of any evergreen/seasonal patterns that may be evident from the time-series of images.

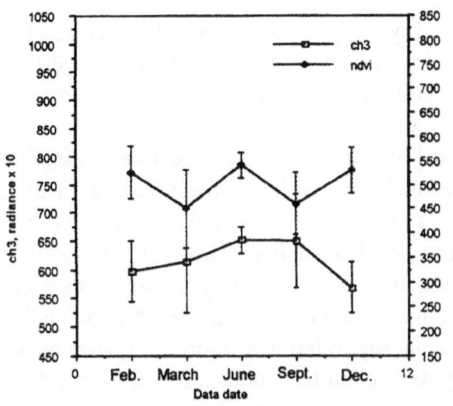

Pattern 1: Closed Evergreen Forest: 70% forest.   Pattern 2: Non-Forest: 85% non-forest.

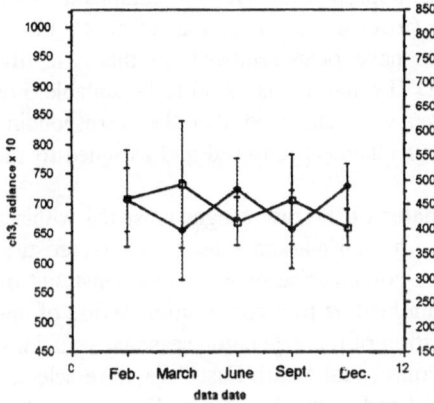

Pattern 3: Mosaics: 65% Forest (swamp and degraded forest)

Figure 2b. Multi-temporal spectral analysis in Sumatra.

While visual analysis often provides quite acceptable results, its subjective and time-consuming nature means that wherever possible, digital classification methods are preferred. Therefore, in the following sections, we will focus on the digitally-based methods for classification. As described in the previous section, where good sequences of good quality images are available, a methodology based on multi-temporal compositing and classification is possible. In other cases, the best parts of individual images have to be classified separately, and the results amalgamated according to predetermined rules.

3.2   MULTITEMPORAL COMPOSITING AND CLASSIFICATION

3.2.1   *Seasonality Considerations.* As stated before, the seasonal behaviour of forests in image time-series can be powerful aids in the labelling of forest type. However, the requirement is on a detailed series of good-quality imagery throughout the dry season-ideally spread from the start to the end of the dry season, so that the progressive drying out of seasonal formations and the increasing defoliation as the dry season progresses can be fully evidenced.

To date, much of the previous focus of tropical forest monitoring by remote sensing has been on closed humid forest, and the dry seasonal has received little attention. This is probably due to its smaller spatial extent and its very complex spatial structure, both in tropical Africa (Achard and Blasco 1990) and in South-East Asia. However, the areas in which these types of formations occur are normally characterised by a well-defined dry season, and this affords the possibility of acquiring a good sequence of high quality images, from which seasonal forests may be reliably separated from more evergreen ones.

In this section, we describe this methodology as applied to continental South-East Asia, for which a good series of cloud-free images from 1990-91 was available.

3.2.2   *Image mosaicking.* The first step of this approach involves the compilation of a set of image mosaics throughout the dry season (from short time periods of images) from which cloud, haze and other unwanted effects have been removed. In this case, the compositing process based on the maximum NDVI criteria was found to be suitable. For each pixel, the maximum NDVI of a set of images was calculated, then the corresponding pre-processed radiometric values of the first four channels retrieved and assigned to the corresponding pixel.

For the mosaicking process, the acquisition dates of the raw images must fall within a relatively short period, particularly for areas with well-marked seasonal characteristics. This is to avoid seasonal artefacts and errors. A compromise between this constraint on the time period and the availability of usable data lead us to a composition period of one month. In order to optimise the sampled information of the vegetation seasonal variations, crucial time periods during the dry season in continental South-East Asia were selected for the mosaicking process: beginning, middle and end of the dry season (December 1990, February 1991, and March 1991). Twelve images were used to make these three composites. Unfortunately, no high-quality images were available for the compilation of a composite image for January 1991.

A field study of the phenology of the different forest types present in the area indicated that in the early dry season all forest types are "green", *i.e.* all trees bear a dense canopy of green leaves. Then in the middle of the dry season the most deciduous types (dry dipterocarp forest and lower mixed deciduous forests) lose their leaves while all other types (upper mixed deciduous forest and evergreen forest) keep theirs. Finally, at the end of the dry season only the evergreen types retain their green leaves. This field-based knowledge proved to be very useful in the determination of a suitable composite image formation, and more importantly, in the subsequent interpretation of the classified image later.

3.2.2 *Unsupervised classification.* After the sea and cloud-covered areas have been masked from the composite images, the next step is to perform image-classification. For large regions it would be impractical and difficult to collect a large and well-distributed homogeneous sample of field data for the purpose of image-classification training, and it is mainly for this purpose that unsupervised methods were initially relied on.

The unsupervised classification procedure used in this case is a recursive clustering, using the radiometric distance criteria. It is initialised with a large number of arbitrary classes (40), and successive iterations are carried out after reprocessing of the grouped cluster means. AVHRR Channels 2 and 3 and the NDVI of each of the three mosaics were used as input in the classification process for continental South-East Asia, and each class was therefore characterised by a mean spectral signature in a 9-dimensional feature space (3 composite dates x 3 spectral values).

3.2.2 *Classification interpretation.* Detailed visual analysis, spectral-temporal feature-space examination, and comparison with ancillary data (or field data where available) then follows. Each of the resulting classes from the unsupervised clustering is examined carefully. Its spatial extent is viewed in relation to any available map and field information and its the spectral/temporal characteristics are scrutinised in order to assign the class to one of the five legend categories (or another one if the class is readily discriminated and distinct enough). In most cases, the extremely contrasted classes (dense evergreen forest, urban areas, non-irrigated agricultural areas) can be allocated easily. More difficulty is encountered in the assignment of the intermediate classes where there is often ambiguous spectral/temporal behaviour (see Figure 3). In these cases, knowledge of the vegetation types and formations in the area, and access to any reliable, accurate ancillary data is paramount. In some cases, where very ambiguous classes result, the classification is re-performed until more separable classes and more satisfactory results are obtained.

The results of the final classification obtained for continental part of South-East Asia are presented in Figure 4 below. A more detailed analysis of this, and a comparison with other data follows in section 4.

294

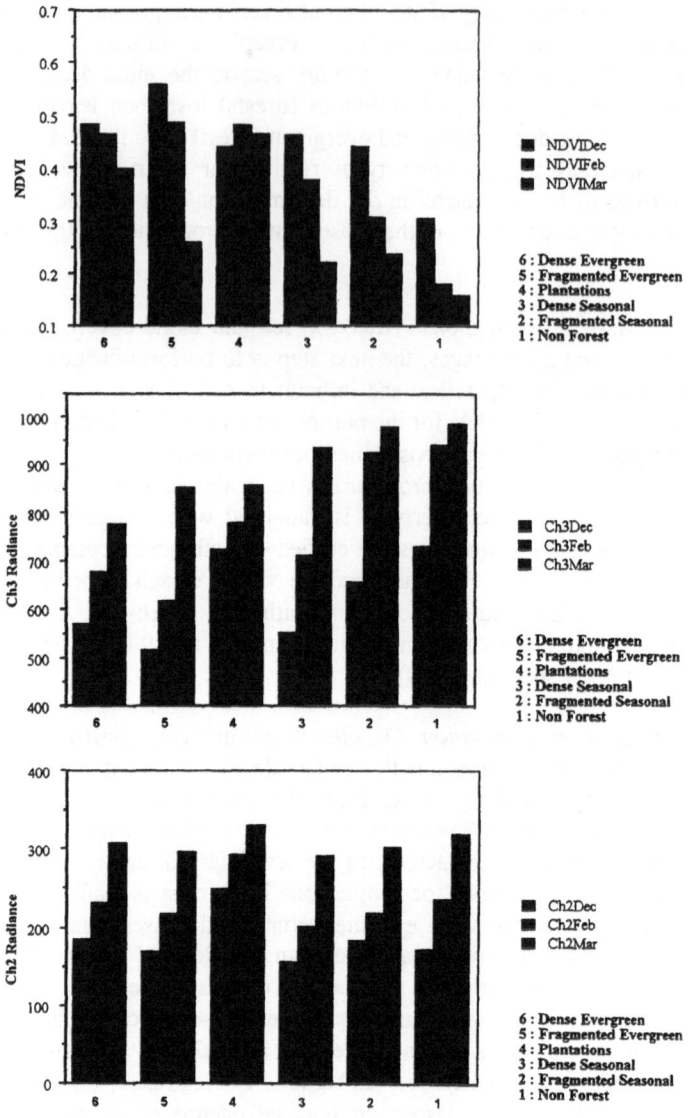

Figure 3. Mean spectral signatures per forest class (NDVI, Ch3, Ch2)..

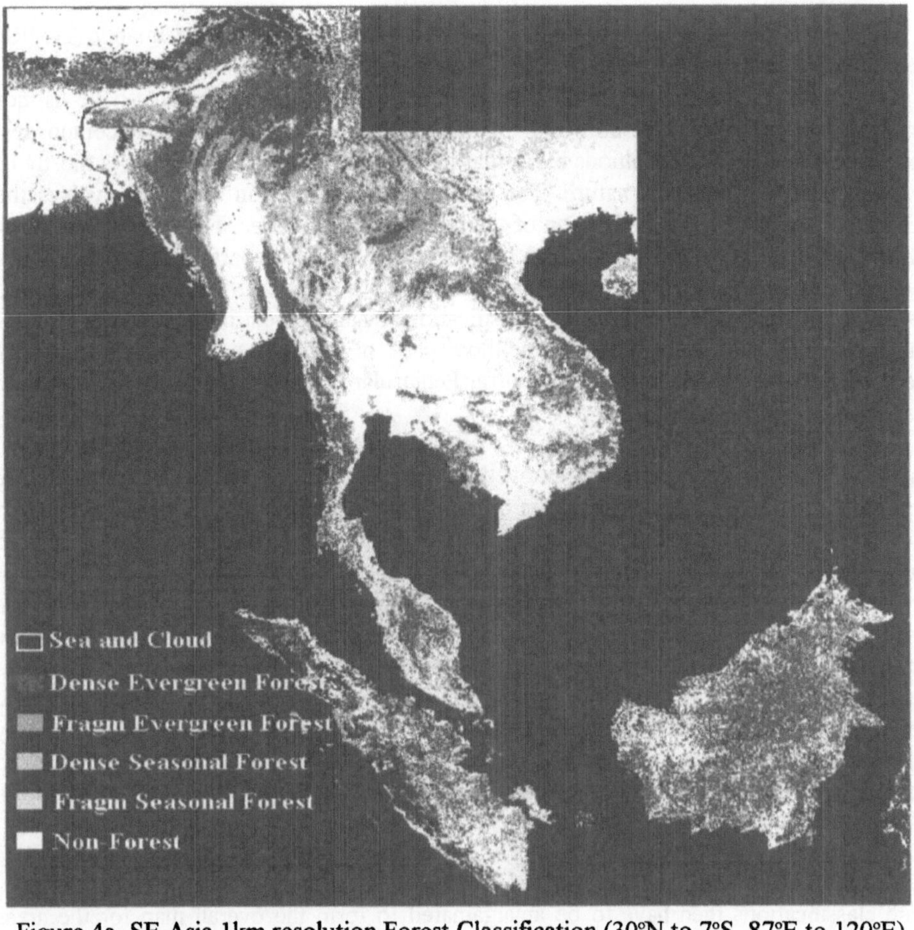

Figure 4a. SE Asia 1km resolution Forest Classification (30°N to 7°S, 87°E to 120°E)
Myanmar, Thailand, Sumatra and Peninsular Malaysia and Borneo

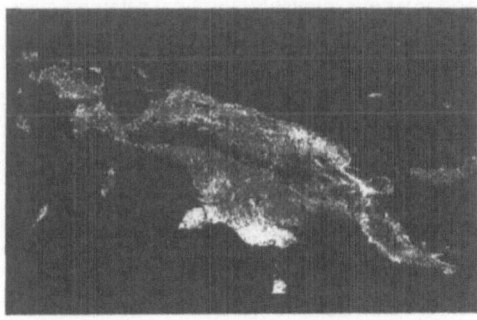

Figure 4b. SE Asia 1km resolution
Forest Classification
(12°S to 0°N, 120°E to 150°E)
Papua New Guinea and Irian Jaya

3.3   INDIVIDUAL IMAGE CLASSIFICATIONS AND AMALGAMATION

3.3.1 *Single-date classifications.* In areas where availability of good quality AVHRR data is very patchy- often the case near the Equator where there is no clear dry season, *e.g.* the insular part of South-East Asia, the methodology described above is not applicable because of the difficulty of making satisfactory cloud-free composites. However, the ever-wet conditions mean that most of the forest in such areas is mostly of an evergreen, not seasonal, nature. The methodology adopted in this case involves the separate (unsupervised) classification of the cloud-free parts of each image independently of the date of its acquisition. Then the results of each separate classification are aggregated together to form the final classification mosaic. Special procedures are necessary in areas where contrasting classifications from different images overlap. This is the approach that has been adopted for those parts of South-East Asia characterised by persistent and ever-present cloud: Sumatra, Peninsular Malaysia, Borneo, Irian Jaya and Papua New Guinea.

The first step of this approach consists of the cloud and sea masking of the individual images. Large clouds were easily removed by use of a Channel 4 brightness temperature threshold (of 290°K), but determination of thin cloud and haze over varying terrain proved to be more difficult. Initial attempts at using published threshold values proved unsatisfactory, so classification and manual methods were adopted instead.

Then for the cloud-free parts of each image, the first four channels plus the NDVI are used in an unsupervised clustering algorithm. The spectral signatures of each cluster are analysed and compared with the spectral typology of the forest/non-forest interfaces established in the examination of spatial transects (such as those in Figure 2a). The resulting preliminary classification is then compared with the vegetation maps and any other ancillary information at hand, which results in the grouping of clusters and labelling of the final classes according to the defined legend, plus additional classes where these are clearly and unambiguously delineated.

3.3.2 *Amalgamation (synthesis) of individual classifications.* Once verified, the single-date classifications then have to be amalgamated to form the overall map for the areas under consideration. For pixels where results from more than one image have been obtained, the final class is based on an assessment of the most reliable *i.e.* the classification that came from the highest quality image is retained irrespective of the way it was classified in other images. If no particularly good image is available, the final retained class is derived from an assessment of which of the classes resulted most frequently for the particular area under consideration.

Some parts of the "*synthesised*" map are then corrected as follows. Areas which are felt to be poorly classified are then extracted from the best single-date image, which is classified iteratively until the contrast apparent visually is retained in the final classification. This extracted image-part is then re-inserted into the synthesised final result. As the amalgamation process can produce an overall map with less spatial detail than in individual-date classifications, areas showing clear and sharp deforestation patterns are also extracted from the best individual date results and re-inserted into the synthesised final result.

The results of the individual classifications are thus amalgamated to form the most complete, cloud-free classification of the whole image window. The application of this methodology resulted in an overall classification of 80% of Sumatra and Peninsular Malaysia, 91% of Borneo, and 82% of Irian Jaya and Papua New Guinea (see Figure 4)

4. South-East Asia classification results and comparisons with other data sets

4.1 GENERAL

Overall, the results of the procedures outlined above provide valuable information relating to the actual distribution of the remaining forest cover, its condition and its seasonal characteristic. For example, in Thailand remnants of dense forest (evergreen, mixed deciduous or dry dipterocarp) are shown to be scarce and restricted mainly to the protected, less accessible or uninhabited areas. Much of the distribution of what was classified as fragmented forest seems to be consistent with the patterns one might expect as a result of shifting cultivation (by far the largest agent of forest conversion in the area).

The ancillary data (*e.g.* published maps) that were used to assist in the classification phase may also be used to check (validate) the results of the classification and/or amalgamation. However, the different methods and criteria used in the compilation of the two quite different thematic results should always be borne in mind when making comparisons- it is unlikely that a high degree of correlation will be obtained, especially as the level of reliability is not available for most conventionally-derived maps, but interesting commission and omission differences in either or both layers of information may be identified. In general one could expect the satellite-derived maps to be more objective and consistent, and certainly more spatially-continuous across country and regional boundaries. However, the conventionally-derived maps will probably have more detailed information about type of forest, especially if they have been compiled from a lot of field data. In this way, the two image sources can be seen as highly complementary.

4.2 CONTINENTAL SOUTH-EAST ASIA

The final classification results obtained showed several classes of diverse spatio-temporal characteristics, probably due to the important topographic and latitudinal variations in the area. Although spectral differentiation between a dry dipterocarp forest and a mixed deciduous forest is notoriously difficult on a single-date image (of low or high spatial resolution), it seems to be more reliably delineated using the multi-temporal approach.

For comparison purposes over continental South-East Asia there exist several different vegetation classification maps at regional scale, for example Royal Forest Dept. of Thailand (1990), Whitmore (1984), Legris and Blasco (1989) and IUCN (1991), but some of these are of variable quality and format. That made by the Royal Forest Department of Thailand (1990) has been adopted in the World Conservation Monitoring Centre

(WCMC) digital tropical forest database and republished in their Atlases (WCMC, 1990). For comparison purposes, the TREES final classification was digitally overlain on this. The resulting comparison details are shown in Figure 5 and Table 1. It can be seen that because the WCMC map omitted the dry dipterocarp forests, part of the AVHRR seasonal forest class coincides with what the WCMC version shows as non-forest. The large discrepancy for the dense evergreen forest also indicates that the two sources of information provide quite different results. Little documentation is available on the level of reliability of the national map, and on close examination, there appeared to be some dubious depiction of forest boundary areas. For this reason, it was decided to validate the overall classification with image classifications derived from interpretations of higher spatial resolution (Landsat TM and SPOT images) at a number of sample locations instead of with conventionally-derived maps.

4.3   INSULAR SOUTH-EAST ASIA

The WCMC digital database and another digital map source for Sumatra (Laumonier 1986) were used for comparison with the classification of the insular part of South-East Asia. In general, all three digital layers showed good general correspondence as far as the depiction of forest/non-forest areas. For example, the classification showed that 30% of Sumatra was covered by closed evergreen forest, while it amounted to 33% in Laumonier's map. When aggregated to forest/non-forest only, the WCMC map and the TREES classification show an overall correspondence of approximately 70% (Table 2 and Figure 6).

Table 1. Tabular WCMC map/AVHRR classification tabular comparison for results over Thailand

WCMC / AVHRR	Rain Forest	Monsoon Forest	Non-Forest	TOTAL	Rain Forest	Monsoon Forest	Non-Forest
Dense Evergreen Forest	30490	3710	21280	55480	54.9%	6.6%	38.3%
Fragm Evergreen Forest	7140	1470	4870	13480	52.9%	10.9%	36.1%
Dense Seasonal Forest	8140	6330	20290	34760	23.4%	18.2%	58.3%
Fragm Seasonal Forest	4000	5790	31500	41290	9.6%	14.0%	76.2%
Non-Forest	13050	7060	260630	280740	4.6%	2.5%	92.8%
TOTAL	62820	24360	338570	425750	14.7%	5.7%	79.5%
Dense Evergreen Forest	48.5%	15.2%	6.2%	13.0%			
Fragm Evergreen Forest	11.3%	6.03%	1.4%	3.1%			
Dense Seasonal Forest	12.9%	25.9%	5.9%	8.1%			
Fragm Seasonal Forest	6.3%	23.7%	9.3%	9.7%			
Non-Forest	20.7%	28.9%	76.9%	65.9%			
TOTAL	100.%	100.%	100.%	100.%			
AVHRR / WCMC	Rain Forest	Monsoon Forest	Non-Forest	TOTAL			

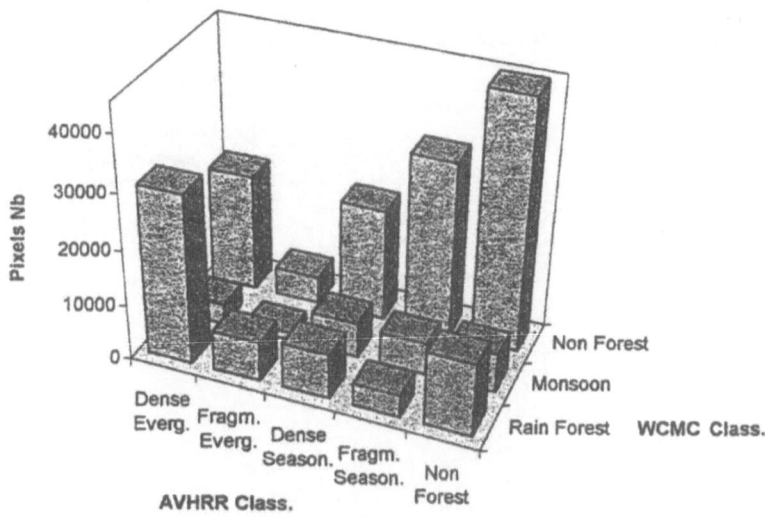

Figure 5. WCMC map/AVHRR classification graphical comparison for results over Thailand

Table 2. Comparison (in percentages) of the 1990 AVHRR-based classification with the 1988 WCMC forest map (from IUCN, 1991) generalised to two classes: forest and non-forest.

1990 AVHRR Classification/ WCMC Forest Map (1988)	Closed Evergreen Forest	Degraded/ Fragmented Forest	Non-Forest
Forest	51	32	17
Non-Forest	19	27	53

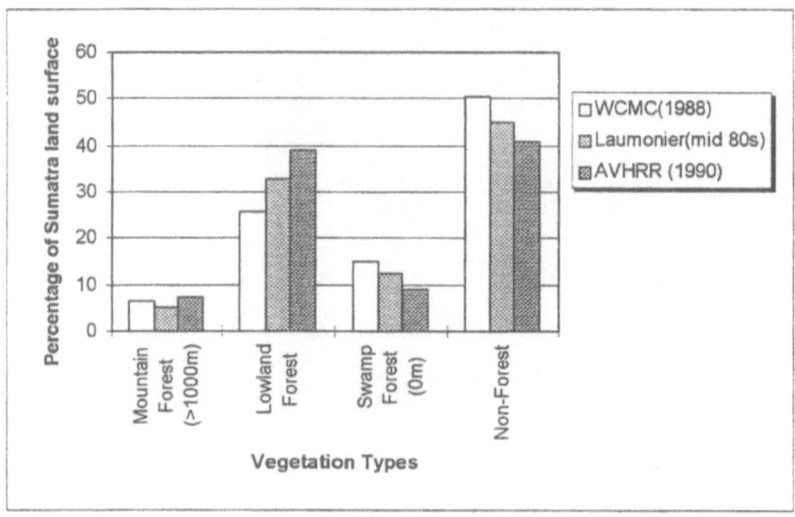

Figure 6. Vegetation classes from Laumonier (1986), WCMC (1988) and the 1990 AVHRR classification, for the island of Sumatra. The classes are expressed as percentages of the total land surface of Sumatra.

Most discrepancy with map data arose from the areas classified as fragmented forest class on the AVHRR-based classification, especially in the lowland areas. Figure 6 summarises the results of the three datasets, grouped according to mountain, lowland, swamp and non-forest. As can be seen, if the AVHRR-derived class of fragmented forest (which by the TREES adopted legend corresponds to forest cover of between 40-70%) is included as forest, a large overestimate with respect to the conventional map sources occurs, especially in the lowland areas. The inclusion of the class in the TREES results is extremely important to the overall mapping, as it indicates areas that may be undergoing some disturbance or degradation, but care must be taken in considering its representation as "forest" or "non-forest" when comparing it with conventionally-derived maps. The difficulty in comparing conventionally-derived thematic maps with satellite-derived ones was highlighted again. In the delineation of a forest polygon on the conventional maps, no indication is provided about the density or distribution or percentage cover within that polygon. On the TREES maps, the legend indicates if the cover is over 70% forest (dense) or between 40-70% (fragmented). Discrepancies in direct comparison without taking these differences into account will undoubtedly occur.

A large part of the ongoing work in the TREES project is to derive region-specific corrective weights to be apportioned to each of the thematic layers, in order to obtain more representative forest area measurements for particular ecosystems and regions. It is planned to do this with classifications based on high spatial resolution satellite imagery from a number of sample locations, probably stratified according to forest fragmentation patterns (see section 5.2 below). This should add considerable interpretative value to the broad classes derived in the global map legend.

5. Calibration and validation with high spatial resolution imagery

5.1 GENERAL

As can be seen above, the processing and examination of AVHRR imagery provides much information about the spectral and temporal behaviour of certain land cover types in the area of interest. Ideally, the classes separated by the classification process will unambiguously correspond to separate and distinct land cover classes, but this should be checked and related to field situations wherever possible. However, field validation of AVHRR-derived results is extremely difficult because of the coarse spatial resolution and the large-area coverage.

Another problem with AVHRR-based mapping is that any area measurements made from a land cover mapping exercise based on its coarse spatial resolution will undoubtedly suffer from some imprecision, arising from scale problems at least. In the example of forest mapping, because of the low spatial resolution of the sensor, several small forest clearings may be missed in the classification, or their area may be grossly overestimated, depending on the arrangement and the spectral contrast of the opening with its surroundings (Townshend and Justice 1988, Malingreau 1991, Cross et al. 1991, Cross 1990). Several studies for different areas have compared forest/non-forest classifications of co-registered AVHRR and high spatial resolution imagery to assess the errors of measurement (e.g. Stone and Schlesinger 1990, Stone et al. 1991, Cross 1990 and Woodwell et al. 1986). Cross (1990) indicated that as the percentage of forest determined from Landsat TM data decreased within corresponding AVHRR pixels, the errors in the AVHRR-based estimates increased. Stone and Schlesinger (1990) reported that where forest cover was less than 50% (within a single AVHRR pixel), the classification of that AVHRR pixels was unreliable, and pixels with less than 35% forest area were classified as non-forest. Cross (1990) also observed that in some circumstances at least 66% of the AVHRR pixel had to be forested before it was reliably classified as forested. This, the amount of over or underestimation of forest area measurements, appears to be highly dependent on the location, nature and extent of forest/non-forest interface, and any area measurements made from AVHRR-derived classifications may not be very reliable

In order to overcome these difficulties in using AVHRR for mapping, the TREES project has adopted a multi-scale, multi-sensor approach for validation of the AVHRR-derived classification, and subsequent calibration of the area measurements therein. For this purpose, a number of classifications based on higher spatial resolution imagery

(Landsat TM and SPOT) have been commissioned at a sample of locations. It is more convenient to carry out field checking of results from the higher spatial resolution imagery, and errors in the AVHRR-based classifications due to scale differences can also be measured by inter-comparisons of the two co-registered sets of data. The two processes (called validation and calibration) are reviewed below.

One point worth mentioning in this context is that both the validation and calibration exercises depend on the important assumption that the data derived from the high spatial resolution imagery is completely accurate and reliable. This is obviously not the case- they will also include some imprecision and perhaps mis-classification, but it is felt that data derived from higher spatial resolution imagery should be more reliable than those derived from coarser spatial resolution, at least for the detailed spatial representation of the different land cover patterns.

Another major difficulty which should be mentioned here is that accurate vegetation classification, particularly that of seasonal forest, is very difficult with single-date images, even if they are of the highest spatial resolution. The acquisition of additional images for multi-temporal analysis at each of the sample sites would, however, be prohibitively expensive. Thus in cases such as these, the two datasets should be seen as complementary- the high spatial resolution imagery providing spatial and spectral detail, with the coarser resolution imagery providing complementary spectral information as well as all-important temporal information, which could be paramount in areas dominated by highly seasonal vegetation.

5.2 VALIDATION

In the TREES project, the process of validation involves the *qualitative* comparison of the class assignments made on the AVHRR-derived classification, with the classifications made from high spatial resolution imagery, which in turn is normally assisted with fieldwork. In some cases, more classes are interpretable on Landsat TM imagery because of the higher spatial and spectral detail, so care has to be taken with comparison and/or amalgamation of different legends. As an example of the validation procedure, we present a pixel by pixel comparison made between the co-registered AVHRR- and Landsat TM-based classifications for an area at the eastern Thailand/southern Myanmar border. The full details of the methodology used and the results obtained in the Landsat TM classification can be found in Stibig (1993). As can be seen in Table 3 and Figure 7, the comparison of the two satellite-image based classifications is much better than the comparison of the AVHRR-based classification with conventionally-derived maps.

The AVHRR 1 k^{m2} pixels, however, may be seen to overlap with several different forest types, in differing proportions, as determined on the TM-based classification. For example, the AVHRR class of *dense evergreen forest* includes 90% of the TM equivalent class, but also 78% of the TM *mixed deciduous forest* class and 70% of the TM *mosaics of evergreen forest/shifting cultivation* class. With the AVHRR-based results (resampled to a TM-equivalent pixel size), a pixel by pixel overlay shows that in total, 23% of the AVHRR *dense evergreen forest* pixels coincide with pixels of the TM equivalent class, but 27% coincide with the TM *mixed deciduous class*, 12% with the TM *mosaics* class,

32% with the TM *dry deciduous* class and only 6% with TM *non-forest* class. Further, 70% of the AVHRR *dense seasonal forest* pixels coincide with the TM *dry deciduous class*, 12% with the TM *mixed deciduous* or *mosaics* class, and only 18% with the TM *non-forest* class. Inter-comparisons of this sort are useful in the assignment of appropriate labels to the AVHRR-derived spectral-temporal classes, and also to make qualitative assessments of their accuracy. However, detailed quantitative comparisons are not possible, because of imprecise geo-matching and scale differences of the two datasets.

Therefore, for a *quantitative* comparison and subsequent correction of forest area measurements a more statistically rigorous method has been defined- that of calibration.

5.3   CALIBRATION

As described in section 5.1 above, area measurements made from AVHRR-based classifications are notoriously imprecise, with the amount of imprecision depending on the pattern of the landscape being mapped. In order to estimate the amount of over- or underestimation inherent in AVHRR-based forest classifications, and to subsequently adjust the area measurements made from them, a calibration procedure has been defined. In this, a sample of high spatial resolution image classifications are co-registered to AVHRR-derived ones, figures of forest/non-forest area measurements from both data sources are made, and corrective coefficients are then derived from regression analyses between the two sets of results.

Table 3   TM/AVHRR classification comparison for Eastern Thailand/Southern Myanmar

Landsat TM AVHRR	Closed Everg'n Forest	Mixed Decid. Forest	Mosaics	Dry Decid. Forest	Burnt Areas	Non-Forest	TOTAL
Dense Evergreen Forest	1635	1921	885	2277	93	309	7120
Fragm Evergreen Forest	97	180	100	595	43	45	1069
Dense Seasonal Forest	65	252	173	2850	420	309	4069
Fragm Seasonal Forest	10	86	68	2686	608	1035	4493
Non-Forest	2	20	27	1213	536	1250	3048
TOTAL	1809	2459	1253	9621	1700	2948	19790
Dense Evergreen Forest	90.4%	78.1%	70.6%	23.7%	5.5%	10.5%	36.0%
Fragm Evergreen Forest	5.4%	7.3%	8.0%	6.2%	2.5%	1.5%	5.4%
Dense Seasonal Forest	3.6%	10.3%	13.8%	29.6%	24.7%	10.5%	20.6%
Fragm Seasonal Forest	0.6%	3.5%	5.4%	27.9%	35.8%	35.1%	22.7%
Non-Forest	0.1%	0.8%	2.2%	12.6%	31.5%	42.4%	15.4%
TOTAL	100. %	100. %	100. %	100. %	100. %	100. %	100. %

Given that the amount of inaccuracy in area measurements is likely to be dependent on the spatial pattern and type of forest/non-forest interface present, the location of the high spatial resolution images for calibration is chosen to provide a representative sample of different forest/non-forest interface patterns. This *"visual stratification"* is based on the AVHRR-derived results, and also on *a priori* knowledge of the deforestation patterns occurring within the area of interest. Each of the validation/calibration samples represents the area covered by a full Landsat TM scene. Suitable images are purchased and their interpretation is commissioned, wherever possible by regional specialists familiar with the area. In most cases, extensive fieldwork is carried out to support their analysis. Currently, 15 sample sites have been identified for the validation and calibration analyses for South-East Asia.

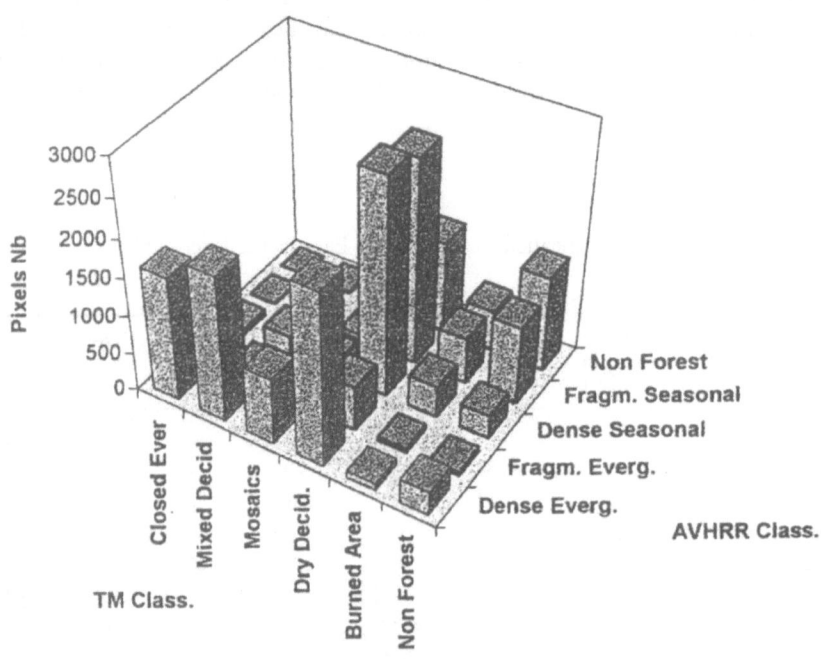

Figure 7. TM/AVHRR classification comparison: Eastern Thailand/Southern Myanmar

Once both TM- and AVHRR-based classifications have been obtained and finalised, and the two datasets co-registered, comparative forest area measurements are made. In order to overcome the low precision in co-registration (due partly to the root mean square error in the geometric correction of the AVHRR imagery, and partly to the difficulty in precisely co-registering data of such different spatial resolutions), a subsampling procedure based on the extraction of forest area percentages is based on 15 km by 15 km boxes, called secondary sampling units (SSUs). (Kleinn *et al.* 1993). Twenty-five of these are distributed systematically over the extent of the Landsat TM image (Stibig 1993). The 25 pairs of forest area measurements from both datasets are then linearly regressed. An example of such a regression is provided in Figure 8. In this particular case, equal-area estimates are found to be in the regions of 20-30% forest cover (within 15 km SSUs). In regions of lower forest cover percentages the AVHRR-based values are underestimated with respect to the TM-based ones, and in regions of more dense cover, the AVHRR-based values are overestimates.

Further analysis of several samples in continental South-East Asia and elsewhere, have shown that the exact form of the regression equation is dependent on the amount of forest cover, its fragmentation and the nature (pattern) of the forest/non-forest interface (Jeanjean *et al.* 1994). As a result, similar regressions are being systematically established for several different areas, which are dominated by several different types of forest/non-forest interface. A result of this work will be a stratification of the whole tropical forest area based probably on the three dominant landscape criteria of: the percentage of forest cover, the degree of forest fragmentation and the geometrical form of the forest/non-forest interface. So far, no automatic stratification system has led to such a map product, so visual analysis is still necessary.

After analysis of all the high spatial resolution images, it is hoped to derive, for each stratum of the whole of the tropical belt, a suitable correction coefficient. This correction coefficient will be used to transform the forest area calculated using the AVHRR-based classification into more accurate forest area measurements where required:

$$\text{AreaForest}_i = \text{AreaForest}_{i,AVHRR} * a_i$$

where:

i   identifies the stratification unit, and

a_i is the calibration coefficient for stratum I, $a_i = \dfrac{AreaForest_{i,LandsatTM}}{AreaForest_{i,AVHRR}}$

6.   Conclusions and future developments

This chapter has presented an example of the use of AVHRR data for land cover mapping and monitoring, in this case for tropical forest determination. It has focused on the TREES methodology for the compilation of a comprehensive pan-tropical forest map at about 1:1 million scale (or 1 km resolution) based on the full spatial and radiometric resolution imagery from the AVHRR instrument. With specific examples drawn from the application to South-East Asia, it has demonstrated the usefulness of this imagery in the determination of forest extent and condition in the tropics. Different methodologies have to be applied, depending on the type and amount of cloud cover in a certain region which impacts the likelihood of obtaining good quality imagery. In areas with a marked dry season, a series of good quality images may normally be obtained, which facilitates easy compositing and multi-temporal classification. In more cloudy areas with no well-marked dry season, the cloud-free parts of single-date images are classified individually and their results are then composited together. The final depiction (Figure 4) provides valuable spatial and temporal information related to the current distribution of the remaining forest cover for the whole of the region, and it compares well, both with conventional maps and with data derived from high spatial resolution imagery, at least in areas where the forest/non-forest interfaces are regular and of low fragmentation.

This presentation has also illustrated methods in which data derived from high spatial resolution imagery may be used in a complementary way for stratification, validation, calibration and improvement of the analysis carried out at coarse spatial resolution.

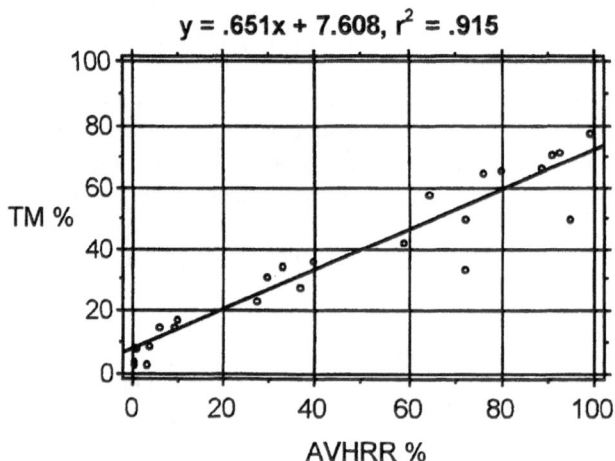

Figure 8. TM-derived versus AVHRR-derived estimates for 15 x 15 km squares.

Encouraged by the positive results obtained in this first TREES project, the Joint Research Centre has proposed a second phase called TREES II (1996-1999). This phase will be geared to the development of a prototype *operational* tropical forest monitoring system. The plan calls for a better streamlining of all remote sensing operations (based on all available high and low spatial resolution sensors), for the full utilisation of the TFIS in the classification process, and for the incorporation of a variety of non-remote sensing data sources in the analysis. Improved linkages with well defined users of tropical forest data will also be organised. Currently, it is expected that the series of AVHRR instruments will remain the prime provider of data for this global forest monitoring exercise. At a later stage, however, other similar instruments such as ATSR-2 (to be flown onboard ERS-2, 1995) and the VEGETATION Instrument to be flown onboard SPOT-4 will also be considered, and the experience gained in the handling and analysis of AVHRR data will undoubtedly be significant in the evolution of their incorporation into the operational system.

7. References and related bibliography

Achard, F. and Blasco, F., 1990. Analysis of Vegetation Seasonal Evolution and Mapping of Forest Cover in West Africa with the use of NOAA AVHRR HRPT data. *Photogrammetric Engineering and Remote Sensing*, **56** (10): 1359-1365.

Achard, F., Malingreau, J.P. and Stibig, H.J., 1993. AVHRR 1 km Resolution Data Analysis for Continental Southeast Asia. *1st TREES Conference*, Belgirate, Italy, 20-21 October 1993.

Blasco, F., 1988. The International Vegetation Map (Toulouse, France), in *Vegetation Mapping*, A.W. Kuchler and I.S. Zonnerveld, (Eds.), Kluwer Academic Publishers, Dortrecht, Netherlands. Chapter 3: pp 443-460.

Buongiorno, A., 1991. SHARP Level 2 - Development Procedures and Format Specifications, ESA EARTHNET Ed., Frascati, Italy, 25 pp.

CEC/ESA, 1990. Tropical Ecosystem Environment Observations by Satellites. Special paper SP-1.90.31, CEC-JRC Publication, Ispra, Italy.

CEC/ESA, 1991. TREES Strategy Proposal 1991-93, Part 1 : AVHRR Data Collection and Analysis, CEC-JRC publication EUR 14026 EN, Ispra, Italy. 20 pp.

Cross, A.M., Drake, N.A., Päivinen, R.T.M. and Settle, J.J., 1991. Sub-pixel Measurement of Tropical Forest Cover using AVHRR Data. *International Journal of Remote Sensing (Letters)*, **12** (5): 1119-1129.

Dickinson, R.E. and Henderson-Sellers, A., 1988. Modelling Tropical Deforestation : A study of GCN land surface parameterizations, *QJR Meteorol. Soc*, 114: 439-462.

D'Souza, G., 1993. Geometric correction techniques. In Advances in the use of AVHRR Data for Land Applications. EUROCOURSE, Ispra, Italy, November 1993.

Estreguil, C., Malingreau, J.P. and Achard, F., 1992. Spectral Contrasts Associated with Forest Types in Tropical Areas as seen on AVHRR Data. Proceedings *IGARSS* 1992, 1:233-235.

Estreguil, C. and Malingreau, J.P., 1993. Vegetation Characterisation and Land Cover Mapping with NOAA-AVHRR Data in the Humid Wet Tropics. *GeoCarto International* (Submitted).

Estreguil, C. and D'Souza G., 1993. Remote Sensing Measurement of Forest Parameters in the Insular Part of Southeast Asia : High/Low Resolution Image Data Analysis. Proceedings *IGARSS* 1993, **2**:733-736.

Estreguil, C., 1993. NOAA AVHRR Studies of Forest Mapping in the Equatorial South East Asia. *1st TREES Conference*, Belgirate, Italy, 20-21 October 1993.

Houghton, R.A., Boone, R.D., Fruci, J.R., *et al.* 1987. The Flux of Carbon from Terrestrial Ecosystems to the Atmosphere in 1980 due to Change in Land-use ; Geographical Distribution of the Global Flux. Tellus 39B, 122-139.

IUCN, 1991. The Conservation Atlas of Tropical Forests : Asia. (N.M. Collins, J.A. Sayer and T.T. Whitmore, (Eds.), Macmillan Press Ltd, London, 256 pp.

Kleinn, C., Dees, M. and Pelz, D.R., 1993. Sampling Aspects in the TREES Project - Global Inventory of the Tropical Forests. JRC Contract 5014-92-10 ED ISP D Final Report, 34 pp + appendices.

Laumonier, Y., Purnadjaya, P. and Setiabudi, S., 1986-87. International Map of Vegetation (1:1,000,000), South Sumatra (83); Central Sumatra (86); North Sumatra (87). BIOTROP, Bogor, Indonesia, ICIV Toulouse, France.

Laumonier, Y., 1992. Sumatra, Environment and development : Its Past, Present and Future. BIOTROP Publication No.46, ISSN 0125-975, ISBN 979-8275-03-9.

Legris, P. and Blasco, F., 1989. Classification and Mapping of Vegetation Types in Tropical Asia. FAO Publication, Rome, 169 pp.

Malingreau, J.P., Stephens, G. and Fellow, L., 1985. The Catastrophic Forest Fires of Kalimantan and North Borneo in 1982-83. *Ambio*, **14**: 314-315.

Malingreau, J.P., 1991. Remote Sensing for Tropical Forest Monitoring : An Overview, in *Remote Sensing and Geographical Information Systems for Resource Management in Developing Countries*, A.S. Belward and C R. Valenzuela, (Eds.), Kluwer Academic Publishers, Chapter 13: 253-278.

Malingreau, J.P., Tucker, C.J. and Laport, N., 1989. AVHRR for Monitoring Global Tropical Deforestation. *International Journal of Remote Sensing*, **10**: 855-867.

Malingreau, J.P., da Cunha, R. and Justice C. (Eds), *Proceedings of the World Forest Watch Conference*, CEC-JRC EUR 14561. En, Ispra, Italy, 84 pp.

Malingreau, J.P. and Achard, F., 1993. Towards an Improved Vegetation Classification Scheme for Global Vegetation Monitoring; Taking into Account the Contribution of Remote Sensing. UNEP/HEM/WCMC/GCTE *"Workshop on Improved Vegetation Classification Scheme for Global Mapping and Monitoring"*, Charlottesville, USA, 24-26 January, 1993.

Myers, N., 1992. Future Operational Monitoring of Tropical Forests : an Alert Strategy. in *Proceedings of the World Forest Watch Conference*, J.P. Malingreau, R. da Cunha and C. Justice, (Eds.), CEC-JRC EUR 14561.EN, Ispra, Italy, pp 9-14.

Päivinen, R.T.M., Pitkänen, J. and Witt, R.G., 1992. Mapping Closed Tropical Forest Cover in West Africa using NOAA/AVHRR-LAC Data, in *Assessment of Tropical*

Forests Using Satellite Data, Sippi Jaakkola *et al.* (Eds.), University of Joensuu Publication, Finland, *Silva Carelica*, **21**: 27-51.

Royal Forest Department of Thailand, 1990. 1:1,000,000 Forest Types Map, Remote Sensing and Mapping Division, Bangkok, Thailand, 1 sheet.

Salati, E., Dourojeanni, M.J., Novaes, F.C., *et al.* 1990. *Amazonia, The Earth as Transformed by Human Action : Global and Regional Changes in the Biosphere over the Past 300 Years*, B.L. Turner (Ed.), Cambridge University Press, Cambridge, Chapter 29: 479-493.

Singh, K.D., 1993. The 1990 Tropical Forest Resources Assessment, *Unasylva*, **174**, (44): 10-19.

Stibig, H.J., Achard, F. and D'Souza, G., 1993. High Resolution Data Integration for Calibration: the TREES Methodology. *1st TREES Conference*, Belgirate, Italy, 20-21 October, 1993.

Whitmore, T.C., 1984. A Vegetation Map of Malesia at Scale 1:5 Million. *Journal of Biogeography*, **11**: 461-471.

USE OF AVHRR DATA FOR THE STUDY OF VEGETATION FIRES IN AFRICA: FIRE MANAGEMENT PERSPECTIVES.

J-M GRÉGOIRE
Joint Research Centre,
Institute for Remote Sensing Applications,
21020 Ispra, Italy.

ABSTRACT: Environmentally, the direct and indirect effect of the increasing anthropogenic pressure on tropical ecosystems are often related to fire. However, fire can be seen in a dual perspective clearly summed up by Wein (1991) in the first issue of the International Journal of Wildland Fire: "Fire is viewed in some instances as the most globally destructive tool used by industrial societies. Yet for many purposes fire is the most environmentally safe, economically sound, and socially acceptable tool that can be used to achieve wildland goals". National and international agencies are more and more convinced that a policy of fire suppression is surely not a sustainable solution in a socio-economic perspective and that it could have negative drawbacks on conservation policies. In such a conflictual context, emphasis should be put on the management of man-made fires and their use in development policies. Still, reliable information on fire dynamics are lacking at local, continental and global scales. Remote sensing technology can contribute to the documentation of this key element in the human-environment interactions. Among the remote sensing tools actually available, the NOAA-AVHRR (Advanced Very High Resolution Radiometer) system, although originally designed for meteorological applications, has proved to be very efficient for the detection and monitoring of vegetation fires.

The situation and dynamics existing in the African continent south of the Sahara will be used here to illustrate how the NOAA-AVHRR system can provide the basic information needed in a fire management perspective.

The general interactions between man-made fires and main components of the environment, will first be analysed. Next, some key domains, in which a management approach to man-made fires is useful/necessary, are defined together with the requested information to support such a management approach. Then, an analysis of how far the NOAA-AVHRR system allows access to the information required for fire management perspectives in the specific domains defined previously will be offered. Finally, some steps towards an operational system for the monitoring of vegetation fires in tropical environments will be discussed.

1. Interactions existing between vegetation fires and the environment

Widely used in South-America, Sub-Saharan Africa and Asia, man-made fires have always been a key element in the human-environment relationships. Fire dynamics can be the result of an equilibrium between environmental conditions and resource management, such as fires used by pastoralists in the savanna zones, or they can be indicators and agents of changes of biomes under increasing anthropogenic pressure, such as the ones occurring in forest ecosystems of Amazonia and Central Africa.

G. D'Souza et al. (eds.), Advances in the Use of NOAA AVHRR Data for Land Applications, 311–335.

Although strongly inter-related, the relationships between man-made fires and vegetation cover, water resources, air and soil will be reviewed separately.

To avoid confusion, it is first necessary to define clearly the terms which will be used in the text (definitions which obviously are not intended to be universal):

- fire: vegetation fire (vegetation fires being one aspect of biomass burning);
- active fire: a fire in a smouldering or flaming state;
- fire event: one or more active fires defined both in time and space;
- fire front: the actively burning part of a fire or a group of fires;
- fire dynamics: the spatio-temporal patterns in fire distribution;
- burnt areas: regions affected by fire (even if not all the vegetation cover is completely burnt);
- burning season: period of the year during which fires can occur for a given region;
- fire calendar: temporal distribution of fires within a burning season;
- fire activity: frequency of active fires for a given region during a given period of time.

1.1   FIRE AND LAND COVER

Existing relationships between fire and vegetation cover do not come down to the simplistic view of fire destroying the vegetation. On the contrary, in tropical regions of the globe there often exists some kind of equilibrium between vegetation cover type and fire. In essence, one can define three typical cases:

- in the first one there is an equilibrium between fire dynamics and vegetation cover type and conditions; in this case vegetation type and conditions are largely controlled by fire. Such a situation is observed in most of the shrub savannas domain. It can be observed also in relatively extreme situations such as the grassy clearings within the Congo forest, for which Menaut (1983) uses the expression "fire-maintained communities";
- in the second case, there is intrusion of fire in ecosystems traditionally immune from fire, which will induce a sudden modification of vegetation cover conditions and characteristics. Such a situation is usually associated with land reclamation in, or at the edges of, forested areas; it can also be the consequence of abnormal meteorological conditions (for example the catastrophic fire event in China or in Kalimantan (Bertault J-G., 1991);
- the third case is characterised by a progressive shaping of vegetation cover under the effect of fire; in this case there is not a sudden break of an equilibrium but a progressive evolution towards a new situation.

To this spatial dimension of relationships between fire and vegetation cover, one should add a temporal one, which might be the most important. This temporal dimension covers two aspects: the date of burning and the frequency of fire occurrence.

Date of burning: for instance in the Miombo woodland of Zambia/Zimbabwe (Menaut, 1983) the vegetation cover is considered to be resistant to fire if burning takes place early during the dry season; such fires do not prevent regeneration of trees; however, late fires can destroy the canopy woodland, reducing it to coppice; openings created which will then facilitate the drying of the standing grass which, in turn, facilitates more intense fires. In this situation establishment or regeneration of woody species becomes more and more difficult.

Frequency of burning: In more open savannas of Senegal, Mali or Burkina Faso the high frequency of fire occurrence has an impact similar to that of overgrazing, which is a progressive eradication of perennial grasses with annuals species becoming subsequently dominant.

Menaut (1983) shows clearly the complexity of the role of fire in the shaping and/or the maintenance of the vegetation cover in Sub-Saharan Africa; temporal dynamics being a key element. In the African context, if one excludes the regions where herbaceous cover is largely dominant (as in the Sahelian zone for instance), one can consider that burning is certainly the major factor determining the distribution and spacing of shrubs and tress. Such observations might not be true everywhere. For example, Bowman *et al.* (1988) have conducted an experiment on the effect of temporal fire regimes on the eucalyptus forest and woodlands of Northern Australia and concluded that, due to a very long history of dry-season burning, vegetation patterns in this part of Australia were primarily determined by edaphic factors and not by fire.

The same type of experiment conducted in Côte d'Ivoire (Dauget and Manaut, 1992) has provided different results. Despite 20 years of annual burning, a woodland savanna plot has shown an increase in shrubs and trees from existing thickets, and almost no new woody species in the herbaceous part of the plot.

It must be stressed that appreciation of the degree of equilibrium, or disequilibrium, existing between fire phenomena and vegetation cover dynamics, depends directly on the scale of observation, both in time and space. This aspect is fundamental in a fire management perspective.

In most cases the most important information concerns on the one hand the spatial distribution of fire events (for instance detection of intrusion of fire activity in ecosystems usually immune from fire), and on the other hand the temporal distribution of fire events, to know for a given burning season (i) if fires occur more than once in a specific location, and (ii) when the fires occur. In any case, the monitoring process must cover long time periods (at least 10 years or more).

1.2.   FIRE AND LAND USE: FIRE PRACTICES.

In Sub-Saharan Africa, man-made fires are found in three landuse contexts: grazing, hunting and agriculture, including land clearance. Depending on the ecological domain, ethnic group, and the time of year, the overall fire activity will derive from one of more of these landuse practices.

Fire and Pastoralism. Among the fire practices in Sub-Saharan Africa, the ones connected with grazing activities appear well known, but there are a lot of misconceptions. The most important one is probably the one which portrays the pastoralists using fire to regenerate grasslands. This is not the case as fire does not regenerate the grass layer. Instead it makes it accessible to cattle for grazing (IEMVT, 1990-a). Such a positive role of fire is not apparent everywhere; for instance, in the Sahelian domain fires have a very negative impact as they destroy the stock of straw remaining from the previous rainy season; straw which is often the main resource of fodder during the dry season. Therefore, fire could be considered at the same time a key tool for grassland management in the shrub and tree savanna domain, but a very harmful phenomenon in the wooded grassland of the Sahelian zone and others (IEMVT, 1990-b). Therefore, environmental management of fire practices must be considered with respect to the ecological region of interest. For instance, fire management plans could be easily formulated for the wet savannas domain, but would prove more problematic in other domains.

One could say that fire practices should not exist in the Sahelian type regions, but it is much harder to define stricter rules in the intermediate or transition zones. In the latter cases, the geographical location of fires may be more important than the burning calendar there (Diallo *et al.* 1990; Langaas, 1992).

Fire and Hunting. Fire is commonly used by hunters in the savanna domain in order to drive and concentrate the game in specific geographical areas, for example in between two gallery forests. Characteristics of these "hunting fires", and their impacts on the environment, can vary widely from one region to another. For instance, fire practices observable in West Africa during the "agoutis" hunting will have impacts very different from the immense fires started by poachers active in Central Africa (at the border between Sudan and the Central African Republic) or from the fires used by cattle thieves in the western region of Madagascar. In terms of fire management and assessment of fire on the environment, then, each type of "hunting fire" must be considered and treated separately.

Fire and Agriculture. Fire is used in agriculture, often to clean plots before the seedtime, at the end of the dry season, or to prepare some new lands for cropping. Such practices end usually in controlled fires limited in extent, but which can last for a much longer time than the fires lit for reach pastoralist or hunting reasons. Moreover, in densely populated agricultural areas, these fires can be very numerous. Even if much less spectacular than fires from pastoralists or hunters, agricultural fires have very strong impacts, both positive and negative, on the environment, and so must not be underestimated.

Finally, fire is used in traditional environmental management, as a cleaning tool; around the villages, as a protection against uncontrolled fires and wild animals, and in the savannas to maintain travel possibilities both for people and cattle.

1.3   FIRE AND WATER/SOIL RESOURCES

Water resources. Relationships between fire and water resources can be observed at different moments of the water cycle at the surface, but most, if not all of them, derive from the impact of fire on the vegetation cover, including modification of key processes such as evapotranspiration, surface runoff, recharge of water tables, contribution to river floods, etc. (Albergel, 1988). As has been mentioned before in the text, fire is a key process in the determination of the respective proportions of grass and shrubs/trees in the overall vegetation cover. Now, some studies, such as the one by Geffard (1991) for West Africa, have shown that the responses of river regimes to interannual changes in rainfall are highly dependent of such proportions of grass and shrubs/trees layers. For the Upper Senegal and the Upper Niger, interannual instability of the river regimes is much higher in the tributaries where the vegetation cover is dominated by the grass layer than in the tributaries where woody formations are important. To a certain extent one could say that, in the savannas domain, fire has an indirect but strong effect on the sensitivity of river regimes to interannual variations in precipitation; river regimes which in turn have an effect on water resources in the same regions.

As far as surface runoff is concerned, and indirectly the recharge of the soil water table, May (1990) has reported a considerable increase in surface runoff after a large fire on a small catchment in Spain. He reported also that hydrological behaviour had reverted to pre-fire conditions three or four years after the fire. Fritsch (1992) has reported, in Guyanne, the strong impact of slash and burn cultivation on the water cycle at the surface, including an increase of the overall run-out volumes, of the peak floods, and of the sediment transports, etc. Such observations might be valid in the African regions or countries situated at the edges of the forest domain, such as Guinee, Cameroun, Congo and Zaire. They may not be valid in the savannas domain in which a kind of equilibrium has been reached between annual fires and vegetation cover type. It must be said however that if burnt, the savannas regions are totally unprotected against soil erosion by heavy storms at the beginning of the rainy season. Very few studies have been carried out to evaluate the effects of fires on surface runoff in Africa. In any case, the main information of interest for fire management here is the interannual repetivity of fire events and the location within the river catchment of the fire events.

Soil resources. Effects of fire on soil may result from direct heating of the soil, such as redistribution and changed availability of nutrients (Moore, 1989; Schmalzer et al. 1991), modification of hydrologic characteristics (Almendros et al. 1990), or from vegetation removal, the third of which has three direct consequences (Gillon et al. 1973):

- a change in soil temperature regime, inducing particularly higher thermal magnitudes, and of the overall soil microclimate, which in turn affects activity of the microbial community;
- an increase in the evaporation which will rapidly exhaust water reserves in the upper layers of the soil;
- an increase in the soil erosion, during the first heavy storms of the rainy season, and consequently of the surface runoff.

The amplitude of soil denudation depends to a large extent, for a given location, on the state of the grass layer at the time of the burning; in effect, on the burning date within the dry season. In a management perspective, the key information here is the burning calendar.

1.4   FIRE AND ATMOSPHERE

It is now generally accepted that biomass burning, with a strong contribution of vegetation fire, is by far the main source of atmospheric pollution over the African continent. This includes atmospheric pollution due to:

- gases and particles with radiative and climatic effects (Cachier *et al.* 1991; Bonsang *et al.* 1991; Delmas, 1992). The exact effects of these atmospheric particles are not completely understood, and the direction of their effect may even sometimes be a source of controversy. For example, Penner *et al.* (1992) tend to say that the effects of aerosol from biomass burning on the global radiation budget are not simple and that, for instance, anthropogenic increases of smoke emission may have actually helped weaken the net greenhouse warming from anthropogenic trace gases (!);
- oxidising effects: production of photochemical ozone and its precursors in the troposphere. It is thought that half of the troposphere ozone over the equatorial Africa is due to vegetation fires of the savanna areas (Cros *et al.* 1988; Lopez *et al.* 1992);
- acidifying effects: organic and mineral acidity of the air and of the precipitation (Lacaux *et al.* 1991; 1992-a-b; Lefeivre, 1993).

In order to evaluate emissions from vegetation fires, both aerosols and gases, one needs to know the amount of biomass burnt and the emission factors which depend on the type and conditions of the burnt material. Emission factors are usually considered to be known with a sufficient precision; the main uncertainty is on the quantification of the burnt biomass (Robinson, J.M., 1989; Delmas *et al.* 1991). This is the biggest uncertainty to be answered in a fire management perspective.

1.5   INFORMATION NEEDED FOR MANAGEMENT OF MAN-MADE FIRES

It is obvious that management of man-made fire is, in Africa, one aspect of environmental management of natural resources such as soil, vegetation (pastures and woody biomass resources), water and air. Such a management approach to fire practices may constitute a key element in operational projects of rural development at national and regional level. National parks management programmes in the North of the Central African Republic (Da Vinci Consulting S.A., 1993) or management of dry forests (Van Wilgen *et al.* 1990) are good examples of actions which integrate fire management in the overall strategy. At the regional level, the ECOFAC project active for five countries of Central Africa is another good example. Despite the complexity of the role of fire in the functioning of African ecosystems, one can define the fire-related information which is of interest in an environmental management perspective. This information is summarised in Table 1: to summarise the requirements, one could say that the key information is *a good description of temporal dynamics, both seasonal and interannual, integrated in a Geographical Information System.* In the next section we will see how far the NOAA-AVHRR system can answer to these requirements.

2.   Use of NOAA-AVHRR data to collect the fire related information

The first attempt to use satellite remote sensing for studying fires in Africa was probably the work of Deshler (1974) who used ERTS imagery for the mapping of burnt areas in Central Africa, north of the equator. More recently Parnot (1988) did an inventory of burnt areas in Burkina Faso, West Africa, during the 1986-1987 dry season; he used Landsat MSS data in a multi-temporal study of bushfire activity for the whole of Burkina.

Table 1. Information of interest for management of man-made fires.

Management domain	Priority information		Expected benefits
Pasture and *"parcours"*	1. 2.	burning calendars location of fires	• impact on percentage of perennial and annual species • control of woody species development • management of fodder during the dry season
Woody biomass resources	1. 2. 3.	burning calendars frequency of burning location of fires	• increase in woody species diversity • improve regeneration capacities of woody populations • improve management of protected areas
Soils resources	1. 2. 3.	location of fires in the topographic ontext burning calendars frequency of burning	• decrease erosion and losses of soils • modify the soil microclimate
Water resources	1. 2. 3.	location of fires in the river basins context burning calendars frequency of burning	• decrease sensivitiy of river regimes to interannual variations in rainfall • regularise the recharge of the soil water table • regularise floods characteristics
Air and precipitation quality	1. 2.	biomass burnt burning calendars	• decrease acidity of rainfall • improve air quality in populated areas

As far as NOAA-AVHRR is concerned, the use of this system for fire detection and monitoring in Africa followed the upgrading of the ESA receiving station at Maspalomas, Canary Islands in 1986 (Arest *et al.* 1984; Fusco and Muirhead, 1987). The availability of good time series data for West Africa then allowed studies such as the one of Langas and Muirhead (1988) for Gambia, or Grégoire *et al.* (1988) for Guinée. For Central Africa, a more recent work carried out in the framework of the US Biodiversity Support Program (World Wildlife Fund, 1993) presents some results on the use of AVHRR data for fire monitoring at the edges of the rain forest domain.

In her review of the current utilisation of satellite remote sensing for studying vegetation fires and their effects on ecosystems, Kennedy (1992) points out four attributes of the NOAA-AVHRR system which make it highly efficient for fire related studies:

- the spectral location of AVHRR channel 3 (3.55 - 3.93 μm) is well suited to the detection of elevated heat sources such as active fires;
- the spatial resolution, 1.1 km in LAC (Local Area Coverage) or HRPT (High Resolution Picture Transmission) configuration and 4 km in GAC (Global Area Coverage) configuration, offers the possibility of obtaining continental and regional coverage of Africa;
- the temporal coverage: depending on the number of satellites in operation, there is the possibility for twice daily and twice nightly coverage of the African continent; when using only one satellite, such as NOAA-11 or NOAA-14, the afternoon overpass time is well suited for fire detection as man-made fire activity is at a maximum sometimes at the beginning of the afternoon;
- the existence of archives of daily AVHRR data (GAC) back to 1981 makes possible the study of general trends in fire activity at the continental level in Africa.

In relation to the requirements of vegetation fires management programmes, NOAA-AVHRR data can also contribute to the documentation of the pre-fire situation, the burning period and the post-fire situation.

- the pre-fire situation; the state and condition of the existing vegetation;
- the detection and monitoring of active fires including the detection of smoke plumes (which will not be covered here in detail, but the reader is referred to the work of Kaufman *et al.* (1989) or Shoji Takeuchi (1993) which show clearly the potential of AVHRR data for the detection and study of smoke plumes to detect active fires);
- the post-fire situation; Mainly the mapping of burnt areas, the characterisation of the vegetation cover which has been affected by fire (type of vegetation cover, regrowth after fire, *etc.*)

2.1   PRE-FIRE SITUATION

One importnat objective of the pre-fire period assessment is to characterise the potential fuel loading. Derived information can then be used to perform risk assessments and to improve the governmental policies regarding early fire practices. Although NOAA-AVHRR data have proved to be very efficient for fire risk assessment (Paltridge and Barber, 1988; Van Wilgen *et al.* 1990), one can understand that fire risk maps (whether derived from NOAA-AVHRR data or other means), are not a priority in most African countries. However, it must be said that governmental policies trying to promote early fires, considered to be less damaging then fires late in the dry season, could gain from the use of pre-fire diagnostics at the country level: population response to governmental directives would be more "positive"!

In any case, the first priority remains the setting-up of monitoring systems which allow the collection of reliable and quantitative information on spatio-temporal dynamics of vegetation fires, and not on pre-fire diagnostics.

2.2   FIRE DETECTION AND MONITORING

2.2.1   *Detection of fire events and characterisation of seasonal dynamics.* The simplest method for fire detection, using AVHRR data, relies on a thresholding of the channel 3 values (3.55 - 3.93 μm). All pixels showing a signal close to the saturation of the sensor in this spectral band are considered as "fire pixels": pixels which contain one or more active fires. This method appears to be very efficient when used to detect fire activity in forested areas of Asia (Malingreau *et al.* 1985) or South America (Malingreau and Tucker, 1988).

When applied to open environments, such as the savanna domains of Africa, this method is not completely satisfactory as surface phenomena other than fire are also found to cause channel 3 saturation (Langaas and Muirhead, 1988; Robinson J.M., 1991, Grégoire *et al.* 1993-a). To avoid confusion between active fires and warm and dry surfaces, a second method (Kaufman *et al*,.1989) uses the difference in the relative responses in channels 3 and 4 (10.4 - 11.3 μm). In the presence of one or more active fires in a pixel, radiance in channel 3 will be increased strongly, compared with only a slight increase in the channel 4 signal. More recent works (Franca *et al.* 1993; Kennedy *et al.* 1993) try to combine thermal and visible data for fire detection and monitoring in the African context.

From the conclusions of a recent workshop (Justice 1993) organised in the framework of the International Geosphere Biosphere Programme (IGBP) Data and Information System (DIS), it seems unlikely that a universal approach to fire detection using AVHRR data sets can be established. It is obvious that particular methodologies must be adapted to meet the specific conditions found in different parts of the African continent, and that the fire detection algorithms must be flexible enough to take into account both the geographical location and the time of the year.

On the other hand, the approach to fire monitoring must allow the combination of satellite derived information and cartographic data related to the geographical context.

Two examples, from results obtained by the MTV unit in the framework of its activities on fire monitoring, are provided here. Although carried out in a research perspective, this work provides results which can be exploited in operational management schemes.

The first example refers to the use of AVHRR-HRPT data for fire detection in Central Africa (Figure 1), in the zone bordering Sudan, Central African Republic and Zaire. Three scenes, situated around the peak of fire activity for this part of Africa, were processed and analysed. Depending on the date of observation, image pixels containing active fires are coded in orange (16.12.1991), red (03.01.1992) and yellow (11.01.1992). These results (Figure 2) bring three general comments:

Figure 1. NOAA-AVHRR 1 km resolution satellite image of the border region between Central African Republic, Sudan and Zaire (11 Jan 1992; 500 x 500 km). The Congo basin forests, bush and tree savanna to the North, and the smoke plumes emanating from several bush fires, driven by the prevailing North Easterly winds, are all indicated, as is one particularly long active fire front.

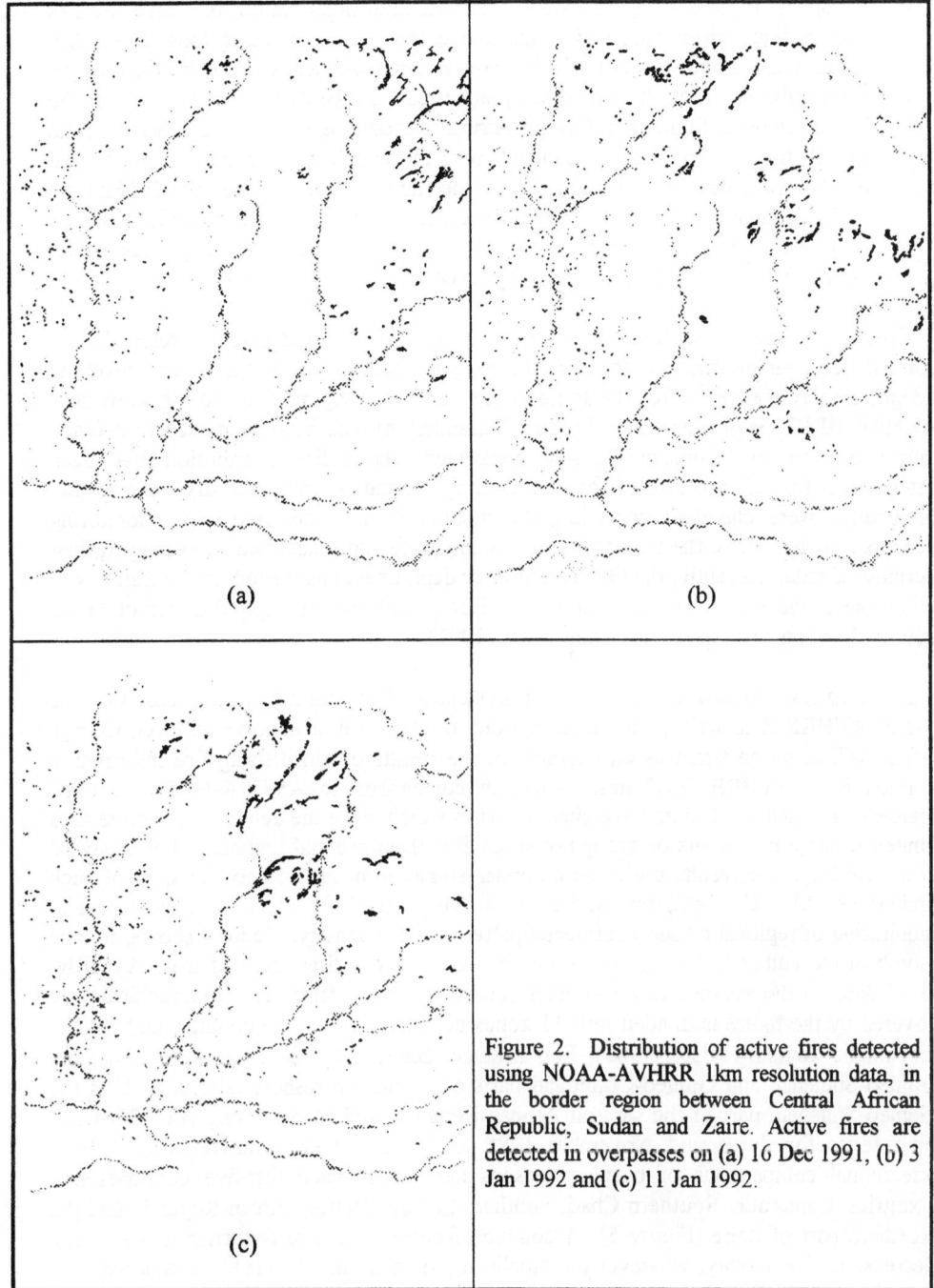

(a)

(b)

(c)

Figure 2. Distribution of active fires detected using NOAA-AVHRR 1km resolution data, in the border region between Central African Republic, Sudan and Zaire. Active fires are detected in overpasses on (a) 16 Dec 1991, (b) 3 Jan 1992 and (c) 11 Jan 1992.

- the spatial organisation, or textural and structural organisation, of active fires, is also evident: within the N-E group active fires are arranged in long, linear fire fronts, while in the Western half fires are scattered with an overall very fine texture;
- some information on the temporal dynamics can also be derived: in the N-E, at the border between Sudan and Central African Republic, a kind of progression of the overall fire activity can be followed from the middle of December 1991 to the middle of January 1992. Progression of specific fire fronts between the 3rd and the 11th of January are evident. In the Western half, temporal patterns are less evident even if a kind of "centrifugal dynamic", a progression of fires in a semi-circular direction around the centre, can be observed.

Operational use of such information, in a management perspective, relies on its combination with information related to the geographical context. At the country level the geographical reference is often the administrative divisions. Figures 3-a and 3-b show how AVHRR-HRPT (High Resolution Picture Transmission) data have been used to provide information to the Guinean forestry department: bush fire distribution has been categorised for 22 prefectures of the country during a complete dry season and prefectures were classified according the number of fires detected. The monitoring process can also derive the temporal patterns and prefectures classified according to their burning calendar. Recently, the Guinean forestry department has started, in the framework of a project funded by the Commission of the European Union, to apply this method at the country level on a routine basis (Anonymous, 1991).

2.2.2. *Regional patterns and interannual dynamics*. Temporal monitoring, based on the use of AVHRR time series, can be carried out at a regional or continental level. Central Africa will again be taken as an example of the results obtained using fire information derived from AVHRR-GAC time series. In comparison to AVHRR-HRPT or LAC imagery, AVHRR-GAC data have characteristics which make the detection of active fires limited to large fire events or groups of small fires (Belward and Lambin, 1989; Belward *et al*, 1993-b). The results are often an underestimation of fire activity. In spite of such limitations, AVHRR-GAC time series remain very useful for the characterisation and monitoring of regional or sub-continental patterns of fire activity. Figure 4 shows, for the month of December 1986, the spatial distribution of active fires, derived from AVHRR-GAC data, in the savanna-forest contact zone of Central Africa. The geographical area covered by the figure is divided into 11 zones depending on (i) the country, and (ii) the bioclimatic domain (White, 1983). For instance, zones I, II and III are the Sudanian, Guineo-Sudanian and Guineo-Congolian parts of Cameroun respectively; zone V is the Guineo-Sudanian part of the Central African Republic, and so on. Using AVHRR-GAC time series for the period November 1985 to January 1988 (Lefeivre *et al.* 1993), interannual comparisons of burning seasons have been made for five countries/sub-countries: Cameroun, Southern Chad, Southern Sudan, Central African Republic and the Northern part of Zaire (Figure 5). A common feature of all five countries is the strong decrease of fire activity, whatever the bioclimatic domain, in 1987-1988, compared with the other two periods.

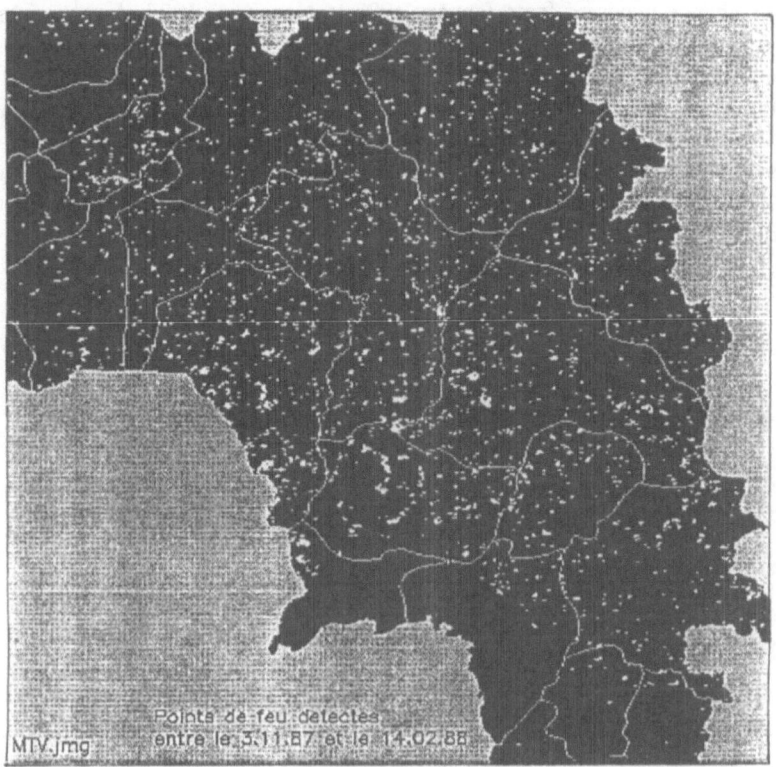

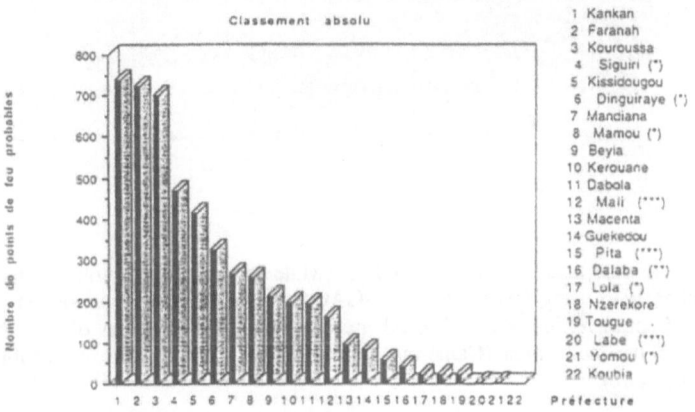

Figure 3. Distribution of active fires, as derived from NOAA-AVHRR-HRPT time series in Guinée during the 1987/1988 dry season.

3a the distribution of fires (white dots) detected between the 3rd November 1987 and the 14th February 1988, and administrative units (prefectures)

3b classification of prefectures on the basis of the number of active fires detected on their territory.

324

This observed difference is unlikely to be due to satellite sensor calibration problems, as no satellite change occurred from January 1987 to November 1987. Another interesting feature is the difference observed in fire activity in Cameroun and Central African Republic. Although at similar latitudes, these two countries show distinct patterns of fire distribution among the Sudanian and Guineo-Sudanian domains. Results from Sudan are probably overestimated in the number of active fires detected, but results over Zaire show clearly the "intrusion" of fire activity into more densely vegetated areas, during the last two decades of January 1986 and January 1987, especially at the border with the Central African Republic and with Sudan.

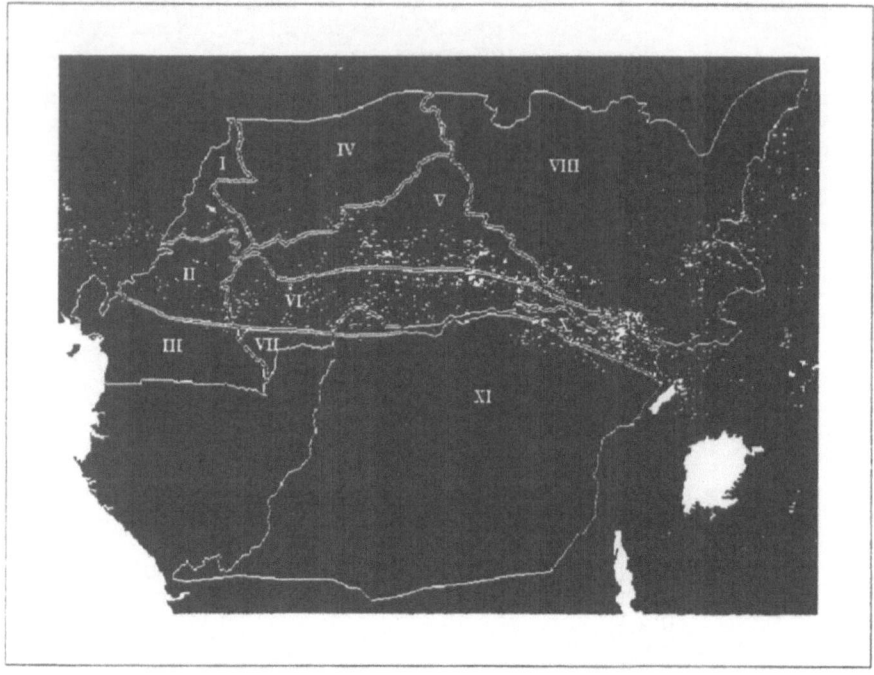

Figure 4. Spatial distribution of active fires (white dots), detected during the month of December 1986 using NOAA-AVHRR-GAC imagery, in the savanna-forest contact zone of Central Africa. Zones I to XI are bioclimatic sub-divisions of the countries present in the study area (Cameroun, Chad, Sudan Central African Republic and Northern Zaire).

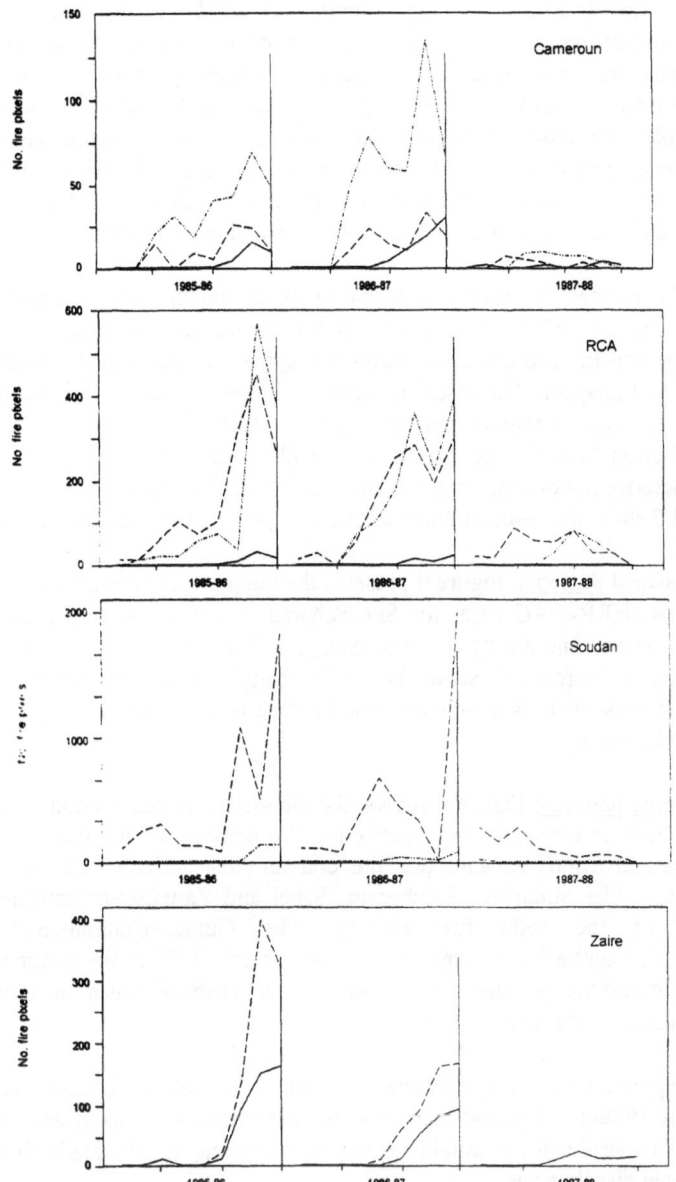

Figure 5. Interannual comparison of the number of active fires detected using AVHRR-GAC time series over Central Africa during 3 dry seasons: 1985-86, 86-87 and 87-88. For each dry season fires have been detected on a daily basis during 3 months (November, December and January) and the resulting numbers accumulated over 10 day periods. Distinction has been made between fires detected in the Sudanian domain (——), Guineo-Sudanian (·····) and Guineo-Congolian (—) zones.

2.2.3. *Trends analysis at the continental level.* At the continental level in Africa, detection of trends in fire activity should focus on regular observations of the bioclimatic domains known to be actually affected by vegetation fires. The Sudanian and Guineo-Sudanian zones (North of the equator) and the Zambezian one (South of the equator) are examples of regions where vegetation fires are highly probable. The spatial requirements (regional approach in a continental context), combined with temporal ones (seasonal dynamics in a multiannual context), make the AVHRR-GAC archives the only source of information currently available for detection and analysis of trends of fire activities.

GAC imagery is, in fact, available at the continental level on a daily basis from 1981. Daily mosaics of the full African continent are created from the unprocessed archive dataset and the fire detection method is applied to the radiometrically and geometrically corrected imagery. The resulting output is in a binary form of fire/non-fire image per day. Daily outputs are then combined on a temporal basis (decade, month and season) and fire distribution "maps" are produced. Finally, these fire distribution maps are used to characterise regions and times of the year in which fire activity is particularly high. Figures 6 and 7 show the results obtained for a one year period: August 1988 to September 1989.

Seasonal Patterns: Figure 6 presents the temporal distribution of active fires, as derived from AVHRR-GAC data, for Sub-Saharan Africa: vegetation fires can be observed at almost every time during the year, except in April; two peaks in fire activity are observed, January to March and September to December. It must be pointed out that during this second peak, there is a 3 month period during which fires are active both North and South of the Equator.

Spatial patterns: Figure 7 shows, for the same one year period, what are the respective contributions within overall continental fire activity, of the main bioclimatic zones. The shrubs and woody savanna domains contain most activity, both North and South of the Equator. The Sudanian, Zambezian, Sahel and Zanzibar-Inhambane zones account for 80% of the total fire activity. The Guineo-Congolia/Sudania and Guineo-Congolia/Zambezia, although accounting for only 10% of the continental fire activity, are very important as they correspond to geographical areas of transition between the savannas and the forest domains.

If applied to the 10-year time series of AVHRR-GAC images, the same method used for the 1988-1989 period may allow the detection of trends, if any, of fire activity during the last decade, for example related to migration of man-made fires and/or changes in seasonal distributions.

2.3   POST-FIRE ANALYSIS

Post-fire analysis covers two main aspects: the quantitative assessment of burnt areas and the evaluation of impacts of fires on the environment.

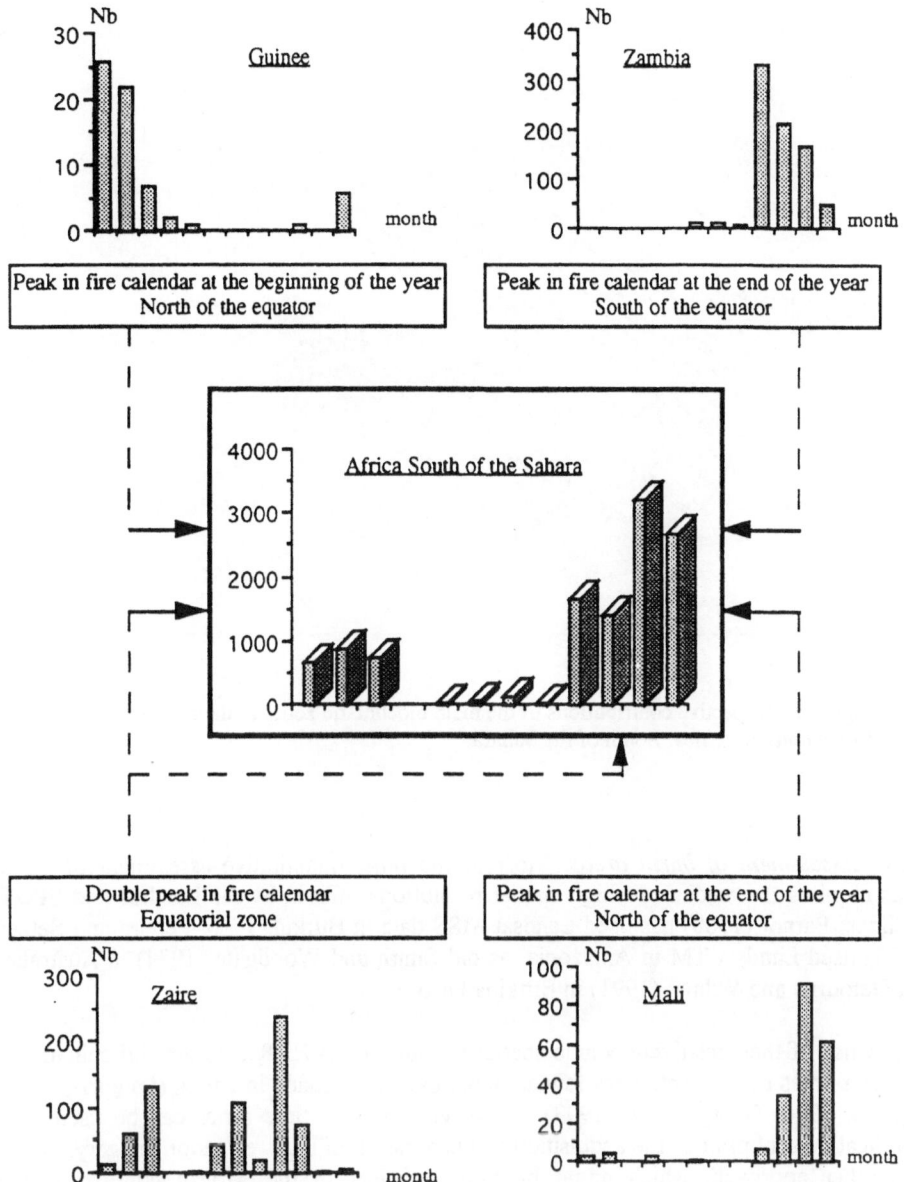

Figure 6. Temporal distribution of active fires, detected with AVHRR-GAC data, for the whole of Sub-Saharan Africa and from some representative countries.

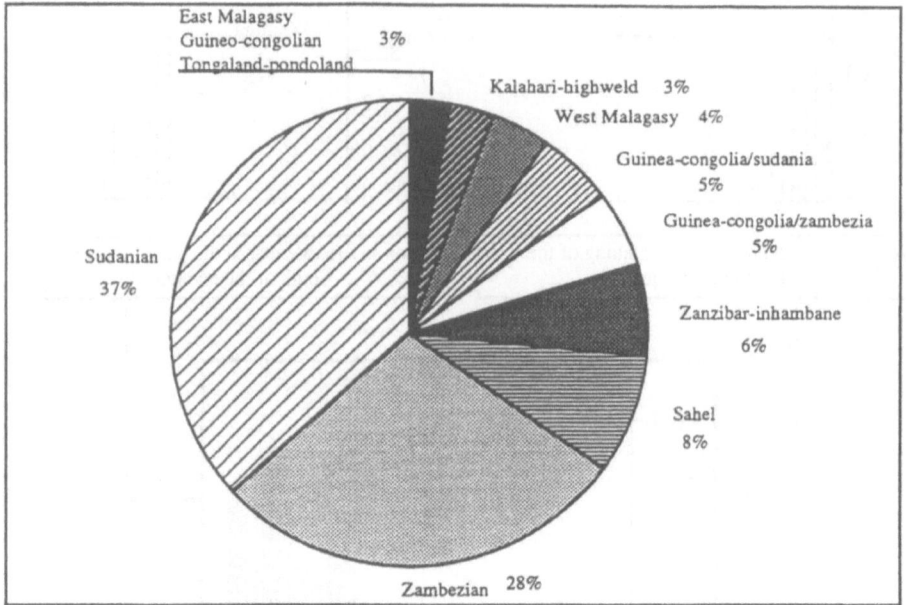

Figure 7. Respective contributions of the main bioclimatic zones to the continental fire activity in Africa, South of the Sahara.

2.3.1 *Assessment of burnt areas.* Most of the time, quantitative assessment of burnt areas requires the medium to high spatial resolutions offered by the Landsat and SPOT satellites: Parnot (1988) has used Landsat MSS data in Burkina Faso, Pereira and Setzer (1993) used Landsat TM in Amazonia, as did Smith and Woodgate (1984) in Australia, and Defourny and Wilmet (1991) in Burkina Faso.

Because of their relatively coarse spatial resolution, AVHRR data are difficult to use for burnt areas assessment. Very few attempts have been made in Africa (Langaas, 1989; Laporte, 1990; George *et al.* 1993). However, AVHRR time series can be used as a stratification tool prior to the acquisition and processing of high resolution imagery.

Another approach, which might be more efficient in the African context, is the combination of AVHRR time series for active fires detection with field knowledge for the qualification of burning efficiency. Depending on the bioclimatic and phytogeographic units in which active fires have been detected, one would derive estimates on the ratio of burnt to total existing biomass. Field knowledge can be built during experimental campaigns, as the one recently carried out in Côte d'Ivoire (Belward *et al.* 1993-a).

Some recent work has also been conducted in the analysis of NDVI temporal patterns, derived from AVHRR-GAC time series, for evaluation of burnt areas at the continental level in Africa (Laport, 1993, personal communication).

2.3.2. *Evaluation of impacts.* As already stated previously, spatio-temporal dynamics of man-made fires can be the result of an equilibrium between environmental conditions and resource management or land use. These dynamics can also be indicators of changing anthropogenic pressure on the environment and/or changing environmental conditions. It is obvious that this notion of "equilibrium" cannot be used if both the spatial and the temporal scales are not well defined (Skarpe, 1992). For instance, in most African countries, changes of the amplitude of fire activity are strongly correlated to fluctuations in climatological conditions and associated movements of populations; fluctuations which can hardly be appreciated on periods shorter than a decade.

In any case, impact studies must rely on (i) temporal sequence of observations long enough to detect multi-annual trends, if any, and (ii) on geographical and meteorological data bases containing the environmental information related to the area under study.

Examples given in 2.2.3 have shown how the NOAA-AVHRR system can help in replacing actual observations in an historical perspective and derive trends specific to different ecological zones.

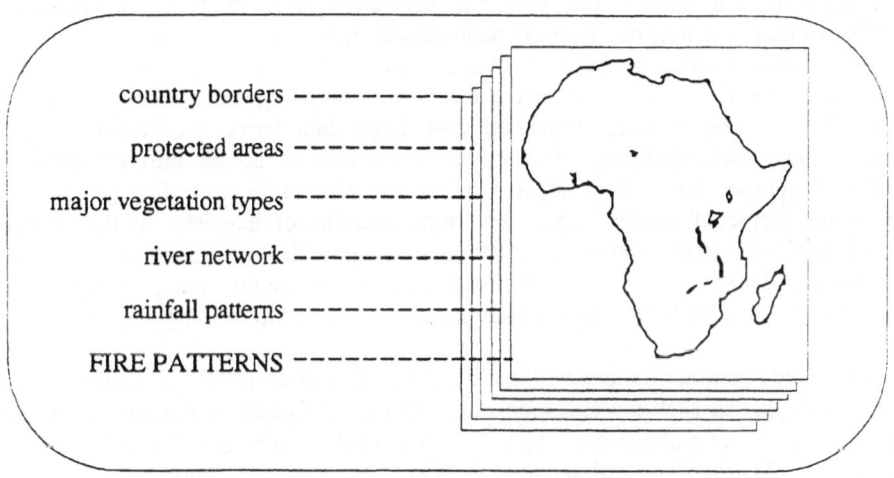

Figure 8. Minimum set of information needed for impact studies at the continental level.

As far as geographical and meteorological data bases are concerned, the situation is much less facilitated for the African continent. Sources of information are scattered among a very large number of institutions and data quality assessment are not always made prior to publication and use. In this context, a minimum set of information can be defined (Figure 8) for impact studies at the continental level. It must be said that even this minimum set is not always accessible: for instance, due to the socio-political perturbations, most of the existing meteorological networks are not operational any more in Africa. In this case, continuous monitoring by AVHRR imagery, or by other similar satellite systems, could become a priority action if the international community wants to keep track of environmental dynamics actually occurring in the African continent.

3. Steps toward an operation system for the monitoring of vegetation fires in Africa

It has been shown how NOAA-AVHRR time series can help in obtaining the information needed for a better understanding of vegetation fires on regional and continental scales in Africa, that is information regarding both the current situation and possible trends during the last decade.

A necessary step, toward an effective use of such information in the fire management perspective, is the design and building of a system which allows the integration of (i) the various sources of information, (ii) the various types of request for environmental management (research, operational management, decision making), and (iii) the various temporal and spatial scales required to operate such a fire management tool. At the country level, the work done by the "Centre de Suivi Ecologique" in Dakar (Federiksen *et al.* 1989) is a good example of a possible national-scale monitoring program.

At the regional or continental level, Malingeau *et al.* (1993) have proposed, during the recent Dahlem Workshop on "Fire in the environment", a Vegetation Fire Information System (VFIS). This system, based on two large data bases (vegetation related information and meteorological information), is intended to access different models, related to vegetation fires behaviour, and to produce the results according to the user requirements (topic of interest, space and time domains of interest). In the overall functioning of the system, satellite observations would provide information on the timing and location of active fires. For reasons already described in the previous paragraphs, the NOAA-AVHRR system is seen as a central element in the remote sensing segment of the VFIS.

A recent study contract, funded by the CEU (Martellaci *et al.* 1993), has evaluated the feasibility of applying such a Vegetation Fire Information System in the context of the African continent. A simplified prototype of the VFIS will soon be installed and tested by the Monitoring Tropical Vegetation unit at the CEU-Joint Research Centre.

To support this effort to make fire management part of a more general approach to environmental management, three priority topics can be defined for research activities:

- a better quantitative description of fire characteristics such as temperature, effective size and smoke plume composition. Together these form information indispensable to improve results of impact studies.

- a better understanding of the interactions existing between socio-economic and ecological aspects of vegetation fires for the main ecological zones of Sub-Saharan Africa, Asia and South America. Without such an understanding, studies of vegetation fires will not go beyond the classical, and (most of the time) unrealistic views on environmental protection.

- the development of impact models, tackling both ecological and socio-economic dimensions, for an objective evaluation of possible damages **and** benefits of biomass burning practices, again for the main ecological zones of the inter-tropical belt. The FIRE project, as proposed by the MTV unit of the Joint Research Centre (Grégoire *et al.* 1993-b) is an attempt to move in this direction.

4.   References

Albergel, J., 1988. Genèse et prédétermination des crues au Burkino Faso. Du m^2 au km^2 - Etude des paramètres et de leur évolution. Thèse de Doctorat, Editions de l'ORSTOM, Collection Etudes et Thèses, Paris, ISBN 2-7099-0903-0, p.341.

Almendros, G., Gonzalez-Vila, F.G. and Martin, F., 1990. Fire-induced transformation of soil organic matter from oak forest: An experimental approach to the effects of fire on humic substances. *Soil Science*, **149** (3), 158-168.

Anonyme, 1991. Suivi de la dynamique spatio-temporelle des feux de brousse en Guinée par télédétection satellitaire. Annexe technique Contract B7-5040/91/003, Ministère de l'Agriculture et des Resources Animales, République de Guinée, p.6.

Arets, J., Berg, A., Fusco, L. and Grégoire, R., 1984. Earthnet MSS data supports CEC hunger-relief projects in the Sahel. ESA Bulletin No. 40, November 1984, pp 66-71.

Belward, A.S. and Lambin, E., 1989. Limitations to the identification of spatial structures from AVHRR data. *International Journal of Remote Sensing*, **11** (5), 921-927.

Belward, A.S., Grégoire, J.M., D'Souza, G., Trigg, S., Hawkes, M., Brustet, J.M., Serça, D., Tireford, J.L., Charlot, J.M. and Vattoux, R., 1993a. In-situ, real time fire detection using NOAA-AVHRR data. *Proceedings of the VI AVHRR Data User's Meeting*, Belgirate, Italy, 28th June - 2nd July 1993.

Belward, A.S., Kennedy, P.J. and Grégoire, J.M., 1993b. The limitations and potential of AVHRR GAC data for continental scale fire studies. *International Journal of Remote Sensing* (in press).

Bertault, J.G., 1991. Quand la forêt tropicale s'enflamme - Près de trois millions d'hectares détruits à Kalimantan. Bois et Forêts des Tropiques, No. 230, 4ème trimestre, 1991, pp 5-14.

Bonsang, B., Lambert, G. and Boissard, C., 1991. Light hydrocarbon emissions from savanna burnings. In *Global Biomass Burning*, Ed. J Levine. MIT Press, Cambridge, pp 155-161.

Bowman, D.M.S, Wilson, B.A. and Hooper, R.J., 1988. Response of Eucalyptus forest and woodland to four fire regimes at Munmarlary, Northern Territory. *Australia Journal of Ecology*, **76**, 215-232.

Cachier, H., Ducret, J., Bremont, M.P., Yoboue, V., Lacaux, J.P., Gaudichet, A. and Baudet J.G.R., 1992. Biomass burning aerosols in a savanna region of the ivory Coast. In *Global Biomass Burning*, Ed. J. Levine. MIT Press, Cambridge, pp 174-180.

Cros, B., Delmas, R., Nganga, D., Clairac, B. and Fontan, J., 1988. Seasonal trends of ozone in equatorial Africa, experimental evidence of photochemical formation. *Journal of Geophysical Research*, **93**, 8355-8366.

Cros, B., Nganga, D. and Fontan, J., 1991. Trophospheric ozone and biomass burning in intertropical Africa. In *Global Biomass Burning*, Ed. J. Levine. MIT Press, Cambridge, pp 143-146.

Dauget, J.M. and Manuat, J.C., 1992. Evolution sur 20 ans d'une parcelle de savane boisée non protégée du fue dans la réserve de Lamto (Côte-d'Ivoire). Candollea 47, pp 621-630.

Da Vinci Consulting S.A., 1993. Carte des formations végétales de al zone pilote de Sangbarapport Da Vinci Consulting S.A., p.8 + annexes.

Defourny, P. and Wilmet, J., 1991. Analyse des brûlis en zone soudano-sahélienne: Présentation d'une méthode de cartographie et de suivi rapide et opérationelle. Rapport Techniques Projet B 947/87 *Evaluation de al biomasse ligneuse en zone soudano-sahélienne*. Université Catholique de Louvain, Louvain-La-Neuve, p 48.

Delmas, R., Lodjani, Ph., Podaire, A. and Menaut, J.C., 1991. Biomass burning in Africa: an assessment of the annually burned biomass. In *Global Biomass Burning*, Ed. J. Levine. MIT Press, Cambridge, pp 126-132.

Delmas, R., 1992. Sources et puits de constituants mineurs atmosphériques d'intérêt climatique. Ecole d'Eté Internationale de Physique Spatiale, *Climats subtropicaux et leur évolution: de l'observation spatiale à la modélisation*. La-Londe-Les-Maures, France, 14-15 September 1992.

Deschler, W., 1974. An examination of the extent of fire in the grassland and savanna of Africa along the Southern side of the Sahara. *Proceedings of the 9th International Symposium on Remote Sensing of Environment*, ERIM, Ann Arbor, pp 23-20.

Diallo, M.B., Perier, J.P. and Watt, A., 1990. Eyude socio-économique-culturelle des populations du Fouta Djallon et de la préfecture de Koundara. Rapport ENDA-TM, Projet République de Guinée - Commission des Communautés Européenes No. 5604-04-94-427, Volume I, *Rapport Général*, p.152, and Volume II *Fiches de village*, p.154.

Franca, J.R.A., Brustet, J.M. and Fontan, J., 1993. A multispectral remote sensing of biomass burning in West Africa. *Journal of Atmospheric Chemistry*, submitted 1993.

Frederiksen, P., Langaas, S. and Mbaye, M., 1989. NOAA AVHRR and GIS-based monitoring of fire activity in Senegal - A provisional methodology and potential applications. . In: *Fire in the tropical biota - Ecosystems Processes and Global Challenges*, Eds. W.D. Billings, F Golley, O.L. Lange, J.S. Olsen and H Remmert. Springer-Verlag: Goldammer, J.G., 1990. pp 400-416.

Fritsch, J.M., 1992. Les effets du défrichement de la fôret amazonienne et de la mise en culture sur l'hydrologie de petits bassins versants. Opération ECEREX en Guyanne française. Editions de l'ORSTOM - Collections Etudes et Thèses, Paris 1992, p.392.

Fusco, L. and Muirhead, K., 1987. AVHRR data services in Europe - The Earthnet approach. ESA Bulletin No.49, pp 9-19.

Geffard, S., 1991. Utilisation de séries temporelles de données satellitaires (NOAA-AVHRR) dans l'étude de la dynamique de systèmes hydrologiques en Afrique de l'Ouest. Rapport de D.E.A. National d'hydrologie, ORSTOM, Université Paris XI, Laboratoire d'hydrologie et de géochimie isotopique, Septembre 1990, p.111 + annexes.

George, H., Touré, A. and Kane, R., 1993. A methodology for correcting over-estimations of burnt area in maps derived from NOAA imagery. Report from the "Centre de Suivi Ecologique", PNUD, Dakar, Sénegal, p.18.

Gillon, D., Monnier, Y. and Vuattoux, R., 1973. Les savanes à ronniers de LAMTO - Fascicule Les feux de brousse. Rapport du Ministère de Education nationale, Côte d'Ivoire, Direction Pédagogie et formation des maîtres, Document No. 546, p.52.

Grégoire, J.M., Flasse, S. and Malingreau, J.P., 1988. Evaluation de l'action des feux de brousse, de novembre 1987 à février 1988, dans la région frontalière Guinée-Sierra Leone. Exloitation des images NOAA-AVHRR - Rapport CCR S.P.I.88.39, October 1988, p.23.

Grégoire, J.M., Kennedy, P., Cambridge, H., Belward, A.S. and Flasse, S., 1992. Documentation et suivi de l'activité des feux de brousse en milieu de transition savane-fôret dans le sud guinéen. Publication Spéciale du Centre Commun de Recherche, S.P.I.92.39, Ispra, October 1992, p.35.

Grégoire, J.M., Belward, A.S. and Kennedy, P., 1993-a. Dynamiques de saturation du signal dans la bande 3 du senseur AVHRR: Handicap majeur ou source d'information pour la surveillance de l'environnement en milieu soudano-guineen d'Afrique de l'Ouest. *International Journal of Remote Sensing*, **14**, (11), 2079-2095.

Grégoire, J.M., Belward, A.S. and Malingreau, J.P., 1993-b. The FIRE Project: Fire In global Resources and Environmental monitoring. *Proceedings of the 1st TREES Conference*, Belgirate, Italy, 20-21 October, 1993. In press.

IEMVT-CIRAD, 1990-a. Les feux de brousse. Fiches Techniques d'Elevage Tropical, IEMVT, Maisons-Alfort, France, Fiche No. 3, mars 1990, p.11.

IEMVT-CIRAD, 1990-a. Les feux de brousse. Fiches Techniques d'Elevage Tropical, IEMVT, Maisons-Alfort, France, Fiche No. 6, juin 1990, p.7.

Justice, C (Ed), 1993. IGBP DIS satellite fire detection algorithm workshop report. NASA-GSFC, Greenbelt, Maryland, USA. February 25-26, 1993. Draft Report, 1993.

Kaufman, Y., Tucker, C.J. and Fung, I., 1989. Remote sensing of biomass burning in the tropics. *Advanced Space Research*, **9**, 265-268.

Kennedy, P., 1992. Biomass burning studies: the use of remote sensing. Ecological Bulletins 42: 133-148. Copenhagen 1992.

Kennedy, P.J., Belward, A.S. and Grégoire, J.M., 1993. An improved approach to fire monitoring in West Africa using AVHRR data. *International Journal of Remote Sensing*, 1993 (in press).

Lacaux, J.P., Delmas, R., Cros, B., Lefeivre, B. and Andrae, M.O., 1991. influence of biomass burning emissions on precipitation chemistry in the equatorial forests of Africa. In *Global Biomass Burning*, Ed. J. Levine. MIT Press, Cambridge, pp 167-173.

Lacaux, J.P., Delmas, R., Kouadio, G., Cros, B. and Andrae, M.O., 1992. Precipitation chemistry in the Mayombe forest of Equatorial Africa. *Journal of Geophysical Research*, **92**, 6195-6206.

Lacaux, J.P., Loemba-Ndembi, J., Lefeivre, B., Cros, B. and Delmas, R., 1992. Biogenic emissions and biomass burning influences on the chemistry of the fogwater and stratiform precipitaitons in the African equatorial forest. *Atmospheric Environment*, **26A**, 541-551.

Langaas, S., 1989. A study of spectral characteristics for burnt areas in Senegal - with recommendations for an AVHRR based bushfire monitoring methodology. Report prepared for Office for Project Services (OPS), UNDP, 1st version, March 1989, 65 pp + annexes.

Langaas, S., 1992. Temporal and spatial distribution of savanna fires in Senegal and The Gambia, West Africa, 1989-90, derived from multi-temporal AVHRR night images. *International Journal of Remote Sensing*, **2**, (1), 21-36.

Laport, N., 1990. Etude de l'évolution spatio-temporelle de la végétation tropicale: Utilisation de données satellitaires NOAA AVHRR sur l'Afrique de l'Ouest. Thèse de Doctorat - Université Paul Sabatier, Toulouse, 4 octobre 1990, p.231.

Lefeivre, B., 1993. Etude expérimentale et par modélisation des caractérisitques physiques et chemiques des précipitations collectées en fôret équatoriale africaine. Thèse de Doctorat - Université Paul Sabatier, Toulouse, 30 mars 1993, 308 pp + annexe.

Lopez, A., Huertas, M.L. and Lacome J.M., 1992. Numerical simulation of the ozone chemistry observed over forested tropical areas during DECAFE experiments. *Journal of Geophysical Research*, **97**, 6149-6158.

Malingreau, J.P., Stephens, G. and Fellows, L., 1985. Remote sensing of forest fires: Kalimantan and North Borneo in 1982-83. *Ambio*, **6**, 314-321.

Malingreau, J.P. and Tucker, C.J., 1988. Large-scale deforestation in the Southeastern Amazon basin of Brazil. *Ambio*, **17**, 49-55.

Malingreau, J.P., Albini, F.A., Andrae, M.O., Brown, S., Levine, J.S., Lobert, J.M., Kuhlbusch, T.A., Radke, L. Setzer, A., Vitouzek, P.M., Ward, D.E. and Warnatz, J., 1993. Quantification of fire characteristics from local to global scales. In *Fire in the Environment: The ecological, atmospheric and climatic importance of vegetation fires*. Eds. P.J. Critzen and J.G. Goldammer. John Wiley and Sons, New York. pp 329-343.

Martellaci, C., Molinari, P., Ligi, R. and Gigotti, R., 1993. Vegetation Fire Information System (VFIS) - Feasibility study. CEC-JRC Contract No. 5245-93-03. ED ISP I. Draft Report, May 1993. 40 pp.

May T., 1990. Vegetation development and surface runoff after fire in a catchment of Southern Spain. In Fire in Ecosystem Dynamics, pp 117-126. *Proceedings of the Third International Symposium on Fire Ecology*, Freiburg, FRG, May 1989. Eds J.G. Goldammer and M.J. Jenkins, SPB Academic Publishing bv, The Hague, NL.

Menaut, J.C., 1983. The vegetation of African savannas. In *Tropical Savannas*, Ed. F. Bourlière, Elsevier, Amsterdam, p.109-149.

Moore, P.D. 1989. Answers that lie in the soil. *Nature*, **342**, 858. December 1989.

Paltridge, G.W. and Barber, J. 1988. Monitoring grassland dryness and fire potential in Australia with NOAA/AVHRR data. *Remote Sensing of Environment*, **25**, 381-394.

Parnot, J., 1988. Inventaire des feux de brousse au Burkina Faso - Saison sèche 1986-1987. *Proceedings of the 22nd Symposium on Remote Sensing of Environment*, Volume II, 20-26 October, 1988, Abidjan, Côte d'Ivoire, p.563-573.

Penner, J.E., Dickinson, R.E. and O'Neill, C.A., 1992. Effects of aerosol from biomass burning on the global radiation budget. *Science*, **265**, 1432-1434.

Pereira, M.C. and Setzer, A.W., 1993. Spectral characteristics of fire scars in Landsat-5 TM images of Amazonia. *International Journal of Remote Sensing*, **14**, (11), 2061-2078.

Robinson, J.M., 1988. The role of fire on earth: A review of the state of knowledge and a systems framework for satellite and ground-based observations. Cooperative Thesis No.112, University of California, Santa Barbara, and National Center for Atmospheric Research, 476 pp.

Robinson, J,M., 1989. On uncertainty in the computation of global emissions from biomass burning. *Climatic Change*, **14**, 243-262.

Robinson, J.M., 1991. Fire from space: Global fire evaluation using infrared remote sensing. *International Journal of Remote Sensing*, **12**, 3-24.

Schmalzer, P.A., Hinkle, C.R. and Koller, A.M., 1991. Changes in marsh soils for six months after a fire. In *Global Biomass Burning*, Ed. J. Levine. MIT Press, Cambridge, pp 273-186.

Skarpe, C., 1992. Dynamics of savannas ecosystems. *Journal of Vegetation Science*, **3**, 293-300.

Shoji Takeuchi, 1993. Monitoring of tropical forest fire by NOAA/AVHRR. RESTEC Newsletter, June 1993, No. 23, 6-7.

Smith, R.B. and Woodgate, P.W., 1984. Assessment of wildlife damage in mountain eucalypt forests in Victoria, Australia. *Proceedings of the 18th International Symposium on Remote Sensing of Environment*, Paris, 1-5 October 1984. Volume 2, pp 1081-1090.

Van Wilgen, B.W., Higgins, K.B. and Bellstedt, D.U., 1990. The role of vegetation structure and fuel chemistry in excluding fire from forest patches in the fire-prone Fynbos shrublands of South Africa. *Journal of Ecology*, **78**, 210-222.

Wein, R.W., 1991. Editorial. *International Journal of Wildland Fire*, **1**, (1).

White, F., 1983. La végétation de l'Afrique. Mémoire accompagnant la carte végétation de l'Afrique Unesco/AETFAT/UNSO. Document ORSTOM - UNESCO. Unesco ISBN: 92-3-201955-8. ORSTOM ISBN: 2-7099-0832-8. 384 pp.

World Wildlife Fund, 1993. Central Africa: Global climate change and development - Overview and Technical Report. Biodiversity Support Program. Corporate Press, Maryland, USA.

CROP PRODUCTION ASSESSMENT FOR THE EUROPEAN UNION: THE MARS-STAT PROJECT INCLUDING THE USE OF NOAA-AVHRR DATA.

P. VOSSEN

Agriculture Information Systems Unit
Institute for Remote Sensing Applications
CEU Joint Research Centre
21020 Ispra (VA) - Italy

ABSTRACT: With its decision of the 26th of September 1988, the Council of Ministers of the E.C. approved the creation of a 10-year research and development Project for the Application of Remote Sensing to Agricultural Statistics. This Project is commonly referred to as the MARS-STAT Project, or "Monitoring Agricultural Statistics with Remote Sensing". Its main activities relate to the quantitative estimation of the acreages occupied by the various crops in a given region or country, vegetation and crop state monitoring, timely crop yield forecasting of mean crop yields per country and the rapid and timely estimation of the E.C.'s total production of the most important crops. Its main users are the E.C.'s Directorate General for Agriculture and the European Statistical Office (EUROSTAT). The various activities of the Project are conceived, developed and implemented on the basis of inputs from approximately 100 institutions from 17 European countries. These institutions provide the required data, models, algorithms and software after having previously validated them for use at the E.U. scale on the basis of site or country specific information. The overall co-ordination and the integration of the various inputs into applications, operational or pre-operational, is carried out by the MARS-STAT Project. Part of the activities, mainly the ones related to crop inventories and rapid estimates of planted areas, are operational and have been transferred to the Member States (crop inventories) and to the E.U's Directorate General for Agriculture (rapid estimates). The crop state monitoring and early yield forecasting activities are still in a validation stage and are carried out on a pre-operational scale by the Joint Research Centre. In this chapter, the methods, results and examples of outputs of the various activities are presented and an evaluation is made of their state of advancement and the needs for improvement or further development, with particular attention on the use of NOAA-AVHRR data in the system. Future perspectives, mainly related to the integrated use of the various Actions into one Advanced Agricultural Information System and for the development of a system for foreign crop production forecasting are also indicated.

1. Introduction

The Ten-Year Programme for the Application of Remote Sensing to Agricultural Statistics was established with the decision of the 26th of September 1988 of the Council of Ministers of the E.C. This Project is commonly referred to as the MARS-STAT Project (Monitoring Agriculture with Remote Sensing) and was initiated by the Directorate General for Agriculture in co-operation with the Statistical Office of the European Communities. The Joint Research Centre is responsible for implementing the

337

G. D'Souza et al. (eds.), Advances in the Use of NOAA AVHRR Data for Land Applications, 337–356.
© 1996 ECSC, EEC, EAEC, Brussels and Luxembourg.

338

programme in close co-operation with national laboratories and organisations. The objective is to develop methods for improving agricultural statistics in the Community with the use of remote sensing techniques. Such methods are to be tested on fairly large areas and are to be developed to a stage where they can be put into operational use. This entails the use of satellite data for which there is a guaranteed continuity. The crops targeted are the ones for which there is the biggest market (excluding fodder crops). Representativeness is sought not only at Community level, but at national and regional and level too (see Figure 1).

In order to meet its overall objectives, the project is organised into five Activities:

A **Activity A:** Regional inventories, which consist of the assessment of crop acreages from a sample of high spatial resolution remote sensing imagery (mainly SPOT and Landsat TM images);

B. **Activity B:** Rapid estimates at the European scale of actual planted areas of the main annual crops, as compared with the areas planted in the previous season;

C. **Activity C:** The Advanced Agricultural Information System, which combines the use of low resolution satellite indices (for example NDVI and surface temperature estimates derived from NOAA-AVHRR data) and agrometeorological crop growth simulation model outputs;

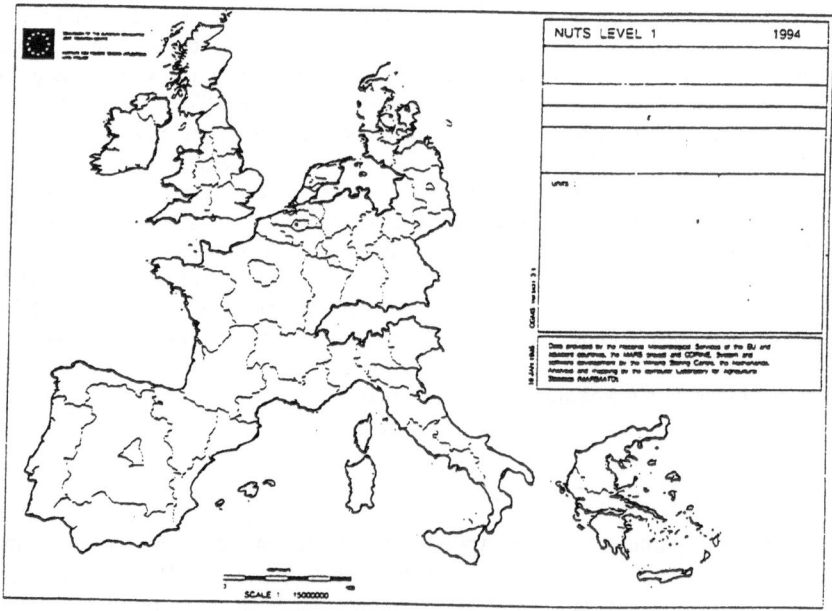

Figure 1. The Member States and statistical regions of the European Union (1994).

D. **Activity D:** The assessment of foreign agricultural production;

E. **Activity E:** The development of new methods and techniques for image analysis and testing of the usefulness for agricultural purposes of new space-borne sensors such as the SAR on ERS-1.

In the context of this book, the most relevant parts of the project are in Activity C. However, for completeness, the other activities are also described briefly. For more details on these activities, the reader is referred to Vossen (1994).

2.   Activity A: Regional Inventories

The aim of this activity is to meet the need for accurate and objective annual information on acreage at the regional level covering the main crops. The method adopted is to establish close links between satellite data and observations on the ground. Development and evaluation have focused on the so-called regression estimator method. This action comprises of two major components (i) objective observations in the field with a sample design established or enhanced by remote sensing; and (ii) automatic classification of the satellite data in order to improve the estimates generated by the ground surveys, using the regression estimator method.

Activity A was implemented with the co-operation and joint financing of national and regional authorities, in administrative regions in Belgium, the Czech Republic, France, Germany, Greece, Italy, Romania and Spain. Figure 2 shows the various regions to which Activity A was applied.

The method used consists of:

*   stratification using satellite images with existing topographical documents and maps in support;
*   selection of a sample of 600 square segments each of 50 hectares;
*   survey on the ground with aerial photographs or hard-copy satellite imagery;
*   simultaneous acquisition of full coverage of the region by both Spot and Landsat-TM;
*   automatic classification of the satellite data in order to improve the estimates generated by the above-mentioned ground surveys, using the regression method;
*   analysis of the results.

Stratification proves to be very economical as, generally speaking, it reduces the cost of the required ground survey by a factor of two for the same degree of precision. High spatial resolution satellite images appear to be the ideal means of detecting homogeneous areas of land use and physical boundaries (such as roads, rivers, paths and the edges of forests) and also for selecting segments and assessing in advance possible difficulties arising from the survey of a number of plots. Since the period in which the images used to

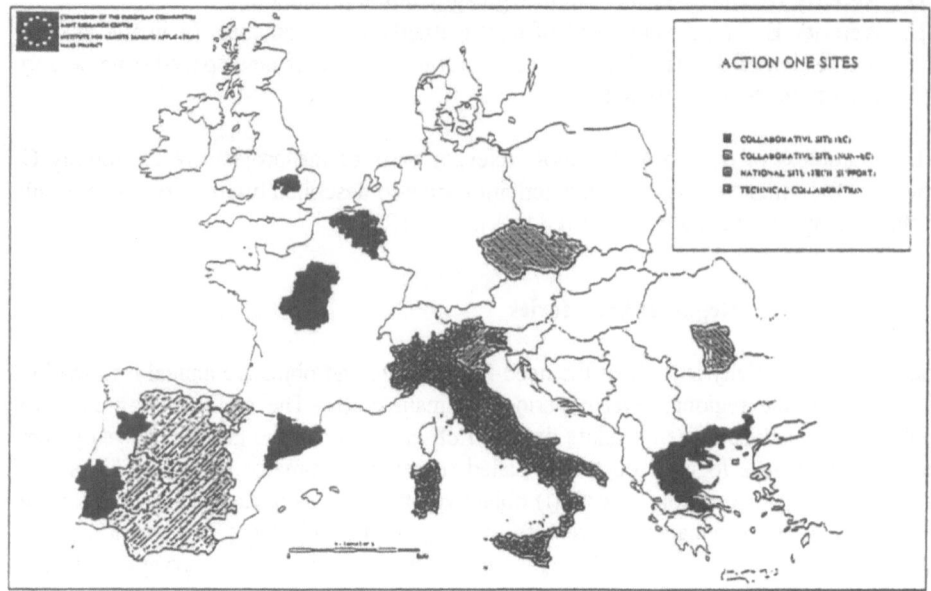

Figure 2. The Regional Inventories study areas, 1988-1994.

assist fieldwork are obtained is not critical, the method has been found to be equally applicable throughout Europe.

The square segments method is quick and effective. The best tool for the enumerator is an enlarged aerial photograph, or a 1/10 000 enlargement of the SPOT images of the segment. As an example, the stratification for the Czech Republic, and the selection of 417 segments within the strata is provided in Figure 3.

SPOT and Landsat TM spatial resolutions proved very suitable for the estimation of crop areas in most cases. A combination of Landsat TM and SPOT imagery was necessary because it was not always possible to cover the whole of a region of 20,000 km^2 within the optimum period, which was about 40 days, with a single type of satellite. Once the crop area estimates were made for the selected segments, the regression estimator method was used to improve and extrapolate the estimates over the whole area of interest. Table 1 provides an example the results for Macedonia, in 1988 and 1989.

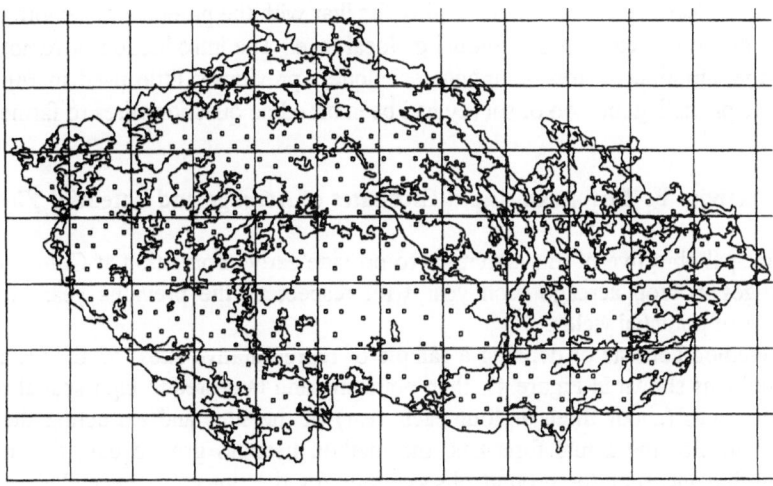

Figure 3. Stratification of the Czech Republic and sample of 417 segments (1992 survey)

Table 1. Results for Macedonia (after stratification)

Culture	Official statistics (x 1000 Ha)	M.A.R.S. estimate (x 1000 Ha)		Relative efficiency: without stratification		with TM-stratification		Crop
	1988	1988	1989	1988	1989	1988	1989	
Blé tendre	195.5	158.5 (11.7)	163.6 (11.0)	1.25	1.39	1.84	1.99	Common wheat
Blé dur	150.3	175.4 (12.3)	161.1 (10.0)	1.17	1.49	2.17	2.13	Durum wheat
Orge		60.4 (6.4)	65.5 (5.5)	1.12	1.29	(*)	1.67	Barley
Légumes secse	5.0	0.6 (0.3)	2.2 (1.0)	1.44	1.72	3.56	(*)	Dried pulses
Maïs	30.4	27.5 (3.3)	22.5 (2.0)	1.70	1.99	1.67	3.11	Maize
Tournesol	16.3	16.6 (3.5)	4.0 (1.2)	1.16	1.09	1.38	1.41	Sunflower
Pomes de terre	4.0	6.0 (3.0)	0.9 (0.7)	0.67	1.10	0.99	7.61	Potatoes
Betteraves	10.3	15.9 (3.8)	13.9 (1.2)	1.36	1.99	1.23	7.16	Sugar beet
Riz	12.0	14.1 (1.1)	9.9 (0.5)	7.04	6.27	6.31	18.86	Rice
Coton	57.7	73.9 (6.2)	60.5 (0.4)	2.26	2.47	1.54	3.19	Cotton
Tabac	20.5	31.7 (4.7)	17.7 (2.2)	1.30	1.60	1.06	2.21	Tobacco

The figures in brackets are standard errors.
(*) The regression correction has not been applied

European statistical services are now fairly familiar with the potential of remote sensing in the area of crop inventories at regional or local level. The introduction of remote sensing entails the adoption of area sampling techniques, previously little used in Europe, and where the prevailing method of survey had been based on questionnaires to farmers.

3.   Activity B: European Rapid Estimates of Acreage and Potential Yield.

The principal objective of this activity is to provide early information at Community level on changes in crop acreage each year with respect to the previous year, as well as indicators of potential yields.

The method consists of defining a sample of representative sites (53 for the European Community as shown in Figure 4), then obtaining and interpreting high spatial resolution satellite images (about three or four each year) of the sites and extracting the required information. For the actual forecasts, the method uses no ground data for the year in question, but incorporates ground observations for the preceding year. This provides a basis for checking the validity of the method retrospectively.

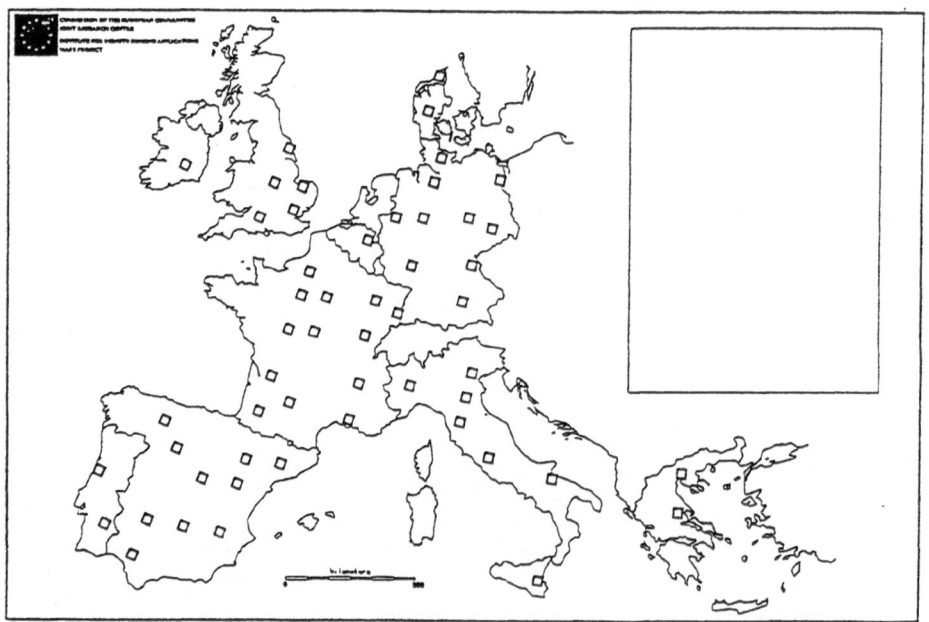

Figure 4. The sample of 53 sites for the rapid estimate of changes in crop acreages.

The Activity depends largely on the SPOT satellite's pointing capability which gives a theoretical frequency for the northern European areas of one image every two to three days. A 40 x 40km sample is selected systematically from the grid of SPOT scene centres. However, the constraints of rendering the sample as efficient as possible are such that, in practice, selection tends to be purposive rather than random. The Joint Research Centre co-operates with the suppliers of the images, SPOT Image and EURIMAGE, in order to fulfil two essential conditions: obtaining three to four images per site over a period of about five months and reducing the delay in obtaining the images to five or ten days between date of acquisition and delivery to the analysis location. The principle is to acquire the earliest available image, whether LANDSAT or SPOT.

Ground surveys of 16 segments of 50 hectares each, selected systematically from the sites, are conducted annually. The survey data for the previous year assists photointerpretation in the current year. To achieve a rapid analysis of a site in three to four days, it is not possible to photointerpret the whole site of 1,600 km². The method therefore consists of photointerpreting some 18 segments selected at random. The site is then classified automatically and the areas obtained are extracted from the classification.

The full scale operational application of the method started in 1992. The deadlines for obtaining the images is kept to about four to five days and three to four days for analysis. All the data acquired, are incorporated into an information bulletin which is sent to the E.U.'s Directorate General Agriculture within 10 days after the image acquisition.

The produced forecasts and acreage estimates are evaluated by comparing the Activity B results with the EUROSTAT statistics. In practice, the EUROSTAT estimates can vary significantly six months after harvest and synthesis is not easy to achieve. The curves on Figure 5 for 1992 durum wheat and common wheat acreages, two crops that are difficult to distinguish in regions where they coexist, show that their percentages were correctly estimated as early as June even though their respective variations between 1991 and 1992 were sharply contrasting. Once some hundred images have been analysed, the variations were not subject to further adjustment.

A comparison of the results on the acreage of the various crops appears to yield better results than a theoretical 2% tolerance, except in the case of sunflowers. A detailed analysis has demonstrated that, for this crop, areas designated as sunflower fields were interpreted as bare ground, probably owing to the very low yield, which explains the very good result achieved by the production forecast on the basis of a very poor acreage forecast.

However, crops that are too localised (e.g., for the E.U.: potatoes) are poorly estimated.

The forecasts are in most cases in good agreement with the final official figures. However, the wheat estimates, mainly those for winter wheat, sometimes show large differences with estimates from other sources.

In general, however, the method developed may be considered operational throughout the European Community. Further improvements, however, could be incorporated, *e.g.* the introduction of the analysis of ERS-1 radar images in order to enhance accuracy and reliability over time. The links with other sources must also be improved, particularly for

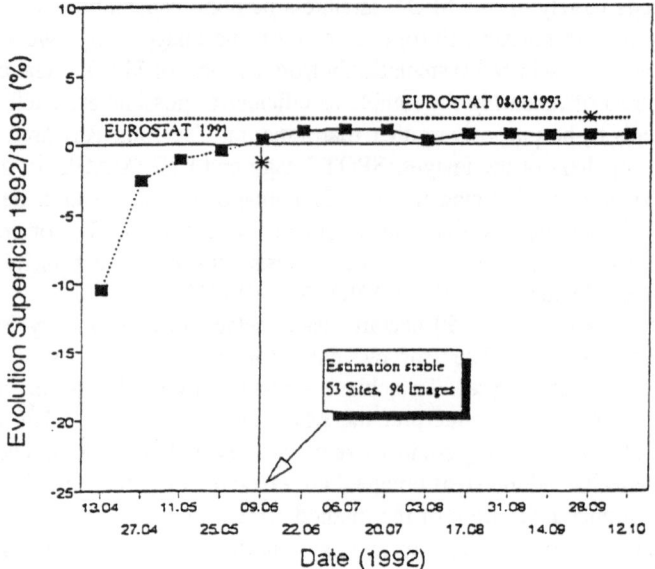

Figure 5. Acreage forecasts for common wheat, and its evolution in 1992.

the yield predictions. A better balance should for example be established between the contribution of Activity C, which is less costly in operational use, and Activity B, for which photointerpretation work will remain an essential, high-cost feature.

4.   Activity C: The Advanced Agricultural Information System.

This Activity consists of the following three components:

- Component 1: The assessment of vegetation conditions and yield indicators with low resolution meteorological satellite imagery (especially NOAA-AVHRR data)
- Component 2: Yield prediction models
- Component 3: The advanced agricultural information system

COMPONENT 1: VEGETATION CONDITION AND YIELD INDICATORS

The objective of this component is to use satellite meteorological data for monitoring vegetation conditions and providing indicators of the yields of the main crops. The method consists of processing NOAA-AVHRR satellite data to generate two indicators: a vegetation index and surface temperature (by methods described elsewhere in this book). Since these indices are directly related to the state of vegetation and crops, a spatial and temporal comparison of these data with other years or areas should make it possible to assess comparative yield levels. In addition to the objectivity of the method, the main attraction is the possibility of providing such indicators at various geographical levels: local, regional, national and European.

In the initial phases, two types of study were carried out independently:

- the development of application methods;
- the pre-processing of AVHRR data.

The development of application methods was carried out at various study sites in the E.U. and Figure 6 shows the map with the various study sites and the corresponding research teams that were involved in the work.

The application methods relate mainly to the three following subjects:

a) Agronomic applications, *i.e.* the study and analysis of two types of data on a multi-annual basis: the vegetation index (NDVI), which is composed from data on visible and near infra-red radiation and its development profile, and thermal infra-red radiation, with the calculation of surface temperatures (T_s) and the search for drought indicators (see for example the chapter by Seguin).

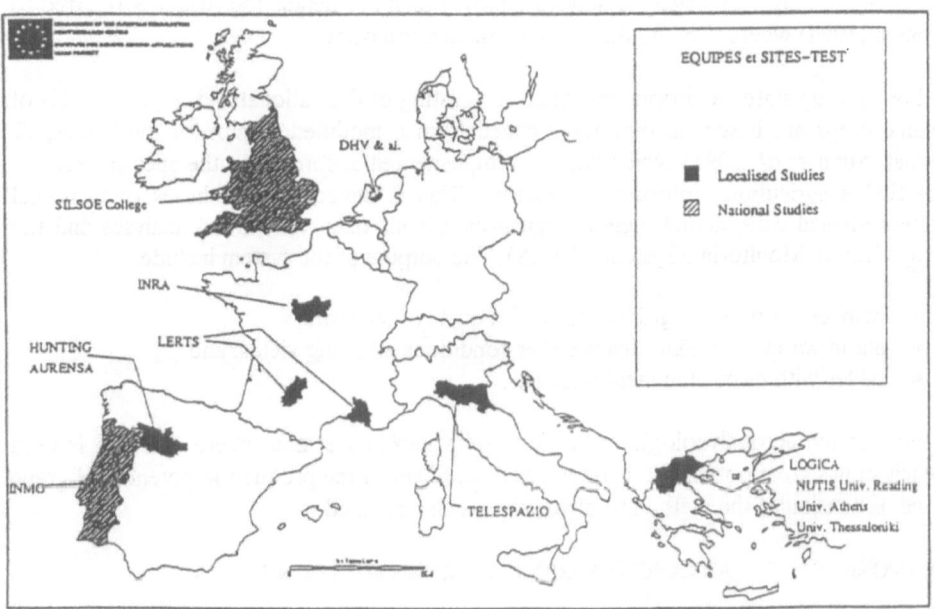

Figure 6. Activity C test sites and research teams (component 1).

b) Generating models of vegetation response on the basis of NOAA-AVHRR data. This is done by incorporating external data, either in the form of high-resolution satellite images, or more traditional data, such as those from weather stations.

c) Definition of useful products for a centralised agricultural information network.

COMPONENT 2: YIELD PREDICTION MODELS

The objective of component 2 is the development, testing and implementation of a system for timely regional crop state monitoring and yield forecasting of the following major E.U. crops: cereals, grain maize, rice, pulses, sunflower, soyabean, potato, sugar beet, colza, wine, olive oil, apple, pear and citrus. The quantitative yield forecasts should be acceptably precise at the national scale and possibly also for large regions.

This component is divided into two separate sub-actions:

a) the development of a semi-deterministic agrometeorological model for predicting annual crop yields;

b) the development of a model for the prediction of vine (and olive) yields based on pollen count methods.

These are only very briefly described below. For more details the reader is referred to Vossen (1994) where they are discussed in much more detail.

For (a) crop state monitoring and yield forecasting of the national and regional yields of annual crops are based on outputs obtained from a modified version of the WOFOST model (Supit *et al*. 1994) which has been implemented according to the special needs of the E.C.'s agricultural information system. This is linked to data bases of historical meteorological data, actual meteorological data, soils data, time trend analyses and the Crop Growth Monitoring System (CGMS). The outputs of the system include:

• mapped outputs of agricultural season quality indicators;
• alarm warning of abnormal weather conditions affecting yields; and
• tables with calculated yield forecasts.

For (b) the aeropalynologic of "capture of pollen in the atmosphere" method is used which shows great potential for the timely assessment of the production potential of grape wine. Extension of the method to olive crops is also planned.

COMPONENT 3: THE ADVANCED AGRICULTURAL INFORMATION SYSTEM.

The objective of this component is to integrate the various actions mentioned above and also to incorporate conventional surveys in order to create a complete information system including the new methods described above.

One of the most important parts of this component of the project is the specially-developed NOAA-AVHRR processing chain *SPACE (Software for the Processing of AVHRR data for the Communities of Europe)*, which generates the following products:

- daily mosaic of the Community generated from NOAA-AVHRR data (level 3);
- database for incorporating data for a homogeneous area or a given period, e.g. ten days, and for generating a more elaborate synthetic product (level 4). These can be time profiles for a given area or cartographic products for a given period.

Presently an archive of seven years of daily AVHRR European coverage is available: 1982 and 1989-1994. These are in the form of pan-European mosaics (made up from several passes) in a standard equal-area projection (Albers). This Level 3 product is the basic archive and consists of the following processed information:

- Channels 1 & 2 calibrated to top-of-atmosphere reflectance, accounting for inflight satellite sensitivity degradation
- Channel 3 calibrated to spectral radiance, and
- Channels 4 and 5 calibrated to brightness temperatures, accounting for non-linear corrections.

For daytime imagery, the solar reflectance component of channel 3 is also calculated by subtracting the emitted part of the radiance signal from the total radiance in channel 3. The latter is approximated using the channel 5 brightness temperature as a "true" value for the physical temperature, and assuming uniform emissivity across the different wavelengths

European statistical services are now fairly familiar with the potential of remote sensing in the area of crop inventories at regional or local level. The introduction of remote sensing entails the adoption of area sampling techniques, previously little used in Europe, and where the prevailing method of survey had been based on questionnaires to farmers.

Additional channels record the presence of clouds, the coastline and other vector geographical features, the pass from which each pixel in the mosaic was extracted, and the pixel view angle. Information on view angle is used to weight the contribution of the pixels to composite products, since data gathered at high view angles will be over-represented, and this unwanted effect can therefore be counter-balance (Sharman, 1989).

Cloud detection is carried out by a modified version of the APOLLO software (see chapter by Kriebel) and geometric correction is carried out by use of a BROLYD orbital model and a number of control points found by automatic correlation matching of images and coastline data bases (see chapter by D'Souza and Sandford). Some correction for atmospheric scatter is made using the "6S" code (see chapter by Vermote *et al.*), but no correction is attempted for atmospheric aerosols since loadings vary locally, and it is not thought possible to eliminate their effect on a continental scale using present techniques.

Routine processing of all European daily passes implies that all geometric, atmospheric and calibration problems must be resolved with little or preferably no action on the part of the operator. However, under extreme conditions (for example where sufficient numbers of coastline control points cannot be found because of cloud cover), some human intervention is required.

The operational processing of data in quasi real time (within three to five days after the acquisition of the images), started in 1993. The qualitative and quantitative results of the studies are particularly good in that they faithfully reflect crop and vegetation conditions. However, location-specific validation of the results is impossible owing to the incompatibility between the low spatial resolution of the NOAA-AVHRR sensors (1 km) and the typically very local detailed yield measurements (field-specific).

However, the variations in the vegetation indices faithfully reflect the variations in the yields of annual crops that occupy fairly large land areas. The results are particularly interesting when an exceptional event occurs, such as a major drought. Figure 7 provides an example for the monitoring of the severe 1992 drought over the south of the E.U (June), resulting in important yield losses.

Parameters such as the thickness of the atmosphere, the viewing angle and the angle of the sun can all influence radiometric quality. Cloud cover in certain areas can reduce the number of ground control points, which can in turn reduce the geometric quality of the images. The results are also very sensitive to the choice of images and to the processing of

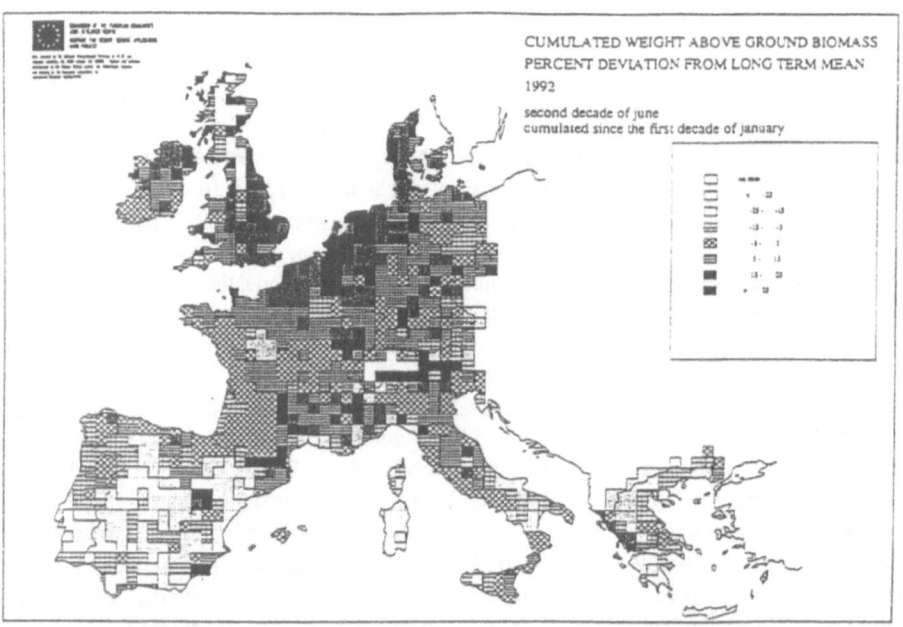

Figure 7. European coverage of the June 1992 NDVI values, averaged for the NUTS level 3 statistical regions, as compared to the same period in 1991.

the data. The quantitative validation in operational terms thus needs further study. Finally, further enhancement of the outputs, by linking them with agrometeorological models or by the establishment of direct relations with crop yield, is ongoing.

The *Advanced Agricultural Information System* produces, since 1993, the monthly *MARS Bulletin*. It is published from March to October and appears within 10 days after the acquisition of the raw satellite and meteorological data. It contains tables with forecasted crop acreages (at European level) and yields (at national level). The acreages are entirely derived from high resolution satellite imagery. The yields forecasts are established in three steps. First the raw outputs of the *CGMS* statistical module (national crop yield in kg/ha) are produced. This output, which is an automatic product, is then analysed by a team of agronomists and statisticians as a function of the growing conditions indicated by the NOAA-AVHRR indicators NDVI and Ts. If necessary, the CGMS forecast is revised, according to the possible recent occurrence of unfavourable conditions depicted by data processed and analysed in SPACE. Specific examples of important depictions by the AVHRR data include the abnormal high surface temperatures in Northern Europe during the teaselling period of grain maize in August 1994 and the excess humidity and low radiation that retarded crop establishment during the spring of 1994.

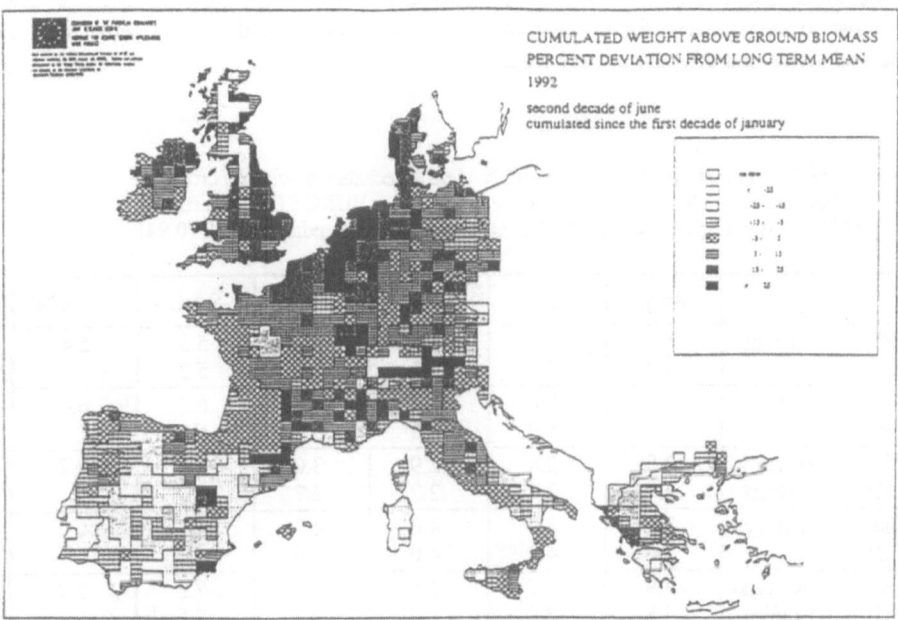

Figure 8. June 1992 simulated biomass production of winter wheat: departures from normal.

Table 2, which summarises the 1994 monthly yield forecasts for selected crops, and Figure 9 which provides the general state of summer crops at the end of September 1994, integrating the map outputs of SPACE (NDVI and Ts) and of CGMS (simulated biomass and grain, state of advancement of the crop development, soil moisture left in the soil, *etc.*).

It is yet too early to make a complete evaluation of the reliability and precision of the forecasts. One consistent result, however, was that for most crops the early forecasts were as from May-June closer to the final, official estimated produced in October-December by the national statistical offices, than the forecasts established by other organisations, often including the national statistical offices themselves.

The MARS automatic processing of NOAA-AVHRR data in near real time and the complete Crop Growth Monitoring System, including the statistical module, are operational for all major annual crops and unique in Europe. The results are published in the Monthly MARS Bulletin, which is provided to the E.U.'s Directorate General Agriculture and to EUROSTAT within 10 days after data acquisition (SPOT and LANDSAT for Activity B, NOAA-AVHRR and meteorological data for Activity C). The MARS Bulletin appears from March to October. It not only contains the acreage estimates, but also quantitative yield forecasts based on the interpretation of the SPACE and the CGMS software outputs.

However, where the acreage estimates at the European scale are operational products, the yield estimates at European and national scales are still in an experimental stage. They will have to be validated for at least two more years (1995 and 1996) against the real observed yields and official national statistics.

Table 2. Summary of the 1994 monthly yield forecasts for selected crops, as elaborated by the MARS Project. (yield = T/ha; E = EUROSTAT figures; M = MARS estimates; the EUROSTAT assessments are the updates on 21.10.94)

Crop		May	June	July	August	Sept.	U (10.94)
(E)	Wheat	5.5	5.4	5.6	5.6	5.5	5.4
(M)		5.2	5.3	5.2	5.2	5.2	
(E)	Common	6.2	6.0	6.2	6.2	6.1	6.0
(M)	wheat	5.9	5.9	5.8	5.8	5.8	
(E)	Durum	2.8	2.8	2.9	3.0	2.6	2.7
(M)	wheat	2.7	2.7	2.7	2.7	2.7	
(E)	Barley	3.9	3.9	4.0	4.0	4.1	4.0
(M)		4.0	4.1	4.0	4.0	4.0	
(E)	Grain	7.9	7.7	7.6	7.4	7.9	7.9
(M)	maize	7.8	7.8	7.6	7.6	7.5	
(E)	Sunflower	1.	1.6	1.6	1.6	1.5	1.6
(M)		1.1	1.4	1.4	1.4	1.4	

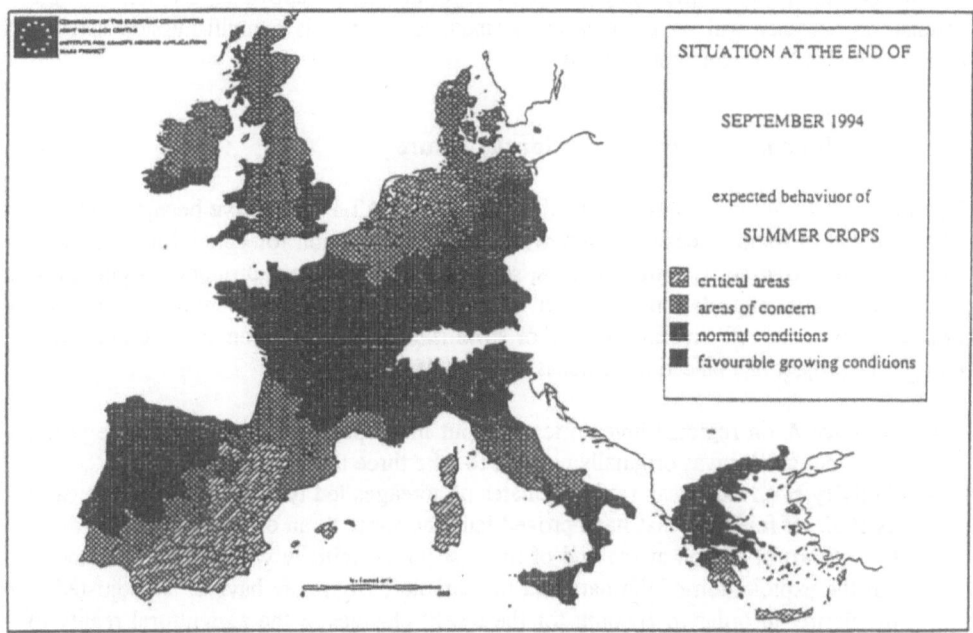

Figure 9. General state of summer crops at the end of the month of September 1994

The systems and approaches of Activity C could have many applications, particularly for environmental problems. As part of the project, there are plans to study applications involving forage crops and forests as well.

5. Activity E: New Methods and Sensors

This Activity focuses mainly on basic research into radar (SAR) techniques and on the prospective integration of the results in the other Activities of the MARS Project, mainly in the Rapid Estimates Activity. Following the launching of the ERS-1 satellite in 1991, work continues using real data for which there is expected to be continuity of service. By mid-1993, it was apparent that radar images are of potential use to the project, for both Activities A and B

This technique is presently being tested on larger areas. But above all, the actual image acquisition capacity will have to be ascertained, as ERS-1 is a multi-mission satellite mainly intended for marine applications.

6.   Conclusions and Orientations for the Future

By the end of 1994, all the basic tools of the MARS-STAT Project have been developed. The potential of remote sensing as a new source of information for agricultural statistics has been fully explored. Certain activities, such as Activity B, have produced results fully in line with initial expectations. In other cases, such as Activity A, remote sensing has proved to be more effective in the area of stratification than in automatic classification, though the project had placed the emphasis on the latter.

- **Activity A** on regional inventories was put into operational use after three years, even though this was originally planned to take three to five years.
- **Activity B** on European rapid estimates of acreages led to the production, as early as 1992, of forecasts that have proved fairly accurate at an early date. It is ready to be put into operation at the end of the first phase, initially without major changes, but the exploitation of the data and the sample design may have to be adjusted in the future, in order to account for the recent changes in the agricultural reality of the Union.
- For **Activity C** on the monitoring of plant growth and agrometeorological yield prediction, a georeferenced database has been established. It can have many applications in the fields of both agriculture and the environment. Cartographic products and trend indicators concerning the state of crops are generated from these models. However, the validation is still ongoing. Nevertheless, they can be used, even though the final development of stable products will take another two or three years. These products must then also be adapted to suit national and especially regional needs. As a by-product, a robust, reliable AVHRR processing suite has been developed, and a high-level dataset of geometrically-corrected and radiometrically-corrected data for the whole of Europe over a period of 7 years is available for detailed analysis and research.

7.   References and selected bibliography (Non exhaustive selection of MARS Publications and Reports)

PUBLICATIONS AND REPORTS RELATED TO ACTIVITY A:

Annoni A., Gallego F.J., 1992. Software for Crop Area Estimation with Remote Sensing (MARS-PED) and Quick Digitization of Sample Units (TTS), Conference on the Appl. of Remote Sensing to Agricultural Statistics (Belgirate). Office for Publications of the E.C. Luxembourg.

Anonyme, 1990. Projet Agriculture, Action 1: Inventaires Régionaux, Régions: France & Italie. Volume 1: Rapport Final Campagne 1989 Région Centre. Study carried out by AQUATER and SODETEG-TAI/SYSAME for the C.E.C., Joint Research Centre, Institute for Remote Sensing Applications. Milan and Paris, 152 pp.

Anonyme, 1990. Regional Inventory of Niederbayern/Oberpfalz. Final Report - Survey 1990. Study carried out by GAF - Gesellschaft für Angewandte Fernerkundung mbH in collaboration with HANSA LUFTBILD GmbH and the INSTITUTE of AGRONOMY of the Technical University of München, for C.E.C., Joint Research Centre, Institute for Remote Sensing Applications. Münster, 166 pp.

Anonyme, 1990. Spain: Castilla-León. Regional Crop Inventory 1990. Final Report Year 3. Study carried out by HUNTING TECHNICAL SERVICES in association with AUXILIAR DE RECURSOS Y ENERGIA SA (AURENSA) for the C.E.C., Joint Research Centre, Institute for Remote Sensing Applications. London and Madrid, 170 pp.

Anonyme, 1991. Projet Agriculture, Action 1: Inventaires Régionaux. Rapport Final: Elaboration de Statistiques des Productions Agricoles au niveau régional. Gréce-campagne 1990. Région Makedonia. Study carried by SYSAME, BRGM and ADK for the C.E.C., Joint Research Centre, Institute for Remote Sensing Applications. Milan and Paris, 242 pp.

Delincé, J., 1990. Un premier Bilan de l'Action 1 "Inventaires régionaux" du Projet Agriculture après deux années d'activité. In: Proceedings of the Conference on The Application of Remote Sensing to Agricultural Statistics, Varese, Italy, 10-11 October 1989, pp.53-58. Publication EUR 12581 EN of the Office for Official Publications of the E.C. Luxembourg, 383 pp.

Gallego, F.J., Delincé, J., Rueda C., 1993. Crop area estimates through remote sensing: stability of the regression correction. *Int. J. Remote Sensing.* **14**, 18, 3433-3445.

González F., López S., Cuevas J.M., 1991. Comparing Two Methodologies for Crop Area Estimation in Spain Using Landsat TM Images and Ground Gathered Data. *Remote Sens. Environ.* **32**, 29-36.

Gallego F.J., 1992. Flächenschatzungen für einjährige Feldfrüchte mit Hilfe Fernerkundung. Neue Wege raumbezogener Statistik. Forum der Bundesstatsistik, volume 20, pp 109-120. Statistisches Bundesamt, Wiesbaden.

MARS Project, 1990. Guidelines for the Action 6 Ground Surveys. Joint Research Centre of the E.U., Ispra (Italy).

PUBLICATIONS AND REPORTS RELATED TO ACTIVITY B:

Sharman M., de Boissezon H., 1991. Action 4 From images to statistics: operational review after two years of rapid estimates of acreages and potential yield of major annual crops throughout the European Community. In: Proceedings of the Second Conference on The Application of Remote Sensing to Agricultural Statistics, Belgirate, Italy, 26-27 November 1991. Publication EUR 14262 EN of the Office for Official Publications of the E.C. Luxembourg, 383 pp.

SOTEMA, 1993. "Estimations rapides au niveau européen des superficies et des rendements potentiels.- Rapport Final Phase 3", Novembre 1993, Toulouse, France. 148pp.

MARS Project, SCOT Conseil and SOTEMA, 1994. Documentation of the MARS Action 4 "European Rapid Estimates of Acreages". Institute For Remote Sensing Applications, Joint Research Centre of the E.U. Ispra (Italy), SCOT Conseil and SOTEMA, Toulouse (France). 650 pp.

PUBLICATIONS AND REPORTS RELATED TO ACTIVITY C:

Besselat, B., Cour, P., 1990. La prévision de la Production Viticole à l'Aide de la Technique 'Capture de Pollen'.In: Proceedings of the Conference on The Application of Remote Sensing to Agricultural Statistics, Varese, Italy, 10-11 October 1989, pp. 261-268. Publication EUR 12581 EN of the Office for Official Publications of the E.C. Luxembourg, 383 pp.

Boons-Prins, E.R., de Koning, G.H.J., van Diepen, C.A., Penning de Vries, F.W.T., 1993. Crop specific simulation parameters for yield forecasting accross the European Community. Simulation Reports N° 32. Centre for Agrobiological Research (CABO) and Joint Research Centre of the E.C. Wageningen (NL), 43pp + annexes

Choisnel, E., de Villele, O., Lacroze, F., 1992. Une approche uniformisée du calcul de l'évapotranspiration potentielle pour l'ensemble des pays de la Communauté Européenne. Publication EUR N° 14223of the Office for Official Publications of the European Communities; Series 'An Agricultural Information System for the European Community'. Luxembourg, 105 pp. + 13 annexes.

De Koning, G.H.J., Jansen, M.J.W., Boons-Prins, E.R., van Diepen, C.A., Penning de Vries, F.W.T., 1993. Statistical validation of crop growth simulation for regional yield forecasting accross the European Community. Simulation Reports N° 33. Centre for Agrobiological Research (CABO-DLO), Wageningen (NL). 77p.

Hooijer A.A., ven der Wal, T., 1994. CGMS Version 3.1 user manual. Winand Staring Centre, Wageningen (The Netherlands) and Joint Research Centre (Ispra, Italy). 247 pp.

Hough, M.N., Parker, J., 1992. The estimation, for the E.C., of global solar radiation and sunshine hours from cloud cover for 1-day and 10-day periods. Final report to the Institute for remote Sensing Applications, MARS Project. London, 49pp.

King, D., Daroussin, D., Jamagne, M., 1994. Development of a soil geographic database from the soil map of the European Communities. Catena, 21: 37-56.

King, D., Le Bas,C., 1994. Programme d'estimation des réserves en eau des sols à partir des paramètres de base de données géographiques des sols de l'Union Européenne. Final report of the Contract N° 5538-93-11 EP ISP F for the Commission of the European Communities, Joint Research Centre, Ispra Establishment. 50pp.

Meteo Consult BV, 1991. AMDaC User Manual. Software package for Actual Meteorological Database Construction. Wageningen, The Netherlands, 60p + 24p of annexes.

Palm, R., Dagnelie, P., 1993. Tendance et effets du climat dans la prévision des rendements agricoles des différents pays de la C.E. Joint Research Centre of the E.C. Publication EUR 15106 F of the Office for Official Publications of the E.C. Luxembourg, Ispra, 128 pp.

Sharman, M., 1991: Action II: monitoring vegetation with AVHRR: purpose, principles and products. In: Proceedings of the Second Conference on The Application of Remote Sensing to Agricultural Statistics, Belgirate, Italy, 26-27 November 1991. Publication EUR 14262 EN of the Office for Official Publications of the E.C. Luxembourg, 383 pp.

Sharman, M., de Boisezzon, H., 1991: Action IV, de l'image aux statistiques: bilan opérationnel après deux années d' estimation rapides des superficies et des rendements potentiels au niveau Européen. Proceedings of the conference on: The Application of Remote Sensing to Agriculture Statistics, Belgirate, J.R.C., Italy.

Supit, Y., 1994. Global radiation. Publication EUR 14767 EN of the Office for Official Publications of the E.C. Luxembourg, 160 pp.

Supit, Y., Hooijer, A.A., van Diepen, C.A., Editors, 1994. System description of the WOFOST 6.0 crop simulation model implemented in CGMS. EUR Publication N° 15956 EN of the Office for Official Publications of the European Communities. Luxembourg,. 168 pp.

van der Voet, P., van Diepen, C., Oude Voshaar, J., 1993. Spatial interpolation of daily meteorological data: a knowledge based procedure for the regions of the European Communities. Report 53/3. T

Vossen, P., 1992. Forecasting national crop yields of E.C. countries: the approach developed by the Agriculture Project. In: Proceedings of the Second Conference on The Application of Remote Sensing to Agricultural Statistics, Belgirate, Italy, 26-27 November 1991, pp. 159-176. Publication EUR 14262 EN of the Office for Official Publications of the E.C. Luxembourg, 383 pp.

Vossen,P., 1994. Early Crop Production Assessment of the European Union. In: Proceedings of the Conference on "The MARS Project: Overview and Perspectives" , Belgirate, Italy, 17-19 November 1993. Publication EUR 15599 EN of the Office for Official Publications of the E.C. Luxembourg.

Vossen, P., Rijks D., 1995. Early crop yield assessment of the E.C. countries: the system implemented by the Joint Research Centre. EUR Publication of the Office for Official Publications of the E.C. Luxembourg, 120 pp. In print.

PUBLICATIONS AND REPORTS RELATED TO ACTIVITY E:

Kohl H.G., King C., H. De Groof, 1993. "Agricultural Statistics: Comparison of ERS-1 and SPOT for the Crop Acreage Estimation of the MARS Project", Proc. of the 2nd ERS-1 Symposium, *ESA SP-361*, **1**, 87-92, Hamburg (Germany), 11-14 Oct. 1993.

H. Laur, 1992. "ERS-1 SAR calibration: Derivation of backscattering coefficient σ° in ERS-1.SAR.PRI products", ESA/ESRIN Technical Note, Issue 1, Rev.0.

Lopes A., Nezry E., Touzi R., Laur H., 1993: "Structure detection and statistical adaptive speckle filtering in SAR images", *Int. J. Remote Sensing*, **14**, 9, 1735-1758.

356

CROP KNOWLEDGE BASES:

Bignon, J., 1990. Agrométéorologie et physiologie du mais grain dans la Communauté Européenne. Publication N° EUR 13041 FR of the Office for Official Publications of the E.C.; Series 'An Agricultural Information System for the E.C.'. Luxembourg, 194pp

Carbonneau, A., Riou, C., Guyon, D., Riom, J., 1992. Agrométéorologie de la vigne en France. Publication N° EUR 13911 FR of the Office for Official Publications of the E.C.; Series 'An Agricultural Information System for the E.C.'. Luxembourg, 165pp

Falisse, A. 1992. Aspects Agrométéorologiques du Développement des Cultures dans le BENELUX et les Régions Voisines. Publication EUR 13910 FR of the Office for Official Publications of the European Communities; Series 'An Agricultural Information System for the European Community'. Luxembourg, 243 pp.

Hough, M.N., 1990. Agrometeorological Aspects of Crops in the United Kingdom and Ireland. Publication EUR 13039 EN of the Office for Official Publications of the E.C.; Series 'An Agricultural Information System for the European Community'. Luxembourg, 310 pp.

MacKerron, D.K.L., 1992. Agrometeorological aspects of forecasting yields of potato within the E.C. Publication N° EUR 13909 EN of the Office for Official Publications of the E.C.; Series 'An Agricultural Information System for the E.C.'. Luxembourg, 247 pp

Narciso. G., Ragni, P., Venturi, A., 1992. Agrometeorological Aspects of Crops in Italy, Spain and Greece. Publication N° EUR 14124 EN of the Office for Official Publications of the E.C.; Series 'An Agricultural Information System for the E.C.'. Luxembourg, 450 pp.

Russell, G., 1990. Barley knowledge base. Publication N° EUR 13040 EN of the Office for Official Publications of the E.C.; Series 'An Agricultural Information System for the E.C.'. Luxembourg, 135 pp.

Wilson, G., Russell, G., 1994. Wheat knowledge base. EUR Publication of the Office for Official Publications of the E.C.; Series 'An Agricultural Information System for the E.C.'. EUR Publication N° 15789 EN of the Office for Official Publications of the European Communities. Luxembourg. 160pp.

THE USE OF AVHRR-DERIVED LAND SURFACE TEMPERATURE ESTIMATES FOR AGRICULTURAL MONITORING.

Bernard SEGUIN.
INRA Bioclimatologie Avignon
BP 91
F.84 143 Montfavet (Cedex)

ABSTRACT.   As seen in some of the earlier chapters in this book, vegetation indices such as the NDVI derived from NOAA-AVHRR satellites are widely used for vegetation monitoring purposes. Land surface temperatures (LSTs), accessible from the thermal infrared channels of the NOAA-AVHRR instrument, may also be considered as a valuable source of information, especially for detecting and quantifying water stress.

A previous chapter by Vogt covered various practical and theoretical ways of retrieving LSTs from NOAA-AVHRR thermal channels. In this chapter we review the physical relationship linking surface temperature to the level of canopy evapotranspiration following a brief analysis of surface energy balance components. Then, the derivation of local methods for monitoring crop water stress is described and the application to satellite-derived data is presented in terms of practical considerations. The results obtained from several experiments in the Sahel, Europe and northern Africa are then presented. Finally, the potential and limitations of land surface temperature estimation from NOAA satellites, particularly for agricultural crop monitoring purposes, are presented and discussed.

1.   Introduction

Monitoring vegetation and assessing crop yield on a regional scale by use of the normalised vegetation index (NDVI) has proven to be feasible in a large number of studies over the past ten years (see preceding chapters in this book, and especially the chapter by Tucker). However, some limits appear in the precise assessment of crop biomass and yield, because of the indirect linkage of these quantities with measured NDVI. Land surface temperature estimates (LSTs) from thermal channels offer an alternative (or, more exactly, a complementary way) for the estimation of these quantities, which has not always been fully exploited up to now, in spite of their availability from the NOAA-AVHRR satellites.

This chapter presents the potential of these thermal infrared channels and illustrates these by presenting the results of several investigations aimed at the monitoring of agricultural crop resources in different regions of the world.

357

G. D'Souza et al. (eds.), Advances in the Use of NOAA AVHRR Data for Land Applications, 357–376.

2.   Surface temperature and energy balance

2.1   BACKGROUND

Surface temperature T_s may be derived from measurement in the thermal infrared channels (as described in detail in the chapter by Vogt). It differs from the air temperature T_a, measured in a meteorological-screen, which attempts to characterise the air mass properties as independently as possible of underlying surface properties (Figure 1).

Access to surface properties (albedo, surface temperature, surface moisture) is a determinant input for the local functioning of the soil-vegetation interface, in relation to the thermal equilibrium governed by the combined processes of energy balance and soil water budget. Thus, a short analysis of this equilibrium, which governs the significance of the surface temperature, is necessary before developing its possible use in agrometeorology (see Seguin, 1989).

2.2.   ENERGY BALANCE OF A THIN SURFACE.

If we consider the daytime case of a thin surface (bare soil or low crop less than 10-20 cm high), the classical equation of energy balance, at a given time,

$$R_n = G + H + LE \qquad (1)$$

expresses the division of net available energy (net radiation R_n) between ground conductive heat flux G and convective fluxes, sensible heat (H) and latent heat (LE) (corresponding to evaporation for a bare soil and evapotranspiration for a vegetation canopy) (Figure 2).

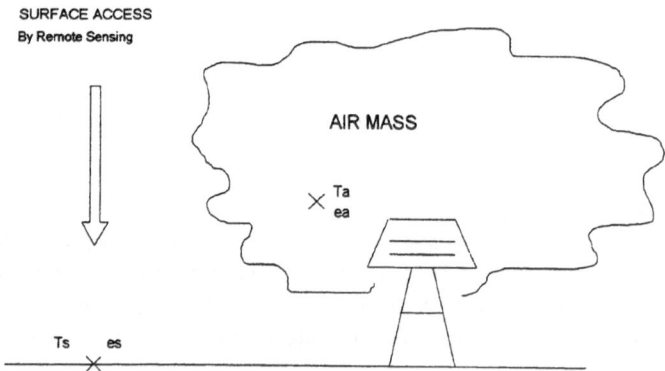

Figure 1. Schematic description of surface characteristics compared with screen measurements.

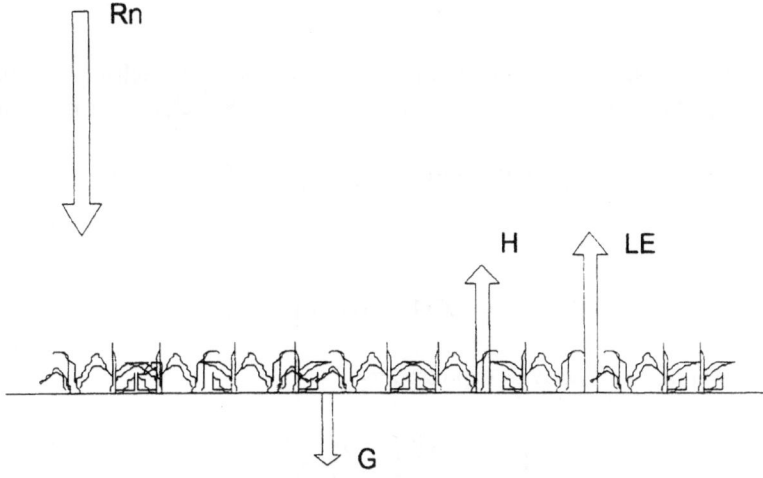

Figure 2. Energy balance components.

As the surface is thin, this equation may be applied directly at the surface level where radiative absorption and emission take place.

The net radiation R_n is then found from the various components of radiation exchanges:

$$R_n^* = (1-a)R_g + R_a - \sigma T_a^4 \tag{2}$$

where the solar global radiation R_g is partly reflected depending on the surface albedo a, σT_s^4 corresponds to the long-wave radiation emitted by the surface (assuming the emissivity being equal to 1.0) and R_a by the atmosphere down to the surface (long-wave atmospheric radiation).

In order to obtain an independent expression for T_s, it is usual to define a "climatic net radiation" R_n^* where T_s is replaced by T_a

$$R_n^* = (1-a)R_g + R_a - \sigma T_a^4 \tag{3}$$

the difference $R_n^* - R_n$ being equivalent to $\sigma T_s^4 - \sigma T_a^4 \approx 4\sigma T^3(T_s - T_a)$.

Using the following expression for H:

$$H = \rho Cph(T_s - T_a)$$ (4)

where the sensible heat flux is related to $(T_s - T_a)$ by the turbulent exchange coefficient h (ρ and Cp being constants, air density and specific heat, respectively), one can obtain

$$R_n^* = R_n + 4\sigma T^3(T_s - T_a) + LE$$

and then

$$R_n^* - G = (\rho Cph + 4\sigma T^3)(T_s - T_a)$$ (5)

so that the deviation $T_s - T_a$ may be expressed by

$$T_s - T_a = \frac{(R_n^* - G) - LE}{\rho Cph + 4\sigma T^3}$$ (6)

2.3.   RELATIONSHIP BETWEEN TS AND EVAPORATION (OR EVAPOTRANSPIRATION)

Taking into account the usual assumption that the ground heat flux G is either negligible (which is the case in wet situations or with a dense canopy) or a fixed proportion of R_n^* (0.1 for example, but it may increase up to 0.2 - 0.3 in the specific case of dry bare soil), we can state from Equation (5) that, for given atmospheric conditions (R_n^* and T_a), T_s is a linear decreasing function of LE. Knowing that $4\sigma T^3 \approx 6 to 8 W.m^{-2}.K^{-1}$ and that ρCph varies between 10 and 100 in the same units, it can be said that the slope of this relationship depends mainly upon the exchange coefficient h.

From the laws of turbulent transfer in the atmospheric surface layer, h can be written as (see Seguin, 1981)

$$h = h_N . F(z/L)$$ (7)

where the exchange coefficient in neutral conditions h_N is

$$h_N = \frac{ku^*}{Log(z/z_{ot})}$$ (8)

with the friction velocity u^* derived from the logarithmic wind profile

$$u^* = \frac{ku(z)}{Log(z/z_o)}$$ (9)

(z_0 and z_{0t} being the aerodynamic roughness and the heat exchange roughness, respectively, which are equivalent for thin surfaces) and F (z/L) the stability effect correction depending on the ratio z/L (L Monin Obukhov length).

So that, on the whole, h depends upon :

- the wind velocity u (z) ;
- the stability effects (L being related to u^* and T_s - T_a) ; and
- the surface aerodynamic roughness z_0 (which is related to the mean height of surface elements h, z_0 being of the order of 0.10-0.15 h).

The relationship between T_s and LE is then globally dependent on :

- the atmospheric variables : R_n^*, T_a, u ; and
- the surface parameters : a and z_0.

2.4.   THE EXTENSION TO TALL CANOPIES.

When applied to canopies, the concept of energy balance and the linked equations are still valid, but problems arise with the definition of the surface level and its properties.

A simplification may occur with the ground heat flux G, because the interception of radiation inhibits the input of energy directly to the soil surface, thus giving a ratio G/R_n less than 0.05 for dense canopies.

Two main differences arise compared with a thin surface :

- The roughness length for heat transfer z_{0t} becomes significantly different from the aerodynamic roughness z_0; corresponding to an excess resistance for heat transfer in the case of rough surfaces (Figure 3) (see Thom, 1975; Garatt, 1978; Brutsaert, 1982). The ratio z_{0t}/z_0 may be estimated as from 0.2 to 0.1 for short to medium-size crops (grass, soybeans, wheat), but there is considerable uncertainty for taller canopies (like corn, fruit trees, forest, *etc.*).
- The corresponding "aerodynamic surface temperature" T_{sa} thus defined, by extrapolation of air temperature profiles down to the level z_{0t}, may differ from the "radiative surface temperature" T_{sr} as measured from the thermal IR radiation. This last parameter results from the radiation emitted by various layers inside the canopy, and may incorporate a significant contribution from the generally colder lower layers (Figure 5). T_{sr} is, therefore, in general lower than T_{sa}. The deviation has been measured as 1 °C in the case of a wheat canopy (Huband and Monteith, 1986). However, it is not well known for taller canopies.

362

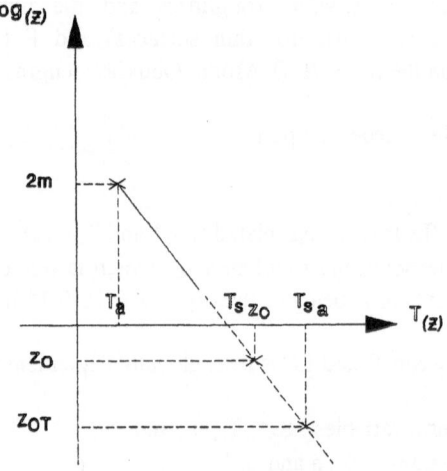

Figure 3. Semi-logarithmic temperature profiles, allowing the separation of z_{0t} and z_0.

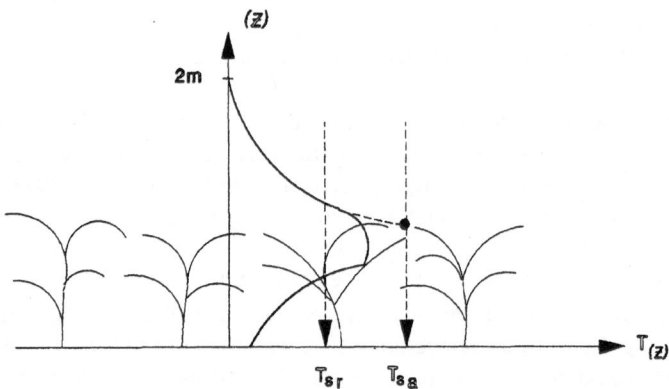

Figure 4. Scheme indicating the case of a developed canopy with T_{sr} significantly different from T_{sa}.

These two points ($z_{0t} = z_0$ and $T_{sr} = T_{sa}$) which are closely linked (for instance if z_{0t}/z_0 is computed from experiments where T_{sa} is measured by T_{sr}, as in Garratt (1978)), lead to large uncertainties on the precise form of the relations between T_s and LE for canopies, as experienced by Hall *et al.* (1992) with the results of the FIFE experiment.

However, both sets of experimental data (like Kohsiek *et al.* 1993) and modelling approaches (like Brunet *et al.* 1991) now provide usable sets of values for the excess resistance and perspectives for its more analytical processing in the near future.

3. Applications to water stress monitoring on a local scale

The combined use of surface temperature measurements and of the energy balance equation (by using Equation (3) for estimating the sensible heat flux term H) has been successfully used for a correct estimation of the instantaneous value of LE in a large range of studies (Bartholic *et al.* 1972, Brown 1974, Stone and Horton 1974, Idso *et al.* 1975, Heilman and Kanemasu 1976, Hatfield *et al.* 1984, Choudhury *et al.* 1986, Kalma and Jupp 1990 *etc.*). Proper assessment of Z_{0t} may reduce the large discrepancies noted by Hall *et al.* (1992), as shown by Kustas *et al.* (1989) or Kohsiek *et al.* (1993).

These micrometeorological applications give an interesting background for the application to crop monitoring in the field of agricultural meteorology, but they evidently need to be adapted to this specific use. Considerable progress has been achieved in this field by the work of R.D. Jackson and his team in Phoenix (see Jackson 1982), for example, the definition of a crop water stress index (CWSI) specifically designed for the scheduling of crop irrigation using portable radiothermometers.

For the purpose of crop monitoring in terms of final yield estimation, however, the need for a regular data acquisition all along the vegetative cycle leads to the consideration of cumulative values. As early as 1977, Jackson *et al.* (1977) have elaborated such a procedure by empirically identifying the cumulative stress degree day (SDD), defined as Σ $T_s - T_a$ which appears to be well related to the final yield when affected by variable water stress. Results of a similar experiment on dry and irrigated fields of wheat, conducted in our laboratory (Steinmetz *et al.* 1991) well illustrate this capacity of SDD (Figure 5).

Although originally empirical, this concept may be directly justified by a parallel suggestion of Jackson *et al.* (1977), assuming that daily evaporation ETd could be linearly related to the instantaneous value of $T_s - T_a$ once a day (near midday preferably)

$$ETd = R_{nd} + a - b\,(T_s - T_a) \tag{10}$$

R_{nd} being the daily net radiation, and a and b being constants of adjustment for a local situation.

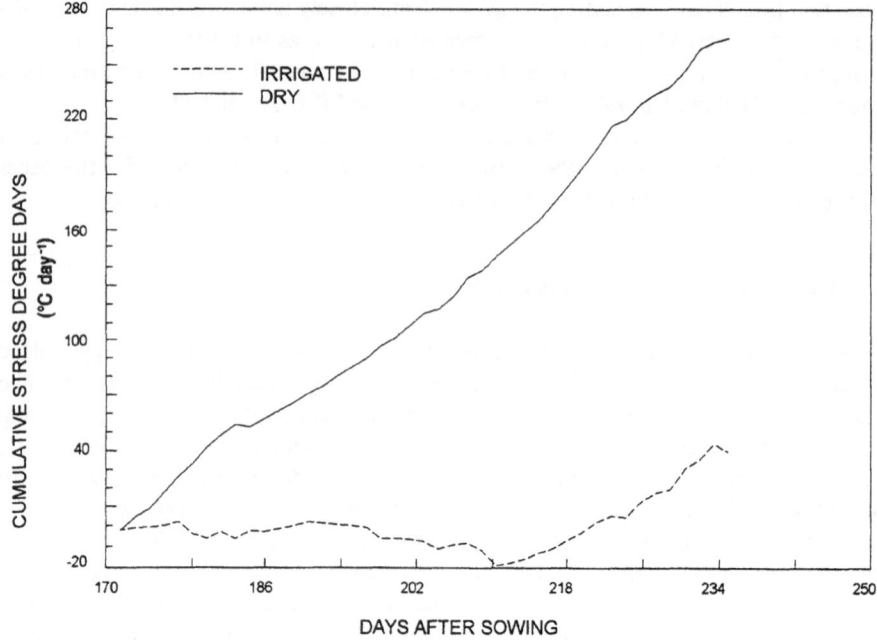

Figure 5. Evolution of Σ (T_s - T_a) along the phenological time of the growing season for adjacent fields of irrigated and dry wheat (from Steinmetz *et al.* 1991).

This approach is effectively a simplified first order estimation, with several underlying questions (for example, is it possible to pass from midday values to daily integrated flux values, in spite of the non-linear effect of wind velocity on the exchange coefficient and possible deviations of the well-organised sinusoïdal temporal variation of radiant fluxes with clouds ? What will be the effect of surface roughness ?, *etc.*). A number of arguments have been presented on this subject, both on a theoretical point of view (Seguin and Itier, 1983, Vidal and Perrier, 1989, Riou *et al.* 1988) and concerning the possible dependance of a and b upon the surface roughness (Lagouarde and Mac Anneney, 1992). These contributions demonstrate that the application of the simplified approach relationship (Equation 9) provides an approximate estimate of daily ET (ETd) accurate to about only ± 1 mm/day. This is obviously very coarse, but good enough to provide useful and reliable information on water availability for crops on a regional level at least. Furthermore, as we have mentioned above, the present basic time scale in agricultural meteorology is 10 days rather than one day, so that the possibility of cumulating ETd values on larger periods allows the dampening of the variability along the main linear relation. The ET computations for 10 days or monthly values (ΣETd) may then be derived from :

$$\Sigma ET_d = \Sigma Rn + a - b\Sigma(T_s - T_a) \qquad (11)$$

with a larger precision in terms of relative error (of about $\pm$ 5 mm for 10 days values and about $\pm$ 10 mm for monthly periods) (Figure 6).

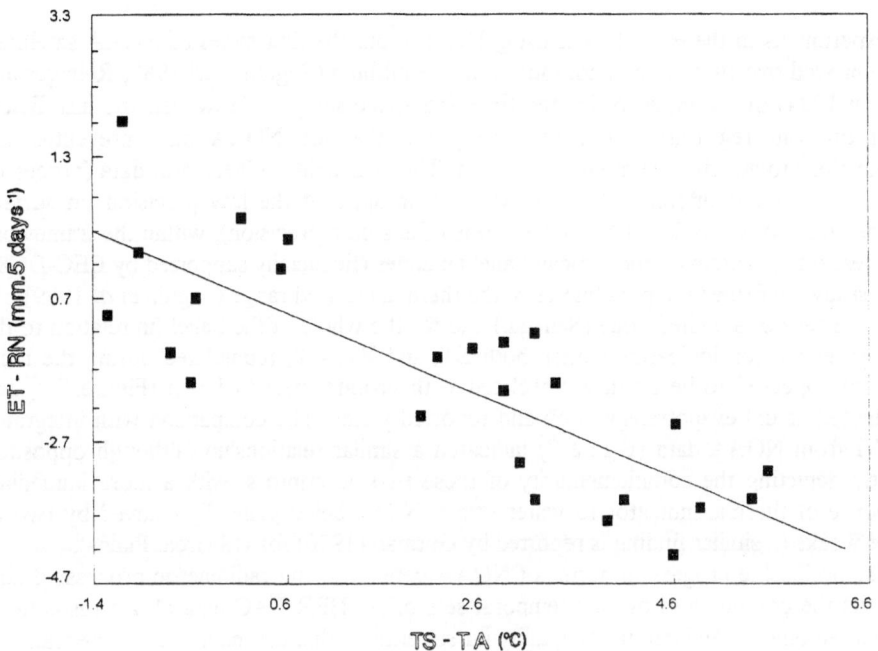

Figure 6. Relationship between ET - R_n and T_s - T_a for five-day periods on a wheat crop (from Steinmetz *et al.* 1991).

4.   Towards the applications with meteorological satellites on a regional scale

As far as surface temperature may also be detected from space using meteorological satellites, these local procedures, established only with ground measurements, may also apply to larger (regional scales), with two conditions:

- that cloud-free conditions permit the collection of good quality images at frequent intervals and that atmospheric influence effects could be corrected in order to obtain a satisfying accuracy in terms of surface temperature estimation precision,
- that the large size of observable pixels (1.1 km for NOAA) would result, in spite of the unavoidable mixing of homogeneous surfaces, in a significant value of surface temperature at this scale.

Experiments in the early 1980s, using HCMM and the first series of NOAA satellites, have proved that these two conditions could be fulfilled (Seguin *et al* 1982, Reiniger and Seguin 1984) in the frame of limited time and space samples. However, the insufficient level of both registration and processing facilities for NOAA data prevented the applications to an effective monitoring effort. The availability of Meteosat data (processed in CMS Lannion) offered a first opportunity, in spite of the low precision on surface temperature (of the order of 5 to 6° in terms of absolute precision), within the framework of a research program on the African Sahelian zones (financially supported by EEC-DG8). This study confirmed the possibilities of the thermal infrared range (Seguin *et al* 1989). In both the case of a limited zone (Senegal) and for the whole of the Sahel (in relation to the Agrhymet meteorological network), both ΣT_s and $\Sigma T_s - T_a$ (cumulated during the rainy season) appeared to be linearly correlated with ground measured rain (Figure 7a), and computed actual evapotranspiration and reported yields. The comparison with integrated NDVI from NOAA data (Figure 7) indicated a similar relationship (although opposite), clearly depicting the complementarity of these two descriptors, with a more immediate response of thermal indicator to water stress, NDVI being generally delayed by two or three weeks (a similar finding is reported by Gutman (1990) for US Great Plains).

The noticeable progress in terms of NOAA geometric and radiometric processing now permits the computation of multitemporal sets of AVHRR LAC data (1 km resolution), with a radiometric and geometric quality far superior to that attainable with Meteosat.

At this stage, a short statement needs to be made about the significance of surface temperature, in relation with the satellite measured radiance (which implies also correct calibration coefficients). The two interrelated points concern the influence of surface emissivity and the atmospheric corrections (see also the chapter by Vogt).

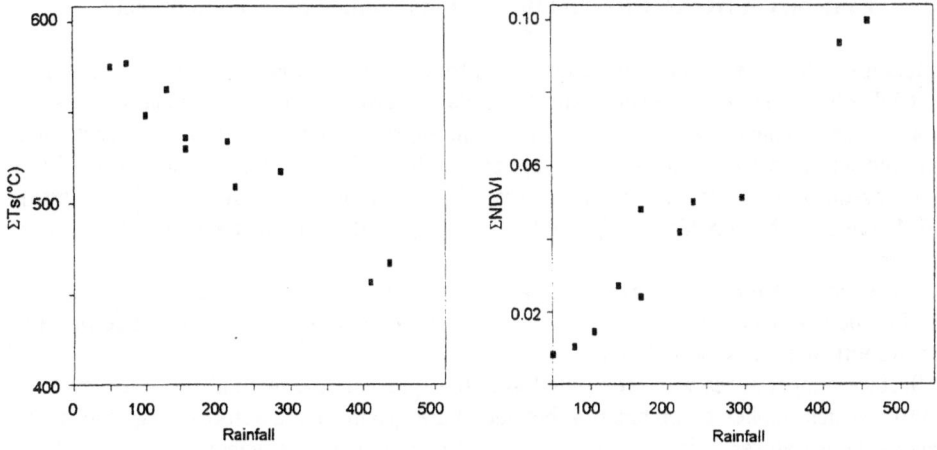

Figure 7. Relationships between rainfall and ΣT_s (a) $\Sigma NDVI$ (b) for 11 stations of Senegal in 1984.

In Equation (1), the thermal-emitted radiation by surface has been written as σTS^4 (assuming the emissivity $\varepsilon = 1$). This approximation is not a serious problem for energy balance considerations, but may create some serious errors for the extraction of surface temperature, since the effective measured radiance (when $\varepsilon < 1$) corresponds to $\varepsilon\sigma TS^4 + (1-\varepsilon) R_a$. The error, when considering $\varepsilon = 1$, evidently depends upon R_a values, but may be roughly approximated to 0.6 - 0.9° for 0.01 error in ε (Becker 1987). On the large scale corresponding to meteorological satellites, and excluding the regions of bare soils (which is acceptable when monitoring vegetation), ε values for the 10.5 - 12.5 µm spectral bands of NOAA are above 0.96 and often approach 0.98. A global underestimation of the order 1 to 2° therefore appears a reasonable first order consequence.

Atmospheric correction may create larger errors, but the generalisation of the split-window method (linearly combining T4 and T5 values derived from channels 4 and 5) provides one way of reducing these. When compared with ground measurements (so that it is impossible to separate the emissivity effect from the atmospheric correction), a global error of about 4° was found with the original coefficients of Deschamps and Phulpin (1980), but recent adaptations by Kerr (1991) and Ottlé and Vidal-Madjar (1992) show improvements in the accuracy up to about ± 2° which is more acceptable for operational purposes.

5. Examples of NOAA thermal infrared data applied to crop monitoring

In relation with these encouraging elements, a three-year set of NOAA has been compiled by CNES-TI in Toulouse, using daily LAC data registered in CMS Lannion. Five-day synthetic scenes have been elaborated (based on the maximum of NDVI) at the scale of 1 km, and analysed (with the financial support of the MARS project of JRC Ispra) by a close association between our laboratory (D. Courault, J.P. Guinot, Ph. Clastre), CEMAGREF in Montpellier (A. Vidal, H. Kerdiles) and LERTS in Toulouse (A. Podaire, Y. Kerr).

The sample of three years was a good opportunity in France, as 1988 was a normal (slightly humid) year, whilst 1989 and 1990 were severely affected by droughts (especially in the south and the west regions).

Surface temperatures were computed with the split-window method proposed by Kerr (1991), which implies a combination between bare ground temperatures, Tsg, and dense canopy temperatures, Tsv, by means of a crop coverage coefficient Cv derived from NDVI

$$Cv = a\,NDVI \tag{12}$$
$$Ts = CvTsv + (1 - Cv)\,Tsg \tag{13}$$
$$Tsg = 3.1 + 3.2\,T4 - 2.2\,T5 \tag{14}$$
$$Tsv = -2.4 + 3.6\,T4 - 2.6\,T5 \tag{15}$$

The main results, already published in Seguin *et al* (1992) and Courault *et al* (1993) may be summarised as follows:

- the identification of dry episodes, as well as the localisation of affected zones, is well depicted by the monitoring of ΣT_s-T_a. The example of seven locations for various contrasted regions in France (Figure 8) displays regional dryness contrasts in space (Figure 9, with dryer conditions in the southern regions) and between years (Figure 10). in this last figure, the difference between 1988 and 1989-1990 is very apparent in Perpignan (early affected), St Laurent and Toulouse (from the beginning of July), whilst Avignon shows only small differences (but the area is mainly occupied by irrigated cultures, thus being almost insensitive to dry episodes). Maps of ΣT_s and T_a may then be used to locate the most affected zones and (considering the differences between the dry year and a reference humid year like 1988) more precisely assess the locations where drought has been severest,

- an assessment of the effect on the agricultural productive regions may be provided as indicated in Figure 11, where the contrast between two years is expressed both in terms of NDVI and ΣT_s-T_a. As previously noted, the two quantities appear at least complementary, the thermal data depicting the water stress earlier than NDVI. At this stage, the complementarity may be used for eliminating possible diagnostic errors (due to cloud contamination or

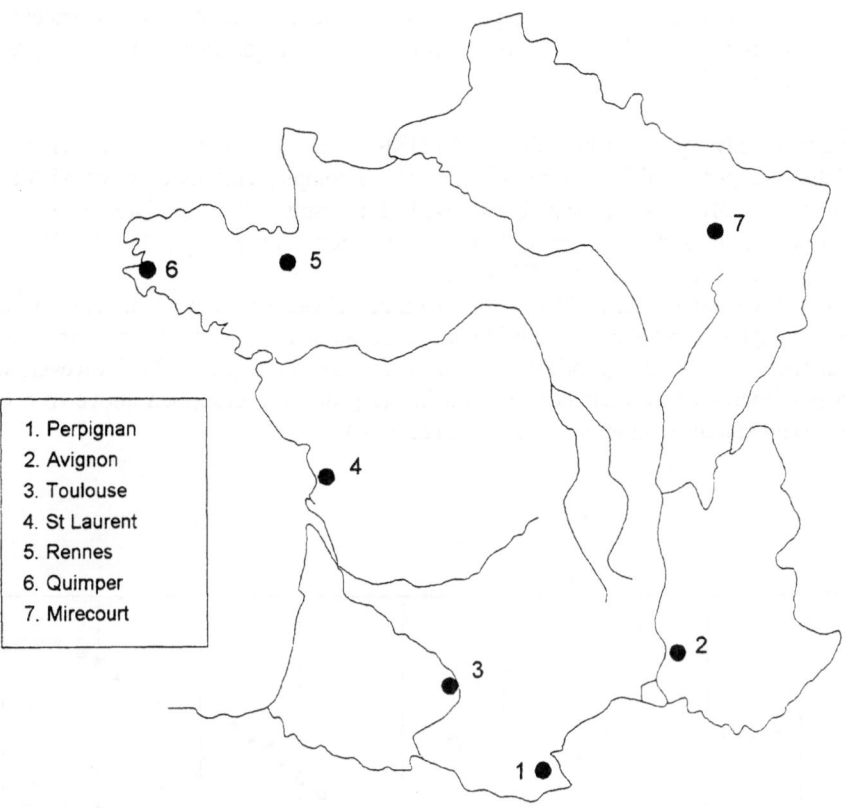

Figure 8. Location of the 7 sites analysed in this study.

atmospheric effects) : when a dry episode occurs, NDVI must diminish and T_s increase simultaneously,

• the use of the linear relationship (9) leads to the computation of ET. In the case of Figure 12, for a large pasture area (140 km^2), the observed difference between 1988 and 1989 in terms of cumulated ET (350 mm for 1989 against 593 mm for 1988) is in good agreement with the observed biomass reduction (about ± 3 ha in 1989 compared with more than ± 5 ha in 1988). Maps of ET could be used as a tool for more quantitative assessments of crop yield, but the large dimension of NOAA pixels (with generally mixed surfaces, except for specific zones like that of Figure 12) is an obstacle to more precise computations.

A parallel study with the same NOAA-AVHRR scenes, but applied to Algeria (Guérif *et al.* 1992) has permitted further investigation in the complementarity between NDVI and T_s. In this study, NDVI has been used for estimating the intercepted PAR (Photosynthetically Active Radiation), then converted into biomass by the Monteith procedure, and T_s (converted into ET values) for introducing a water stress reduction in the efficiency conversion term. The relation with cereal yields (as always affected by some noticeable imprecision) may be considered as acceptable (due to all the encountered uncertainties) : $r = 0.53$ for NDVI, $r = 0.41$ for ΣT_s-T_a (Figure 13). However, it is significantly increased by their combination in the procedure described above ($r = 0.78$ between yield and the computed biomass) (Figure 14).

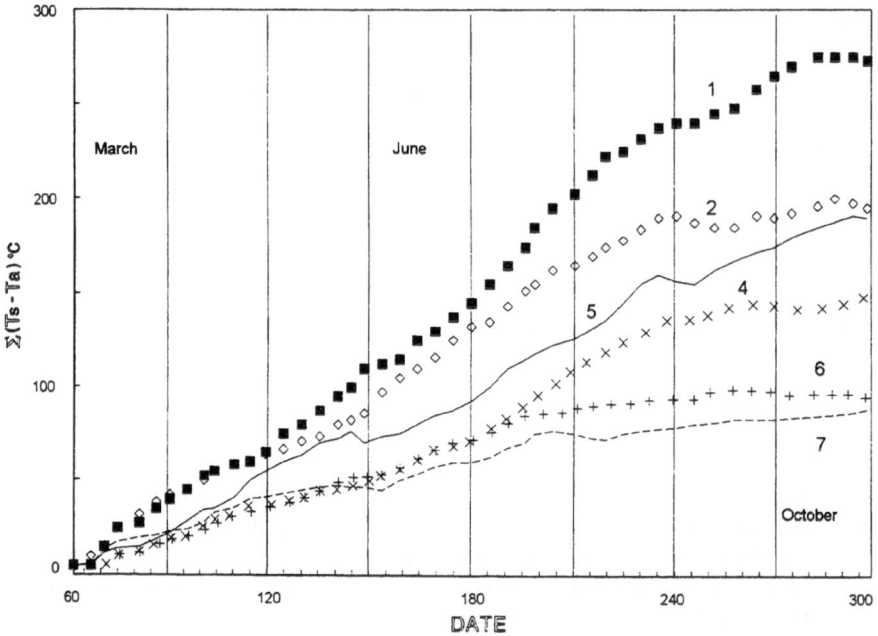

Figure 9. $\Sigma (T_s - T_{amax})$ curves for the first six locations in 1988. Their location is shown in Figure 8.

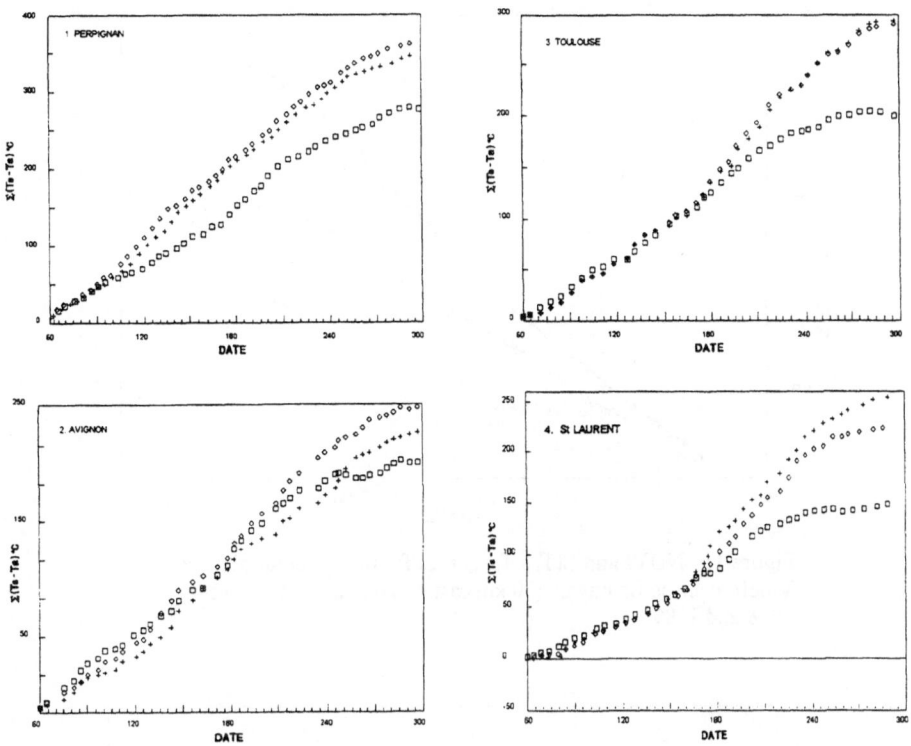

Figure 10.

Analysis of the differences between the three years for the four south and west sites.
1. Perpignan, 2. Avignon, 3. Toulouse, 4. St Laurent.

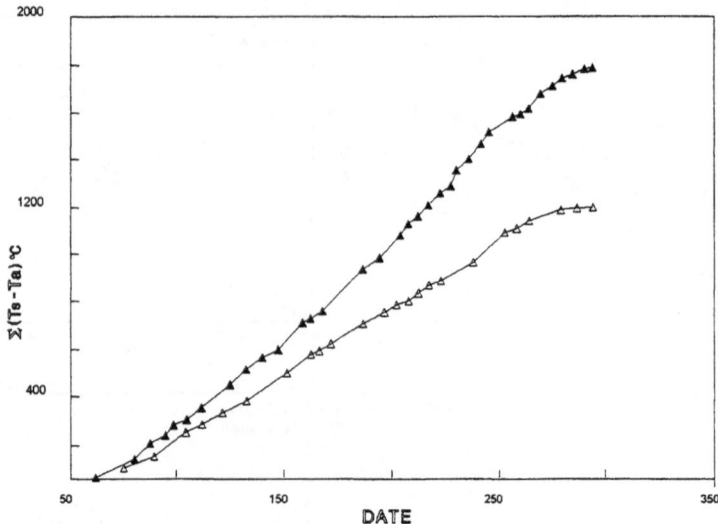

Figure 11. NDVI and $\Sigma(T_s - T_a)$ curves for the agricultural area of Villefranche de Lauragais (30 km east of Toulouse) for the years 1988 and 1989.

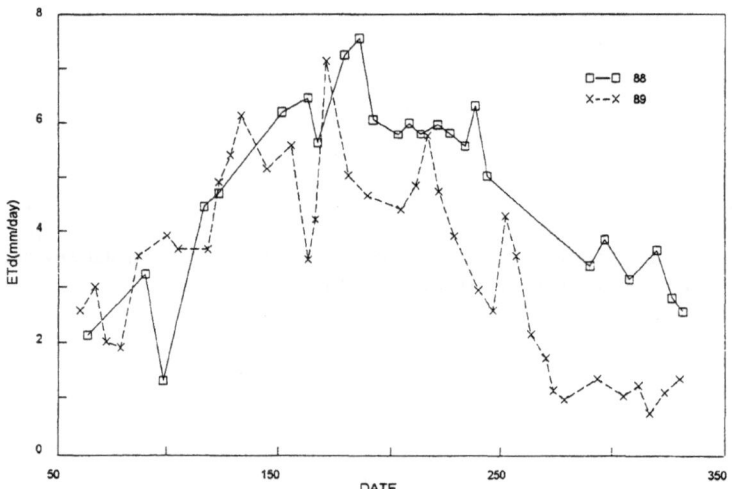

Figure 12. Comparison of ET values for the St Laurent site (4) in 1988 and 1989.

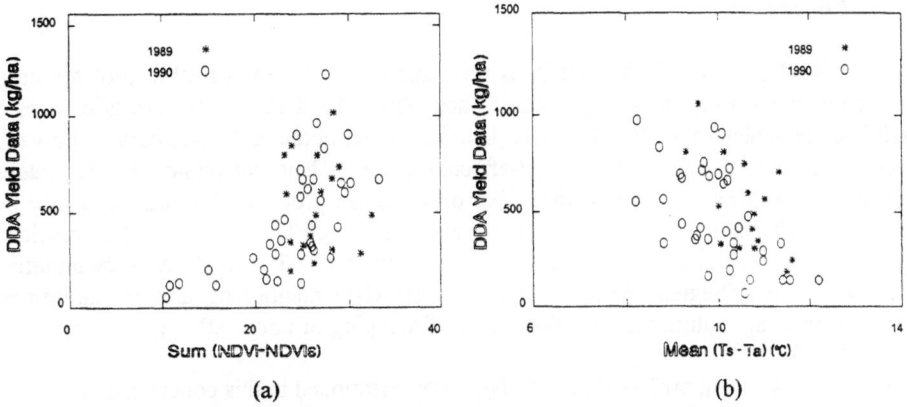

Figure 13. Relationship between cereal yield for the same examples.
a) Σ (NDVI - NDVI$_s$)
b) Σ (T$_s$ - T$_a$)

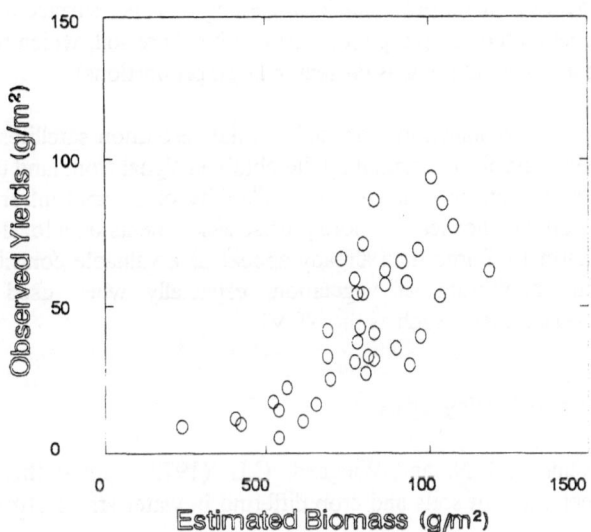

Figure 14. Relationship between observed yield and estimated biomass.

6. Conclusion

Thermal IR data from NOAA satellites then appear as a very valuable tool for crop production monitoring on a large scale. Their ability to detect water stress ensures a straight complementary with NDVI data. In a first stage, obtained ΣT_s-T_a curves allow the detection and delimitation of drought-affected zones. Their conversion to ET values provide perspectives to compute the effect of stress on produced biomass, and may be combined with NDVI values for an analytical processing. They may also be combined with agrometeorological models (by way of computed ET) to allow a quantitative estimate of yield (Seguin 1992). It is also worthwhile mentioning another interesting contribution to agricultural meteorology, that of mapping of zones affected by frosts (see Seguin 1989).

However, two main problems need to be clearly mentioned in this conclusion

- one is related to a technical aspect of NOAA satellites : the drift in the overpass time, which may reach one to two hours after some years (as experienced with NOAA-9). Surface temperature is sensitive (in relation with the energy balance aspects explained at the beginning of this contribution) to the time of data acquisition. Approximate corrections may be applied as a first solution, but a better control of the overpass time is clearly needed in the future.

- the other aspect is related to the spatial scale. The general occurrence of mixed pixels creates difficult problems for assessing a proper surface temperature for a given vegetation or crop type (especially when bare soil, which temperature may increase in dry conditions, is present in large proportions).

As for the NDVI, a combination with high spatial resolution satellites (like SPOT or Landsat) will be necessary for decorrelating the obtained signal from land use. However, it alone will not be sufficient, and the regular availability of thermal infrared data at high spatial resolution would be needed for more precise assessments on a local scale.

Nevertheless, thermal infrared data already appear as a valuable contribution in global and regional scale monitoring of vegetation, especially when used together with vegetation index measurements such as the NDVI.

7. References and Bibliography

Bartholic, J.F., Namken, L.N. and Wiegand, C.L. (1972). Aerial thermal scanner to determine temperatures of soils and crop differing in water stress. *Agronomy Journal*, **64**; 603.

Becker, F., (1987). The impact of spectral emissivity on the measurement of land surface temperature from satellite. *International Journal of Remote Sensing*, **8**, 1509-1522.

Brown, K.W. (1974). Calculations of evapotranspiration form crop surface temperature. *Agric. Meteor.*, **14**, 199.

Brunet, Y., Paw U., K.T., Prevot, L. (1991). Using the radiative surface temperature in energy budget studies over plant canopies. *Proc. 5th. Coll., Physical measurements and signatures in remote sensing*, ESA SP-319, 557-580.

Choudhury, B.J., Reginato, R.J. and Idso, S.B. (1986). An analysis of infrared temperature observations over wheat and calculation of the latent heat flux. *Agric. For. Meteor.*, **37**, 75-88.

Deschamps, P.Y., and Phulpin, T. (1980). Atmospheric correction of infrared measurements of sea-surface temperature using channels at 3.7μm, 11μm and 12 μm. *Boundary Layer Meteorology*, **18**, 131-143.

Guérif, M., de Brisis, S. and Seguin, B., (1992). Complementarity of SPOT-HRV and NOAA-AVHRR data for crop yield assessment in semi-arid environments. EARSEL workshop on agrometeorology. Florence (Italy) 13-14 April 1992. To be published in Advances in Remote Sensing.

Gutman, G.G., (1990). Towards monitoring droughts from space. *Journal of Climate*, **3**, 282-295.

Hall, F.C., Huemmrich, K.F., Goetz, S.J., Sellers, P.J. and Nickeson, J.E. (1992). Satellite remote sensing of surface energy balance : success, failures and unresolved issues in FIFE, *J. Geoph., Res.*, **97**, D17, 19061-19090.

Hatfield, J.L., Reginato, R.J., and Idso, S.B. (1984). Evaluation of canopy temperature-evapotranspiration models over various crops. *Agric. Meteorol.*, **32**, 41-53.

Heilman, J.L., and Kanemasu, E.T. (1976). An evaluation of a resistance form of the energy balance to estimate evapotranspiration. *Agronomy Journal*, **68**, 607-611.

Idso, S.B., Schmugge, T., Jackson, R.D. and Reginato, R.J. (1975). The utility of surface temperature measurements for the remote sensing of water status. *J. Geophys. Res.*, **80**, 3044.

Jackson, R.D., Reginato, R.J., and Idso, S.B., (1977). Wheat canopy temperature : a practical tool for evaluation of water requirements. *Water Res. Res.*, **13**, 651-656.

Jackson, R.D., (1982). Canopy temperature and crop water stress. *Advances in Irrigation*, 1, 43-85.

Kalma, J.D. and Jupp, D.L.P. (1990). Estimating evaporation from pasture using infrared thermometry, evaluation of a one layer resistance model. *Agric. For. Meteorol.*, **51**, 223-246.

Kerr, Y.H., (1991). Corrections thermiques dans l'infrarouge thermique : cas de l'AVHRR. Proc. 5th Coll. Physical Measurements and Signatures in Remote Sensing. ESA, SP 319, 29-34.

Kohsiek, W., de Bruin, H.A.R., The, H. and Van den Hurk, B. (1993). Estimation of the sensible heat flux of a semi-arid area using radiative temperature measurements. *Boundary Layer Meteorology*, **63**, 213-230.

Kustas, W.P., Choudhury, B.J., Moran, M.S., Reginato, R.J., Jackson, R.D., Eay, L.W. and Weaver, H.L. (1989). Determination of sensible heat flux over sparce canopy using thermal infrared data. *Agric. For Meteorol.*, **44**, 197-216.

Lagouarde, J.P. and Mac Anneney, K.J. (1992). Daily sensible heat flux estimation from a single measurements of surface temperature and maximum air temperature. *Boundary Layer Meteorology*, **59**, 341-362.

Ottlé, C. and Vidal-Madjar, D. (1992). Estimation of land surface temperature with NOAA9 data. *Remote Sensing of Environment*, **40**, 27-41.

Reiniger, P. and Seguin, B., (1986). Surface temperature as an indicator of evapotranspiration and soil moisture. *Rem. Sens. Review*, **1**, 277-310.

Riou, Ch., Itier, B., and Seguin, B., (1988). The influence of surface roughness on the simplified relationship between daily evaporation and surface temperature. *International Journal of Remote Sensing*, **9**, 1475-1481.

Seguin, B. and Itier, B., (1983). Using midday surface temperature to estimate daily evaporation from satellite thermal IR data. *International Journal of Remote Sensing*, **4**, 371-383.

Seguin, B., (1989). Use of surface temperature in agrometeorology. In "Applications of Remote Sensing of Agrometeorology". Kluwer Ed., Amsterdam, 221-240.

Seguin, B., Assad, E., Freteaud, J.P., Imbernon, J., Kerr, Y.H. and Lagouarde, J.P., (1989). Use of meteorological satellite for water balance monitoring in sahelian regions. *International Journal of Remote Sensing*, **10**, 1001-1017.

Seguin, B., Lagouarde, J.P., Savane, M., (1991). The assessement of regional crop water conditions from meteorological satellite thermal infrared data. *Remote Sensing of Environment*, **35**, 141-148.

Seguin, B., (1992). Utilisation combinée de données satellitaires et de modèles agrométéorologiques. *Proceedings of the second conference on remote sensing applied to agricultural statistics*, Belgirate (Italy), 26-27 November 1991. Ed by JRC Ispra, 199-208.

Seguin, B., Fischer, A., Kerdiles, H., Louahala, S. and Podaire, A., (1992). Suivi agroclimatique des cultures en France à partir des données NOAA, METEOSAT ans Spot. *Proceedings of the second conference on remote sensing applied to agricultural statistics*, Belgirate (Italy), 26-27 November 1991. Ed. by JRC Ispra,339-342.

Steinmetz, S., Lagouarde, J.P., Delecolle R., Guerif, M., and Seguin, B., (1991). Evapotranspiration and water stress using thermal infrared measurements : a general review and a case study on winter durum wheat in Southern France. Proc. ICARDA-INRA Symposium. Montpellier (France) 3-6 July 1989. Ed. by INRA, 55, 89-114.

Stone, L.R. and Horton, M.L. (1974). Estimating evapotranspiration canopy temperatures : field evaluation, *Agronomy Journal*, **66**, 450-454.

Vidal, A. and Perrier, A., (1989). Analysis of a simplified relationship for estimating daily evapotranspiration from satellite thermal IR data. *International Journal of Remote Sensing*, **10**, 1327-1337.

INTEGRATION OF AVHRR AND FINE SPATIAL RESOLUTION IMAGERY FOR TROPICAL FOREST MONITORING

RICHARD M. LUCAS
Department of Geography
University College of Swansea
Singleton Park
Swansea SA2 8PP
United Kingdom

ABSTRACT. A multi-sensor approach to tropical forest monitoring is discussed whereby data from the NOAA AVHRR is used to identify changes in forest cover on a regional to continental scale and to direct more detailed assessments using satellite sensor imagery with a finer spatial resolution. Temporal, spatial and spectral attributes of optical sensors and the procedures of stratified sampling applied to different forms of forest disturbance are outlined. Regression techniques, linear mixture models and fuzzy classification algorithms (which can be used to estimate the proportions of land cover types within AVHRR pixels and which require fine resolution imagery for development and validation) are also reviewed in the context of tropical forest monitoring. The use of thematic information derived from fine resolution data for increasing the monitoring capabilities of the AVHRR is considered.

1. Introduction

A multi-sensor approach to tropical forest monitoring has been advocated (Nelson and Holben, 1986) as integration of data from different satellite sensors allows a more complete representation of the spatial and temporal changes occurring within the tropical forest biome. In tropical forest monitoring systems (TFMS), the NOAA AVHRR (Advanced Very High Resolution Radiometer) provides an overview of tropical regions and can be used to detect specific disturbance events, large scale changes in forest cover, and adverse trends. The AVHRR defines areas where further investigation with finer resolution sensors is required. A more detailed representation of tropical forests can then be obtained by using imagery from fine resolution sensors such as the Landsat MSS (Multi-spectral Scanner System) and TM (Thematic Mapper) and the SPOT (Satellite Probatoire pour l'Observation de la Terre) HRV (Haute Resolution Visible).

This 'top-down' approach has been suggested for monitoring tropical forests (Malingreau and Laporte, 1988). However, due to the common occurrence of cloud, smoke and haze over tropical regions and the infrequent temporal coverage of Landsat and SPOT sensors, the acquisition of fine resolution imagery of sufficient quality is limited. Furthermore, the routine use of such imagery within monitoring systems has been deterred by the high cost and the large quantity of data involved (Nelson and Horning, 1993). Consequently, attention has focused recently on deriving more comprehensive

G. D'Souza et al. (eds.), Advances in the Use of NOAA AVHRR Data for Land Applications, 377–394.

information from the AVHRR by estimating the proportions of different land cover types within the AVHRR pixels (Cross *et al.*, 1991). In this 'bottom up' approach, fine resolution imagery are used in the development and validation of techniques, such as regression analysis, linear mixture models and fuzzy classification algorithms, which estimate proportions and which may play an important rôle in tropical forest monitoring in the future.

This chapter rationalises the importance of a multi-sensor approach to tropical forest monitoring, reviews the relative merits of the different sensors for detecting and describing forest disturbance, and investigates the use of both 'top-down' and 'bottom-up' approaches to integrating data from different satellite sensors. Although a number of satellite sensors routinely observe tropical regions, only the NOAA AVHRR and the Landsat and SPOT sensors will be considered in this chapter.

2. Requirements of a tropical forest monitoring system

Primarily, a TFMS should relate observations from a range of satellite sensors to the nature and extent of disturbance within the forest. In many areas of the tropics, change within the forest is gradual. In regions of South East Asia and Africa, for example, shifting cultivation and selective logging have resulted in structural alteration rather than wholescale clearance of forests (Green and Sussman, 1990, Miller and Williams, 1978, Gilruth *et al.*, 1990). Periods of drought have also led to the gradual degradation of large tracts of forest in several regions of South East Asia (Malingreau, 1986). Observations at a range of spatial scales and comparisons of both multi-date AVHRR and fine resolution imagery over periods extending from months to decades are therefore necessary in order to identify and assess the impact of these changes.

In other tropical regions, large areas of forest may be cleared and change is both rapid and dramatic. For example, extensive tracts of forest in the Brazilian states of Para, Mato Grosso and Maranhaõ were replaced by cattle pastures (Fearnside, 1991, Malingreau and Tucker, 1988). Similarly, massive forest clearance has resulted from large scale logging operations in South East Asia and West Africa. In these situations, more frequent satellite sensor observations are required to detect specific disturbance events and short-term changes. Detailed observations by fine resolution sensors may then be necessary to assess the nature and magnitude of change.

Changes in forest extent over smaller areas result from activities such as mining, road construction, rubber tapping, oil exploration, hydro-electric power generation, and plantation forestry. Although the AVHRR may on occasion be able to give evidence of clearance, fine resolution imagery alone may only provide the capability to detect and determine the extent of intrusion into the forest

In all the cases mentioned, a warning system is required followed by an assessment programme. A TFMS should recognise anomalies in the seasonal and interannual phenological characteristics of tropical forests at scales ranging from local to continental and identify changes in the actual extent of forests by comparing alterations in land cover observed by satellite sensors on two or more dates. Such change-detection techniques should highlight areas of new deforestation, assess the expansion of existing forest clearances, and determine rates of change. Specific events, such as burning activity, road construction, and recent forest clearance, should also be identified.

TFMS therefore have to consider the spatial, temporal and spectral aspects of the disturbance process in relation to the equivalent characteristics of the satellite sensors used. The nature of change within the forested area will decide the optimal sensor for identifying, quantifying and monitoring the processes of forest conversion. Selection of the most appropriate sensor and stratification of the sampling procedure will also be determined by the particular characteristics of the sensor. The principal features of both the AVHRR and fine resolution sensors should therefore be considered in order to optimise the integration of these systems.

3. Principal features of the AVHRR and fine resolution satellite sensors of importance for forest monitoring

3.1 SPATIAL DIFFERENCES

The NOAA AVHRR, with a spatial resolution of 1.1 km at nadir, provides wide area observations of tropical regions (Malingreau and Belward, 1992). The Landsat MSS and TM sensors, with spatial resolutions of 80 m and 30 m respectively, and the SPOT HRV, with a spatial resolution of 20 m (multi-spectral mode) and 10 m (panchromatic mode), provide more detailed local and regional observations. The thermal channel (band 6) of the Landsat TM has a spatial resolution of 120 m but will not be considered further in this chapter. As no single spatial resolution will satisfy all the requirements for a TFMS (Townshend and Justice, 1988), selection of the most appropriate sensor will depend upon the extent of disturbance and the type of forest conversion.

3.2 TEMPORAL VARIATIONS

With two NOAA satellites currently in orbit, the AVHRR is capable of observing the tropical belt four times per day. The mid-afternoon overpass allows observation in both the spectral and thermal wavebands while the remaining three overpasses are best suited to thermal observation. Although the temporal resolution of the Landsat and SPOT sensors is 16 days and 26 days respectively, these sensors generally overpass an area on different days thereby increasing the frequency of observations and the likelihood of cloud free imagery. Furthermore, two Landsat and two SPOT satellites are currently in orbit allowing observation of targets every six to nine days and 13 days respectively. Both SPOT HRV sensors also have off-nadir viewing capabilities thereby allowing images to be recorded as frequently as every three days. The temporal frequency of satellite sensors needs to be related to the time-scales of change within forest areas in order to optimise the use of imagery for detection of disturbance activity in forested regions.

3.3   SPECTRAL CHARACTERISTICS

The NOAA AVHRR senses in five wavebands (Table 1). AVHRR channel 3 is optimal for defining forest extent as the strong thermal and reflectance differences between the forest and non-forest areas are emphasised particularly during the dry season and late in the afternoon (Malingreau, 1991; Malingreau and Tucker, 1988). AVHRR channels 1 (visible) and 2 (near infrared) can also be combined to generate the NDVI (Normalised Difference Vegetation Index) which may be used to delineate areas of deforestation (Malingreau *et al.*, 1989)

Biomass burning is detected by identifying pixels that are saturated or close to saturation in AVHRR channel 3 as this waveband is sensitive to elevated heat sources (Malingreau and Laporte, 1988). Flaming fires as small as 10 m x 10 m and smouldering fires as small as 30 m x 30 m can be detected using this channel (Kaufman *et al.*, 1990) and the size and temperature of fires may be derived by using a combination of AVHRR channels 3 and 4 if the pixels are not saturated (Pereira and Setzer, 1993a). AVHRR channels 1 and 2 are also used to detect smoke palls resulting from burning activity (Pereira and Setzer, 1993a).

Table 1. Wavebands of the AVHRR and fine resolution sensors

Band No.	NOAA AVHRR	Band No.	SPOT HRV
1	0.58 - 0.68 µm		*(Panchromatic)*
2	0.72 - 1.10 µm	1	0.51 - 0.73 µm
3	3.55 - 3.93 µm		*(Multi-spectral)*
4	10.5 - 11.5 µm	1	0.50 - 0.59 µm
5	11.5 - 12.5 µm	2	0.61 - 0.68 µm
		3	0.79 - 0.89 µm
Band No.	Landsat TM	Band No.	Landsat MSS
1	0.45 - 0.52 µm	1	0.50 - 0.60 µm
2	0.52 - 0.60 µm	2	0.60 - 0.70 µm
3	0.63 - 0.69 µm	3	0.70 - 0.80 µm
4	0.76 - 0.90 µm	4	0.80 - 1.10 µm
5	1.55 - 1.75 µm		
6	10.4 - 12.5 µm		
7	2.08 - 2.35 µm		

Changes in vegetation phenology and structure are described using multi-temporal AVHRR NDVI at various spatial resolutions ranging from 1.1 km resolution LAC (Local Area Coverage) or HRPT (High Resolution Picture Transmission) and 4 km GAC (Global Area Coverage) data to 15 km GVI (Global Vegetation Index) data (Malingreau and Laporte, 1988, Malingreau *et al.*, 1989). The thermal AVHRR channels may also indicate change in the forest canopy (Malingreau *et al.*, 1989).

The Landsat and SPOT sensors observe in the optical wavebands with the exception of Landsat TM band 6 (Table 1). Recently deforested areas are identified using the visible channels of these sensors although revegetated areas are generally indistinguishable from closed forest. Secondary vegetation is, however, detected using the near and middle infrared wavebands. With the Landsat TM sensor, channels 4 and 5 are optimal for discriminating fire scars from other cover types, while smoke palls are detected using channels 1, 2 and 3. Channel 7, which is able to penetrate smoke, is used for locating active fires which are not observed within the other channels (Pereira and Setzer, 1993b).

The NDVI may also be derived using the visible and near infrared bands of the Landsat and SPOT sensors, and used to monitor changes in vegetation phenology and canopy structure.

3.4   TIME SERIES INFORMATION

Many of the changes in tropical forests have been chronicled by optical sensors which have been observing the tropics for over 20 years. The NOAA VHRR (Very High Resolution Radiometer) was deployed from 1972 while later generation satellites supporting the AVHRR sensor observed the Earth from 1978 to the present (Hastings and Emery, 1992). Landsat MSS data has been available for much of the tropics from 1972, and from 1982 the Landsat TM sensor provided an opportunity to observe the tropics at a finer resolution and in seven wavebands (Sader *et al.*, 1991). SPOT HRV data has been available since 1986 (National Remote Sensing Centre, 1992). This immense archive of photographic and digital Landsat and SPOT sensor imagery for the tropical regions is held by various institutions worldwide and provides an important accessible data source for deriving baseline information and detecting change at different spatial and temporal resolutions.

4.   Integration of AVHRR with fine resolution imagery - the "top-down" approach.

4.1   FOREST/NON FOREST

Data from the AVHRR sensor may be used for the discrimination and estimation of the spatial extent of forest and non-forest areas (Nelson and Horning, 1993). However, the ability to detect forest clearances using the AVHRR depends upon the size of the cleared areas in relation to the spatial resolution of the sensor, the spatial arrangement of the target, and the spectral contrast between the deforested or degraded land and the original forest (Townshend and Justice, 1988). For example, linear patterns of disturbance, such as roads, and large discrete clearances with distinct geometric boundaries, such as cattle ranches in eastern Amazonia or logged areas in south-east Asia, are often distinct within 1 km resolution AVHRR imagery, even in cases where the spectral contrast is limited (Malingreau, 1991, Townshend and Justice, 1988). However, close to the margins of clearance and where a complex arrangement of forest and non-forest areas or small clearances exist (e.g., in Rondonia, Brazil), the true spatial extent of forest is often unclear within the AVHRR image (Townshend and Justice, 1988).

As a result of the observation of subpixel elements and forest/non-forest boundaries and the saturation of AVHRR channel 3 by fires, the area of deforestation is often over-estimated when using data from the AVHRR (Cross *et al.*, 1991, Cross, 1990). For example, Skole and Tucker (1993) suggested that the AVHRR over-estimated the deforested area in the Legal Amazon by as much as 50% in comparison to estimates derived using Landsat TM data. Even so, a number of studies (Justice *et al.*, 1985, Malingreau and Tucker, 1990, Woodwell *et al.*, 1987, Nelson and Holben, 1986) have suggested that estimates of forest cover using the AVHRR are in fact comparable with those derived using Landsat MSS data.

Several studies have compared forest/non-forest classifications of co-registered NOAA AVHRR and fine resolution imagery to assess the errors of measurement (e.g. Stone and Schlesinger, 1990, Stone *et al.*, 1991, Cross, 1990, and Woodwell *et al.*, 1986). For example, Cross (1990) indicated that as the percentage of forest determined from TM data decreased within corresponding AVHRR pixels, the errors in the AVHRR estimates increased. Stone and Schlesinger (1990) reported that where forest cover was less than 50%, the classification of AVHRR pixels was considered unreliable and ground pixels with less than 35% forest failed to be classified as such. Cross (1990) also observed that at least 66% of the AVHRR ground pixel needed to be forested for inclusion in a forest category within the classified image.

Due to the limitations in the classification of forest and non-forest using AVHRR data, fine resolution imagery should be used to estimate the actual deforested area with a higher level of accuracy, especially for areas where disturbance activity is proceeding rapidly. These active deforestation fronts may be indicated within AVHRR data by fire activity, noticeable expansion of the deforested area, and linear patterns of clearance which may suggest road construction (Malingreau and Laporte, 1988). In the state of Rondonia, Brazil, for example, a "fish bone" pattern of side roads and associated settlements developed along the main BR-364 highway which links the Brazilian cities of Porto Velho and Cuiaba. Burning activity and the overall pattern of development was evident within full spatial resolution AVHRR imagery but only fine resolution Landsat TM and MSS data were able to describe the road systems, river networks, urban areas and agricultural partitions in detail (Woodwell *et al.*, 1987, Justice *et al.*, 1985).

4.2   EVIDENCE OF BURNING ACTIVITY

In tropical areas, and especially during dry seasons where burning activity is greatest, the AVHRR has the potential to provide daily observations of fire activity. Although the

AVHRR sensor is often unable to give details of the burning event and associated changes in forest cover, the spatial arrangement and variations in the location of fires may suggest the spatial dynamics and magnitude of deforestation events. For example, regular patterns of fire activity may indicate organised clearance (Malingreau, 1991) and linear fire patterns or multi-temporal observations may suggest movement of the fire front (Qingxi and Jiyuan, 1992, Matson *et al.*, 1987) or road construction (Malingreau, 1990). Multi-temporal analysis of AVHRR fire images over several years may also indicate the rotation cycles for shifting cultivation. However, fires often saturate the AVHRR channel 3 response leading to imprecise location of burning activity within the ground pixel and little indication of the impact of fire on the forest environment (Cross, 1990). This may have accounted for the low correlations observed between fire activity recorded by AVHRR channel 3 and MSS estimates of deforested area (Nelson *et al.*, 1987)

Fine resolution imagery can therefore be used to improve the description of fire events and supplement information on the nature and extent of clearance. Pereira and Setzer (1993b) recommended the use of a combination of supervised single cell and maximum likelihood algorithms applied to Landsat TM channels 3, 4 and 5 to distinguish fire scars from secondary forest, closed forest and degraded pastures. The age of fire scars was determined using channel 5 and also a time-series of Landsat data while channel 7 pixels, which were saturated or approaching saturation, were used to detect active fires and fire fronts.

As fires identified by the AVHRR sensor are major indicators of forest disturbance a monitoring system should be capable of guiding more detailed observations of affected areas using fine resolution satellite data. However, in many tropical regions, burning activity is widespread. In Amazonia alone, 350 000 fires were recorded in 1987 (INPE, 1990). The intensity of fire activity should therefore be defined as a criteria to direct further sampling by fine resolution sensors (Nelson *et al.*, 1987).

4.3   CHANGES IN VEGETATION PHENOLOGY AND THE STRUCTURAL COMPOSITION OF
       FORESTS

TFMS should distinguish changes in the phenology of vegetation canopies induced by drought or climatic variation from structural alteration of the forest as a consequence of shifting cultivation, wood collection, or selective logging. Both phenomena are reflected in the AVHRR NDVI but the scale and magnitude of change is markedly different.

Broad continental scale changes can be monitored using a time series of MVC (Maximum Value Composite) GVI and GAC NDVI data. For example, changes in the phenology of tropical forests in northern Thailand and East Kalimantan, which occurred in response to droughts in the early 1980s, were reflected in the GAC NDVI (Malingreau, 1986, Malingreau *et al.*, 1989). However, regional or local scale alterations in the structural composition of tropical forests can often only be assessed by comparing AVHRR NDVI values at full spatial resolution from different observation periods

TFMS therefore require identification of anomalies in the interannual or seasonal AVHRR NDVI at a range of spatial scales with further investigation using fine resolution data to identify the cause and extent of change. A stratification of the sampling procedure for multi-resolution analysis is necessary to optimise this integration. GVI or GAC NDVI data can be analysed routinely to monitor continental changes in vegetation phenology. Where anomalies are observed, full spatial resolution AVHRR data can be used to focus on specific areas at risk. These sites could include the disturbance hotspots identified by Myers (1992) which are likely to undergo greatest change as a result of widespread drought or variations in climate conditions. Fine resolution imagery can then be used to further investigate areas where specific events (e.g. fires) or additional anomalies are identified within full spatial resolution AVHRR NDVI imagery. For determining structural alteration of the forest, GAC NDVI or full spatial resolution AVHRR NDVI can be monitored to identify trends or more dramatic changes in the forest canopy. Fine resolution imagery should then be used to determine the cause and magnitude of change.

4.4   CHANGES IN FORESTED AREA

Loss of forest cover can be determined through quantification or spatial representation of differences in the spatial extent of forest between observation periods. As AVHRR imagery is often classified into forest and non-forest for monitoring purposes, changes in forest extent will only be recorded when the level of clearance between image dates is of sufficient magnitude for pixels to be reallocated within the classified image. Identification of change therefore depends upon the ability to define forest/non-forest boundaries using AVHRR data and also upon the rate and extent of forest conversion.

In areas where extensive forest loss occurs and the movement of the deforestation front is rapid, AVHRR imagery may be adequate for detecting change although mensuration of the boundary change is frequently inaccurate (Cross, 1991). Consequently, finer resolution imagery is often necessary to detect slow and localised depletion of forest cover and to define more precisely the boundary change (Townshend and Justice, 1988). The capability of fine resolution imagery to detect change is also improved due to the higher accuracies of image-to-image registration compared to the AVHRR.

5.   Integration of AVHRR with fine resolution imagery - the "bottom-up" approach.

5.1   INTRODUCTION

Techniques for extracting more detailed information from AVHRR data have received considerable attention in recent years. This section reviews several of the procedures which have been developed in order to estimate the proportions of land cover within each AVHRR pixel thereby improving the use of this sensor for monitoring tropical forests in the future. The rôle of fine resolution imagery in supporting the development of these techniques is also reviewed.

Classification or density slicing of AVHRR data often fails to define accurately the extent of forest cover due mainly to the assumption that each ground pixel (i.e., the area of ground that is viewed by an image pixel) consists of one land cover type (e.g., forest or non-forest) which exhibits a distinctive spectral signature, and that any one pixel is constrained to belong to only one cover type (Fisher and Pathirana, 1990). Consequently, a single image is produced which allocates only one of a range of thematic classes to each pixel (Settle and Drake, 1993).

In reality, the ground pixel will often contain a mixture of cover types, subpixel features (e.g. forest fragments within cleared areas or small clearances in forest expanses) and boundaries such as the forest/non-forest interface (Figure 1; Cross *et al.*, 1991). Estimating the land cover composition of these 'mixed' pixels is one of the challenges which has been confronted using techniques such as regression, linear mixture modelling and fuzzy classification. As yet, most of these techniques have been applied to obtain more accurate differentiation of forest and non-forest classes using the AVHRR. Many of these approaches also employ fine resolution data for their development and validation.

5.2   REGRESSION TECHNIQUES

Regression techniques such as that described by Iverson *et al.* (1989) determine empirical relationships between the percentage forest cover (independent variable) derived from fine resolution imagery and AVHRR digital values (dependent variables). Iverson *et al.* (1989) obtained percentage forest cover within individual AVHRR pixels through comparison with a co-registered binary map of forest and non-forest generated using an unsupervised classification of Landsat TM imagery for an area in Illinois, USA. The distribution of five percentage cover classes was ultimately mapped for the entire region using a linear combination of AVHRR bands 1 and 2 with an adjusted r^2 value which best predicted the proportion of forest cover. The final AVHRR map showed 24.1% forest cover compared with 23.2% established using forest inventory data.

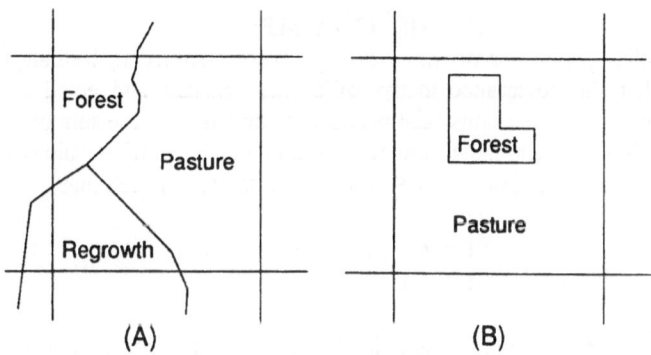

Figure 1. Examples of the mixed pixel problem: a) a boundary pixel; b) a subpixel object (after Fisher and Pathirana, 1990).

5.3   LINEAR MIXTURE MODELS

Linear mixture modelling represents an alternative approach to estimating proportions. This technique assumes that, due to minimal multiple scattering between different cover types, each photon that reaches the sensor has only interacted with one cover component on its path between the Sun and sensor. Consequently, the reflectance values of each pixel are assumed to be equal to a linear mix of component reflectances (Settle and Drake, 1993, Cross et al., 1991, Bierwirth, 1990).

As most ground pixels consist of a number of different cover types, mixture models represent an advancement on standard classification procedures as several images, each indicating the proportion of different land cover types (e.g., percentage forest cover) within the scene, are generated (Cross et al., 1991, Settle and Drake, 1993).

The linear mixture model assumes that the spectral signature x for a pixel is given by

$$\mathbf{x} = \mathbf{Mf} + \mathbf{e} \tag{1}$$
$$= \mathbf{e} + f_1\mathbf{u}_1 + f_2\mathbf{u}_2 \dots f_c\mathbf{u}_c$$

where, for n spectral bands and c components, $\mathbf{M}$ is an $n \times c$ matrix (where the columns are end-member spectra), $\mathbf{f}$ is a $c \times 1$ vector of unknown fractions, and $\mathbf{e}$ represents a zero noise term (Cross et al., 1991, Quarmby et al., 1992). The end-member spectra $(\mathbf{u})$, or columns of $\mathbf{M}$, represent the pure spectral signatures of the component cover types (end-members). In the multi-spectral feature space, a pixel with a mixture of two components (e.g., forest and non-forest) will occupy a line joining the two end-member spectra while a pixel with three components (e.g., forest, non-forest, shade) will occupy points within a triangle defined by the three end-member spectra (Settle and Drake, 1993).

The unmixing process estimates the proportions vector $\mathbf{f}$, given an observation $\mathbf{x}$, by minimising the sum of squares of the errors (Adams et al., 1985, Holben and Shimabukuro, 1993). Cross et al. (1991) minimised the quadratic

$$(\mathbf{X} - \mathbf{Mf})^{\mathrm{T}} \mathbf{C}^{-1} (\mathbf{X} - \mathbf{MF}) \tag{2}$$

as the statistical properties of the error term were assumed to be independent of the mixture and that the covariance matrix of $\mathbf{e}$ was constant and equal to the constant variance matrix $\mathbf{C}$. A linear constraint needs to be added since the sum of the proportions for any resolution element must sum to one and the proportion values must be non-negative (Holben and Shimabukuro, 1993, Cross et al., 1991) such that:

$$f_1 + f_2 + \dots f_c = 1 \tag{3}$$
$$f_i \geq 0, i = 1, \dots c.$$

Although linear mixture modelling fails to provide information on the exact location of forest/non-forest boundaries and sub-pixel forest, the accuracy of forest cover estimation is often improved. Colour composite or classified fine resolution imagery can be used as a reference with which to compare the results of mixture modelling. For example, Cross et *al.* (1991) compared binary forest/non-forest images generated using a maximum likelihood classification of Landsat TM images for Ghana and Rondonia, Brazil, to the result of a mixture model applied to AVHRR imagery of the same areas. Considerable improvements in the classification of forests in Ghana were reported although in Brazil, the model failed to improve estimates of forest cover due mainly to fires which saturated AVHRR channel 3 pixels.

Fine resolution imagery are also used for determining the location of 'pure' end-member pixels for each land cover component (e.g., forest or non-forest) within the AVHRR image (Foody and Cox, 1992, Settle and Drake, 1993). Alternatively, the end-member spectra can be determined from a scattergram of the first two principal components as the end-member signatures occupy the extremes of the distribution of pixel signatures in the principal component feature space. The corresponding digital values in the feature space of the original image can then be extracted as the end-member signatures (Cross *et al.*, 1991).

For the discrimination of forest and non-forest and the estimation of forest proportions, Cross *et al.* (1991) indicated that AVHRR channels 1, 2, 3 and 4 may be used in the model although the use of channels 3 and 4 assumed that each cover type was thermally distinct. However, the radiometric response of the AVHRR sensor may not be represented accurately by a linear mixture model leading Holben and Shimabukuro (1993) to consider only the reflective component of AVHRR channel 3 in combination with AVHRR channels 1 and 2. Ratios such as the NDVI are unable to be used for mixture modelling.

In general, mixture models applied to AVHRR data enhance the classification of forest/non-forest where discrete areas of forest are cleared (e.g., cattle ranches) but the models are less successful where a complex type of forest clearance exists (e.g., shifting cultivation) and in areas where only slight variations in the spectral response of forest canopies indicate forest disturbance. The models also perform better where a non-forest category maintains a similar spectral response. Mixture modelling improves the extraction of information on forest cover from AVHRR data and with further development this technique is likely to become more important to TFMS in the future.

5.4   FUZZY CLASSIFICATION ALGORITHMS

Fuzzy classification algorithms have only recently been considered in the context of tropical forest monitoring using the AVHRR. These algorithms generates fuzzy membership values which indicate the strength of membership of each pixel within an image to land cover classes. Values range from 0 (weak membership) to 1 (strong membership), and sum to 1 (Bezdek *et al.*, 1984). Pixels where no cover type approaches 1 can be assumed to represent a mixed pixel with the proportions of each cover type indicated by the fuzzy membership values.

The fuzzy classification may be unsupervised (e.g. the Fuzzy C-Means or FCM; Bezdek *et al.*, 1984) or supervised (Wang, 1990). In an unsupervised FCM classification, the initial allocation of *n* pixels to *c* fuzzy groups is random. The mean digital value, v_i, of each fuzzy group is calculated and a measure of dissimilarity, d_{ik}, is determined for the ith class and the kth attribute on the basis of the distance between a pixel observation x_k (which represents a vector the length of which is the number of image bands or attributes p used) and the class mean such that

$$(d_{ik})^2 = (x_k - v_i)^T A(x_k - v_i) \qquad (4)$$

where A is a weight matrix which determines the norm to be used (e.g. the Mahalanobis norm).

The FCM algorithm then iteratively relocates pixels to other classes so as to minimise the generalised least squared errors functional J_m where

$$J_m(U, V) = \sum_{k=1}^{n} \sum_{i=1}^{c} (u_{ik})^m (d_{ik})^2 \qquad (5)$$

and U is a fuzzy c-partition of n pixels to c fuzzy groups, V is a matrix of size c x p whose elements, v_{ik}, represent the mean of the kth attribute in the ith class, m is a weighting component ($1 < m < \infty$) and u_{ik} represents the fuzzy membership value of x_k to the ith class (Foody and Cox, 1992).

The fuzzy classification algorithm ultimately generates a fuzzy partition of the spectral space where membership grades indicate the level of association of each pixel with different cover types (Figure 2). This is in contrast to hard classifications which produce a strict partition of spectral space such that each pixel is assigned to only one class (Figure 3).

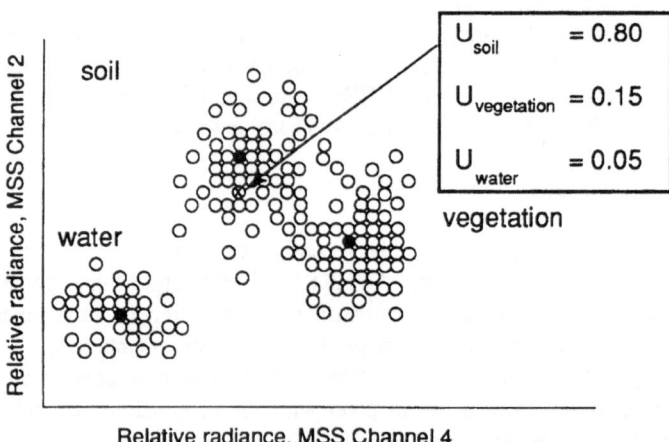

Figure 2. Membership grades of a pixel in a fuzzy partition of spectral space (after Wang, 1990).

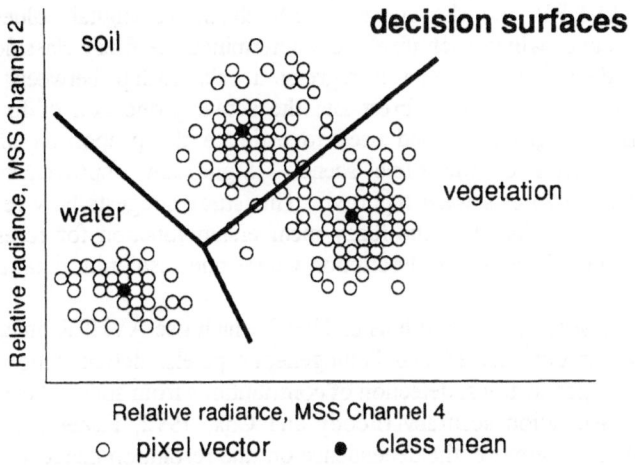

Figure 3. A hard partition of spectral space and decision surfaces (after Wang, 1990)

Fine resolution imagery can be integrated to determine the number of clusters required for an unsupervised fuzzy classification of multi-spectral AVHRR data. For example, comparison with co-registered Landsat TM scenes for a tropical area may indicate five dominant land cover types (e.g. closed tropical forest, secondary forest, croplands, bare soil and water) within the area of interest thereby suggesting that at least five clusters should be requested from the algorithm.

Supervised fuzzy classification algorithms, such as that described by Wang (1990), may however be preferred for tropical forest monitoring as the proportion of specific classes within AVHRR pixels (e.g., forest, non-forest and regrowth stage) can be determined using either training data selected by comparing co-registered fine resolution imagery or proportion images generated from a preliminary unsupervised fuzzy classification. The membership allocations for AVHRR pixels can also be validated through comparison with fine resolution data of the same area.

Several images indicating the fraction of cover types within the scene are generated from the fuzzy classification. Fraction images derived from a fuzzy classification of AVHRR data may be used to improve the definition of forest and non-forest and other thematic classes, the detection of small scale disturbance such as road construction, and the description of burning activity. Multi-temporal comparison of fraction images may also indicate subpixel changes in forest extent.

Few studies, however, have examined the potential of the fuzzy classification algorithm to improve tropical forest cover assessment. One study by Foody and Cox (1992) classified a Landsat MSS channel 3 image extract of Ghana into forest and non-forest categories. The classified and original image were then degraded to a spatial resolution equivalent to the AVHRR and the average MSS channel 3 digital values and the proportion of forest cover within each pixel were determined. A fuzzy classification was performed on the degraded data and a regression relationship between the fuzzy membership function associated with forest and the actual proportion of forest within selected pixels was then generated and used to estimate the proportion of forest in remaining degraded pixels of the same scene. Significant improvements in the classification accuracy of forests were reported. This study suggests how relationships between actual AVHRR values and the fuzzy membership function for forests can be established for improved forest cover description where fine resolution data are used to provide baseline information.

The use of fuzzy classification algorithms in TFMS which use AVHRR imagery would allow differentiation of both mixed and homogeneous pixels, determination of mixed cover classes and their proportions, detection of contributions from subpixel elements, and improvements in classification accuracy (Foody and Cox, 1992, Fisher and Pathirana, 1990, Wang, 1990). Furthermore, the dependence on fine resolution imagery for detailed assessment of disturbance processes would be reduced.

6.   The use of thematic information for improving the description of forest cover using AVHRR

Fine resolution sensors are able to provide thematic information on the spatial distribution of land covers, such as regenerating forest and agriculture, which can be used as reference data to improve both the description of forest types and the nature of conversion using the AVHRR and hence the monitoring capabilities of this sensor.

For example, Lucas *et al.* (1994) outlined a procedure for describing regrowth forest in an area north of Manaus in the Brazilian Amazon using a time series of Landsat MSS and TM data. Landsat TM images from 1985, 1988, 1989 and 1991 were classified separately into closed forest, agricultural land and secondary forest classes using a supervised minimum distance algorithm. The age of secondary forest was deduced by comparing the occurrence of this class within the classified images from all years. A summary image was then produced which defined the spatial distribution of secondary forest that was either <2 years, 2 - 3 years, 3 - 6 years or 6 - 14 years of age in 1991. The maximum possible age of regrowth on cleared land in 1991 was at most 14 years as the oldest clearance supported no regeneration in 1977. Subsequent comparison with co-registered AVHRR images of the same area has suggested that the AVHRR NDVI may be used to determine different growth stages of secondary forest and that the age of regrowth may also be determined by comparing NDVI values from different years.

7.   Conclusions

The AVHRR represents an important sensor for TFMS due to daily coverage of the tropical regions in both spectral and thermal wavebands. The AVHRR is capable of providing up-to-date information which identifies active deforestation fronts, anomalies in the phenology of vegetation over time, and large scale changes in forest extent. However, when used independently of other satellite sensors, the utility of the AVHRR for forest monitoring is restricted. For this reason, the integration of AVHRR with finer resolution data has been encouraged.

Fine resolution imagery, although more limited in terms of spatial and temporal coverage, allows more precise estimation of forest extent and location of forest/non-forest boundaries, assessment of the nature and extent of forest conversion, and description of forest and other land cover types. Furthermore, the extended archives of data allow an improved description of long term processes associated with tropical forest conversion. Reference data for the development, validation and improvement of techniques for assessment of forest extent and disturbance using AVHRR data are also provided by fine resolution sensors.

Fine resolution data also plays an important rôle in development of advanced procedures for estimating proportions within AVHRR pixel, such as mixture modelling and fuzzy classifications, which may improve the use of AVHRR and ultimately reduce dependence on fine resolution data for monitoring tropical forests in the future.

8. Acknowledgements.

Professor Paul Curran, Dr. Giles Foody and Dr. Gintautas Palubinskas are thanked for their helpful comments on the manuscript.

9. Bibliography

Adams, J.B., Smith, M.O., and Johnson, P.E., 1985, Spectral mixture modelling: A new analysis of rock and soil types at the Viking Lander I site. *Journal of Geophysical Research*, **91**, 8098-8112.

Bezdek, J.C., Ehrlich, R., and Full, W., 1984, FCM: The fuzzy c-means clustering algorithm. *Computers and Geosciences*, **10**, 191-203.

Bierwirth, P.N., 1990, Mineral mapping and vegetation removal via data-calibrated pixel unmixing using multi-spectral images. *International Journal of Remote Sensing*, **11**, 1999-2017.

Cross, A., 1990, Tropical deforestation and remote sensing: The use of NOAA AVHRR data over Amazonia. *Final Report to the Commission of the European Communities*. Article B946/88, Natural Environment Research Council (NERC), Reading, 60p.

Cross, A., 1991, Tropical forest monitoring using AVHRR: Towards an automated system for change detection. *Final Report to the United Nations Environment Programme (UNEP) Global Resources Information Database (GRID)*, Natural Environment Research Council (NERC), Reading, 51p.

Cross, A.M., Settle, J.J., Drake, N.A., and Paivinen, R.T.M., 1991, Subpixel measurement of tropical forest cover using AVHRR data. *International Journal of Remote Sensing*, **12**, 1119-1130.

Fearnside, P.M., 1991, Deforestation in Brazilian Amazon. *In* G.M. Woodwell (ed.) *The Earth in transition: patterns and processes of biotic impoverishment*, Cambridge University Press, Cambridge, 211-238.

Fisher, P.F. and Pathirana, S., 1990, The evaluation of fuzzy membership of land cover classes in the suburban zone. *Remote Sensing of Environment*, **34**, 121-132.

Foody, G.M. and Cox, D.P., 1991, Estimation of sub-pixel land cover composition from spectral mixture models. *Spatial Data 2000*, Remote Sensing Society, University of Nottingham, 186-187.

Gilruth, P.T., Hutchinson, C.F., and Barry, B., 1990, Assessing deforestation in the Guinea Highlands of West Africa using remote sensing. *Photogrammetric Engineering and Remote Sensing*, **56**, 1375-1382.

Green, G.M. and Sussman, R.W., 1990, Deforestation history of the eastern rainforests of Madagascar from satellite images. *Science*, **248**, 212-215.

Hastings, D.A. and Emery, W.J., 1992, The Advanced Very High Resolution Radiometer (AVHRR): A brief reference guide. *Photogrammetric Engineering and Remote Sensing*, **58**, 1183-1188.

Holben, B.N. and Shimabukuro, Y.E., 1993, Linear mixing model applied to coarse spatial resolution data from multi-spectral satellite sensors. *International Journal of Remote Sensing*, **14**, 2231-2240.

INPE, 1990, INPE Space News, **1**, 8 p.

Iverson, L.R., Cook, E.A., and Graham, R. L., 1990, A technique for extrapolating and validating forest cover across large regions. Calibrating AVHRR data with TM data. *International Journal of Remote Sensing*, **10**, 1805-1812.

Justice, C.O., Townshend, J.R.G., Holben, B.N., and Tucker, C.J., 1985, Analysis of the phenology of global vegetation using meteorological satellite data. *International Journal of Remote Sensing*, **6**, 1271-1318.

Kaufman, Y.J., Tucker, C.J., and Fung, I., 1990, Remote sensing of biomass burning associated with deforestation. *In* J.A. Smith (ed.) *Remote Sensing of the Biosphere*, Florida, 2-13.

Lucas, R.M., Honzak, M., Curran, P.J., Foody, G.M., and Corves, C., 1994, Characterising tropical secondary forests using multi-temporal Landsat sensor imagery. *International Journal of Remote Sensing*, in press.

Malingreau, J.P., 1986, Global vegetation dynamics: Satellite observations over Asia. *International Journal of Remote Sensing*, **7**, 1121-1146.

Malingreau, J.P., 1990, A tropical deforestation alarm system. *Draft report*, Joint Research Centre, Ispra, 12 p.

Malingreau, J.P., 1991, Remote sensing for tropical forest monitoring: An overview. *In* A.S. Belward and C.R. Valenzuela (eds) *Remote Sensing and Geographical Information Systems for Resource Management in Developing Countries*, Kluwer Academic, Dordrecht, 253-278.

Malingreau, J.P. and Laporte, N., 1988, Global monitoring of tropical deforestation. AVHRR observations over the Amazon Basin and West Africa. *Proceedings, Forest Signatures Workshop*, Ispra, 32 p.

Malingreau, J.P. and Tucker, C.J., 1988, Large-scale deforestation in the southeastern Amazon Basin of Brazil. *Ambio*, 17, 49-55.

Malingreau, J.P. and Tucker, C.J., 1990, Ranching in the Amazon Basin: Large scale changes observed by AVHRR. *International Journal of Remote Sensing*, 11, 187-189.

Malingreau, J.P. and Belward, A.S., 1992, Scale considerations in vegetation monitoring using AVHRR data. *International Journal of Remote Sensing*, 13, 2289-2307.

Malingreau, J.P., Tucker, C.J., and Laporte, N., 1989, AVHRR for monitoring global tropical deforestation. *International Journal of Remote Sensing*, 10, 855-867.

Miller, L.D. and Williams, D.L., 1978, Monitoring forest canopy alternation around the world with digital analysis of Landsat imagery. *Proceedings, International Symposium on Remote Sensing for Observation and Inventory of Earth Resources and the Endangered Environment*, 3, 1721-1761.

Myers, N., 1992, Future operational monitoring of tropical forests: An alert strategy. *Report for the Joint Research Centre, Commission of the European Community*, Joint Research Centre, Ispra, 81 p.

Nelson, R. and Holben, B., 1986, Identifying deforestation in Brazil using multi-resolution satellite data. *International Journal of Remote Sensing*, 7, 429-448.

Nelson, R., and Horning, N., 1993, AVHRR-LAC estimates of forest area in Madagascar. *International Journal of Remote Sensing*, 14, 1463-1475.

Nelson, R., Horning, N., and Stone, T.A., 1987, Determining the rate of forest conversion in Mato Grosso, Brazil, using Landsat MSS and AVHRR data. *International Journal of Remote Sensing*, 8, 1767-1784.

National Remote Sensing Centre, 1992, A guide to Earth observing satellites, 79 p.

Pereira, M.C. and Setzer, A.W., 1993a, Spectral characteristics of deforestation fires in NOAA/AVHRR images. *International Journal of Remote Sensing*, 14, 583-597.Pereira, M.C. and Setzer, A.W., 1993b, Spectral characteristics of fire scars in Landsat-5 TM images of Amazonia. *International Journal of Remote Sensing*, 14, 2061-2078.

Qingxi, T. and Jiyuan, L., 1992, Monitoring forest cover change using remote sensing techniques in China. *Proceedings, International Space Year (ISY) World Forest Watch (WFW) Conference*, Sao Jose dos Campos, Brazil, 71.

Quarmby, N.A., Townshend, J.R.G., Settle, J.J., White, K.H., Milnes, M., Hindle, T.L., and Silleos, N., 1992, Linear mixture modelling applied to AVHRR data for crop area estimation. *International Journal of Remote Sensing*, 13, 415-425.

Sader, S.A. and Joyce, A.T., 1988, Deforestation rates and trends in Costa Rica. *Biotropica*, 20, 11-19.

Settle, J.J., and Drake, N.A., 1993, Linear mixing and the estimation of ground cover proportions. *International Journal of Remote Sensing*, 14, 1159-1177.

Skole, D., and Tucker, C., 1993, Tropical Deforestation and Habitat Fragmentation in the Amazon: Satellite Data from 1978 to 1988. *Science*, 260, 1905-1909.

Stone, T.A., Brown, I.F., and Woodwell, G.M., 1991, Estimation by remote sensing of deforestation in central Rondonia, Brazil. *Forest Ecology and Management*, **38**, 291-304.

Townshend, J.R.G. and Justice, C.O., 1988, Selecting the spatial resolution of satellite sensors required for global monitoring of land transformations. *International Journal of Remote Sensing*, **9**, 187-236.

Wang, F., 1990, Fuzzy supervised classification of remote sensing images. *IEEE Transactions on Geoscience and Remote Sensing*, **28**, 194-201.

Woodwell, G.M., Houghton, R.A., and Stone, T.A., 1986, Deforestation in the Brazilian Amazon Basin measured by satellite imagery. *In* G. T. Prance (ed) *Tropical rainforests and the world atmosphere*, American Association for the Advancement of Science, Selected Symposium, **101**, 23-32.

Woodwell, G.M., Houghton, R.A., Stone, T.A., Nelson, R.F., and Kovalick, W., 1987, Deforestation in the Tropics: New Measurements in the Amazon Basin using Landsat and NOAA Advanced Very High Resolution Radiometer Imagery. *Journal of Geophysical Research*, **92**, 2157-2163.

AVHRR DATA ACQUISITION, PROCESSING AND DISTRIBUTION AT THE EUROPEAN SPACE AGENCY (ESA)

O. ARINO
ESA/ESRIN,
C.P. 64, via Galileo Galilei,
00044 Frascati, Italy.

ABSTRACT. This paper describes the operational environment from which the European Space Agency (ESA) delivers AVHRR data from the NOAA satellites to the user community. It concentrates on that part of the community concerned with land applications, for which the Earthnet Coordinated Tiros Network (ECTN) (incorporating operational receiving stations in Europe, Africa and South-East Asia) was specifically established in 1986. The network delivers information which is then converted by the ESA in-house processing system to a number of output products in a common format. The data archive, catalogue and set of image Quick Look products is located and maintained at ESRIN, in Frascati. The processing, format and archiving of these products, as well as access to them, are all described in this chapter.

The development of higher level and Fast Delivery Products, as well as more refined tools to browse and/or place an order electronically for all archived data and products are in progress. ESA/ESRIN is also participating in the international "Global 1 km Land Cover AVHRR Data Set" project in collaboration with NOAA, NASA, USGS and CSIRO. This project, requested by the scientific community including the International Biosphere Geosphere Program (IGBP) and the Commission of the European Union (CEU) as well as the Moderate Resolution Imaging Spectrometer (MODIS) team amongst others, is directly related to the studies of the Global Change affecting our planet. The project, incorporating as it does the co-ordinated collection and processing of large volumes of data from diverse data sources, provides ESA with a good opportunity to install procedures and systems to cope with handling large amounts of data in an operational context. Lessons learnt from the participation in this project will serve to support the development of a primed ground segment for future missions and, in particular, for the Medium Resolution Imaging Spectrometer (MERIS) instrument to be flown on the future satellite system ENVISAT.

1. Introduction

This chapter provides an overview of ESA's activities related to the processing, archiving and distribution of AVHRR data, with specific emphasis on those capabilities in relation to land surface applications. It includes discussion on: the data acquisition performed by ESA's co-ordinated network of receiving stations; the processing systems of ESRIN, including its hardware and software facilities, and data format and transmission considerations; and the ESA products (including the Quick Looks) which are made available to users in an operational manner.

With regard to data provision, ESA's user service with its central catalogue and acquisition and sales activities will also be described briefly. The "Ionia" concept (Quick

395

G. D' Souza et al. (eds.), Advances in the Use of NOAA AVHRR Data for Land Applications, 395–432.
© 1996 *ECSC, EEC, EAEC, Brussels and Luxembourg.*

Looks and ordering facilities available on CD-Browser and Internet) will also be described and some examples provided. The special procedures put in place at ESRIN to deal with data for the "Global 1 km Land Cover" project are also highlighted. With regard to processing activities, the current development at ESRIN to prototype the processing of higher level products (cloud classification and detection, atmospheric correction, land surface temperature, remapping and compositing) are detailed below. The collaboration with the CEU-Joint Research Centre to establish a real-time AVHRR channel 1 and channel 2 onboard calibration by repeated viewing of a desert site is also described. Finally, details about other developments of interest to the reader, for example the integration of the AVHRR processing in a common processing system with the SeaWiFS instrument, and the development of the Fast Delivery Product line for land applications (Pitella 1993), are also provided.

2. The Acquisition Network

As described in the chapter by Kidwell, AVHRR data acquired by the sensors onboard the NOAA satellites are either down-linked in real-time on L-band transmission, or recorded on-board and downloaded to one of the NOAA receiving stations in Wallops Island, Virginia or Fairbanks, Alaska. Our concern within the ECTN is the acquisition and processing of the down-linked near real-time full resolution data referred to as High-Resolution Picture Transmission (HRPT).

Several receiving stations exist inside and outside Europe operated by national entities carrying out local observation programmes (HRPT, 1991). With the purpose of co-ordinating the European activities for NOAA/AVHRR and in order to provide the European users with uniform access to wide-coverage data through a common archive and a central catalogue, ESRIN, in 1986, established a network of ground facilities called the ESRIN Coordinated Tiros Network (ECTN) composed of a number of existing acquisition stations and a central facility at Frascati (Fusco and Muirhead, 1987, Fusco 1989, Fusco *et al.* 1989).

The major tasks of the ECTN are:

- the collection of data at a number of European and non-European receiving stations;
- the generation of standard compatible products;
- the creation and the maintenance of a central archive and a catalogue at ESRIN;
- the distribution of products to users; and
- the development of algorithms and processing software to serve the user community needs and developments;

All contributing stations have been equipped with standardised hardware and software systems, which are described in the next section. Data have been collected, processed and archived within the ECTN since early 1986 and currently, the ECTN receives and archives

AVHRR data from 10 stations covering Europe and Africa as well as South-East Asia. The stations participating with (or having an agreement to contribute to) the ECTN central archive in Frascati are the following: Tromsö in Norway, Oberpfaffenhoffen in Germany, Scanzano in Italy, Cairo in Egypt, Maspalomas in the Canary Islands, Niamey in Niger, Nairobi in Kenya, La Réunion in the La Réunion islands, Manila in the Philippines and Terranova in the Antarctic. The station coverage is shown in Table 1 which also lists the start time of the operation at the different stations and the presence or otherwise of a local archive.

For most of the stations, the processing is performed at the receiving station itself using the ESA-developed "Station HRPT Archiving and Reprocessing Kernel" (SHARK) software, (Pittella and Bamford, 1989). Other stations may send data directly to ESRIN in the raw HRPT format which is then converted to Standard Family HRPT Archive Request Product (SHARP) format and archived at ESRIN. Since April 1992, HRPT and LAC data have been exchanged with NASA/USGS to ensure that the full archive of the "Global 1 km Land Cover AVHRR Data Set Project" is available both at ESRIN and the Eros Data Center (EDC) (10,000 passes per project year, Buongiorno et al. 1993). As most of this data set was collated in raw HRPT format, some of the ECTN station were requested to send to ESRIN HRPT data in addition to the normal supply of standard SHARP products.

Table 1. ECTN participating stations

STATION	LATITUDE	LONGITUDE	START OF OPERATION	SHARP LOCAL ARCHIVE
Cairo	30.01	31.24	Sep. 1992	NO
Dundee	56.46	-2.98	Jun. 1991	YES (historical archive from '78)
La Réunion	-20.87	55.48	Aug. 1990	NO
Manila	14.39	121.04	Aug. 1993	NO
Maspalomas	27.77	-15.63	Mid. 1986	YES
Nairobi	-1.25	36.75	Aug. 1991	NO
Niamey	13.53	2.08	Mid. 1990	NO
Oberpfaffenhofen	48.05	11.16	Jan. 1990	YES
Scanzano	37.90	13.35	Jan. 1993	YES
Terranova	-74.41	164.07	Seasonal activity	NO
Tromsö	69.66	18.94	Jan. 1989	YES

3.   The Processing System

3.1.   OVERVIEW

SHARK is a hardware and software system that has the capability of extracting AVHRR imagery from the raw HRPT data stream, to process it, and to write the processed data to storage media in SHARP format. The SHARP format is an archive and distribution format for AVHRR and TOVS data which conforms to the Landsat Technical Working Group's "Standard Family Tape Format".

The SHARK system has been designed to fulfil the ECTN requirements for a compact and easy to operate system providing the following functions:

- HRPT data acquisition;
- Display of acquired data and selection of scenes to process;
- Generation of SHARP products and digital Quick Looks (Q/L);
- Generation of entries in the catalogue of all archived data;
- On-line quality control;
- Archiving of products on optical disk;
- Archive management and CCT/exabyte user product generation;
- Hard copy (laser) Q/L generation and distribution via *fax*.

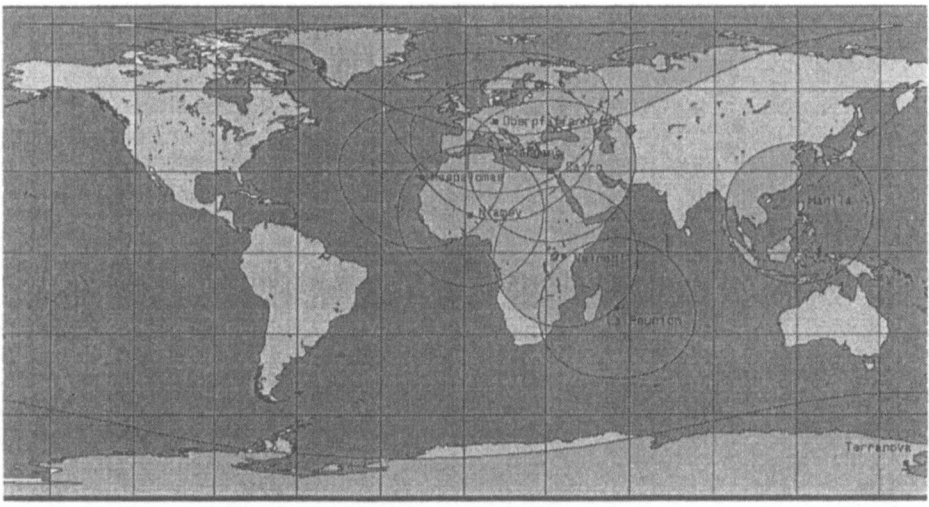

Figure 1. Location and coverage of participating ECTN receiving stations

3.2. THE EXCHANGE OF FILE INFORMATION BETWEEN ESRIN AND OTHER ECTN STATIONS

During data processing and archiving several files containing various pieces of information are created or accessed at the different stages of the processing. SHARK provides interactive/automatic tools in order to help the operator handle those files and related directories. Of particular importance to the correct processing are the satellite ephemeris data (TBUS). These are generally located in a directory of files updated regularly and automatically across networks. A set of tools is available to allow listing, checking and editing of the TBUS data within any directory. A certain number of functions, both automatic and interactive, are also available for TBUS and other data management which allow:

- satellite ephemeris collection (from NOAA network bulletins);
- satellite ephemeris transmission (ESRIN to ECTN receiving station);
- catalogue entries transmission (ECTN receiving station to ESRIN);
- browse/Quick Look products transmission (ECTN receiving station to ESRIN);
- reports transmission (ECTN receiving station to ESRIN);
- Fast delivery product (FDP) data transmission from ECTN receiving station to end user *via* networks.

3.3. SHARK HARDWARE AND SOFTWARE ENVIRONMENT

SHARK is a Unix-based system using a SUN workstation and a number of standard peripherals (Fusco and Pitella 1991). It includes a magnetic disk with a minimum capacity of 500 Mbytes and, if it is used to archive SHARP products, at least one 12" optical disk drive. Other peripherals recognised by the system include 1/2" and exabyte tape drives, PostScript laser printers and at least one type of network connection.

The SHARK software is installed in a tree-structured hierarchy of UNIX directories which contain data files, UNIX scripts, executable and source code written in Fortran and C. The entire directory structure occupies 80 Mbytes. Two areas of the disk are reserved by SHARK to ensure that there is sufficient working space available. One area is designated to hold raw HRPT data and a second one is used for SHARP products generated during a SHARK session. HRPT data from a single pass occupies about 115 Mbytes. Data from each HRPT pass, with their associated SHARK header files, are then stored in a reserved area on disk known as "passage". The passages are initially set up by the system administrator. Users can then write HRPT passages to disk and delete data from them, but creation of new ones or deletion of existing ones are functions restricted to the administrator.

The total number of passages that can be held on a system simultaneously depends on the size of the disk. If the disk is large enough to support more than nine then they are grouped into passage sets, each set containing nine passages. SHARK can only access one passage set at a time.

SHARP products are stored in reserved areas on disk known as "slots". There are usually at least ten slots available but the exact number varies from system to system depending on how much disk space is available (see SHARK Users' Manual 1993, and SHARK Technical Reference Manual, 1993).

3.4.   SHARK UTILITIES

SHARK contains two main utilities for creating SHARP products from HRPT data sets. Both are interactive and use a graphical user interface (GUI) which runs under Sun View. The first generates level 1 products. In this, the operator has to perform two tasks interactively before the SHARP scenes are created. First, to check the image navigation and, if necessary, correct it by matching the position of the coastline that is visible on the image with that calculated from the scan times and satellite orbit parameters. Second, the operator must define the extent of the scenes to be created, as a single HRPT pass acquisition can be as long as 16 minutes (5760 lines), but each of the SHARP scenes are limited to four minutes duration (1440 lines). The default action is to segment the pass into such scenes starting from the first scan line, but the operator can override this. Generally, operators try to maintain a consistency in the segmentation relevant to the main applications (*i.e.* if land applications were the piority, the segments would be made over the sea wherever possible). The operator can also modify the default action to exclude parts of the image with excessive cloud cover or large numbers of missing or corrupted scan lines of data.

Once the coastline correction has been made and the limits of the four-minute scenes defined, the display program either generates the level 1 products directly or creates requests for the defined products to be generated by a background job. This, once started, runs without operator intervention. SHARK's second interactive utility provides facilities for managing SHARP archives on optical disk or magnetic tape, and for creating level 2 and Quick Look products. A different set of tools is available to handle the Optical Disk (OD) operation (formatting and labelling, mounting/ unmounting, content listing, *etc.*).

3.5.   OPERATING SHARK

3.5.1. *Reading HRPT data.* The data sets processed by SHARK are typically large; a single HRPT pass contains about 120 Mbytes and a single SHARP level 1 product occupies about 34 Mbytes. For convenience and speed, both raw data and SHARP products are kept on reserved areas of the magnetic disk area during processing. HRPT data can be ingested into the SHARK software environment either directly from a ground station's acquisition system or, indirectly, from magnetic tape or over a network connection. Direct ingestion writes HRPT data from the reception equipment directly into a passage on disk. By the time the whole pass is acquired the data are ready for processing by SHARK. Indirect ingestion from magnetic tape or network is slightly more complex. Each ground station writes HRPT data to magnetic tape in a slightly different format, so a separate utility is provided to read each station-specific format, to reformat the HRPT data and to load it into a common format passage.

3.5.2. *Displaying HRPT data.* Once loaded, the data in any of the HRPT passages on magnetic disk can be displayed by running SHARK's "display" program. Each passage contains data from all four or five bands of the AVHRR. Any of these bands can be loaded into either of the display's two viewing panels. Once an image has been loaded its contrast can then be interactively stretched to emphasise particular features. Images have to be sub-sampled to fit them into the viewing panels. By default only every fifth pixel and every eighth scan-line are displayed, but a smaller part of the image can be selected and re-displayed at full resolution, should the user want to examine some of the data more closely.

3.5.3. *Navigating an HRPT passage.* In addition to the satellite image data, SHARP products contain information on the geographical location of the imagery. This additional information has to be calculated from the timing information carried in the HRPT data stream and a knowledge of the satellite's orbit. Up-to-date values of the orbit parameters (which change frequently) are distributed by the NOAA in the "TBUS bulletins" which are archived on SHARK and kept up to date by the system administrator. The process of calculating the position of pixels within an image is known as "image navigation".

To allow an operator to judge the accuracy of the navigation and, if necessary, to correct it, SHARK (under user definition) extracts a small part of the pass, displays it, calculates the geographical location of each pixel, and overlays a vectorised coastline and country border on top of the displayed extract. The quality of the calculated geographical positions can then be judged from the accuracy with which the superimposed coastline fits those parts of the actual coastline which are visible in the image. If the fit is not sufficiently accurate it can be improved by "sliding" the overlaid coastline/country border vector over the image. When the user is satisfied with the fit, the system relates the defined shift to corrections to the calculated satellite Equator crossing time and longitude. These corrections are then stored with the image, and are used later to re-navigate SHARP scenes extracted from the pass with improved accuracy.

3.5.4. *Extracting a SHARP scene.* A full HRPT passage can contain all of the data transmitted by the satellite and recorded at the receiving station during the time that the satellite is above the station's horizon. The longest passages are collected when the satellite passes very nearly directly above the station. On these occasions the satellite is visible for up to 16 minutes if there are no obstructions to the line of sight between the satellite and the receiving antenna. During that time the AVHRR acquires nearly 6000 scan lines of image data. The SHARP format imposes a limit of 1440 scan-lines in a single scene which corresponds to a transmission of four minutes. Thus, SHARK's display tool can be used to define, interactively, up to four separate SHARP scenes to be created from each full HRPT passage. SHARP products contain not only all the image data (all 5 bands) and the raw telemetry, but also geolocation information (latitude and longitude, sun and satellite angles at selected pixels in the scene), coastline, state boundaries and latitude/longitude overlays, calibration data and histograms of the raw pixel values for each band. While the image data in level 1 products is recorded in raw pixel values that in level 2 products is converted to physical quantities such as reflectance and brightness temperature.

A SHARP scene can be of any size from 720 to 1440 lines (two to four minutes). Times are used to describe the size of a scene because each line of raw HRPT data contains the time at which it was acquired. This time stamp is used by SHARK to identify and sequence individual scan lines. An HRPT pass may be either northbound or southbound, but, SHARK displays all images with north toward the top of the screen. Because of this, and because the display program uses the scan time to identify scan-lines, the four minute extracts may be defined from either the top or the bottom of the viewing panel, depending on the direction of the original satellite pass.

3.5.5. *Displaying SHARP scenes and creating SHARP level 2 format data.* This section describes how to use the "Shark" program to display ESA standard products (including SHARP level 1 and level 2, and the Quick Look). Within the Shark program, the user can create the level 2 products, and archive and retrieve SHARP products to and from optical disk. The Shark program is an interactive tool, run from a GUI, which can be used for a wide variety of tasks. Its primary purpose is to create SHARP level 2 products from SHARP level 1 products. It can also create "Quick Looks", which are classified, single-band, reduced-resolution versions of SHARP scenes suitable for inclusion in browse-files to indicate the coverage, cloud-cover and data quality of archived products. These Quick Looks have also been designed so that they can be printed in black and white and faxed without significant loss of information.

As well as generating level 2 products the Shark program provides all the facilities needed to manage archives of SHARP products. SHARK systems normally use optical disks to archive SHARP level 1 products and the Shark program includes functions to format new disks, and to copy and retrieve SHARP level 1 products to and from archive. To make the task of creating level 2 products easier the Shark program is also able to display SHARP products on the workstation screen. The program displays any of a product's five bands and the Quick Look (if it has already been generated). Then extracts of the image data can be displayed at full resolution and an interactive contrast stretch facility is available to emphasise particular features of interest.

3.5.6. *Archiving, retrieving and distributing SHARP products.* All SHARP level 1 products can be archived on optical disk. The operator may archive SHARP products from either the Shark program of the display process, and a product can be archived directly from its slot on magnetic disk to the optical disk. It is not necessary to load and display the product before archiving it.

Apart from creating and archiving SHARP products the SHARK software also makes it very easy to copy products to magnetic tape for distribution to other users. Some of this software has been designed to work without operator intervention, so it is well suited to deal with customer requests for all scenes covering a specified area over several days, producing either Level 1 or Level 2 products according to the users requests.

403

3.5.7. *Creating Quick Look Products.* Quick look images, which are composed of classified images (land/sea/cloud) with various parameters assigned (*e.g.* NDVI for land, surface temperature for sea or cloud, *etc.*) are automatically generated in the course of creating level 1 SHARP products. One can examine a scene's Quick Look on the workstation display, print it in black and white or colour, create a version suitable for transmission by *fax* or geometrically transform it to match a Universal Transverse Mercator map projection. All of these Quick Look functions are provided in the Shark program.

4. The Products

4.1. INTRODUCTION

All SHARP products consist of segments of AVHRR imagery that contain no fewer than 720 scan lines and no more than 1440 scan lines. These size limits correspond to two and four minutes transmission time respectively. In SHARP format, the data are arranged with the northern-most scans first, irrespective of the direction in which the satellite was flying when the image was scanned. This allows easier orientation when viewing.

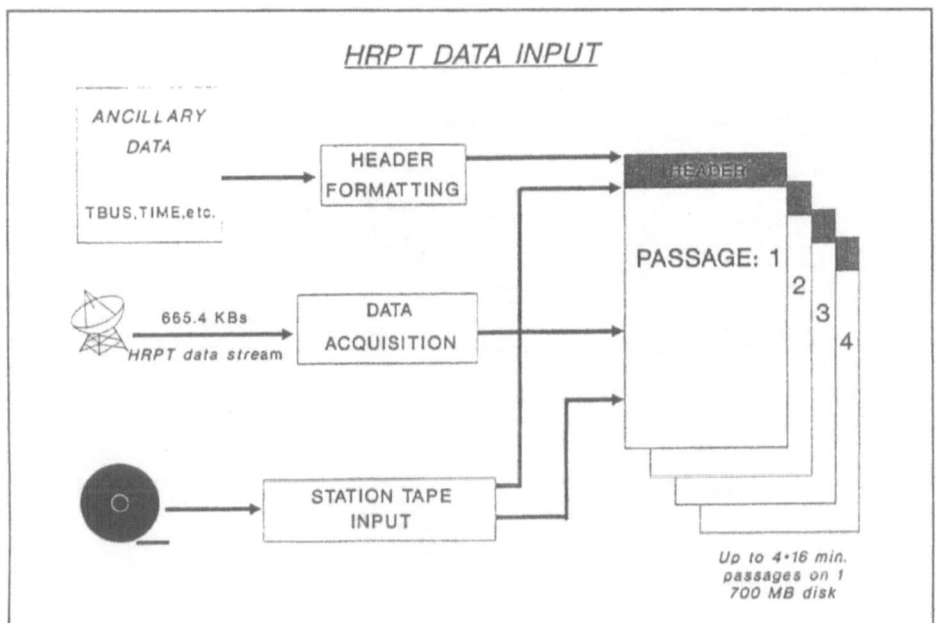

Figure 2. HRPT data input to the SHARP software.

4.2.   SHARP LEVEL 1 PRODUCTS

The ESRIN AVHRR products are archived under the SHARP 1B format (SHARP, 1989). These products are made of four minutes of raw data in the original satellite projection from all 4 or 5 channels. Each SHARP scene contains the unmodified 10-bit AVHRR pixel values as quantised onboard the satellite, with derived calibration information to convert pixel values for the infrared bands into equivalent brightness temperatures. The SHARK software calculates the onboard thermal channel calibration coefficients for each scan line as it reformats the raw data. At the same time, it calculates the histograms of the raw pixel values in each band and also adds these to the SHARP output product.

To aid the interpretation of the imagery, SHARP products include the geographical position of the centre of each pixel on a subsampled lattice grid within the image. The grid consists of every 32nd pixel on every 16th scan-line. SHARP products also include the azimuth and elevation of the sun and satellite from the centre of each pixel of the lattice.

Finally, graphical overlays of the coastline, state boundaries and a latitude-longitude grid are also included in each SHARP product (in the higher order bits of the 16-bit words), so each overlay can be plotted over the image data independently as required.

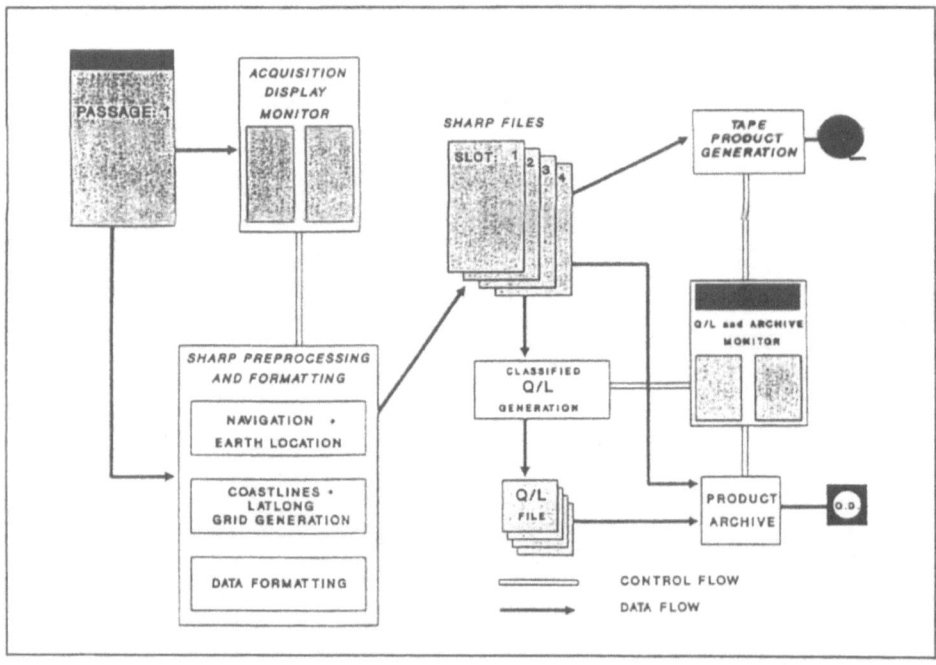

Figure 3. Overview of SHARP software capability and organisation.

4.2.1. *Processing from the raw data to SHARP Level 1 data product.* The input to the SHARP Level 1 processing system is the HRPT raw data stream as output from the frame-synchroniser and stored on magnetic disk by real-time acquisition programs, the so-called "level 0" data. The processing from level 0 data to the level 1 data product then consists of:

- the reformatting of the HRPT 10-bit words into 16-bit words by adding 6 empty bits to the most significant byte (MSB);
- the checking of the frame (time sequence, synchronisation words, *etc.*) and detection, labelling and/or replacement of bad or missing lines;
- the extraction and calculation of the calibration coefficients;
- the computation of the image navigation data using the satellite ephemeris information available, *i.e.* TBUS message for NOAA satellites;
- the generation of the coastlines, boundaries, late-long and land/sea mask files extracted from a reference database (a generalised version of the CIA World Data Base II);
- the refinement of image navigation accuracy by matching image data with computed coastlines (under operator control);
- the generation of the browse (Quick Look) product for the full pass (actually the classified black and white Quick Look, but see later the three-bands colour composite "Ionia" Quick Look) and of its hard-copy file (in postscript format);
- the generation of the catalogue entry relative to the full pass;

Thus, the resultant level 1 data product generated as output on magnetic disk will be composed of the SHARP products reformatted and corrected for bad/missing lines including:

- the calibration coefficient file;
- the geolocation points file;
- the coastlines, boundaries, lat-long and land/sea mask file;
- the browse (Quick Look) product file with its hard-copy file;
- the catalogue entry file, and
- the processing log file.

4.2.2. *Algorithm for SHARP Level 1 data processing.* In the generation of level 1 products, two algorithms are needed: one for the generation of image navigation data and another for the superimposition of coastlines, geographic grids and land/sea mask on the image.

The image navigation algorithm consists of:

- the reading (and eventual conversion) of satellite orbital elements;

- the propagation (by orbit computation) of the satellite position from image data start time to image data stop time in step of pre-defined size;
- the computation, at each propagation step, of the intercept of the instrument line-of-view with the earth ellipsoid at regularly spaced scan angles;
- the computation at each of the above determined ground points of the following parameters: geodetic co-ordinates, satellite pointing angles, sun pointing angles; and
- the generation of a file containing the computed parameters forming a grid of parameters at even-spaced pixel/line positions

The existing ESA implementation of this algorithm consists of about 2200 lines of Fortran code and needs about 20 seconds total time on a SUN Sparcstation2 to generate data for a typical AVHRR 10-minutes acquisition.

The coastline generation algorithm consists of:

- the reading of the earth location file generated by the navigation program;
- the extraction from the geographic data set (World Data Base II) of the portion of information (coastlines, state boundaries and land-sea mask) relative to the latitude-longitude range interested by the image to process;
- the generation of a coastline, state boundaries and land-sea mask file to be superimposed on the image by means of (bilinear) interpolation using as interpolation grid the content of the earth location file.

Both algorithms are satellite-independent.

4.2.3. *Archiving of SHARP Level 1 products*. SHARP level 1 data products temporarily stored on magnetic disk need to be copied together with complementary data to Optical Disk for long-term archiving. This product is called the SHARP level 1 archive product. The archives are maintained at each ECTN receiving station (for national usage) as well as at the central facility at ESRIN. The catalogue entry and the Browse product are sent to ESRIN via mail/network in order to update the central catalogue and the Browse products database. After a period of about one month, all SHARP Level 1 products are then sent using appropriate media (OD, exabytes, network), from the ECTN receiving station to ESRIN where they are then archived on Optical Disks and/or exabyte cartridge.

The input is the level 1 data product on magnetic disk including:

- the updating of the catalogue entry file (to include the OD label for OD archived products);
- the copying of the level 1 data product to the optical disk and/or to the Exabyte tape;

- the copying of the browse product and catalogue entry files to a telecom mail-box area in order to send them to ESRIN central facilities; and
- the copying to OD and/or to Exabyte of the auxiliary data corresponding to the level 1 data (if available).

In order to perform the archiving, the following functionality is available:

- handling of the Optical Disk (OD) operations, OD formatting, OD mounting and unmounting, OD content listing;
- handling of the Exabyte tape operations, tape positioning, tape inspect;
- handling the auxiliary data, uploading the data from CCT, exabyte, CD-ROM to magnetic disk, checking of the data, removing of the data from magnetic disk;

A centralised long-term archive is located at ESRIN based on the SHARP Level 1 archive products received from the Station.

4.3.   LEVEL 2 PRODUCTS

A level 2 product can be provided on request (SHARP, 1992a and 1992b). This product consists of data that have been further processed from level 1. Two level 2 products have been defined, known as 2A and 2B. Each consists of 5 bands with the same restrictions on size and the same ancillary data as level 1 products. All daytime passages acquired by the ECTN stations can be processed to generate SHARP level 2 products. A passage is considered to be day-time if the AVHRR band 1 histogram read from the SHARP level 1 trailer file has a mean value greater than a certain threshold. Because of the possibility of making changes in the data calibrations procedures or in the geophysical data retrieval algorithms, the SHARP level 2 data are not operationally archived, but generated on request.

The SHARP 2 format (SHARP, 1992a) has the same structure as the SHARP level 1 data, in that SHARP level 2 is a 5 band image in band-interleaved-by-line (LINN) format. Together with the image data, other information can be found in the prefix and suffix data of the image records. SHARP level 2 pixel is organised in 2-byte words (SHARP 2, 1992). Starting from the Least Significant Bit (LSB) each data word consists of:

- 10 bits used for image data;
- 3 bits used for state boundary, coastlines and Lat/Long grids flags; and
- 3 bits used for classification flags. Currently, the classification of Muirhead and Malkawi (1989) is used.

However, instead of the raw pixel values, which are written to level 1 products, each band in a level 2A product contains calibrated data from the corresponding AVHRR channel. Thus, bands 1 and 2 of a level 2A product contain calibrated reflectance of channel 1 and 2 using the coefficients given by Kaufman and Holben (1993), while band 3 contains radiance of channel 3 and bands 4 and 5 the brightness temperatures from channel 4 and 5 (Lauritson *et al.* 1979).

Level 2B products are more complicated. Bands 2, 3 and 4 are identical to those in the level 2, but the type of data in bands 1 and 5 varies depending on the nature of the target. Each pixel in a level 2 is classified as either land, sea, cloud, snow/ice or sun-glint. There is also an extra, unclassified class. The class to which each pixel is assigned determines which type of data it holds, for example:

1. Pixels which are classified as land contain, in band 1, a scaled value of the Normalised Difference Vegetation Index (NDVI). All other pixels contain the AVHRR band 1 reflectance as in level 2A product.
2. Similarly, pixels which are classified as either sea or sun-glint contain, in band 5, a scaled value of the Sea Surface Temperature (SST). This SST has been calculated using the non-linear split-window algorithm from McClain *et al.* (1990). This algorithm accounts for the satellite zenith angle variations within the image. The brightness temperature are corrected from the non-linearity effects of the detectors. All other pixels contain the AVHRR channel 5 brightness temperature.

All the data in SHARP Level 2 format are coded on 10 bits: the reflectance values are coded in 1000 levels with a resolution of 0.001 percent reflectance units; the spectral radiance in band 3 is multiplied by 100; and the brightness temperatures are coded from 0 to 1023 representing respectively 223 K to 323.3 K with 0.1 K resolution. The SST is coded from 0 to 500 representing 0 °C to 50 °C. The NDVI is multiplied by 1,000.

4.4.   QUICK LOOK PRODUCTS

Each SHARP product also includes a "Quick Look" image which is intended to indicate to the user the geographical area covered by a particular scene and the amount of cloud cover within it. Quick Looks are 16-bit single-band images consisting of 480 lines of 512 pixels each. They cover the entire length of the original image but only the middle 75% of the swath. The outermost parts of each scan are distorted because of the AVHRR's viewing geometry and are consequently excluded from the Quick Looks.

For day-time acquisition, Quick Looks contains a classification of the original scene into land, sea, cloud, snow/ice and sun-glint, and a residual unclassified category (Muirhead and Malkawi, 1989). The classified image can be printed using only four grey levels (one per major cover class) so that it can be faxed satisfactorily. Quick Looks for night-time scenes are generated from AVHRR band 4 data converted into brightness temperatures. A look-up table converts this to an appropriate grey-scale for hard-copy output.

The classified image Quick Look contains a 5-bit composite image with pixel values which represent either the brightness temperature or NDVI depending on the class to which the pixel has been assigned. Both types of Quick Look also include the coastline, state boundary and latitude-longitude overlay.

4.5.   THE NEW QUICK LOOK PRODUCTS: THE IONIA CONCEPT

Recently, a new Quick Look generation algorithm has been developed (Melinotte and Arino 1993) in order to:

- provide a better visual impression of the image quality;
- be accurate for any NOAA satellite at any latitude and acquisition time; and
- highlight specific interesting features (*e.g.* cloud height, water temperature, sun glint, vegetation strength, active fire, ice and snow).

4.5.1. *Quick Look generation.* The processing uses as input the standard SHARP 1B product at full resolution from the ESA archive, and is based on the following:

- Reflectance in channel 1 (R1) and 2 (R2) are computed using the calibration coefficients provided by Kaufman and Holben (1993), for odd-numbered satellites and NOAA-14 and pre-flight coefficients for other even-numbered satellites. The normalisation by the cosine of the solar zenith angle is performed for every 32 x 16 pixels.
- Brightness temperatures in channel 3 (BT3) and 4 (BT4) are computed by inverting the Planck Function using the in-flight calibration coefficients as described in Lauritson *et al.* (1979).
- Active fire detection is performed at full resolution using an ESA/ESRIN developed algorithm. This algorithm consists of a thermal threshold for night images and spectral analysis of R1, R2, BT3 and BT4 for day-light images to avoid erroneous detection of fires. This algorithm has been tested on more than one thousand images and was found to be both coherent, competent and probably conservative *i.e.* detected fires can be considered as highly likely and underestimated, if anything. The fire detection is applied only for latitudes between 30 degrees South and 15 degrees North to fully cover the African savannah and forest zones where biomass burning is of particular importance (Cahoon *et al.* 1992). Two different image processing techniques are applied to day-time and night-time images, the selection of which is based on the sign of the sun elevation angle at the centre of the scene. In both cases, a 3 x 3 pixel average is performed to obtain a 512 x 480 Quick Look. The cells containing at least one pixel identified as fire at full resolution are set to fire. For day-time imagery, a partial classification is independently performed on each band, resulting in four classes for R1, three for R2 and three for BT4. Each class is then defined by its physical properties in each spectral band (Arino, 1993). The four classes identified are water, bare soil, vegetation and cloud. Each class is attributed to a segment of the colour intensity spectrum. To enhance the contrast, histogram equalisation is performed on each segment. The colour composite is made by attributing R1 to the red colour, R2 to the green and BT4 to the blue.
- For night time the BT4 values are coded in grey scale after histogram equalisation.

4.5.2. *Quick Look Display*. A 24- to 8-bit colour compression is performed using a clustering algorithm. Coastlines, state boundaries and latitude/longitude grid are set to violet, while fire pixels are set to red. The scaling of the image is such that:

- Cloud-coded regions range from white for thick cloud to yellow for low cloud with the exception of cold semi-transparent clouds which have blue tones;
- The water surfaces have different blue tones depending on their temperature: dark blue for warm water to light blue for cold water;
- Sun glint patterns are identified by red or yellow tones over the sea;
- Land surfaces are represented from yellow to brown for bare soil and from light green to dark green for low to highly vegetated surfaces;
- Ice and snow can generally be separated from cloud based on their yellow aspect.

The AVHRR Quick Look obtained with this algorithm shows a good separation of the clouds from the land and sea surfaces, whilst maintaining the contrast of surface characteristics. The colour code best represents the natural visual tones. This Quick Look extends the standard geographical and cloud cover information to additional qualitative, but also to quantitative information on the features that appears in the higher level application products (*e.g.* land monitoring, oceanography, biomass burning, clouds type and form, *etc.*)

5. The ESRIN User Service

5.1. CENTRAL CATALOGUE

The ESRIN central catalogue also called "LEDA" for AVHRR is fully described in the LEDA Users Guide (LEDA, 1989). This makes use of the Inventory Exchange Format (IEF) information collected through the various receiving stations and centralised at ESRIN. A full description of the IEF is given in CEOS (1992). LEDA allows users to interrogate the catalogue remotely about the availability of a scene within a specific time and/or location of interest. As well as time and space, the query can also refer to other types of information such as the percentage of cloud cover per image.

5.2. ECTN DATA ACQUISITION AND DISTRIBUTION STATISTICS

The sale of AVHRR data in comparison with other experimental missions from 1 January to 30 September 1993 are given in the Table 2. As can be seen, the amount of AVHRR data sold far outweighs the amount from the other missions shown. The total number of images acquired by each station from the beginning of 1993 to September 1993 is shown in Table 3.

Table 2: Sale of products by source in Jan-Sept 1993

Source	Number of products
Seasat	8
Nimbus	278
NOAA	4342
MOS	0
Meteosat	2521

Table 3: Number of NOAA-11 and NOAA-12 acquisitions by receiving station for Jan- Sept 1993.

Station	NOAA-11	NOAA-12
Maspalomas	1995	1620
Oberfaffenhofen	2580	1005
Tromsö	2650	96
Niamey	1126	673
Nairobi	186	0
Scanzano	1276	0
Manila	120	0
Tennanova	70	32
Cairo	76	5

5.3.   EURIMAGE

Eurimage is the commercial partner for ESA product distribution and, as such, plays an important role in providing data to users. The company provides remote sensing products and services in Europe, North Africa and Middle East. Created in 1989, Eurimage is a service-oriented company with British Aerospace (UK), Dornier (Germany), Telespazio (Italy) and Satellitbild (Sweden) equal shareholders. The company provides remote sensing users with a centralised multi-mission service, supplying data products from several satellites. Its headquarters are in Rome where the marketing and sales, administration and finance, business development and public relations departments are based. The customer services offices are located at ESA/ESRIN in Frascati, Italy. For details on the other services and products and for conditions of sale and prices, please refer to the product and service guide of Eurimage.

6   The Ionia CD-Browser

6.1.   INTRODUCTION

Recently, ESA has developed a tool called "CD-Browser" for digital Quick Look consultation and product selection from CD-ROMs. As a first demonstration, the AVHRR CD-Browser Ionia CD-ROM has been produced and widely distributed (Arino *et al.* 1993). ESA wishes to encourage and receive the maximum user community feedback on this demonstrator before initiation of a routine CD-ROM Quick Look production service. The AVHRR Quick-Look images provided on this CD-ROM give a partial view of the ESA AVHRR SHARP data archive. Clearly, Quick Look image consultation is a fundamental step before ordering products. CD-Browser is a generic image base consultation system specifically designed for CD-ROM. CD-Browser provides a complete service, from the Query definition to the generation of product Orders. It uses "standard" formats, the Inventory Exchange Format (IEF) for the catalogue information and the Graphics Interchange Format (GIF) for the Quick look images. The AVHRR CD-Browser is the second member of the ESA CD-Browser family, OCEAN CD-Browser, dedicated to CZCS data, being the first realised a few years ago.

6.2.   THE QUERY

A normal CD-Browser session starts with the Query Definition (see Figure 4a). CD-Browser allows searches through inventory information associated with the Quick Look images using a combination of criteria such as: earth location, day/night, maximum % of cloud, acquisition station and satellite number. In a few seconds, CD-Browser builds the list of images corresponding to the Search criteria (Figure 4b). CD-Browser then allows you to SAVE and RESTORE your queries for future utilisation.

6.3.   THE SELECTION PROCESS

Once the search is performed, CD-Browser allows you to browse through the Quick Look images that satisfy the user selected criteria. Together with the lists of image products "Retrieved" and "Selected", CD-Browser can then display, for each image product, the catalogue information available and the mini Quick Look associated with them (see Figure 4c). The list of selected products can then be saved and/or loaded from session to session as required.

6.4.   THE DISPLAY

Next, CD-Browser can display any Quick Look image from both lists at full size. The Quick-Look image is a three-colour composite of channel 1 reflectance in the blue channel, channel 2 reflectance in the green channel and channel 4 brightness temperature in the red channel. A classification into water, vegetation, bare soil, cloud has been performed on each channel. The Quick-Look is a 512 x 480 pixels (3 x 3 pixels average) image coded on 8-bit representative of the SHARP image product described in a previous

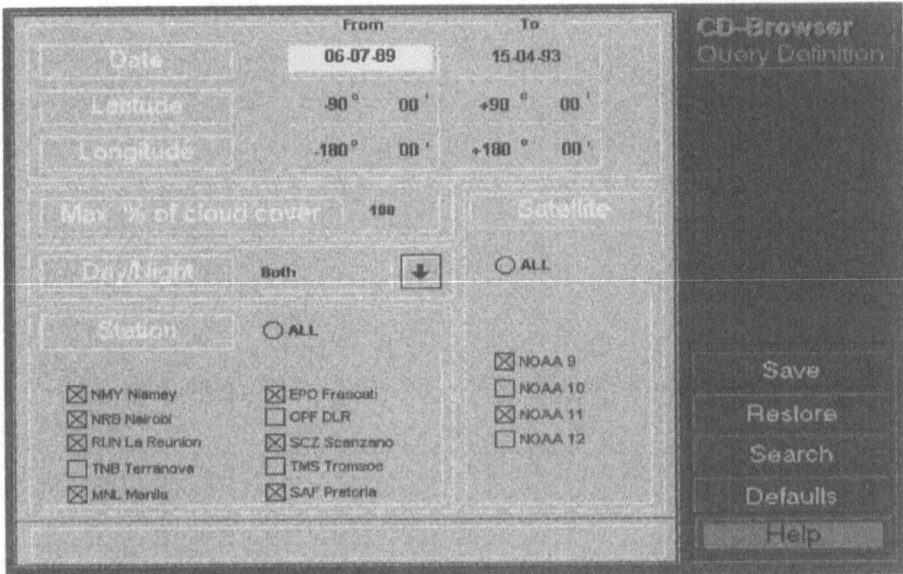

Figure 4a. The CD-Browser session normally starts with the query definition. Dates and areas of interest can be specified, as well as other parameters such as maximum percentage cloud cover, day/night imagery, receiving station and satellite.

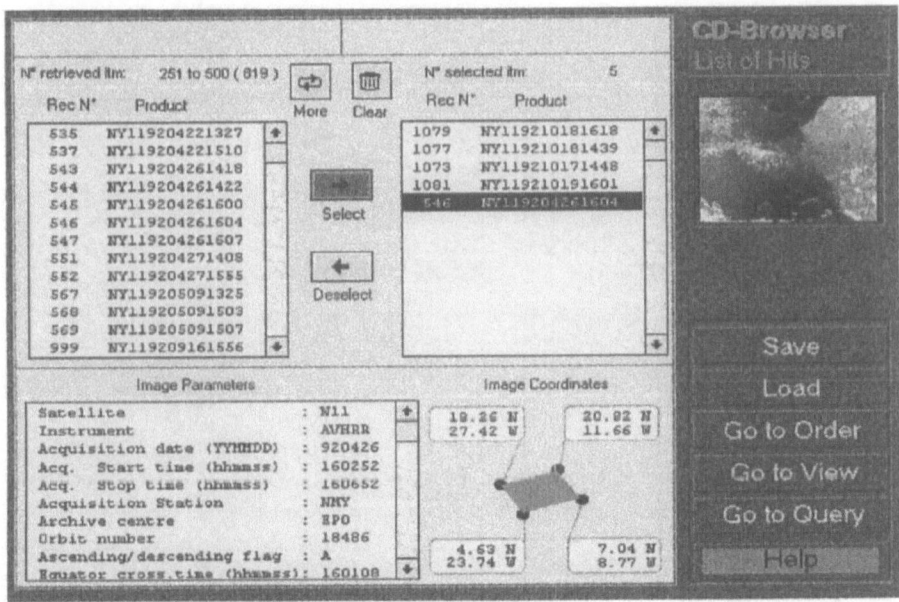

Figure 4b. The list of images, their details and quick-looks that satisfy the defined query. are then made available for browsing. CD-Browser then allows you to SAVE and RESTORE your queries for future utilisation

414

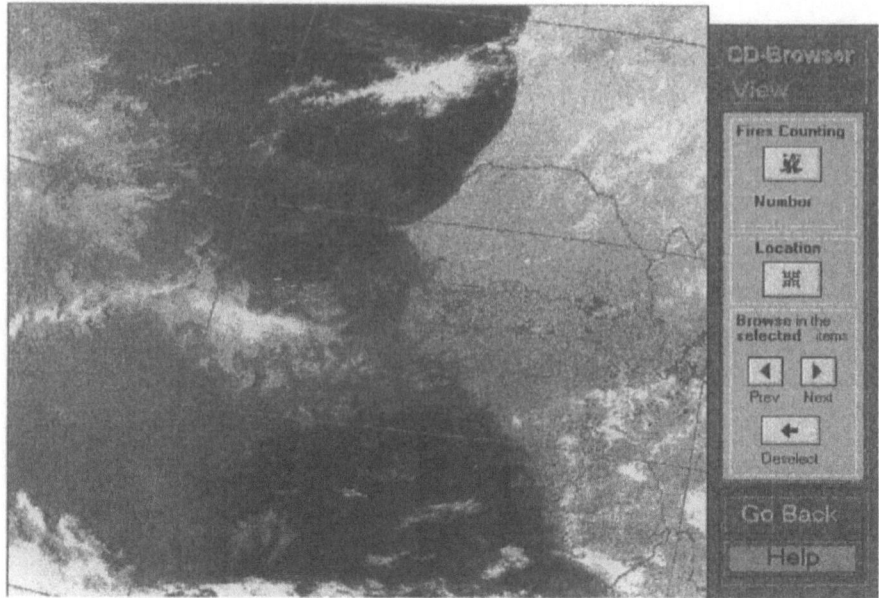

Figure 4c. Each quick-look can also be displayed at better resolution. Latitude and longitude graticules and country borders and coastlines are also provided.

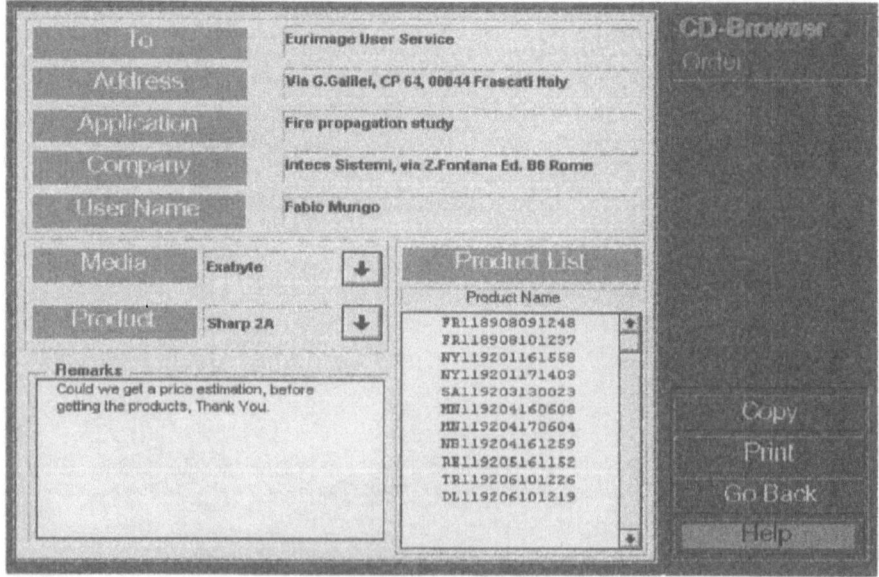

Figure 4d. Various products can be ordered, specifying the level of product and desired media. The order form can be printed or mailed on floppy disc.

NB. In reality all displays are in full colour.

section. Furthermore the red pixels mark the active fire detected by the Quick-Look algorithm, CD-Browser allows you to count the pixels with a fire in any part of the image you may define. CD-Browser then allows you to SELECT or DESELECT the images directly from this screen.

6.5. THE ORDER

CD-Browser also has the capability to generate the products ORDER using the list of products you have selected (see Figure 4d). It also allows you to customise your order by defining the level of products and the media requested. As output, you can print the order form to be sent by fax or save your order on floppy disk. ESA will then use the product list saved on the floppy for automatic delivery tape generation.

6.6. THE NET VERSION

An internet network version of the CD-Browser, with more or less the same functionality as the CD-Browser is also available on URL http://shark1.esrin.esa.it. This is specifically designed to allow the remote browsing and ordering of products collected in the framework of the global 1 km land cover project, which is described in the next section. The operation of the Ionia "1 km" net browser is covered in section 7.5.

7.   "TheGlobal Land Cover 1 km" Project

7.1.   SUMMARY

The increasing interest shown by the scientific community for world-wide remotely sensed data set at a nominal 1 km spatial resolution, led to the definition of the requirements for the compilation of the "Global Land Cover 1 km AVHRR data set", whose collection started operationally on April 1992. A co-operation plan was established between NOAA, NASA, ESA, USGS and CSIRO (Australia) with the aim of ensuring the acquisition over all land surfaces using NOAA "afternoon" satellites. USGS/EDC and ESA/ESRIN made available their facilities to set up the necessary infrastructure and implemented dedicated software and hardware for data collection, archiving, prototyping and processing. This section contains a brief description of the current status of the operational aspect of the project and the result achieved after the first phase completion (see also Buongiorno et al. 1993).

7.2.   INTRODUCTION

In the past years, various groups including the International Geosphere Biosphere Programme (IGBP), the Commission of the European Union (CEU), the Science Team of the Moderate Resolution Imaging Spectrometer (MODIS), and several other groups have identified the need for the acquisition and the compilation of a global 1 km resolution data set based on a compilation of daily AVHRR images for all land surfaces of the Earth. The scientific investigations carried out by these groups confirmed the importance of using the AVHRR data to derive significant information for a number of global phenomena involving the Earth system. As a result, the "Global Land Cover 1 km AVHRR Data Set" project was initiated with the purpose of collecting, archiving and processing of full-resolution HRPT and LAC AVHRR daily coverage of all of the world's land surfaces. This project is recognised by the scientific community to be very important for ongoing global change studies. It is also recognised that access is needed as soon as possible to this data set.

The project involves the following major participants: NOAA, with its HRPT stations and LAC recording facility; ESA with its HRPT ground receiving station network; USGS and NASA, with the USGS/EDC HRPT ground station network augmented by the CSIRO co-ordinated Australian HRPT stations (see Figure 5); and IGBP and CEOS for the definition of the Science requirements, the processing methods and the relevant formats (see Townshend *et al.* 1992 and following chapter by Belward).

7.3.   THE DATA SET

The Global Land Cover 1 km AVHRR data set is composed of 5 channels, 10-bit raw AVHRR data at full radiometric and spatial resolution (1.1 km at nadir view angle). The data set is made up of all the afternoon passages (usually two or more per acquisition station) collected daily from the NOAA polar-orbiting satellites. The data collection began on April 1 1992, at the outset for a term of 18 months, but the collection phase continued until the failure of the NOAA-11 AVHRR instrument.

7.4.   THE DATA ACQUISITION, ARCHIVING AND MANAGEMENT

Currently, the main priority for this project is the acquisition and archiving of the global data set. Each participating organisation has assumed responsibility for their own data collection and archiving, and for co-ordinating the data acquisition through their network of acquisition stations. Once the collection and archiving is performed, the data are copied to allow their exchange among the participants. In this way, a complete global land cover AVHRR data set is available at each major facility.

During the ingestion of the data, quality checks are performed both at ESRIN and at EDC using the time stamp information extracted from the HRPT data stream. The results from the quality checks are archived together with the data and referred to during the subsequent processing. For each passage the CEOS_IEF metadata lines describing the AVHRR data are generated and archived as well.

The archiving at ESA/ESRIN is made on OD media, each having about 75-pass storage capability. The data archive format consists of the HRPT data stream with each of the 10-bit words stored in 2 bytes. The full data set is managed by a OD Juke-Box which provides the file system mount of the requested data. All the CEOS_IEF metadata information generated is made available to an ORACLE database to allow efficient local browsing of the global data set.

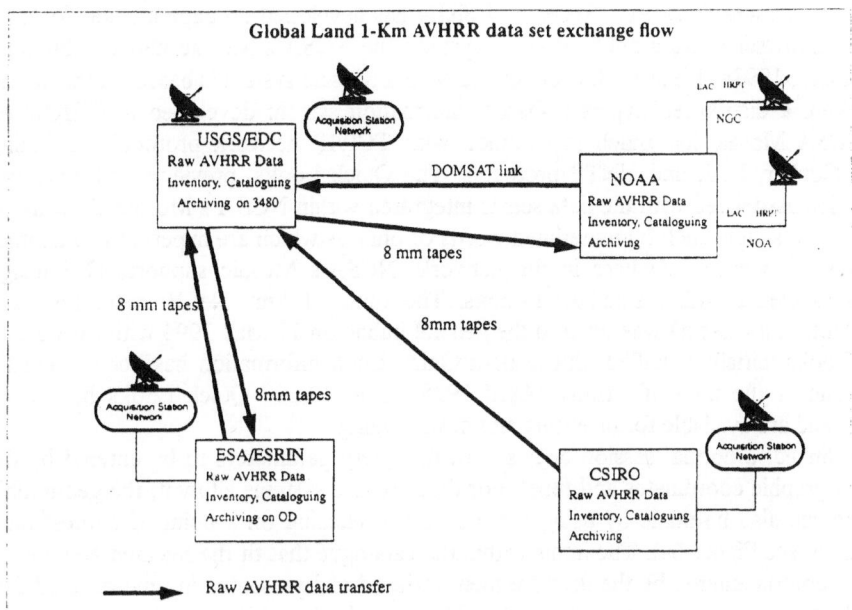

Figure 5   Global land cover 1 km AVHRR data set exchange flow

7.5.   THE DATA VISIBILITY

7.5.1. *Quick Look Products.* In consultation with a wide range of users, the Quick Look product for this project has been identified as follows: colour composite in GIF format subsampled to approximately 6 km spatial resolution using the Ionia algorithm (Arino, 1993, Melinotte and Arino, 1993). The Quick Look size is based on the archived HRPT pass length, which is variable depending on its origin. The width is fixed to 462 pixels and takes into account the satellite viewing angle so that each pixel is representative of an area of 6 x 6 km. The maximum size of this Quick Look actually processed in ESRIN is around 300 kbytes.

For reasons of processing time, active fires are not included on the Quick Look product, and neither are coastlines, state boundaries nor Latitude/Longitude gridlines. Nevertheless, it is easy to locate the area in the image due to the length and area of coverage of each pass. These Quick Look products will eventually be collated on CD-ROMs for the overall project.

7.5.2. *The Browse Software.* Recently, ESRIN has developed an experimental Ionia "1 km" Net Browser using a public domain system: the NCSA's Mosaic network browser (Andreessen, 1993a, 1993b). The core of the NCSA Mosaic system is based on the World Wide Web, a distributed hypertext-based information system developed at CERN. As such NCSA Mosaic has much in common with TCP/IP network protocols, including HTTP, Gopher, FTP, and NNTP protocols. The Quick Looks, previews and inventory information associated with the data sets is integrated within NCSA's Mosaic allowing its retrieval by a simple click on underlined words or phrases which are hyperlinks to another document or picture anywhere in the network. NCSA's Mosaic supports GIF image formats as well as other standard formats. The Ionia "1 km" Net-Browser (at URL http://shark1.esrin.esa.it) was open to the general public on 17 May 1994 with over 3,300 Quick Looks initially installed. Subsequent Quick Look information has been gradually added, and at the time of writing (April 1995), over 13,200 Quick Looks have been installed and are available for browsing and downloading.

Once linked, a normal session begins with the query parameters to be entered by the user (geographic coordinates and time). For those whose systems allow it, the geographic selection can also be made by using the mouse and clicking on a point of interest on a world map (see Figure 6a). The items within the catalogue that fit the selected geographic and time criteria entered by the user are then retrieved and presented in time-ordered lists of products coming from the overall accessible Quick look archive (an example of which is provided in Figure 6b).

Figure 7 shows, by month in the last year of access, the number of queries and images (previews and full Quick Looks) retrieved by mainly external users (ESRIN internal users represent only 2-8% of the accessees). Apart from the first peak (May 1994) which represents initial interest and curiosity during the time when the facility was included in the NCSA "What's New" page, the number of accesses and transfers has steadily increased. A simple analysis of all the accesses shows that more than a third of the interest is generated by users performing effective browse activity (submitting queries and retrieving Quick Looks); 15% of the traffic is generated by users having a "professional" interest, including users from groups such as: INRA, JRC, Meteo France, DLR, CNES, ERICSSON, NRSC, NERC, RAL, SSC, CNR, INRIA, *etc.*; and more than 55% of the overall accesses are generated by users within the European Community and about 20% by users in the US (about 10% educational and 10% commercial users). Overall, the analysis clearly demonstrates that this type of on-line accessible browse service is of considerable and growing interest to the user community, and should lead to better user service.

7.6.   THE HARDWARE EQUIPMENT

The present hardware configuration at ESRIN assigned for this project includes three systems; the first one is dedicated to ingesting and archiving operation and consists of a SUN 4/470 with 64 Megabytes RAM and 1.3 Megabytes of disk space, 8 mm exabyte cartridge tape drive, CD-ROM player and ATG Optical disk drive. The other system, dedicated to the generation of prototype products, is a SUN SparcStation10 with 128

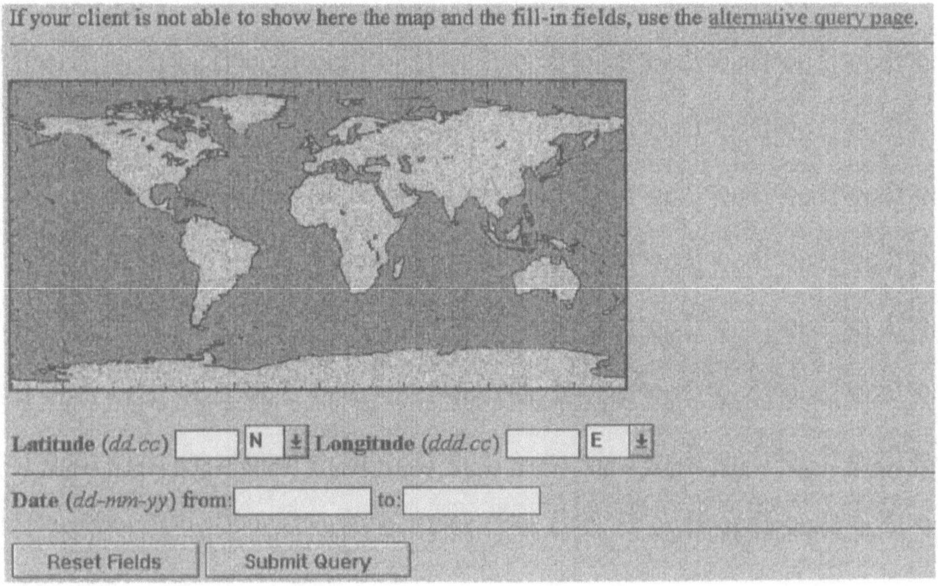

If your client is not able to show here the map and the fill-in fields, use the alternative query page.

Latitude (*dd.cc*) [ ] N ± Longitude (*ddd.cc*) [ ] E ±

Date (*dd-mm-yy*) from: [ ] to: [ ]

[Reset Fields] [Submit Query]

Fig 6a. The image query/selection page. For those whose systems allow it, the geographical selection may be made by mouse click on the world map. Latitude/longitude and date ranges can be entered via the keyboard.

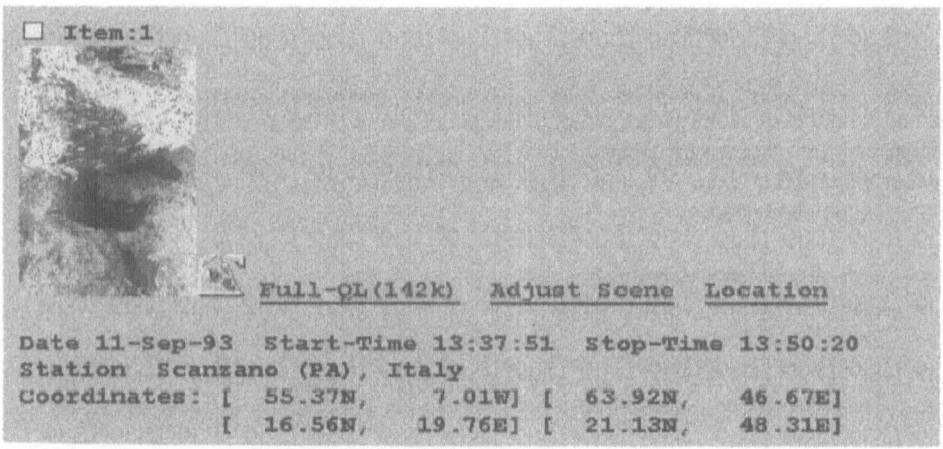

☐ Item:1

Full-QL(142k)   Adjust Scene   Location

Date 11-Sep-93   Start-Time 13:37:51   Stop-Time 13:50:20
Station   Scanzano (PA), Italy
Coordinates: [55.37N,   7.01W] [63.92N,   46.67E]
             [16.56N,   19.76E] [21.13N,   48.31E]

Fig 6b. The details and quicklooks of the images within the catalogue that fit the selected geographic and time criteria entered by the user are then retrieved for browsing by the user. An example of one of these is provided above. The pages are actually in full colour.

420

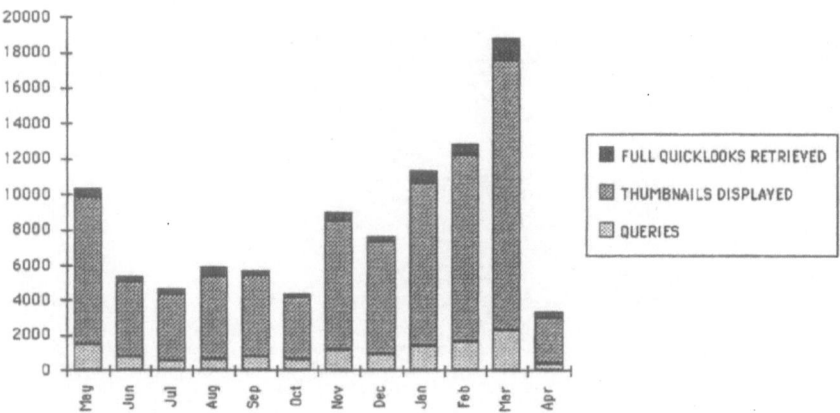

Figure 7. Accesses of queries and retrievals from the Ionia "1 km" Net Browser for May 1994-April 1995.

Megabytes RAM, 3 Gigabytes of disk space and it is equipped with two 8 mm tapes, an Optical Disk Juke box including two ATG Optical disk drives capable of handling 200 ATG Optical Disks for a total amount of 1.8 Terabytes of data. A Silicon Graphics machine has also been recently acquired to further process the data set.

7.7. DATA PROCESSING

The main goal of the processing phase is to generate products that are optimised for the Science applications following the requirements defined by IGBP (see Townshend (1992a, 1992b and the chapter by Belward later in this book). At present the following data processing steps (Figure 8) have been identified and prototype products are being implemented and tested :

- Orbital segments assembling;
- Radiometric calibrations;
- Atmospheric corrections;
- Geometric correction;
- Maximum-value compositing.

USGS/EDC, who will integrate the reference product generation software, have implemented the above mentioned steps generating a prototype 10-day composite global product consisting of ten output bands (see below for details). The Orbital segment

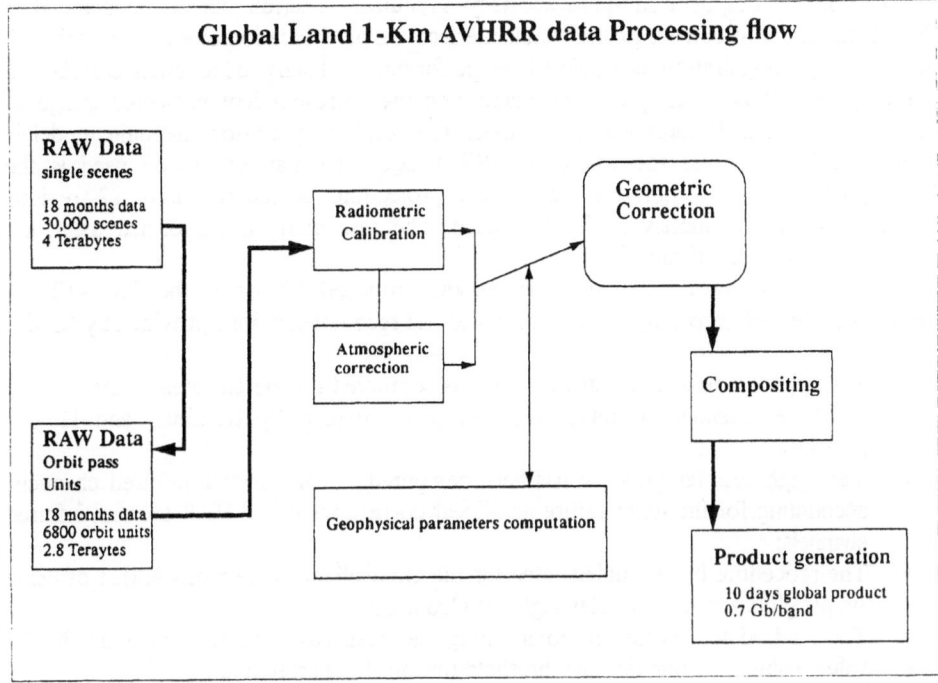

Figure 8. Global land cover 1 km AVHRR data processing flow

assembling processing provides a sensible data reduction and improved quality from the world-wide daily acquired passages to 14 daytime orbital strips. In this way the input data volume for the subsequent processing is reduced by the 30%. The method used can be summarised in the following steps:

- Sort by time the passages belonging to the same orbit;
- Determine the overlap regions between passages using time stamp information;
- Copy the image data to the output orbital image starting from the earliest time;
- In overlap regions, check for bad lines in the current segment replacing them with good lines from the next segment;
- Fill with zero filled lines the regions with no data.

The geometric correction process is the most time-consuming (about six hours to process one orbit element of AVHRR data with 10 output bands) but sensible multi-temporal image composite sub-pixel accuracy is required. A terrain correction is added to the orbital model to correct for the parallax effects on pixel location due to mountains. A precise image registration is obtained by performing a binary edge cross-correlation, matching band-2 AVHRR projected image with the correspondent projected image of coastlines and inland land/water boundaries. The resulting precision mapping model is then applied to process the whole AVHRR image. The map projection used is the Interrupted Goode Homolosine equal area projection, segmented into 12 regions consisting of approximately 5,000 lines each. The resampling is performed using the nearest neighbour algorithm.

The global data product includes among the proposed 10 bands, the five AVHRR channels calibrated according to the procedures and recommendations provided by IGBP:

- The apparent reflectance at the sensor is computed for the channels 1 and 2 using the time-dependent calibration coefficients provided by Kaufman and Holben (1993);
- The brightness temperature has been computed for the thermal infrared channels, accounting for the temperature non-linear corrections of AVHRR thermal infrared channels;
- The procedure being implemented for the atmospheric corrections is that of using simple global models for Rayleigh and Ozone correction;
- The method for the image compositing has been based on the maximum NDVI value, using only one thermal threshold for the cloud screening.

ESRIN is giving a high priority to its participation in this project, and is preparing the needed infrastructure in view of hosting the processing capabilities.

8.   Ongoing Developments

8.1.   THE PROTOTYPING OF HIGH LEVEL PRODUCTS

As well as the activities listed above, ESA/ESRIN has investigated the possibility of deriving higher level products in its operational chain using AVHRR and Along-Track Scanning Radiometer (ATSR) data simultaneously. To offer these new services, ESA/ESRIN has prototyped for inclusion new, but well-accepted and tested, scientific algorithms within the ESA processing chain. The project is currently at the end of the prototyping phase and it includes algorithms for the following functions: cloud detection, cloud classification, atmospheric correction and objective analysis (temporal interpolation) of time-series data. A set of specifications for an operational phase has now been generated from the tests of the prototypes.

Cloud detection, together with classification into more than 21 meaningful classes including land, water, ice, not ice, snow, not snow, totally cloudy, cloud free and partially cloudy, is performed using an extended version of APOLLO (Saunders and Kriebel, 1988) modified to cope with tropical conditions and to incorporate information from ATSR channels, Kriebel *et al.* 1994.

Atmospheric correction for all AVHRR and ATSR channels is performed by inverting 6S, (extended version of 5S, Tanre *et al*, 1990) for the visible channels and LOWTRAN 7 (Kneizys *et al.* 1988) for the infrared channels. Given knowledge of surface emissivity, two methods to generate Land Surface Brightness Temperature (LSBT) are implemented for the cloud free pixels: a single channel correction by inversion of LOWTRAN 7, and a split-window algorithm correction (Ottle and Vidal Madjar, 1992, Ottle and Francois, 1994).

Intercalibration of the ATSR and AVHRR sensors is carried out to take advantage of the precise ATSR calibration (0.1 K accuracy in brightness temperature), Arino *et al.* 1995.

The objective analysis (Santoleri *et al.* 1991) allows the retrieval of SST fields when satellite measurements are missing due to clouds or unrepresentative due to diurnal warming. The algorithm is based on optimum interpolation of time series of SST maps. Correlation times and lengths have been found to be region-dependent so the final software has been designed to take this into account (Santoleri *et al*, 1994).

8.1.1. *Background.* In response to user requests for standardised high level products with different levels of complexity, ESA/ESRIN has begun to incorporate, in a modular scheme, different recognised and approved scientific algorithms and software from the research community (Arino, 1992 and Arino *et al.* 1993). Given the AVHRR and ATSR global purposes and possible complementarity, the set of considered software addresses both types of data and generate, in certain cases, complementary cross-products.

Data output have been standardised following the recommendations of the Committee on Earth Observation by Satellite (CEOS). In that respect, one objective of the prototyping exercise was to tackle data format standardisation in input as well as in output for the ATSR and the AVHRR. Maintaining the coherence between formats allows the users to re-use the same software to address the new data products. The ERS-1 low bit rate data (*i.e.* ATSR data) is currently received at the Kiruna, Gatineau, Prince Albert and Maspalomas receiving stations. ATSR data are processed by the Rutherford Appleton Laboratory (RAL) up to level 2 using their "in-house" software called Synthesis of ATSR Data Into Sea-surface Temperatures (SADIST), (Bailey, 1993-a) and then converted to CEOS format, (ESRIN, 1993).

It is worth noting that currently no critical point exists between the different software to be interfaced with the ESRIN products, although a global synergy between some of them is preferable. The output format will remain equivalent to SHARP 2, ATS.IBT and ATS.SST formats in size and structure with atmospherically corrected radiometry, LSBT and SST information provided in the Imagery File.

8.1.2. *Cloud detection and classification.* The APOLLO software (Saunders and Kriebel, 1988) makes use of 5 threshold algorithms to detect cloud free pixels and two threshold algorithms for fully cloudy pixels. APOLLO was developed for AVHRR over temperate regions, but it will be adapted to cope with tropical regions over sea as well as over land. The thresholds will be set automatically as a function of location (latitude) and time (in the year). The implementation of APOLLO within SHARK is fully described by Gesell *et al.* (1993). The classification into more than 21 meaningful groups of pixels (including land/sea, cloud, snow, ice) is reported in the spare bits of the imagery file for any level 2 generation.

8.1.3. *Atmospheric correction and LSBT generation.* The correction of the infrared channels of the AVHRR is performed by creating Look Up Tables (LUT's) from LOWTRAN 7 with a 10 degree increment for the viewing angles. Standard atmospheric profiles of the LOWTRAN library can be used as input as well as radiosonde data entered by the user.

The visible and near infrared channels of the AVHRR are corrected by inverting an improved version of 5S, (Tanre *et al.* 1990) where the Rayleigh scattering and the gaseous absorption calculation have been improved. Two aerosol models will be available: maritime and continental. Three levels of corrections are planned: Rayleigh only, Rayleigh + gaseous absorption by O_2, O_3, H_2O, CO_2, Rayleigh + gaseous absorption + aerosol scattering. In operation, the software will run using either ECMWF data for pressure, temperature and water vapour or climatological atmospheric contents for ozone, water vapour, and visibility or user defined values. The validation of the implementation of the software using in situ measurements over the HAPEX SAHEL site in Niger has been performed (Ottle and Francois, 1994). The output formats are identical to the input formats: SHARP 2A with the corrected radiometry replacing the top-of-atmosphere (TOA) measurements and comments of the applied processing in the Leader File of the product.

Over land, the LSBT (land surface brightness temperature) can be derived from the 11 micrometer channel by inverting LOWTRAN 7 given the surface emissivity at this wavelength. Another method described in Ottle and Vidal-Madjar, (1992) is a split-window method where the coefficients are dependent on angle, emissivity and type of atmosphere (polar, temperate, tropical). The output format will be identical to the SHARP 2B where, over land, the LSBT could be put in channel 5 for AVHRR with comments on the applied processing in the Leader File.

8.1.4. *Possible extensions.* A set of potentially useful other products has been identified, and the possibility of merging multi-sensor data is under investigation. A set of operational constraints on format descriptions, order handling procedures and operational processing limitations has been identified. This will serve to establish an early specification for an operational system capable of generating and distributing the identified high level products. The modular concept for the incorporation of the different software algorithms and packages into the ESA operational processing chain will allow the future evolution of the package in line with the needs of the potential users and future missions. In the short-term, a Fast Delivery Production (FDP) line is also being designed, and this will include calibrated TOA and atmospherically corrected NDVI, SST, LSBT, Active Fire and Quick Looks. These FDPs are described in the section 9 following. The inclusion of the classification scheme derived from APOLLO will be integrated first within the level 2 processing scheme.

8.2. AVHRR CALIBRATION

8.2.1 *The calibration of the short-wave channels.* The short-wavelength channels of the AVHRR (channels 1 and 2) are not re-calibrated in flight, although their sensitivity is known to change, sometimes considerably, over the lifetime of the satellite. Studies whose results are available in the literature use a limited number of methods to estimate the change in sensitivity from the characteristics of the data received by the satellite. These studies provide often substantially different calibration coefficients at irregular and unpredictable intervals. Typically the delay between the acquisition of the imagery and the publication of the results is of the order of two to four years. This is clearly unacceptable for an operational near-real time project requiring calibrated data.

ESA has developed a Level 2 AVHRR data product which lacks absolute calibration in real-time. As ESA has the intention to market its Level 2 product more widely than the raw data, the Level 2 data will have to be calibrated in near-real time. The CEU-JRC MARS project (see chapter by Voessen earlier in this book) requires calibrated data from NOAA-7, -9, -11 -14 and future afternoon-pass satellites. It also requires them in as near-real time as possible, with each calibration estimate accurate to within about 5-10%, allowing a combined estimate whose precision increases over the lifetime of the satellite. It requires the calibration to be absolute, so that the true reflectance of terrestrial targets can be calculated.

Thus this project intends to set up an operational system for continuous calibration of the currently operational AVHRR sensor using two target types: desert (dark target) and sun-glint (bright target). This system will run automatically, with the data being extracted and the calibration coefficients being calculated with little or no human intervention. Some parts of the Saharan desert which are within view of the ECTN network will be used for calibration targets.

8.2.2 *The AVHRR and ATSR intercalibration.* The characterisation of the AVHRR and ATSR instrument's relative performances and the intercalibration of their data products has been carried out. The intercalibration of two different infrared sensors shall be planned as an operational exercise in the future with automatic software tools. This study gives the foundation for the identification of the main problems relevant to this intercalibration. Each instrument is characterised in terms of geometry, radiometry and their capacity to detect Sea Surface Structures (SSS). At Brightness Temperature (BT) level, the ATSR has been found to be more sensitive to the SSS than the AVHRR. Even with geometric resampling already applied to the distributed products, the ATSR has a better radiometric resolution due to its precise calibration and to its 12- bit digitisation in comparison with the 10-bit precision of the AVHRR. The processing up to SST obviously reduces the sharpness of SSS. The impact of the use of the double view for processing the ATSR data to Sea Surface Temperature (SST) was not fully tested: only two of the four specific data sets analysed use the double view capacity.

Time series of images taken at different hours of the days have been analysed. The normalisation of these time series to a given hour is possible. This allows the use of information coming from other acquisition times to be used together in the reconstruction of a given product: AVHRR and ATSR data can be merged to improve the temporal coverage. This issue is relevant when the adjustment of the two data sets is needed for global change analysis of long time series.

8.3.   THE SEASHARK SYSTEM

The SeaStar-1 satellite, which includes the Seawifs instrument, is a mission handled jointly by NASA, for the scientific utilisation of the data, and Orbital Sciences Corporation (OSC), USA, for the commercial exploitation of the SeaWiFS instrument. The present SeaStar mission is planned to be launched in early 1996 and SeaWiFS data will then be made available to foreign station two to three months later. The data stream (High Resolution Picture Transmission) down-linked from SeaWiFS instrument will be similar to the one of the already operational TIROS/AVHRR and, in general, the SeaWiFS system has been designed in such a way that existing AVHRR HRPT receiving facilities can be used, with the only difference being the encryption of the SeaWiFS data for commercial reasons.

8.3.1. *Assumptions*. In order to set-up the needed ground segment system for handling SeaWiFS data, ESA is procuring and implementing a new software package, called "SeaShark", for the simultaneous processing of both SeaWiFS and AVHRR data, making maximum re-use of the existing AVHRR dedicated hardware.
The system will be installed at some of the already existing ECTN stations.
Candidates are the following acquisition stations:

- TSS station in Tromsö (Norway);
- DLR station in Oberpfaffenhofen (Germany);
- Telespazio station in Scanzano (Italy);
- and at ESRIN Frascati (Italy).

with a possible addition of Maspalomas, Dundee, La Réunion, and others which may be added at a later stage.

8.3.2 *System*. The SeaShark system (Figure 9) will provide for both SeaWiFS and AVHRR data, the functionality for :

- generation of Level 1 Products;
- generation of Fast Delivery Products;
- generation of Browse Products;
- generation of catalogue entries;
- archiving of Level 1 Products;

- distribution of products;
- quality control on data.

The systems installed at acquisition stations and the one installed at Frascati will be connected on a network via an X25 link. In the above scheme, several types of

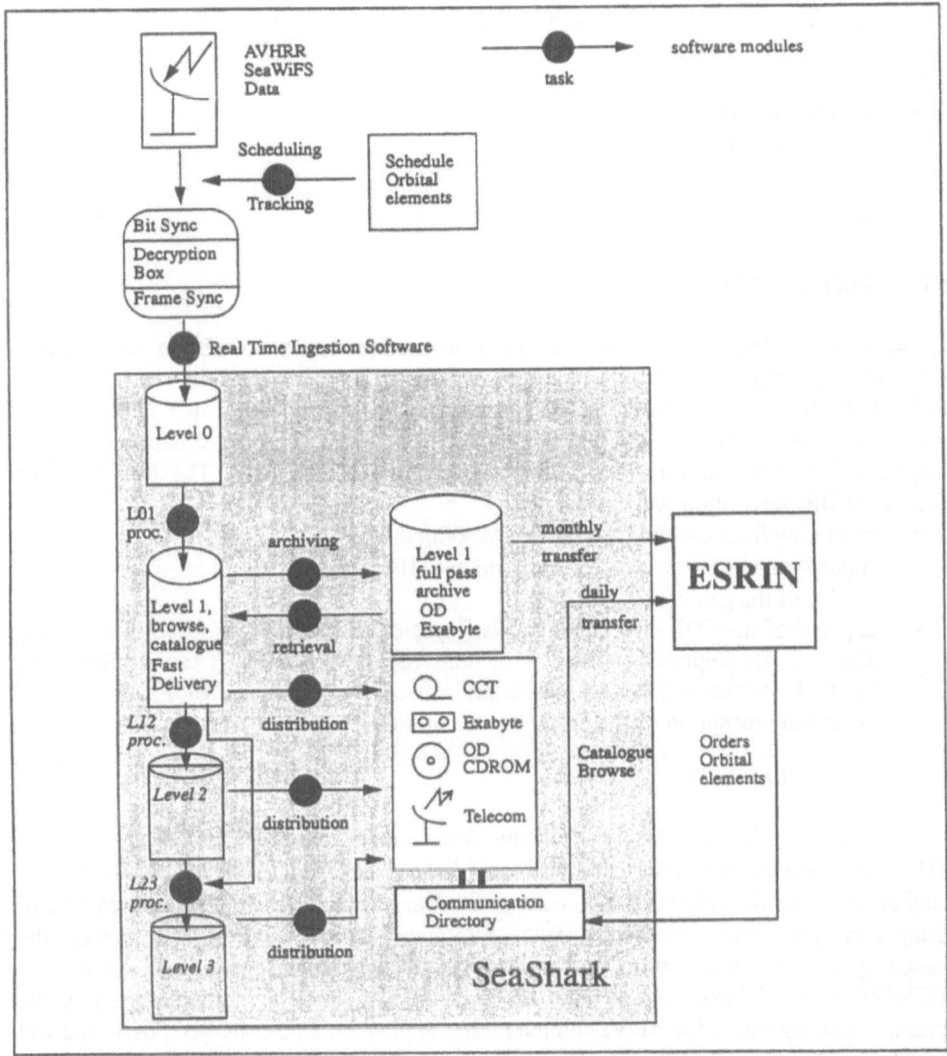

Figure 9. Flow diagram of the SeaShark receiving stations capability.

information need to be exchanged between ESRIN and the receiving stations. The handling of this information will also be part of SeaShark and will include real-time transmission from the various receiving stations to ESRIN of:

- catalogue entries and browse products (daily);
- reports of delivered products (monthly);
- operations reports (monthly);
- as well as the transmission from ESRIN to the various receiving stations of:
- satellite orbital elements;
- decryption key;
- acquisition plan;
- production orders.

9. The Fast Delivery Products

9.1. BACKGROUND

A Fast Delivery Product is a full resolution product (normally covering a well-defined small area) generated in near real-time using a simple algorithm (*e.g.* for land: NDVI or FIRE algorithm). The parameters for generation can be provided either automatically within an order file or interactively. The input can be either the level 0 data product on magnetic disk (MD) or the level 1 data product on MD or OD. The Fast Delivery Product (FDP) generation will consist of:
- extraction from the full pass of the area of interest;
- application of the selected FDP algorithm to the image;
- display of the generated FDP;
- copying of the FDP either to a media (tape) or to a telecommunication dedicated directory, for shipment to the user or transmission to ESRIN. It is considered that the FDP service will have a subscription basis for the daily, twice daily or four times daily reception of the products on a limited given area.

9.2. AVHRR 'LAND' FDP

'Land' FDP consists of the Top Of the Atmosphere calibrated NDVI (Normalised Difference Vegetation Index) using calibrated band 1 and band 2 values on pixels over land areas. A Surface NDVI is also considered. This surface NDVI should be calculated using atmospherically corrected surface reflectance. To derive these reflectances, the necessary auxiliary information on the atmospheric constituents can either come from ECMWF data or climatological data, or be entered by the user before request of the product. The algorithm for NDVI computation is applied on a pixel-by-pixel basis and will be integrated into the SeaShark system.

9.3.   AVHRR 'FIRE' FDP

This product provides the detection of active fires and their locality for daytime. The algorithm will be integrated into the SeaShark system. The Ionia CD-Browser (Melinotte and Arino, 1993, Arino, 1993, Arino *et al.* 1993) includes a set of thousand images with an active fire detection algorithm. The SeaShark software will provide the active fire product as a real-time service. This will be specially useful for the atmospheric chemistry community as well as the land management services of the national countries (see Cahoon 1992, Malingreau 1990, and chapter by Grégoire earlier in this book).

The following example illustrates the importance of the active fire in the west Africa Savannahs during the dry season. Nearly all savannah surfaces are burned annually. Figure 10 is the compilation of all afternoon images acquired by NOAA-11 by the Niamey station (40 SHARP images) for the month of October 1992 and (49 SHARP images) for the month of January 1993. It is noticeable how the number of fire events increases while the dry season progresses. The location of the fires are also seen to progress from North to South from September to April.

10.   Conclusion

Since the establishment of the ECTN in 1986, ESA has continued to develop and enhance its activities and capabilities for real-time data access, data supply and high-level product development to serve the growing Science, research and operational monitoring user community for AVHRR data for land applications.

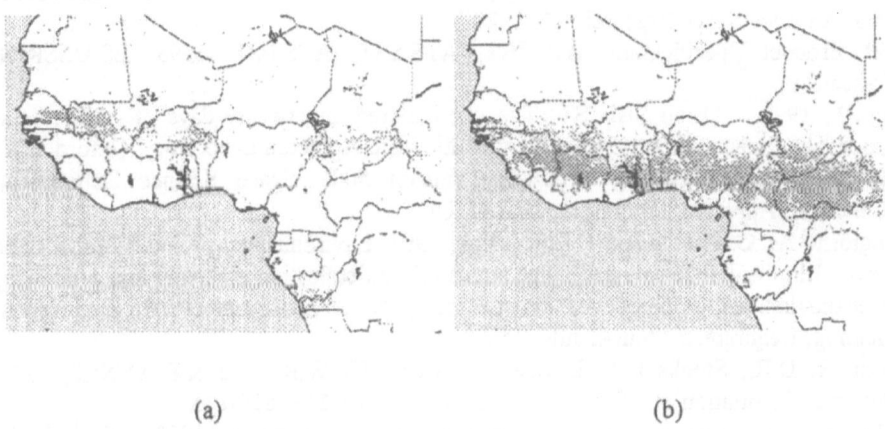

(a)   (b)

Figure 10. Summary of detected fires for (a) October 1992 and (b) January 1993.

Acknowledgements

This chapter is based on the work carried out by a team led by G. Pittella and composed of A Buongiorno, J.M. Melinotte and O. Arino within the ESA/ESRIN Earth Observation Division of the Department of Project and Engineering. It relies also on the experience acquired by many people: station operators, contractors and the user service. Finally thanks to L. Fusco who initiated the ECTN activity.

References

Andreessen M., 1993-a: NCSA Mosaic Technical Summary. Revision 2.1., May 4, 1993, National Center for Supercomputing Applications, 605 E. Springfield, Champaign IL 61820.

Andreessen M., 1993-b: Getting started with NCSA Mosaic. June 1, 1993, National Center for Supercomputing Applications, 605 E. Springfield, Champaign IL 61820.

Arino O., 1992: Technical Specification Document, SHARK research study: Tools for generating and merging geophysical products from the AVHRR and the ATSR instruments in the ESA SHARK environment. ESA internal document, ref. EPO/9206OA00, Frascati.

Arino, O., 1993: Specifications for AVHRR Quick Look Processing and "Active Fire" Detection. ESA internal document, March 1993, 10 pages, ESA/ESRIN, Frascati.

Arino O, J.-M. Melinotte and G. Calabresi, 1993: Fire, cloud, land, water: the "Ionia" AVHRR CD-Browser of ESRIN. *Earth Observation Quarterly*, **41**, July 1993, ESA/ESTEC, Noordwijk.

Arino O. *et al*, 1993. Software prototypes to derive high level products within ESA operational context. *Proceeding of the 6th European AVHRR Data Users' Meeting*, Belgirate 29 June-2 July 1993.

Arino, O., A. Buongiorno and P. Goryl, 1995: Intercalibration of AVHRR and ATSR data. Advance Space Research, **17**, 1, 29-38.

ATSR Product Specification: ATS.IBT, ATS.SST, ATS.PST, 1993. ESA/ESRIN, Frascati.

Bailey P., 1993-a: Using SADIST: The one-hundred and twenty days of ATSR data-processing. Space Science Department, Rutherford Appleton Laboratory, Oxford, UK.

Bailey P., 1993-b: SADIST Products (Version 500). Space Science Department, Rutherford Appleton Laboratory, Oxford, UK.

Buongiorno A., G. Pittella and J. Eidenshink, 1993: Environmental project "Global Land Cover 1 km AVHRR data set": The status of data collection archiving and processing after the first year of experience. *Proceeding of the 6th European AVHRR Data Users' Meeting*, Belgirate, 29 June-2 July 1993.

Cahoon Jr, D.R., Stocks B.J., Levine J.S., Cofer III W.R., and K.P. O'Neill, 1992: Seasonal distribution of African savanna fires. *Nature*, **359**, 812-815.

CEOS Inventory Exchange Format, Version 2.1.1, 7 February 1992, CEOS WGD Catalogue Subgroup.

Earthnet TIROS AVHRR Catalogue user guide, 1990, ESA/ESRIN, Frascati.

Fusco L. and K. Muirhead, 1987: AVHRR Data Services in Europe. *ESA bulletin*, **49**, February 1987, ESA/ESTEC, Noordwijk.

Fusco L., 1989: The European Coordinated TIROS Network: Operational Experience and Future Developments. *Proceeding of the 5th European AVHRR Data Users' Meeting*, Rothemburg, 5-8 September 1989.

Fusco L., Muirhead K. and G. Tobiss, 1989: Earthnet's Coordinated Scheme for AVHRR Data. *International Journal of Remote Sensing*, **10**, 625-636.

Fusco L. and G. Pittella, 1991: CEOS (Committee on Earth Observation Satellites) AVHRR Work Plan: An International Effort for the Coordination of AVHRR Related Activities. Proceeding of the 5th European AVHRR Data Users' Meeting, Tromsö, 25 June-28 June 1991.

Gesell G., T. Koenig, K.T. Kriebel and H. Mannstein, 1993: SHARK-APOLLO: Quantitative satellite data analysis based on ESRIN/SHARP-2A and DLR/APOLLO. *Proceeding of the 6th European AVHRR Data Users' Meeting, Belgirate*, 29 June-2 July 1993.

Harris A.R. and I.M. Mason, 1992: An extension to the split-window technique giving improved atmospheric correction and total water vapour. *International Journal of Remote Sensing*, **13**, 881-892.

HRPT Ground Station Database, 1991: World-wide review of High Resolution Picture Transmission Ground Stations. Release June 1991, ESA/ESRIN, 1991.

Kaufman Y.J. and B.J. Holben, 1993: Calibration of the AVHRR visible and near-IR bands by atmospheric scattering, ocean glint and desert reflection. *International Journal of Remote Sensing*, **14**, 21-52.

Kidwell K.B., 1991: NOAA Polar Orbiter Data Users Guide. NOAA, Washington D.C.

Kneizys *et al.* 1988: users guide to LOWTRAN 7. Report AFGL-TR-88-0177, AFGL, Hanscom AFB, MA 01731, USA, 137 p.

Kriebel, K.T., G. Gesell and H. Mannstein: 1994: SHARK research study on cloud detection, final report, ESA Contract No 10271/93/YT-I-(SC).

Lauritson L., G.J. Nelson and F.W. Porto, 1979: Data extraction and calibration of TIROS-N/NOAA radiometers. *NOAA Technical Memorandum* **NESS 107**, NOAA, Washington D.C.

Malingreau J.-P., 1990: The contribution of Remote Sensing to the Global Monitoring of Fires in Tropical and Subtropical Ecosystems. *Fire in the Tropical Biota, Ecosystem Processes and Global Challenges*, 337-370, edited by J.G. Goldammer.

McClain E.P., Walton C.C. and L.L. Stowe, 1990: CLAVR cloud/clear algorithms and non linear atmospheric corrections for multi-channel sea surface temperatures. *Fifth Conference on Satellite Meteorology and Oceanography*, 3-7 September 1990, London, 133-138.

Melinotte J.M. and O. Arino, 1993: AVHRR CD-Browser "Ionia" *Proceeding of the 6th European AVHRR Data Users' Meeting*, Belgirate, 29 June-2 July 1993.

Muirhead K. and O. Malkawi, 1989: Automatic Classification of AVHRR Images. *Proceeding of the 5th European AVHRR Data Users' Meeting*, Rothemburg, 5-8 September 1989.

Ottle C. and D. Vidal-Madjar, 1992: Estimation of land surface temperature with NOAA 9 data. *Remote Sensing of the Environment*, **40**, 27-41.

Ottle,C. and Francois,C., 1994: SHARK research study on atmospheric correction, final report, ESA Contract No 10272/93/YT-I(SC).

Pittella G. and C. Bamford, 1989: The EARTHNET "SHARK" system for TIROS data acquisition processing and archive. *Proceeding of the 5th European AVHRR Data Users' Meeting*, Rothemburg, 5-8 September 1989.

Pittella, G., 1993: Using Non-ESA Missions to prepare for ENVISAT. *ESA bulletin*, **75**, In print, ESA/ESTEC, Noordwijk.

Santoleri L., S. Marullo and E. Bohm, 1991: An objective analysis scheme for AVHRR imagery. *International Journal of Remote Sensing*, **12**, 681-693.

Santoleri, L. and R. Leonardi, 1994: SHARK research study on Objective Analysis, final report, ESA Contract No 10180/93/YT-I-(SC).

Saunders R.W. and K.T. Kriebel, 1988: An improved method for detecting clear sky and cloudy radiances from AVHRR data. *International Journal of Remote Sensing*, **12**, 123-150.

SeaShark: Development of the SeaShark software system for the SeaWiFS European Data Information System (SEDIS). Invitation To Tender AO/1-2489/93/I-HGE, 10 June 1993, ESA/ESRIN, Frascati.

SHARK Users' Manual, 1993. ESA/ESRIN, Frascati.

SHARK Technical Reference Manual, 1993. ESA/ESRIN, Frascati.

SHARP level 1 Technical Specification of CCT Format, Release 1.1, 1989. ESA/ESRIN, Frascati.

SHARP level 2 Technical Specification of CCT Format, Release 1.0, 1992a. ESA/ESRIN, Frascati.

SHARP level 2 User Guide, Release 1.0, 1992b. ESA/ESRIN, Frascati.

Tobiss G. and K. Muirhead, 1989: Earthnet's AVHRR Catalogue Service. *Proceeding of the 5th European AVHRR data users' meeting*, Rothemburg, 5-8 September 1989.

Tanre *et al.* 1990: Description of a computer code to Simulate the Satellite Signal in the Solar Spectrum, the 5S code. *International Journal of Remote Sensing*, **11**, 659-668.

Townshend J.R.G. *et al.* 1992: The Global 1 km AVHRR Data Set: Further Recommendation. IGBP-DIS Working Paper number 3, June 10 1992.

Townshend J.R.G. *et al.* 1992: Improved Global Data for Land Applications. Report of the Land Cover Working Group of IGBP-DIS. *IGBP Report Number* **20**.

AVHRR DATA ACQUISITION, PROCESSING AND DISTRIBUTION AT NOAA

KATHERINE B KIDWELL
National Oceanic and Atmospheric Administration,
National Environmental Satellite Data and Information Service,
National Climatic Data Center,
Satellite Data Services Division,
Washington, D.C. 20233.

ABSTRACT. NOAA/NESDIS acquires Advanced Very High Resolution Radiometer (AVHRR) data from the NOAA Polar Operational Environmental Satellites (POES). This paper fully describes the AVHRR instrument and includes its history, channel configurations, data types and resolutions, as well as the orbital characteristics of the TIROS-N satellite series and the diverse applications for which the AVHRR have been used. The manner in which these data are acquired is described, in addition to the methods used for appending navigation and calibration information to the Level 1b data sets. NOAA's archive of AVHRR data is maintained by both the Satellite Data Services Division (SDSD) and the Satellite Active Archive (SAA). SDSD is the traditional archive where digital data can be obtained on a physical media such as cartridge or round tape, whereas SAA is an on-line archive that can supply users with small samples of AVHRR data electronically through the Internet.

1. Introduction

This paper describes the manner in which the National Oceanic and Atmospheric Administration's (NOAA) National Environmental Satellite Data and Information Services (NESDIS) acquires, processes and disseminates Advanced Very High Resolution Radiometer (AVHRR) data from the current series of NOAA Polar Orbiting Environmental Satellites (POES). This description includes a brief history of the AVHRR instrument, as well as characteristics of the sensor itself and some of the many and diverse applications for which these data have been used. The method of navigation and calibration applied to the data is also described, as are the different AVHRR formats.

NOAA's archive of these data include the traditional distribution of the data on electronic media (round tapes and cartridges) through NESDIS' Satellite Data Services Division (SDSD), as well as the recently formed NESDIS data center, the Satellite Active Archive (SAA), which can electronically provide small samples of data directly to any user on the Internet. A short description of the NOAA Bulletin Board is also included, plus future plans for the AVHRR.

G. D'Souza et al. (eds.), Advances in the Use of NOAA AVHRR Data for Land Applications, 433–453.
© 1996 *ECSC, EEC, EAEC, Brussels and Luxembourg.*

History

The Very High Resolution Radiometer (VHRR) was the forerunner of the AVHRR instrument and was carried aloft on the ITOS series (NOAA-1 to NOAA-5) satellites from 1972 to 1978. VHRR was a scanning radiometer with only two channels: one in the visible (0.6 - 0.7 micrometers) region of the spectrum and one in the infrared (10.5 - 12.5 micrometers) region. As a result of the primitive scanning system on the ITOS series, it was necessary to carry two VHRR instruments onboard each satellite. The normal operating mode was the direct transmission of alternating visible and infrared scan lines from both VHRRs (visible from one radiometer and infrared from the other).

Immediately following the ITOS series came a third generation of NOAA polar orbiters that began with the launch of TIROS-N in October 1978. TIROS-N was followed by NOAA-6, NOAA-7, NOAA-8, NOAA-9, NOAA-10, NOAA-11, NOAA-12, NOAA-13 and NOAA-14. This series of satellites is referred to as the TIROS-N series. The AVHRR instrument is an integral part of the payload on the TIROS-N series and contributes data required to meet a number of operational and research oriented environmental objectives.

2. AVHRR instrument

The AVHRR instrument is a broad-band, four or five channel (depending on the model) cross-track scanning radiometer, sensing in the visible, near-infrared (IR), and thermal infrared portions of the electromagnetic spectrum. These channels were specifically selected to permit multispectral analyses of meteorological, oceanographic and hydrologic variables.

Table 1 shows the spectral band widths of the AVHRR (in micrometers), its Instantaneous Field of View (IFOV), measured in milliradians and how each channel is typically utilised.

Table 1. Spectral band widths, IFOVs and primary applications of the AVHRR sensors.

Ch #	TIROS-N	NOAA-6, 8, 10	NOAA-7, 9, 11, 12, 14	IFOV (mr)	Primary Applications
1	0.55-0.90	0.58-0.68	0.58-0.68	1.39	Daytime cloud/surface mapping
2	0.725-1.10	0.725-1.10	0.725-1.10	1.41	Surface water delineation, ice and snow melt
3	3.55-3.93	3.55-3.93	3.55-3.93	1.51	Sea surface temperature, night cloud mapping
4	10.5-11.5	10.5-11.5	10.3-11.3	1.41	Sea surface temperature, night cloud mapping
5	Ch. 4 repeated	Ch. 4 repeated	11.5-12.5	1.30	Sea surface temperature, day and night cloud mapping

There are three types of AVHRR data: High Resolution Picture Transmission (HRPT), Local Area Coverage (LAC) and Global Area Coverage (GAC). HRPT data are full resolution (1 km) image data that are transmitted to a ground station as they are collected (in real time). LAC are also full resolution data, but are recorded with an on-board tape recorder over selected areas for subsequent transmission to a ground station during the next overpass. GAC data are low resolution data (4 km) that provide subsampled global coverage recorded on the satellite's tape recorders which are then transmitted to a ground station (see chapter by Tucker).

NOAA/NESDIS operates two Command and Data Acquisition (CDA) stations at Wallops Island, Virginia and Fairbanks, Alaska. The CDA stations process and rebroadcast the data via a DOMSAT link to NOAA's computing facility in Suitland, Maryland. All LAC and GAC data acquired are permanently archived by NOAA/NESDIS.

2.2   ORBITAL CHARACTERISTICS

The TIROS-N series satellites are designed to operate in a near-polar, sun-synchronous orbit. The sun-synchronous orbit precesses eastward about the Earth's polar axis 0.986 degrees per day (at the same rate and direction as the Earth's average daily rotation about the Sun). The orbit has an inclination angle of 98 degrees which gives it a near-polar orbit (Rao *et al.* 1990).

The orbital period is about 102 minutes which produces approximately 14.1 orbits per day. Because the number of orbits per day is not an integer, the sub-orbital tracks do not repeat on a daily basis, although the local solar time of the satellite's passage is essentially unchanged for any latitude, at least early in the operating lifetime of the satellite.

The TIROS-N series was designed as a two spacecraft system of circular sun-synchronous orbits with nominal altitudes of 833 km and 870 km. The altitudes are deliberately staggered to minimise observing and acquisition conflicts. Their orbital planes are approximately 90 degrees apart with one spacecraft having a southbound equator crossing time of approximately 0730 Local Solar Time (LST) (morning spacecraft) and the other a northbound equator crossing time of approximately 1340 LST (afternoon spacecraft). These orbits were designed so that the equator crossing times would always occur at the same LST for consistent scene illumination.

However, the satellite's orbits drift over time (Price 1991). This drift causes a systematic change of illumination conditions and local time of observation which is the major source of non-uniformity in multi-annual satellite time series.

Table 2 contains the approximate times of the ascending node (northbound Equator crossing) and the descending node (southbound Equator crossing) in LST for the TIROS-N series when the satellites were launched. This table also contains the ascending and descending nodes as of March 1995 for the active POES satellites.

Table 2. Approximate equatorial crossing times of the TIROS-N satellite series.

Satellite	Asc. Node (Launch)	Dsc. Node (Launch)	Asc. Node (3/95)	Dsc. Node (3/95)
TIROS-N	1500	0300		
NOAA-6	1930	0730		
NOAA-7	1430	0230		
NOAA-8	1930	0730		
NOAA-9	1420	0220	2116	0916
NOAA-10	1930	0730	1753	0553
NOAA-11	1330	0130	1723	0523
NOAA-12	1930	0730	1915	0715
NOAA-13	1340	0140		
NOAA-14	1330	0130	1330	0130

The general degradation of NOAA-11's orbit can be seen in Table 2. When NOAA-11 was launched in September 1988, it had a 1330 LST equator crossing. By 1995, its orbit had degraded to 1723 LST.

Table 3 summarises the important dates for the satellites which have been launched from the TIROS-N series. The date range in this table is at best an approximation. There may be scattered data sets available before or after these dates. Users of NOAA-11 data should note that NOAA-11's AVHRR failed on Sept. 13, 1994, but sounding products continued to be archived. Tentative planned dates for future launches are also shown, illustrating the guaranteed long-term continuity of the system and data supply.

2.3   APPLICATIONS

NOAA generates several operational products from the AVHRR data. Products include Mapped GAC data, Radiation Budget data, Sea Surface Temperature (SST) data and Global Vegetation Index (GVI) data.

Mapped GAC data consist of mosaics of GAC passes in Mercator and polar stereographic map projections. The mapped mosaics are of daytime visible and infrared and nighttime infrared imagery.

Radiation Budget data are generated from GAC data. There are several different products, but the basic data consist of nighttime and daytime longwave radiation, available solar energy and absorbed solar radiation collected on a daily, monthly or seasonal basis, in Mercator and polar stereographic projections.

Table 3. Key dates for NOAA series launches.

Satellite	Launch Date	Date Range		
TIROS-N	Oct. 13, 1978	Oct. 19, 1978	-	Jan. 30, 1980
NOAA-6	June 27, 1979	June 27, 1979	-	March 5, 1983
		July 3, 1984	-	Nov. 16, 1986
NOAA-B	May 29, 1980	Failed to achieve orbit		
NOAA-7	June 23, 1981	Aug. 19, 1981	-	June 7, 1986
NOAA-8	Mar. 28, 1983	June 20, 1983	-	June 12, 1984
		July 1, 1985	-	Oct. 31, 1985
NOAA-9	Dec. 12, 1984	Feb. 25, 1985	-	Nov. 7, 1988
NOAA-10	Sept. 17, 1986	Nov. 17, 1986	-	Sept. 16, 1991
NOAA-11	Sept. 24, 1988	Nov. 8, 1988	-	present
NOAA-12	May 14, 1991	May 14, 1991	-	present
NOAA-13	Aug. 9, 1993	Aug. 9, 1993	-	Aug. 21, 1993
NOAA-14	Dec. 30, 1994	Dec. 30,1994	-	present
NOAA-K	April 1, 1996			
NOAA-L	Dec. 1, 1997			
NOAA-M	April 1, 1999			
NOAA-N	May 1, 2000			
NOAA-N'	June 1, 2002			

GAC infrared and visible data are processed into a basic set of SST observations at 8 km resolution over the global oceans. All observations are values which have been integrated over an 8 km diameter spot. However, they have variable spacing, ranging from 8 km contiguous in the U.S. coastal waters to 25 km in the open oceans. This database is further processed to generate gridded SST analyses at the regional, global and local scales.

Various combinations of AVHRR Channels 1 and 2 have been found to be sensitive indicators of the presence of chlorophyll in green vegetation and are referred to as vegetation indices. GVI data are produced from GAC data, orbit by orbit, and then mapped to a standard base projection. Channels 4 and 5 are also used (in later generations of the Vegetation Index) for cloud screening.

AVHRR Channels 1 and 2 are used to observe vegetation, clouds, land-water boundaries, snow and ice extent and, when data from both channels are compared, they indicate ice/snow melt inception. Channels 3, 4, and 5 are used to measure cloud distribution and to determine the temperature of the radiating surfaces (land, water, sea and clouds). Channels 3 and 4 are used to compute the sea surface temperature. Channel 5 (when available) is used for removing radiant contributions from water vapor when determining surface temperatures.

AVHRR data have been used for many diverse applications. In general, AVHRR applications encompass meteorological, climatological and land use. Obvious meteorological and climatological applications include: detection of cold fronts; tracking plumes; weather systems; cloud movement; squall lines; boundary clouds; jet stream; cloud climatology; flooding and hurricane tracking. In addition, land use applications of the AVHRR include monitoring of: food crop evaluation; volcanic activity; forest fire detection; deforestation; vegetation evaluation; snow cover; sea ice location; desert encroachment; iceberg tracking; oil prospecting and geology applications. Other miscellaneous AVHRR applications include the monitoring of: migratory patterns of various animals; animal habitats; environmental effects of the Gulf War; oil spills; locust infestations and nuclear accidents such as Chernobyl.

In a recent analysis of user requests for data it was found that over 30% of the orders were for land applications.

3. Data Acquisition

As the satellite orbits the Earth, data are both broadcast continually (direct readout mode) and recorded onboard the satellite for later playback. The two CDA stations receive both recorded and direct readout image data from the satellite and transmit these data to Suitland, Maryland via satellite relay. However, during two (sometimes three) sequential orbits of the Earth, the satellite remains out of contact with both of these sites.

The amount of HRPT data received during one pass of the satellite over the ground station is limited to the acquisition range of the station. A satellite pass directly over an antenna site is within view of that antenna (horizon to horizon with no obstruction) for about 15.5 minutes when the satellite is at an altitude of 833 km and 16 minutes when it is at 870 km.

The TIROS-N series satellites carry five digital tape recorders, each with a single electronic module and dual tape transport, to record data for subsequent transmission through the CDA to the data processing facility. Each transport has the capacity to record either 115 minutes (slightly more than a full orbit) of GAC or 11.5 minutes of LAC/HRPT data.

3.1 REQUESTING LAC DATA TO BE SCHEDULED

Users may request scheduling of LAC data which are recorded outside the direct readout range of Wallops Island, Virginia or Fairbanks, Alaska. Because recorder space and transmission time must be shared by many requestors, requests must be received at least one month prior to the data acquisition period. Requests are considered on a first-come, first-served basis, and according to the following priority considerations:

- National emergencies, as specified in the various national emergency plans;
- Situations where human life is in immediate danger (*i.e.*, search and rescue operations);
- U.S. strategic requirements;
- Commercial requirements;
- Scientific investigations and studies;
- Other miscellaneous activities.
- Requests must also be accompanied by the following:
- Brief description of application;
- Geographical area (*i.e.*, East Greenland, Korea Straits, *etc.*);
- Latitude and longitude coordinates bounding the area of interest;
- Desired frequency of coverage (*i.e.*, once weekly, *etc.*);
- Spectral channels required for image processing - Channel 1 or 2 = visible; Channel 3, 4, or 5 = infrared. Include range of expected brightness values or temperatures for image enhancement purposes.
- Type of data - digital data available on computer compatible tape, and/or analog data, available as photographic prints.
- Beginning and ending dates of the study period.
- Satellite preference.
- Name, address, and telephone number of requestor.

Failure to provide this information at the time of the request may cause a delay in scheduling of the LAC data. Requests for AVHRR LAC data may be phoned in, but must be followed by written documentation. Requests must be submitted to the LAC Coordinator, Data Collection and Direct Broadcast Branch, NESDIS, Room 806, NOAA Science Center, Washington, D.C. 20233.

Every effort is made to accommodate each request, for example, by combining requests of overlapping areas. However, because the number of requests for LAC coverage always surpasses scheduling resources, NESDIS does not guarantee complete or even partial fulfillment of LAC requirements. When lack of scheduling resources severely limit the acquisition of LAC coverage, requestors will be notified by LAC scheduling personnel. Users are not charged a fee for scheduling services.

Users should be aware that a request for LAC scheduling is not an implicit request for data. Users must also contact SDSD or SAA and meet all prepayment requirements before the actual processing of a data request can begin. Special procedures are made for international collaborative projects such as the IGBP Global 1 km Land Cover project.

4.   Processing of recorded data

The processing of recorded AVHRR data is a complex operation. Basically, the processing is divided into three main categories: preprocessing, navigation and calibration. The methods for generating the Level 1b data sets are described in more detail below.

4.1   DATA PREPROCESSING

NOAA/NESDIS has a data preprocessing computer system known as the Central Environmental Satellite Computer System (CEMSCS). The CEMSCS (located in Suitland, Maryland) is an operational satellite data handling hardware/software system designed for both the NOAA POES and geostationary (GOES) environmental satellites. Its major functions include satellite data ingest, data preprocessing, image processing, data transmission to external users, data archival, and environmental and atmospheric product processing.

NESDIS Satellite Operations Control Center (SOCC) receives the instrument data via communications satellite from the CDA stations. These data are sent from SOCC to the CEMSCS. The ingest activities on the CEMSCS are data driven and automatically begin the preprocessing steps when the raw data are received. The data are buffered and repacked into the Level 1a data format. (According to FGGE terminology, Level 1a data are raw data; Level 1b data are raw data with earth location and calibration appended, but not applied). Each AVHRR data type (GAC, LAC and HRPT) is stripped out into separate Level 1a data sets. These data sets are then passed on as input to the Level 1b preprocessing software called DEFER.

Each day about 14 to 15 orbits (per satellite) of GAC data are processed along with LAC and HRPT data. The following procedures are used to process each of the AVHRR data sets on the CEMSCS.

- Data are ingested.
- Data are quality checked.
- Raw data are placed in the Level 1a format.
- Earth location parameters are appended.
- Calibration coefficients are appended.
- The quality checked data and the appended information are placed in the Level 1b format.
- Level 1b data are provided to internal and external users for product processing.

The DEFER process contains all information necessary to calibrate and earth locate the instrument data. The calibration and earth location information are appended to the Level 1a data in fields already provided. Quality indicators are set to denote certain conditions such as no earth location data present, no calibration data present, time code errors, and ascending

/descending indicator flags. The Level 1b data set and calibration history files are created by the DEFER process.

4.2   NAVIGATION

NOAA/NESDIS receives a high precision orbital element set for each operational POES satellite from the U.S. Space Command (USSC). This element set contains an inertial position and velocity vector with corresponding orbit number, ballistic coefficient, solar flux, average solar flux, planetary index, and other parameters. The accuracy of this vector is reported to be at least one kilometer. A Cowell numerical integrator and algorithms to model atmospheric effects are used to predict this vector ahead in time and create a User Ephemeris File (UEF). The UEF contains a header record with the original USSC vector and data records with predicted inertial position and velocity vectors at 60-second intervals for a 10-day period. The UEF is used in generating the earth location data for the Level 1b files, TBUS bulletins (ephemeris data), and Search and Rescue (SAR) ephemeris data for the polar satellites.

The 10-day UEF is used by the Advanced Earth Location Data System (AELDS). AELDS is software that has been integrated into the DEFER process. Using AELDS, DEFER computes earth location parameters for the AVHRR data. The data time codes for each scan along with the instrument scanning parameters are used to obtain an interpolated satellite position and velocity vector from the UEF and the desired earth location values for that particular scan. Fifty-one earth location parameters per scan line are provided in the Level 1b AVHRR data. The physical algorithm used to compute earth locations is described in Appendix C of Lauritson *et al.* (1979).

Currently, the earth location data do not contain the one kilometer accuracy that is found in the UEF. Future enhancements to the overall earth location process will be geared toward reaching a 1.1 kilometer accuracy at nadir to meet AVHRR specifications.

4.3   CALIBRATION

All of the AVHRR instruments flown on the TIROS-N series have undergone extensive pre-launch calibrations at the instrument manufacturer's facilities in order to establish the instrument's stability, linearity of response and sensitivity. AVHRR Channels 1 and 2 (visible and near-infrared) are calibrated against lamps where output is viewed through the aperture of an integrating sphere, whereas Channels 3, 4 and 5 (infrared) are calibrated against precision blackbody sources (Lauritson *et al.* 1979; Planet 1988; and Paris 1994).

Although the pre-launch calibration procedures are quite extensive, it is not sufficient to rely on these calibration data alone to achieve the desired accuracy from AVHRR data. The instrument characteristics cannot be expected to remain the same in orbit as they were before launch! This situation occurs primarily because the thermal environment varies with the satellite's position in orbit, causing the output in digital counts to vary. Initially, Channels 1 and 2 are observed to degrade in orbit because of the outgassing and launch associated contamination (Rao and Chen 1993). Continued exposure to the harsh space environment (Brest and Rossow 1992) is also a contributing factor. In addition, the instrument's components age in the years that elapse between the pre-launch tests and actual launch. Furthermore, this aging process continues during the two or more years that the instrument is typically operational.

The TIROS-N series AVHRR was designed to view cold space and an internal warm blackbody as part of its normal scan sequence in orbit. This provides calibration data in the thermal infrared channels for determining signal-to-noise and radiometric slopes and intercepts. Unfortunately, there are no onboard calibration sources for the visible channels and the pre-launch calibration must be used or the user must rely on ground-based experimental techniques for deriving the calibration equations.

NOAA and the National Aeronautics and Space Administration (NASA) recognised the inherent problems with the AVHRR data and collaborated to form the NOAA/NASA AVHRR Pathfinder program. The main objective of the Pathfinder program is to reprocess and rehabilitate the long term records of AVHRR and AVHRR-derived geophysical products from 1981 to the present. As part of this program, the AVHRR Pathfinder Calibration activity has determined the in-orbit degradation of the AVHRR visible and near-infrared channels (Rao *et al.* 1993). After applying the appropriate formulae to account for in-orbit degradation, most (if not all) spurious trends are removed from the long term records of AVHRR and AVHRR-derived geophysical products. Currently, these formulae exist for the AVHRRs flown on NOAA-7, NOAA-9 and NOAA-11 spacecraft.

4.3.1 *Visible Channel Calibration.* Calibration for the visible and near infrared AVHRR channels (Channels 1 and 2) is performed before the launch of the satellite. The calibration source is a large aperture integrating sphere equipped with twelve calibrated quartz-halogen lamps. These lamps were carefully selected to match each other as closely as possible in spectral output and operating current. The sphere is then calibrated with all twelve lamps on against a National Institute of Standards and Technology (NIST) secondary standard of spectral irradiance. The ratio of the output of n lamps to that of twelve lamps is also determined. This yields the spectral output of the sphere when any number of lamps is on. By varying the number of bulbs which are turned on, a calibration curve from dark level to a maximum of twelve lamps output is obtained.

4.3.2 *Thermal Channel Calibration.* In orbit calibration of the AVHRR IR channels (Channels 3, 4 and 5) is possible because the instrument output is linear with input energy. During every scan line, the instrument views cold space (0 radiance) and its housing (approximately 290 K). The housing portion of the instrument has been designed to be a blackbody target to be used in orbit for instrument calibration. Four Platinum Resistance

Thermometers (PRTs), whose output values are included in the data stream, are embedded in the housing and monitor the temperature of the target. By determining the instrument output while viewing the known warm target and cold space (which is also included in the data stream), it is possible to determine the instrument response curve.

Pre-launch calibrations of the AVHRR infrared channels are carried out in a thermal vacuum chamber to minimise absorption of radiation in the path between the source and the radiometer and to simulate conditions in space. The radiometer sequentially views the warm calibrated laboratory blackbody (in place of the earth "target"), a blackbody cooled to approximately 77 K (representing the cold space view), and its own internal blackbodies. Temperatures of all blackbodies are sensed with thermistors or PRTs. Radiances for each channel can be computed from those temperatures. Data are collected as the laboratory blackbody is cycled through a sequence of temperature plateaus approximately 10 K apart between 175 K and 320 K, which spans the entire range of earth target temperatures. The entire procedure is carried out independently for several instrument operating temperatures (*e.g.* 10, 15 and 20°C for the AVHRR) that encompass the range of operating temperatures encountered in orbit. The operating temperature is represented by the temperature of the instrument's baseplate, which is also approximately the same as the temperature of its internal warm blackbody.

There are other coefficients necessary for in-orbit calibration that must be derived from pre-launch test data. These include the coefficients to account for the non-linearity in the AVHRR's response. The non-linearity may be accounted for by adding a correction term to the brightness temperature of the scene. The correction term is determined by interpolation in a table of correction terms vs. scene brightness temperatures specified at 10 degree intervals between approximately 200 and 320 K. The corrections, also functions of the AVHRR's internal calibration target temperatures, are made available for each spacecraft for internal calibration target temperatures of 10, 15, and 20 °C for each channel (Kidwell 1991). The appropriate correction is determined by interpolation on the internal calibration target temperature and the scene temperatures.

4.4   LEVEL 1b FORMAT

Recalling that Level 1b data are defined as raw data that have been quality controlled and to which Earth location and calibration information have been appended, Level 1b data are present on the data base as a collection of data sets. Each data set contains data of one type for a discrete time period. Thus, there are separate HRPT, LAC and GAC data sets. Time periods are arbitrary subsets of orbits, and may cross orbits (*i.e.* may contain data along a portion of an orbital track that includes the ascending node, the reference point for counting orbits). Generally, GAC data sets have a three to five minute overlap between consecutive data sets and begin approximately at the ascending node.

4.4.1 *GAC.* The processor onboard the satellite samples the real-time AVHRR data to produce reduced resolution GAC data. GAC data contain one out of three original AVHRR lines and the data volume and resolution are further reduced by averaging every four adjacent samples and skipping the fifth sample along the scan line. GAC resolution at the subpoint is actually 1.1 km x 4 km with a 3 km gap between pixels across the scan line, but is usually referred to as 4 km resolution. All of the GAC data computed during a complete pass are recorded onboard the satellite for transmission to Earth on command. The 10-bit precision of the AVHRR data is retained.

Each GAC data set contains an individual satellite recorder playback (or a portion of a playback if there is an interruption in the data due to noise, *etc.*, in which case a single playback may be fragmented into a number of data sets). Data within each GAC data set is in chronological order with one logical record per scan. Two logical records are written per 6440-byte physical record. Typically, a GAC data set is approximately 40 Megabytes in length. Each logical record contains 3220 bytes written in binary format as shown in Table 4.

As shown in Table 4, the GAC data record contains descriptive information for each scan. This information includes the scan number, the time, quality indicators (errors detected during processing plus Data Acquisition and Control Subsystem [DACS] quality errors), calibration coefficients, scan geometry, telemetry and the GAC video data. For more detailed information on the GAC Level 1b format, see Kidwell 1991.

The GAC video data consist of five readings (one for each channel) for each of the 409 points in a scan. They are packed as three (10-bit) samples in four bytes, right-justified. The 2045 samples (409 points x 5 channels) are ordered scan point 1 (Channel 1, 2, 3, 4, 5), scan point 2 (Channel 1, 2, 3, 4, 5), *etc.*, (also known as Band Interleaved by Pixel [BIP]). When

Table 4. Overview of the GAC data record.

Byte #	# Bytes	Content
1 - 2	2	Scan line number
3 - 8	6	Time code (year, day, hour, minute, second)
9 - 12	4	Quality indicators
13 - 52	40	Calibration coefficients
53	1	Number of meaningful zenith angles and earth location points appended to scan
54 - 104	51	Solar zenith angles
105 - 308	204	Earth location data
309 - 448	140	Telemetry (HRPT minor frame format)
449 - 3176	2728	GAC video data
3177 - 3220	44	Spare (zero filled)

there is no sensor for Channel 5 (on TIROS-N, NOAA-6, NOAA-8, and NOAA-10), Channel 4 data is repeated in the Channel 5 position. The video data are stored in binary.

Level 1b GAC data are also available in an unpacked format which consists of the video data represented in 16-bits instead of 10-bits. The unpacked format is more voluminous than the previously described packed format, but it greatly facilitates software development.

4.4.2 *LAC/HRPT.* The AVHRR data are digitised to 10-bit precision. The digitised data are both transmitted from the satellite in real-time as HRPT data, and selectively recorded onboard the satellite for subsequent playback as LAC data. A maximum of 11.5 minutes of LAC data may be recorded per orbit. LAC and HRPT have identical formats.

Each LAC/HRPT data set contains the LAC/HRPT data from one CDA contact. The data within each data set are in chronological order with one scan contained in two records. Typically, LAC data sets are approximately 46 Megabytes in length, while HRPT data sets are usually 57 Megabytes. The records are written in binary and contain 7,400 bytes in the format shown in Table 5. The logical record size is 7,400 bytes and the physical record size is 14,800 bytes.

The content of the LAC/HRPT records is very similar to the GAC data record, and includes the same descriptive information. The LAC/HRPT video data contain five values (one for each channel) for each of the 2048 points in a scan (*i.e*, 2048 points x 5 channels = 10240 samples). The data are packed as three (10-bit) words in four bytes, right-justified.

The video data are stored in binary and ordered scan point 1 (Channel 1, 2, 3, 4, 5), scan point 2 (Channel 1, 2, 3, 4, 5), *etc.,* (BIP).

Table 5. Overview of the LAC data record.

Rec.#	Byte #	# Bytes	Content
1	1 - 2	2	Scan line number
	3 - 8	6	Time code
	9 - 12	4	Quality indicators
	13 - 52	40	Calibration coefficients
	53	1	Number of meaningful zenith angles and earth location points appended to scan
	54 - 104	51	Solar zenith angles
	105 - 308	204	Earth location
	309 - 448	140	Telemetry
	449 - 7400	6952	LAC/HRPT video data
2	1 - 6704	6704	LAC/HRPT video data
	6705 - 7400	696	Spare (zero-filled)

446

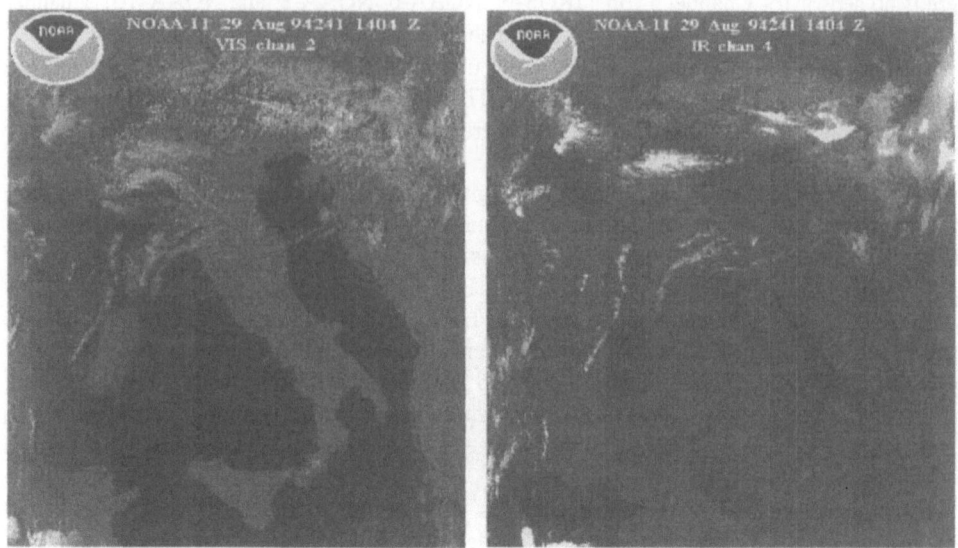

Figure 1. NOAA-11 AVHRR quick-look images over Italy, 29 August 1994

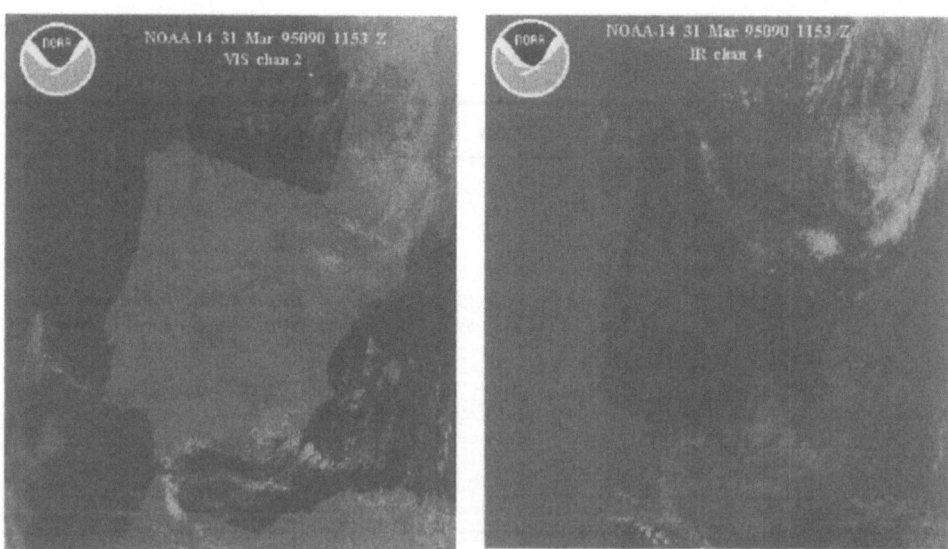

Figure 2. NOAA-14 AVHRR quick-look images over Spain, 31 March 1995

Level 1b LAC/HRPT data are also available in the unpacked format which is analogous to the GAC unpacked format previously described.

5. Archiving methods

The SDSD of the National Climatic Data Center (NCDC), under the auspices of NOAA/NESDIS manages a database of environmental satellite data and information derived from NOAA operational environmental satellites, certain NASA research satellites, Department of Defense meteorological satellites, and certain foreign government satellites. The bulk of the data received by SDSD is in digital form and is archived on tape cartridges. The mission of SDSD is to preserve the digital and analog data and to produce copies, extracts or derived products from the database for a wide range of U.S. Government and private customers.

5.1 CURRENT AVHRR DATA HOLDINGS AT NOAA

The SDSD maintains the digital and analog archive of all AVHRR Level 1b and product data beginning with TIROS-N data in 1978 to the current operational polar orbiter, NOAA-14. Consult Table 3 for specific operational dates for each TIROS-N series satellite. Because of spatial constraints, SDSD keeps only the most current year of POES data at its archive facility in Suitland, Maryland. The older data are shipped to an offsite location in Asheville, North Carolina where they are stored for ready accessibility.

5.2 SATELLITE ACTIVE ARCHIVE

The SAA serves as both a data centre and a system designed to provide easy access to data from NOAA's satellites. SAA contains descriptive information about those data sets, and permits users to search inventories of data holdings for availability based on geographic and date requirements. In addition to providing advanced on-line data query and product request capabilities, SAA also provides an on-line graphical browse tool which can assist the user in determining geographic coverage of individual data sets and display on-line digital representations of those data sets. Once the data requirements have been determined, an order may be placed electronically, and data may be delivered either electronically (SAA) or through physical medium (SDSD).

The SAA resides on several IBM RS/6000 UNIX Workstations at NOAA/NESDIS in Suitland, Maryland and can be reached from the Internet using the following address.

telnet saa.noaa.gov or telnet 140.90.232.101

Users who require further assistance from SAA should contact:

NOAA/SAA User Assistance
5627 Allentown Road
PES, Suite 100
Camp Springs, MD 20746
United States

e-mail: saainfo@nesdis.noaa.gov

Phone: (301) 763-8400 or (301) 457-5100
Fax: (301) 763-8443 or (301) 457-5105

Currently, SAA contains Level 1b AVHRR (HRPT, LAC, and GAC) data available from March 1994 forward. SAA's future plans include the availability of the following datasets: TOVS Level 1b data; Pathfinder Level 3 Products: AVHRR Atmosphere, TOVS Path C; DMSP: SSM/I and SSM/T; and Derived Products: aerosol, clouds, ozone, SST and winds.

In addition to providing advanced on-line metadata query and product request capabilities, SAA also offers a number of graphical aids to users accessing SAA through one of the supported graphical interface platforms. Among these are on-line digital image browse services for selected satellite image data sets.

These image browse services are primarily intended to support data set selection for ordering by allowing users to visually judge overall image quality, determine the extent of cloud cover, and verify geographic coverage. The AVHRR Level 1b data sets are the first satellite image data sets to be supported with browse services from SAA.

On January 14, 1994, SAA began operationally capturing browse images from each of the approximately 30 GAC passes it acquires on a daily basis. These browse images are obtained by capturing AVHRR Channel 2 and Channel 4 data as the data are acquired and passed to the Mass Storage System. They are generated by truncating the raw 10-bit data to eight bits before being transferred to the SAA Server system for storage. Thus, a full 100 minute GAC pass, for example, reduces to approximately a five Megabyte browse image on the SAA Server.

SAA's main menu is tailored to allow users to choose whether to learn a great deal about the platforms, instruments and data in the SAA through the Directory and Guide options, or to go directly to the Inventory option to search for, view and then order their data of interest. In addition, the menu has options providing news of general interest (such as the addition of a new data set), linking to remote computer systems, and personalised user information where registered user's addresses and such things as preferred format for specifying coordinates are stored.

5.3   DISTRIBUTION TO USERS

A user may place an order for AVHRR data from either SAA or SDSD. Currently, the maximum amount of data that can be acquired per day (free of charge) from SAA is five Megabytes (although this is expected to increase to 10 Megabytes in the near future and eventually to a whole dataset). If a user exceeds the specified daily maximum and has an account set-up with sufficient funds, they will be notified and their order will be transferred to SDSD to fill using traditional physical media such as cartridge or round tape. The user will be charged accordingly for this data on physical media.

When ordering data from SDSD, the following parameters should be specified to insure the selection of the proper data:

- Type of data;
- Dates and times of data;
- Channels;
- Satellite name;
- Day, night, or both;
- Area (latitude/longitude box);
- Orbit numbers if known;
- 3480 cartridge/9 track 1600 or 6250 bpi computer compatible tape;
- Packed or unpacked (Level 1b only).

When ordering data directly from SAA, the user must follow the on-line directions to guarantee proper selection of data.

6.   NOAA Bulletin Board

NOAA also operates a multi-line electronic bulletin board (EBB) system capable of disseminating a large number of computer files and messages to the environmental satellite user community. This electronic bulletin board is called the NOAA Environmental Satellite Information System, hereby referred to as NOAA.SIS. A maximum of eight dial-in phone lines (2400 bps) can be supported simultaneously serving up to 500 calls daily.

The purpose of NOAA.SIS is to provide technical information to the global satellite user community having express interest in the management and operation of NOAA's earth-observing environmental satellites. This information and other associated satellite activities are routinely posted on NOAA.SIS and may be accessed via telephone/modem from any place on the globe.

NOAA.SIS is menu-driven and permits users to select messages from a variety of topics. Special user help assistance is also available. The Satellite Services Division (not to be confused with SDSD), an organisational component of NESDIS, is principally responsible for the operation of this electronic bulletin board. The European Space Agency and the Japanese Meteorological Agency, members of the World Meteorological Organisation, also contribute information to NOAA.SIS.

Topics posted on NOAA.SIS include: Weekly Space Craft Events; Polar and Geostationary Orbital Elements; TBUS Messages; Eclipse Schedules; WEFAX Transmission Schedules; Special Operational Notices; News Items and other helpful instructive information.

The minimum hardware needed to access NOAA.SIS is:
- Microcomputer/work-station;
- 1200 bps modem (CCITT standard V.42 or MNP 2-5) or 2400 bps modem;
- Normal voice grade telephone line.

The minimum software required to logon WITHOUT color menus is:
- Communication package which supports ASCII character terminal emulation, supports No Parity, 8 Start Bits, and 1 Stop Bit transfer mode, and supports ASCII and XMODEM data transfer protocols.

The minimum software required to logon WITH color menus is:
- Communication package which supports full ANSI color terminal emulation, supports No Parity, 8 Start Bits, and 1 Stop Bit transfer mode, and supports ASCII and XMODEM data transfer protocols.

The recommended dial-in configuration is a PC/XT/AT compatible computer with a 2400 bps MNP 5 modem using a commercially available communication software package. The communication package should be set-up to dial the telephone number 301-763-8500 using 2400 bps speed, N/8/1 transfer mode (default), and color ANSI terminal emulation (default).

After setting up the hardware/software as described above, the initial logon screen should appear. The system will ask the user for their first name, last name and desired password for future logon. The user will also be required to answer a brief questionnaire. After answering the questionnaire, the user will be shown the BULLETIN/NEWS MENU. Users should follow the on-line instructions for additional details.

NOAA.SIS is now also available on the Internet. The Internet Uniform Resource Locator (URL) address will be: http://psbsgi1.fb4. noaa.gov:8080/MOSDEV.html, click on EBB/ NOAASIS.

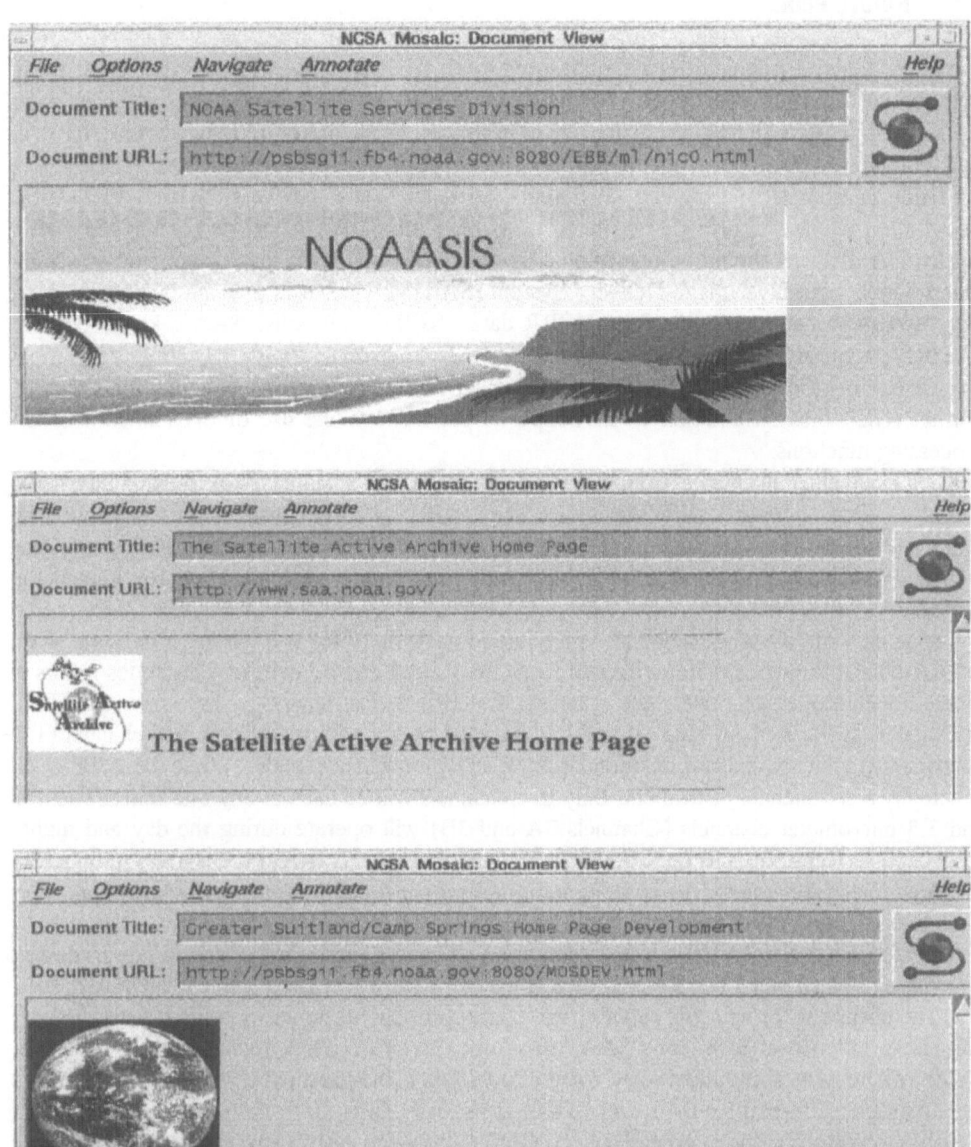

Figure 3. Some important NOAA WWW home-pages

7.   Future Plans

The SAA plans to populate their database with retrospective Level 1b AVHRR data, from October 1986 through March 1994. This means that users will be able to access these data on-line and have a sample electronically transmitted to them (or if they choose, have a magnetic tape shipped to them from SDSD).

SDSD is currently designing a Satellite Archive Data Delivery System (SADDS) workstation which will provide output to users on 4 mm (DAT tapes) and 8 mm (Exabyte) tapes. The SADDS should be operational by mid-1995. SDSD is also developing a system called Satellite Archive Image Delivery System (SAIDS) which will provide customers with high resolution image products for AVHRR data. SAIDS will utilise SAA 8 km resolution AVHRR browse functions to provide SDSD personnel with a low resolution browse/hardcopy capability. SAIDS will also provide high resolution Graphics Interchange Format (GIF) files, in addition to hardcopy output through the use of SAA's work order processing functions.

SDSD currently maintains a Home Page under NCDC's Home Page on the Internet's World Wide Web (WWW). The URL for this Home Page is http://www.ncdc.noaa.gov/ncdc.html. Future plans for WWW include the addition of the NOAA Polar Orbiter Data User's Guide and other user's guides on-line for immediate access.

The launch of NOAA-K (currently scheduled for April 1996) will initiate a new series of NOAA polar orbiting satellites, referred to as the NOAA-K,L,M series. This series marks a moderate improvement over the current AVHRR/2 instrument. The new AVHRR instrument, designated AVHRR/3, will be upgraded by the addition of one new channel (1.6 micrometer) which will allow discrimination of snow/ice versus clouds. While the AVHRR/3 is actually a six-channel instrument, only five channels will be used at any one time. The 1.6 and 3.8 micrometer channels (Channels 3A and 3B) will operate during the day and night, respectively. Capability has been built into the instrument to switch at the Earth's terminator, or specific latitudes, or by stored command tables or manually from the CDA stations. The final determination is yet to be decided.

The changes in the AVHRR instrumentation will have a significant effect on the processes which are downstream. The formats of all Level 1b data will change with the NOAA-K,L,M era. The change in formats will require modifications in the ingest and archive software which will affect software at SDSD and SAA. In addition, a new NOAA Polar Orbiter Data Users Guide will be written and distributed for the NOAA-K,L,M spacecraft.

The proposed format for the Level 1b data is radically different from the current format. In addition to embedded documentation and a different header record structure, the new format includes attitude corrections, more orbital elements, calibration coefficients, cloud classification, Earth/Sun distance ratio, more telemetry data and record lengths in multiples of

512 bytes for the convenience of VAX users. The new format will be much more flexible than the old format and will facilitate future expansion of the data, if needed.

All of these improvements and future plans are part of NOAA's continuing commitment to the ever-increasing number of NOAA-AVHRR data users.

8.   References

Brest, C.L. and W.B. Rossow, 1992, Radiometric Calibration and Monitoring of NOAA AVHRR data for ISCCP, *International JOURNAL Remote Sensing*, **13**, 235-273.

Kaufman, Y.J. and B.N. Holben, 1993, Calibration of the AVHRR visible and near-IR bands by atmospheric scattering, ocean glint, and desert reflection. *International Journal of Remote Sensing*, **14**, 21-52.

Kidwell, K.B., July 1991, NOAA Polar Orbiter Data User's Guide, U.S. Dept. of Commerce, NOAA/NESDIS, Washington, D.C., 302 pp.

Kidwell, K.B., December 1994, Global Vegetation Index User's Guide, NOAA/NESDIS, 129 pp.

Lauritson, L.; Nelson, G.J., and Porto, F.W., November 1979, Data Extraction and Calibration of TIROS-N/NOAA Radiometers, *NOAA Technical Memorandum NESS 107*, 73 pp.

Paris, C.A., February 1994, NOAA Polar Satellite Calibration: A System Description, *NOAA Technical Report NESDIS 77*, Washington, D.C., 61 pp.

Planet, W.G. (Editor), revised October 1988, Data Extraction and Calibration of TIROS-N/NOAA Radiometers, *NOAA Technical Memorandum NESS 107 Revision 1*, 130 pp.

Price, J.C., 1991, Timing of NOAA afternoon passes, *International Journal of Remote Sensing.*, **12**: 193-198.

Rao, C.R.N. and J. Chen, 1993, Calibration of the visible and near-infrared channels of the Advanced Very High Resolution Radiometer (AVHRR) after launch, *Proceedings of the SPIE Conference on Recent Advances in Sensors, Radiometric Calibration and Processing of Remotely Sensed Data*, 1993, 56-66.

Rao, C.R.N., J. Chen, F.W. Staylor, P. Able, Y.J. Kaufman, E. Vermota, W.R. Rossow and C. Brest, 1993, Degradation of the visible and near-infrared channels of the Advanced Very High Resolution Radiometer on the NOAA-9 spacecraft: Assessment and Recommendations for corrections, *NOAA Technical Report NESDIS 70*, U.S. Dept. of Commerce, Washington, D.C.

Rao, P. K., S.J. Holmes, R.K. Anderson, J.S. Winston, and P.E. Lehr, editors, 1990, Weather Satellites: Systems, Data, and Environmental Applications, American Meteorological Society, Boston, Mass.

Schwalb, A., March 1978, The TIROS-N/NOAA A-G Satellite Series, *NOAA Technical Memorandum NESS 95*, 75 pp.

Schwalb, A., Feb. 1982, Modified Version of the TIROS-N/NOAA A-G satellite series (NOAA E-J) - Advanced TIROS-N (ATN), *NOAA Technical Memorandum NESS 116*, 23 pp.

AVHRR DATA SETS FOR GLOBAL TERRESTRIAL ECOSYSTEM MONITORING

A. S. BELWARD

European Commission's Joint Research Centre,
Institute for Remote Sensing Applications,
21020 Ispra (Varese) Italy.

1. Introduction

The availability of daily data, at a resolution of 1 km for the entire globe is often cited as an important attribute of the NOAA / AVHRR system for global terrestrial ecosystem monitoring exercises. However, the idea of global 1 km resolution AVHRR archives spanning more than a decade is sadly false. Prior to April 1992 there was no daily global 1 km data archive. Length of archive and geographical area covered varies considerably from region to region. For much of the Earth the only AVHRR data archives offering systematic, long term coverage are of sampled AVHRR data. Firstly there is the Global Area Coverage (GAC) archive which is the full resolution data spatially sampled to give a nominal resolution of 4 km. Secondly there is the Global Vegetation Index (GVI) data set, which has a resolution of around 19.5 km and is the result of spatial, radiometric and temporal sampling of the GAC data. Figure 1 summarises the GAC and GVI sampling schemes.

Recognition of the limited availability of full resolution AVHRR data (High Resolution Picture Transmission - HRPT; Local Area Coverage - LAC) eventually led to the International Geosphere Biosphere Programme's Data and Information System (IGBP-DIS) initiative, the global 1 km Land Cover project (Townshend 1992a). Essentially this was a two phase programme, with the first phase consisting of the collection and archiving of global HRPT and LAC resolution AVHRR data, followed by phase II, the processing of these data and preparation of derived land cover products.

The 1 km efforts of IGBP-DIS have been paralleled by an independent project, the NASA Pathfinder GAC programme (James and Kalluri 1994). This aims to process *all* GAC data archived back to July 1981.

A number of continental scale processed AVHRR data sets are also available. These include those produced at the Global Inventory Monitoring and Modelling Studies (GIMMS) group at the Goddard Space Flight Center (see Tucker, chapter 1), and the daily, multi-channel multi-annual GAC time series created for Africa and South East Asia by the Monitoring of Tropical Vegetation Unit (MTV) at the European Commission's Joint Research Centre (Malingreau and Belward 1994).

This chapter describes these large area AVHRR data sets, and provides information on their availability.

G. D'Souza et al. (eds.), Advances in the Use of NOAA AVHRR Data for Land Applications, 455–470.
© 1996 ECSC, EEC, EAEC, Brussels and Luxembourg.

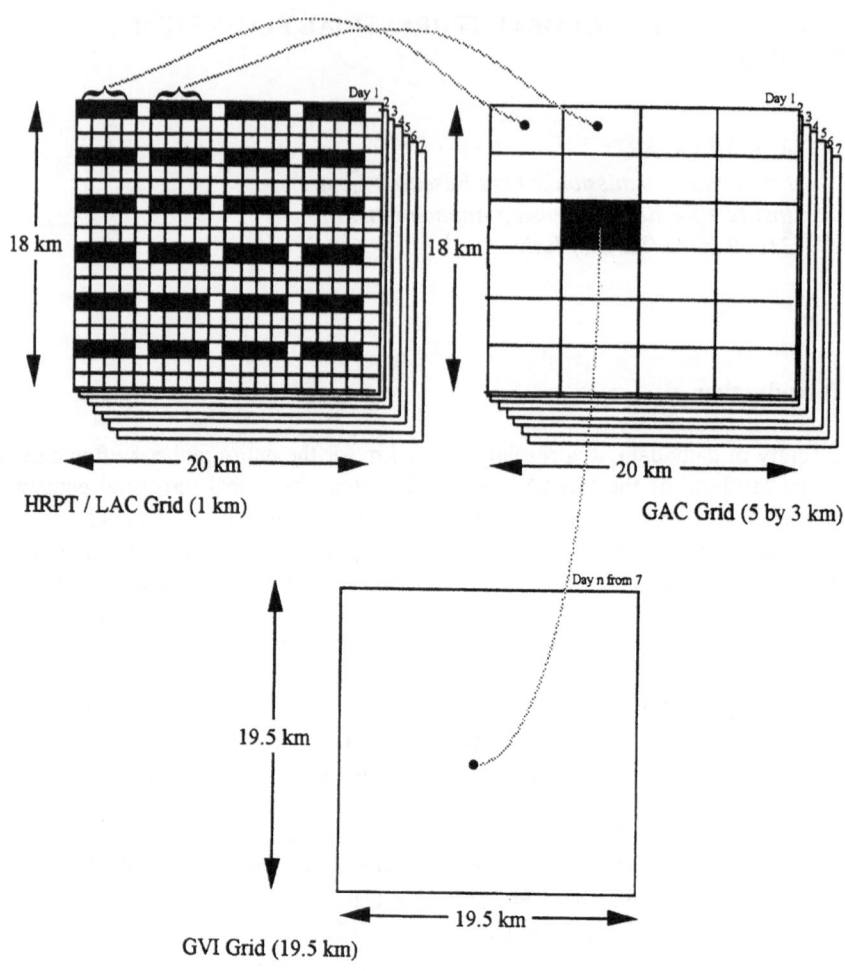

18 km

20 km

HRPT / LAC Grid (1 km)

Day 1

18 km

20 km

GAC Grid (5 by 3 km)

Day n from 7

19.5 km

19.5 km

GVI Grid (19.5 km)

Pixels sampled to create the GAC and GVI data sets

Figure 1. The relationship between the AVHRR full resolution data sets (High Resolution Picture Transmission [HRPT] and Local Area Coverage [LAC]), and the sampled data sets Global Area Coverage (GAC) and Global Vegetation Index (GVI)

2. The Global Vegetation Index data

The GVI is only available as mapped data, with a radiometric precision of 8 bits, rather than the 10 bits of the original AVHRR data. Data collection and processing started in 1982 (Tarpley *et al.* 1984) and has continued ever since. However, along the way there have been a number of changes to this product. From May 1982 to March 1985 the GVI data set consists of the difference Vegetation Index (VI) which is simply AVHRR channel 2 minus AVHRR channel 1, the Normalised Difference Vegetation Index (NDVI), plus channels 1 and 2, truncated to 8 bit precision. From April 1985 on, channels 4 and 5 (as 8 bit "GOES counts" corresponding to temperature values) were also included along with solar zenith angle and sensor view angle information (Kidwell 1990).

The NDVI is calculated from the day-time passes of all 14 daily orbits of GAC data, though users should be aware that this computation uses instrument digital counts or numbers (DN), not calculated radiances. This has two important consequences; firstly NDVI based on DNs alone is sensitive to the irradiance and brightness of the target as well as spectral reflectance (Price 1987). This means that in a single scene a bright target, such as grassland can give higher DN-based NDVI values than dark targets such as forest, even though the two may have identical NDVIs, if these are computed from reflectance values (Goward *et al.* 1991); the second consequence is that the sensor degradation reported in the earlier chapters by Vermote *et al.* are not accounted for, hence GVI NDVI time series will contain trends which are artefacts of sensor performance, not due to vegetation dynamics.

From May 1982 to March 1985 the data were mapped onto a polar stereographic projection with a resolution cell size of 15 km at the equator, and from April 1985 to the present a Plate Carrée projection (16 km at the equator), and a Mercator projection (19.5 km at the equator) were used. Only one GAC pixel is selected for each GVI pixel. This single pixel is simply the last GAC value mapped to each grid cell of the output map, however, in the GVI experimental phase, from 10th May 1982 to 27th March 1983, a number of different sampling methods were tried (Kidwell 1990). This makes the early part of the time-series difficult to compare with later data. The final product is the result of compositing the daily arrays over a seven day period such that the maximum VI value, (and equivalent channel 1, 2, 4 and 5 values) are retained for each location.

Because GVI data are produced by radiometric, spatial and temporal sampling of the GAC data they are best suited to terrestrial monitoring applications of a general nature. Indeed, Goward *et al.* (1994a) show that the GVI becomes considerably more representative of the GAC data from which it is derived if a three three by three pixel spatial average is used. This would reduce the effective GVI resolution to approximately 60 by 60 km.

Goward *et al.* (1991) and Goward *et al.* (1994a) provide a very complete review of the limitations of the GVI product. In essence the authors conclude that the original GVI NDVI component should not be used for vegetation studies. The recommendation is for computation of a new NDVI based on calibrated channel 1 and 2 reflectances, where post-launch calibration coefficients are used (see earlier chapters by Vermote *et al.*).

Observations with solar zenith angles above 70° should not be used, a two-week compositing period using NDVI, rather than seven-day VI composites is preferred and an equal area map projection should be used. A GVI data set following these recommendation is being prepared (Goward *et al.* 1994b). With the support of the NASA Pilot Land Data System and NOAA Data Diskette Project the Goward GVI will be released on CD-ROMS. The data set is currently going through the alpha and beta test phases, and minor modifications are being made. For information on the status of the Goward *et al.* GVI data set and for access to the CD-ROMS, contact *The NOAA Geophysical Data Center, Boulder, Colorado, USA*.

3.   The GAC data

The problems encountered with the GVI data set highlight the drawbacks of sampling. Unfortunately the GAC data too are sampled, though to a lesser extent than the GVI. The data are left at full radiometric resolution and there is no temporal compositing. The sampling is restricted to the spatial domain where the following scheme is applied to the full resolution data, on-board the satellite; for a given scan line the first four pixels out of every five are averaged, and only every third scan line is processed. This gives a resolution at nadir of approximately 1.1 km by 4 km, with a gap of around 3 km between pixels across the scan lines, though the data are conventionally referred to as having 4 km resolution (Kidwell 1991). This unusual sampling strategy seems to have been chosen purely on engineering grounds, as it provides ease of on-board data processing (Justice *et al.* 1989).

The suitability of this sampling scheme was studied by Justice *et al.* (1989) who applied different sampling strategies to five by three blocks of HRPT/LAC pixels. The conclusion was that the GAC sampling strategy gave a relatively poor representation of the full resolution data. In fact, out of five different sampling procedures tested, only the random selection of a single pixel from the block of 15 gave a poorer result than the method adopted by NOAA.

The sampling scheme used to generate the GAC data loses some of the spatial and spectral information contained in non-sampled AVHRR. However, if the GAC data themselves are spatially averaged to give a pixel size of around 12 km (a 3 x 3 GAC pixel average) then the GAC provide the same information as is found in 12 km pixels produced by degrading AVHRR HRPT or LAC data (Belward 1992, Malingreau and Belward 1992). Because of the sampling limitations the GAC data are best suited for terrestrial ecosystem monitoring on large areas at coarse scales, rather than for local studies.

3.1.   THE CONTINENTAL DATA SETS

Much of the pioneering work concerning preparation of GAC time series was undertaken by the GIMMS group (see Tucker, chapter 1). The first data set GIMMS produced was for Africa. This was a ten-day NDVI maximum value composite (MVC) product (Holben, 1986) covering the period July 1981 to the present. The GIMMS group also have an

NDVI MVC data set for the Indochina peninsula (Justice *et al.*, 1991), a comparable African time-series prepared by the FAO's ARTEMIS project exists (Snijders, 1989) and monthly NDVI MVCs for the Australian Continent are available (Graetz *et al.*, 1992). The ARTEMIS Africa product is a 10-day NDVI MVC at a resolution of 8 km. Data from August 1981 to December 1992 have been released by FAO on a CD - ROM along with PC based software for extracting time-series curves. More information on the current status of this CD can be obtained from *Da Vinci Consulting, Chaussée de Huy 230 B-1325 Chaumont-Gistoux, Belgium.*

Though the study of the continental NDVI datasets went a long way towards highlighting the potential of the AVHRR for terrestrial applications it is known that the index is sensitive to factors other than vegetation dynamics (Gutman 1991). Furthermore, this geophysical parameter alone does not realise the full potential of the GAC archives as a source of data for terrestrial ecosystem monitoring. For example, biomass burning studies require channel 3 and 4 brightness temperatures for the detection of active fires, vegetation indices and channel 2 reflectance for burned area assessments and fuel loading studies. Land cover classifications use vegetation indices for studying vegetation seasonality, and channels 1, 2 and 3 for discrimination between vegetation types (Townshend *et al.* 1991, Grégoire *et al.* 1993). The thermal channels of course are needed for surface temperature measurements and for cloud masking (see the earlier chapters by Vogt, Seguin and Kriebel). The individual AVHRR channels are also vital for assessing changes in forest canopies, determining seasonal characteristics of forest communities, detecting fire in forest ecosystems and identifying new deforestation fronts (see the earlier chapter by Malingreau *et al.*). Recognising this, the JRC's Monitoring of Tropical Vegetation unit (MTV) has established pre-processing chains to prepare continental multi-channel AVHRR time series.

In the framework of a NASA - JRC collaborative research agreement, the MTV unit acquired GAC data for Africa, S.E. Asia and Europe. The initial efforts in data processing concentrated on Africa. The MTV's Africa GAC archive covers the period 9th July 1981 to mid-1992. There are over 10,000 scenes where each scene is a part-orbit between 40° N and 40° S. Data are available in all 5 channels at full 10 bit radiometric resolution, around 12.5 Megabytes per scene. Even though we are now in the age of multi-Gigabyte cartridges, much of the AVHRR data are still found on computer compatible tapes (CCTs); the 125 Gigabytes (GB) of raw Africa data arrived at the MTV unit on over 1, 200 individual tapes. The GAC processing chain was originally built for this specific data set. The number of scenes per day that can be handled, the map projection, image size, co-ordinates and resampling method are all fixed, as are the output channels. It is a fully automatic chain that starts with level 1b data (see the earlier chapter by Kidwell) on computer tapes and finishes with a multi-channel daily mosaic for the continent archived to magneto-optical media. Figure 2 summarises the data processing steps, a detailed description is given by Malingreau and Belward (1994). Example images from the MTV Africa data set can be found via the World Wide Web (WWW) Centre for Earth Observation Home Page using Uniform Resource Locator (URL) http ://ceo-www.jrc.it/.

460

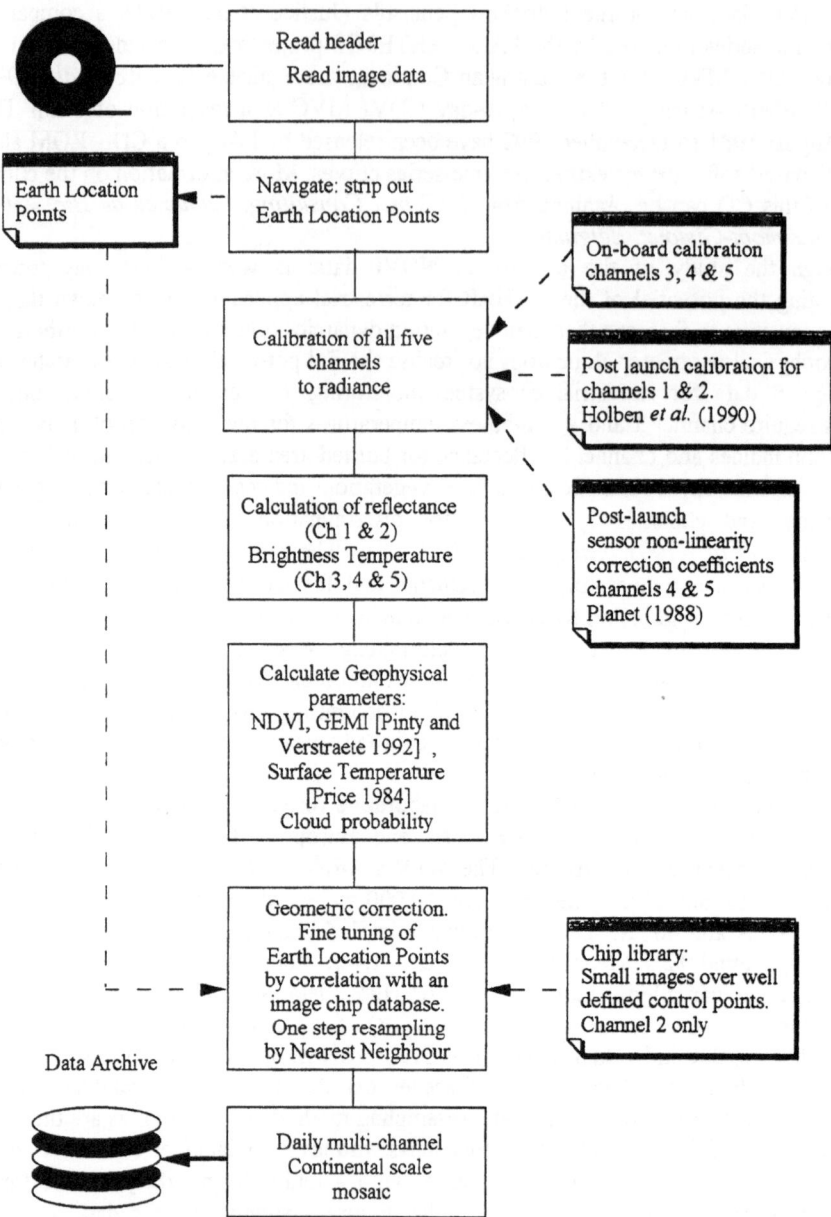

Figure 2. Data processing steps for the JRC's MTV Africa and South-East Asia GAC archives (see Malineau and Belward 1994).

3.2.   THE GAC PATHFINDER LAND DATA SET

Though continental GAC data sets are now a well established product, the bulk of the 30,000 CCTs making up the GAC archive remained untouched until the NASA AVHRR Pathfinder project was initiated in 1990.

The pathfinder GAC land data set consists of global, NDVI with associated reflectances, brightness temperatures, solar and satellite viewing geometry and cloud estimation. There are three products; a daily, global, land surface NDVI + coincident channels at 8 km; a global *quasi* 10-day NDVI Maximum Value Composite plus coincident channels at 8 km; a 10-day NDVI MVC only at a resolution of 0.5 - 1 degree. The raw data are stored on 420 6 GB WORMs. (Some idea of the work that goes into preparation of extensive AVHRR data sets can be gleaned from the fact that the transcription from tape to WORMs alone took three shifts working seven days/week from eight to 10 weeks to transcribe each year's data.)

The data processing steps are outlined in Figure 3. Full details of the pathfinder data set can be found in James and Kaluri (1994), and from the World Wide Web using URL http://xtreme.gsfc.nasa.gov/.

The data are archived in the EOS standard format, Hierarchical Data Format (HDF) at the Earth Observing System (EOS) Distributed Active Archive Center (DAAC). Because of the size of each daily global data set (227 MB), the DAAC is planning to implement software which will allow users to request data for specific geographical regions in addition to specific dates. Users should note that the DAAC will also supply software for reading the HDF data. For more information on the project contact: *Goddard DAAC User Services Office (USO), Earth Observing System Distributed Active Archive Center, Global Change Data Center, NASA Goddard Space Flight Center, Code 902.2, Greenbelt MD 20771, Email daacuso@eosdata.gsfc.nasa.gov.*

The DAAC has an on-line catalogue. This allows you to search through the processed global GAC data set, and to order data. Access is via the following: *telnet 192.107.190.75, login: daacims, passwd: gsfcdaac.*

Processing began in early June 1993, beginning with data from January 1987. Processed data from January 1986 to December 1990 are now available and the complete GAC archive should be processed by the end of 1995.

4.   The HRPT / LAC data sets

The earlier chapters by Saint and Malingreau *et al.* emphasised the importance of high resolution satellite data sets for land cover classification. It is axiomatic from sections 2 and 3 above that the sampling used in the creation of the GVI and GAC exclude these data sets from such a role. Indeed as Townshend and Justice (1990) point out, even the use of the full resolution AVHRR for land cover change is open to question, with a resolution closer to 250 metres being optimum. Nevertheless, the gap between the 250 m resolution required and the existing global land cover data sets' resolutions (in the order of 50 - 100 km) is so great that there is a clear requirement for 1 km, unsampled AVHRR

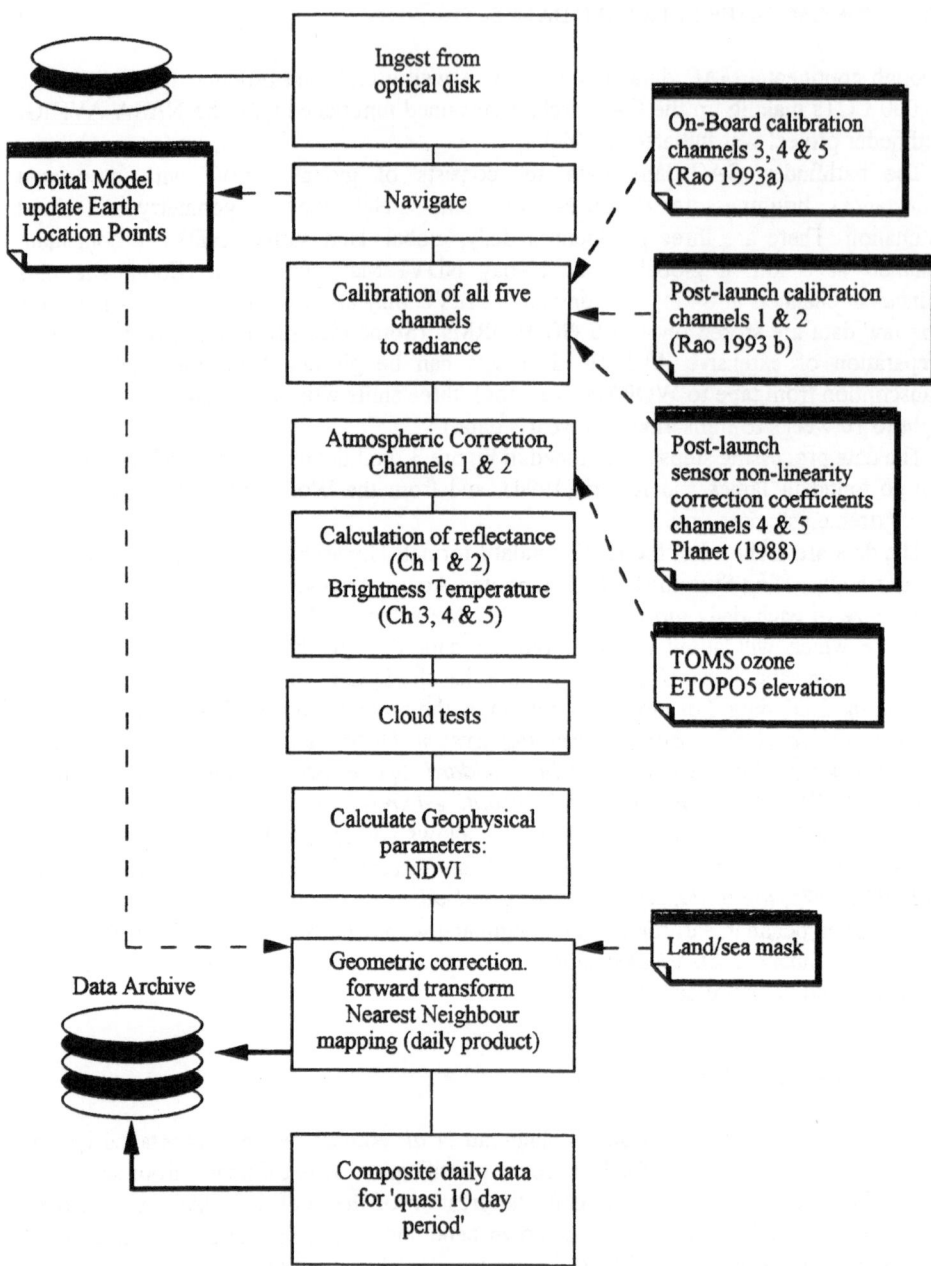

Figure 3. Data processing steps for the GAC Pathfinder Land Area Set (see James and Kaluri 1994).

data. Unfortunately, as stated in the introduction, systematic global 1 km archival of AVHRR data only began in 1992.

The 1 km data are available in two formats; HRPT in which case they are recorded at ground receiving stations for the area in which the satellite is in direct line of sight of the station's antenna. This will vary depending on the position of the station and the antenna elevation used; LAC in which case they are recorded on board, though there is only on board recorder space for 10 to 12 minutes data from each 100 minute orbit for these data (hence the GAC product in the first place). The combination of LAC acquisitions and the network of HRPT receiving stations (at least 48 of these around the world are known to have digital archives of AVHRR data) has for long had the potential for real global 1 km data coverage, but there was no mechanism for the co-ordinated data acquisition.

This changed in April 1992 when the IGBP-DIS 1 km project got under way. Simply put, this has the goal of collecting, archiving and processing AVHRR 1 km data for the entire planet surface, every day and then deriving related land cover products. This is all set out in IGBP report 20 "Improved Global Data for Land Applications" (Townshend 1992a). The initial 18 month period ended September 1993, though the NASA pathfinder project has agreed to extend funding to complete at least 24 months data acquisition. Indeed, regular archiving of data from NOAA 14 (successfully launched on 30th December 1994) has begun, though data processing is still only guaranteed for the first 18 months of data collected in the IGBP-DIS project framework.

4.1   THE IGBP-DIS 1 KM DATA SET

The United States Geological Service's EROS Data Center (EDC), NASA, ESA and the Australian CSIRO, acting in concert with a number of independent receiving stations (30 in total) such as the Louisiana State University facility at Baton Rouge and the Hartebeesthoek receiving station in South Africa, co-ordinated the process of global collection of the 1 km data. The data are routinely sent to EDC where each incoming scene is catalogued. A copy of the entire global 1 km raw data set is also held at the ESA ESRIN facility, Frascati (see the earlier chapter by Arino and WWW URL http://shark1.esrin.esa.it/).

The data processing overview is given in Figure 4. The first task is assembly of a coherent global data set, such that the thousands of individual scenes (for the initial 18 months of the project the total would be 43,000) are combined to give 14 complete orbits per day (7,560 in total). For each orbital pass a north-to-south progression is followed, the time stamps are used to identify overlap, the northern scene is used until lines of bad data start to appear, this is usually at the edge of the stations acquisition horizon. At this point acquisition is transferred to the next scene in sequence, regions with no data are zero filled.

464

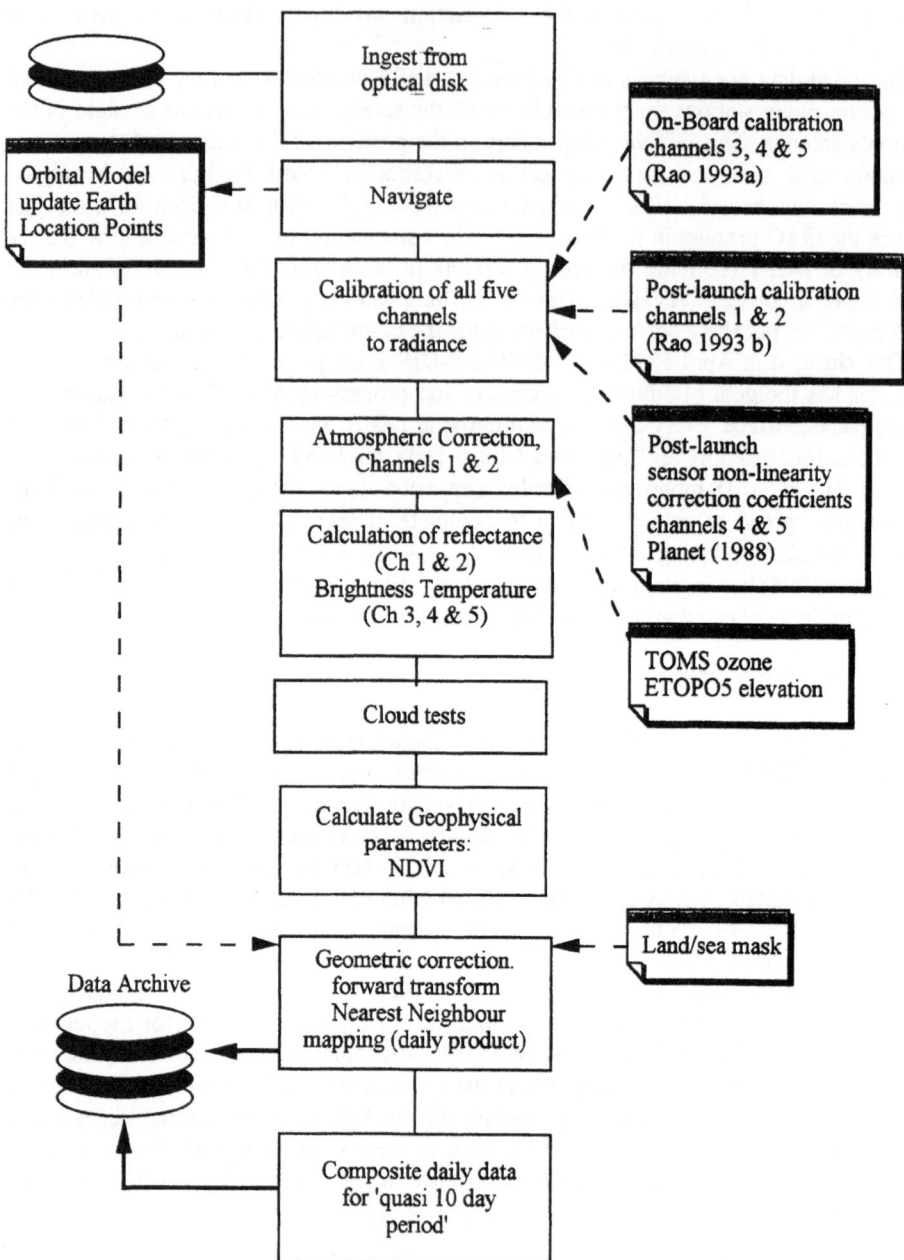

Figure 4. Data processing overview of the IGBP-DIS Global 1 km Land Cover Data Set.

Channels 1 and 2 of the AVHRR are converted to radiances using calibration coefficients recommended by Teillet and Holben (1994), also see the earlier chapters by Vermote *et al.* Channels 3, 4 and 5 use the on-board calibration coefficients.

The target accuracy for the final mapped product is a root mean square error of 0.8 pixels in relation to known ground control points (Townshend 1992b). This requires the use of an orbital model for first order image correction and then correlation with a library of pre-selected ground control points for fine correction. For the fine correction the project currently uses coastlines, rivers and inland water bodies from the Digital Chart of the World and World Vector Shoreline. Early experiments with image 'chip' libraries from AVHRR or from Landsat were found to be impractical on a global scale.

All data are remapped to the Goodes Interrupted Homosoline Equal Area Projection (Steinwand 1994) at a resolution of 1 km. Nearest Neighbour resampling is used via the inverse method. i.e., once the map location is known, the original image pixel closest to that location is chosen. The final dataset is around 1.4 Gigabytes, per channel per day.

The Maximum Value Composite method is used to create 10-day composite images. For each image/pixel in the compositing period the Maximum NDVI value is retained. The date for each pixel is noted and the equivalent individual channels and sun angles are selected. The final product is a 10 channel product. The 10 bit precision channels are stored as 2 byte data (around 1.4 gigabytes per day per channel), the 8 bit channels are correspondingly smaller.

The full product is as follows. Each of the 10 channels are listed in order.

1   AVHRR Channel 1 (reflectance, 10 bit precision)
2   AVHRR Channel 2 (reflectance, 10 bit precision)
3   AVHRR Channel 3 (radiance, 10 bit precision)
4   AVHRR Channel 4 (apparent brightness temperature, 10 bit precision)
5   AVHRR Channel 5 (apparent brightness temperature, 10 bit precision)
6   The NDVI (calculated from reflectances, 8 bit precision)
7   Solar Zenith Angle (8 bit precision)
8   Satellite Zenith angle (8 bit precision)
9   Sun Satellite relative azimuth (8 bit precision)
10 Day of pixel selected within the composite

Note that there is no cloud mask.

Rayleigh scatter and Ozone correction are applied to the data *after* the compositing. This is something of a deviation from the recommendations set out in Townshend (1992, 1992b). The rationale behind atmospheric correction after compositing are twofold; firstly, if non-atmospherically corrected radiances for channels 1 and 2 are used in the MVC process these will tend to favour smaller satellite and solar angles and thus reduce any directional effects (see the earlier chapters by Verstraete and Flasse and by Dedieu for a discussion of directional reflectance as a factor in analysis of AVHRR data); secondly correcting after compositing means that new approaches can be tried without reprocessing the entire data set (Cihlar and Huang 1994).

All raw data collected under the auspices of the IGBP-DIS have been included in the EDC catalogue. These are thus available through the normal EDC data request system. The raw data archive, and the processed data can be accessed via the WWW using URL http:/sun1.cr.usgs.gov/landdaac/landdaac.html.

Data transfer via ftp is under development. The global data set is available either at full resolution or reduced 2 km, 4 km, 8 km or 16 km. All bands are available, though have to be retrieved one at a time. A standard Unix compression program may be used if required to speed-up data transfer.

5. Conclusions

The AVHRR was designed as a meteorological instrument, with recognised potential for oceanographic applications too. However, soon after the first HRPT data became available, workers interested in terrestrial applications, rather than marine or meteorological, started to use the data. They soon discovered that, whilst the AVHRR data offered unique attributes for terrestrial studies, they were not easy to work with. Resolution varies across the swath, local solar time changes across the swath, sensor response drifts with time, the time of satellite overpass drifts with time, the satellite attitude is not well known, the atmosphere is not entirely transparent at the wavelengths the AVHRR visible and near infrared channels use, and the characteristics of the entire NOAA AVHRR system results in considerable spatial and temporal variation in the geometry of illumination and observation.

Since Townshend and Tucker (1981), and Schneider *et al.* (1981) first highlighted the use of AVHRR data for land applications the AVHRR data users have reached a consensus on many data processing issues. The use of post-launch coefficients for channel 1 and 2 calibration is now almost universal (though as can be seen from the three processing chains shown in Figures 2 - 4 above, there is still variation in the exact coefficients used). Geometric correction usually employs nearest neighbour resampling. Resampling occurs after data calibration and computation of geophysical parameters. Indeed, not only do the data products conform reasonably well from one chain to the other, but so too do thematic products derived in turn from these (e.g., Belward *et al.* 1995).

However, there is still significant diversity in the approaches to atmospheric correction. The three processing chains discussed in detail above deal with this issue in different ways. The MTV data processing carries out no atmospheric correction, but generates the GEMI vegetation index which accounts for many atmospheric effects directly in its computation. The NASA Pathfinder data processing chain atmospherically corrects before NDVI maximum value composites are generated, the IGBP-DIS 1 km data processing chain atmospherically corrects after compositing. Arguments in support of each approach can be made, yet consensus has yet to be reached.

Other data processing issues, such as cloud detection, have been adequately resolved on a regional scale (see the earlier chapter by Kriebel), but there is still no consensus on a universally applicable global cloud detection routine. Correction for directional reflectance factors too is an ongoing area of research. Variations in viewing and illumination geometry, inherent in the AVHRR system, can result in reflectance measurement differences from vegetation which are attributable to the anisotropic nature of the vegetation reflectance field, rather than variations in the biophysical properties of the vegetation itself. The development of corrections for angular variations over the highly complex canopies of terrestrial ecosystems has been the subject of much research (Verstraete et al. 1990, Pinty et al. 1990). Though operational corrections are not yet available, the approaches outlined in the earlier chapter by Verstrate and Flasse suggests that these will not be long in coming.

Though a number of data processing issues remain unresolved, data sets are being produced, and more importantly are being used. The role of earth observation data, and in particular data from AVHRR is now widely accepted for global terrestrial ecosystem monitoring. Indeed, it is on the basis of past success with the AVHRR data that *data users* (and not space agencies) in Europe have sponsored the development of the Vegetation instrument that will fly on SPOT 4 in October 1997. The Vegetation instrument is similar to the AVHRR insofar as it has visible and near infrared channels operating at a resolution of 1 km. However, as the name of the sensor suggests, the system has been designed for operational vegetation monitoring from the outset. Details can be found via the World Wide Web using URL http://ceo-www.jrc.it/.

Other new sensors include the Moderate Resolution Imaging Spectroradiometer (MODIS) scheduled to fly on the NASA Earth Observing System (EOS) satellites by 1998. MODIS will have a spatial resolution of from 250 m to 1 km and 36 channels. The 1 km resolution Along Track Scanning Radiometer (ATSR) which flew on ERS -1 will be followed by ATSR 2 on ERS-2 in 1995, which will have visible and near infrared channels. This in turn will be followed by the Advanced ATSR (AATSR) on ESA's ENVISAT in early 1999. ENVISAT will also carry the Medium Resolution Imaging Spectrometer (MERIS) with 300 m resolution and a choice of 15 channels at any one time out of a total of 200 available. The Sea-viewing Wide-field-of-view Sensor SeaWiFS also offers 1 km resolution data, and though the visible and near infrared channels have been developed for ocean colour applications, the system design is such that the sensors will not saturate over land.

Just as this multiplicity of new sensors has profited from the last decade and more of research based on AVHRR data so too has the AVHRR itself. As Kidwell et al. describe in the earlier chapter, improved AVHRRs will continue to fly on the NOAA missions till well into the next Century.

References

BELWARD, A. S., 1992, Spatial attributes of AVHRR imagery for environmental monitoring. *International Journal of Remote Sensing* **13,** 193-208.

BELWARD A. S., HOLLIFIELD, A. and JAMES, M. E., 1995, The potential of the NASA GAC Pathfinder product for the creation of global thematic data sets: the case of biomass burning patterns *International Journal of Remote Sensing* (in press).

CHE, N. and PRICE, J. C., 1992. Survey of radiometric calibration results and methods for visible and near-infrared channels of NOAA - 7, -9 and -11 AVHRRs, *Remote Sensing of Environment*, **41**, 19 - 27.

CIHLAR, J. and HUANG, F., 1994, Effects of atmospheric correction and viewing and restriction on AVHRR data composites. *Canadian Journal of Remote Sensing*, **20**, 132 - 137.

EIDENSHINK, J. C., and FAUNDEEN, J. L., 1994, The 1 km AVHRR global land data set: first stages in implementation, International Journal of Remote Sensing, **15**, 3443 - 3462.

GOWARD, S. N., MARKHAM, B., DYE, D. G., DULANEY, W. and YANG, J. 1991, Normalized Difference Vegetation Index measurements from the Advanced Very High Resolution Radiometer, *Remote Sensing of Environment*, **35**, 257 - 277.

GOWARD, S. N., DYE, D. G., TURNER, S. and YANG, J. 1994a, Objective assessment of the NOAA Global Vegetation Index data product, *International Journal of Remote Sensing* **14**, 3365 - 3395.

GOWARD, S. N., TURNER, S., DYE, D. G., and YANG, J. 1994b, The University of Maryland improved Global Vegetation Index product, *International Journal of Remote Sensing* **15**, 3365-3395.

GRAETZ, D, WILSON, M. and FISHER, R. 1992, *Looking back: 1972 - 1992 The Australian Continent*, (CSIRO: Canberra)

GREGOIRE, J-M, BELWARD, A. S., and KENNEDY, P. J., 1993, Saturation du signal dans la bande 3 du senseur AVHRR: Handicap majeur ou source d'information pour la surveillance de l'environnement en milieu soudano-guinéen d'Afrique de l'Ouest? International *Journal of Remote Sensing* **14**, 2079 - 2095.

GUTMAN, G. G., 1991, Vegetation Indices from AVHRR: An Update and Future Prospects, *Remote Sensing of Environment*, **35**, 121-136.

HOLBEN, B., 1986, Characteristics of Maximum Value Composite images from temporal AVHRR data, *International Journal of Remote Sensing*, **7**, 1417 - 1434.

HOLBEN, B. N, KAUFMAN, Y. J., and KENDALL J. 1990, NOAA-11 AVHRR visible and near IR in-flight calibration, *International Journal of Remote Sensing*, **11**, 1511 - 1519

JAMES M. E. and KALLURI S. N. V. 1994, The Pathfinder AVHRR land data set: An improved coarse resolution data set for terrestrial monitoring (Special edition of the *International Journal of Remote Sensing*, Creating Global Remote Sensing Data Sets edited by J. Townshend vol 15 no 17 20th November 1994)15, 3347-3363

JUSTICE, C. O., MARKHAM, B. L., TOWNSHEND, J. R. G., and KENNARD, R. L., 1989, Spatial degradation of satellite data, *International Journal of Remote Sensing*, **10**, 1539-1561.

JUSTICE, C. O., TOWNSHEND, J. R., G., and KALB, V. L., 1991, Representation of vegetation by continental data sets derived from NOAA-AVHRR data, *International Journal of Remote Sensing*, **12**, 999-1021.

KAUFMAN, Y. J., and HOLBEN, B. N (1992), Calibration of the AVHRR visible and near IR bands by atmospheric scattering, ocean glint and desert reflection, *International Journal of Remote Sensing*, (in press)

KIDWELL, K. B., 1990, Global Vegetation Index users guide, NOAA NESDIS, National Climate Data Center, Washington, D.C.

KIDWELL, K. B., 1991., NOAA polar orbiter data user's guide, revised July 1991, NOAA NESDIS, National Climate Data Center, Washington, D.C.

MALINGREAU J-P., and BELWARD A. S., 1992, Scale considerations in vegetation monitoring using AVHRR data. *International Journal of Remote Sensing* **13**, 2289-2307.

MALINGREAU J-P., and BELWARD A. S., 1994, Recent activities in the European Community for the creation and analysis of global AVHRR data sets. *International Journal of Remote Sensing* **15**, 3397-3416.

PINTY, B., VERSTRAETE, M. M. and DICKINSON, R. E., 1990, A physical Model of the Bidirectional Reflectance of Vegetation Canopies, 2. Inversion and Validation, *Journal of Geophysical Research*, **95**, 11,767 - 11,775.

PINTY, B. and VERSTRAETE, M. M., 1992, GEMI: A non-Linear Index to monitor global vegetation from satellites, *Vegetatio*, **101**, 15 - 20.

PLANET, W.G., 1988, Data extraction and calibration of TIROS-N-NOAA radiometers, NOAA Technical Memorandum NESS 101 - Rev. 1, U.S. Department of Commerce, National Oceanic and Atmospheric Administration, Washington, D.C.

PRICE, J.C., 1984, Land Surface Temperature Measurements from the Split Window Channels of the NOAA - 7 AVHRR, *Journal of Geophysical Research*, **89**, 7231 - 7237.

PRICE, J. C., 1987, Calibration of satellite radiometers and the comparison of vegetation indices, *Remote Sensing of Environment* **21**, 15 - 27.

RAO, C.R.N. (editor), 1993a, Nonlinearity corrections for the thermal infra-red channels of the AVHRR. NOAA Technical report NESDIS 69 (NOAA: Washington).

RAO, C.R.N. (editor), 1993b, Degradation of the Visible and Near-infrared channels of the AVHRR. NOAA Technical report NESDIS 70 (NOAA: Washington DC).

SCHNEIDER, S. R., MCGINNIS, S. R., Jr., and GATLIN, J. A., 1981, Use of NOAA/AVHRR visible and near infra-red data for land remote sensing. NOAA Technical report, NESS 84, USDC, Washington, D. C. 48pp

SNIJDERS, F. L., 1989, Operational monitoring of environmental conditions relevant to crop production and desert locust plague prevention using NOAA-AVHRR data, *In the proceedings of the 4th AVHRR data users' conference, held at Reichsstadthalle, Rothenburg ob der Tauber, Federal Republic of Germany, on 5th - 8th September 1989.* (Darmstadt-Eberstadt: EUMETSAT) pp 179 -184.

STEINWAND, D. R., 1994, Mapping raster imagery to the Interrupted Goode Homosoline projection, *International Journal of Remote Sensing* **15**: 13463 - 3471.

TARPLEY, J. D., SCHNEIDER, S. R., and MONEY, R. L., 1984, Global vegetation indices from the NOAA-7 meteorological satellite. *Journal of Climate and Applied Meteorology*, **23**, 491 - 494.

TEILLET, P. M. and HOLBEN, B. N., 1994, Towards operational radiometric calibration of NOAA AVHRR imagery in the visible and infrared channels, Canadian Journal of Remote Sensing, **20**, 1-10.

TOWNSHEND, J.R.G, and TUCKER, C.J., 1981, Utility of AVHRR of NOAA 6 and 7 for vegetation mapping. in *Matching Remote Sensing Technologies and their applications* - Proceedings (London: Remote Sensing Society), pp 97 - 109.

TOWNSHEND, J. R. G., and JUSTICE, C. O., 1990, The spatial variation of vegetation changes at very coarse scales, *International Journal of Remote Sensing* **11**: 149 - 157.

TOWNSHEND, J.R.G., JUSTICE, C.O., LI, W., GURNEY, C., and McMANUS, J., 1991, Global land cover classification by remote sensing: Present Capabilities and Future Possibilities, *Remote Sensing of Environment*, **35**, 243-255.

TOWNSHEND, J. R. G., 1992a, editor, Improved global data for land applications, IGBP report number 20. (IGBP secretariat, Royal Swedish Academy of Sciences, Box 50005, S-10405, Stockholm, Sweden). 87pp.

TOWNSHEND, J. R. G., 1992b, editor, The Global 1 km AVHRR data set, further recommendations, IGBP working paper #3, (IGBP-DIS, University of Paris VI, 4, place Jussieu, 75252 Paris Cedex 05, France).

VERSTRAETE, M. M., PINTY, B. and DICKINSON, R. E., 1990, A physical Model of the Bidirectional Reflectance of Vegetation Canopies, 1. Theory, *Journal of Geophysical Research*, **95**, 11,755 - 11,765.

SUBJECT INDEX

478

EURO

COURSES

REMOTE SENSING

1. A.S. Belward and C.R. Valenzuela (eds.): *Remote Sensing and Geographical Information Systems for Resource Management in Developing Countries.* 1991   ISBN 0-7923-1268-6
2. F. Toselli and J. Bodechtel (eds.): *Imaging Spectroscopy: Fundamentals and Prospective Applications.* 1992   ISBN 0-7923-1535-9
3. V. Barale and P.M. Schlittenhardt (eds.): *Ocean Colour: Theory and Applications in a Decade of CZCS Experience.* 1993   ISBN 0-7923-1586-3
4. J. Hill and J. Mégier (eds.): *Imaging Spectrometry – a Tool for Environmental Observations.* 1994   ISBN 0-7923-2965-1
5. G. D'Souza, A.S. Belward and J-P. Malingreau (eds.): *Advances in the Use of NOAA AVHRR Data for Land Applications.* 1996   ISBN 0-7923-3911-8

KLUWER ACADEMIC PUBLISHERS – DORDRECHT / BOSTON / LONDON